PRINCIPLES OF
GEOTECHNICAL ENGINEERING

THE PWS SERIES IN ENGINEERING

Anderson, *Thermodynamics*
Askeland, *The Science and Engineering of Materials*, Third Edition
Bolluyt, Stewart, and Oladipupo, *Modeling for Design Using SilverScreen*
Borse, *FORTRAN 77 and Numerical Methods for Engineers*, Second Edition
Bronzino, *Biomedical Engineering and Instrumentation*
Clements, *68000 Family Assembly Language*
Clements, *Microprocessor Systems Design*, Second Edition
Clements, *Principles of Computer Hardware*, Second Edition
Das, *Principles of Foundation Engineering*, Second Edition
Das, *Principles of Geotechnical Engineering*, Third Edition
Das, *Principles of Soil Dynamics*
Fleischer, *Introduction to Engineering Economy*
Gere and Timoshenko, *Mechanics of Materials*, Third Edition
Glover and Sarma, *Power Systems Analysis and Design*, Second Edition
Janna, *Design of Fluid Thermal Systems*
Janna, *Introduction to Fluid Mechanics*, Third Edition
Kassimali, *Structural Analysis*
Keedy, *Introduction to CAD using CADKEY*, Second Edition
Knight, *The Finite Element Method in Mechanical Design*
Logan, *A First Course in the Finite Element Method*, Second Edition
McGill and King, *Engineering Mechanics: Statics*, Second Edition
McGill and King, *Engineering Mechanics: An Introduction to Dynamics*, Second Edition
McGill and King, *Engineering Mechanics: Statics and an Introduction to Dynamics*,
        Second Edition
Raines, *Software for Mechanics of Materials*
Reed-Hill and Abbaschian, *Physical Metallurgy Principles*, Third Edition
Reynolds, *Unit Operations and Processes in Environmental Engineering*
Sack, *Matrix Structural Analysis*
Schmidt and Wong, *Fundamentals of Surveying*, Third Edition
Segui, *Fundamentals of Structural Steel Design*
Segui, *LRFD Steel Design*
Shen and Kong, *Applied Electromagnetism*, Second Edition
Sule, *Manufacturing Facilities*, Second Edition
Vardeman, *Statistics for Engineering Problem Solving*
Weinman, *FORTRAN for Scientists and Engineers*
Weinman, *VAX FORTRAN*, Second Edition

PRINCIPLES OF GEOTECHNICAL ENGINEERING

Third Edition

BRAJA M. DAS

Southern Illinois University at Carbondale

PWS Publishing Company

Boston

PWS Publishing Company
20 Park Plaza
Boston Massachusetts 02116

PWS Publishing Company is a division of Wadsworth, Inc.

International Thomson Publishing
The trademark ITP is used under license.

  This book is printed on recycled, acid-free paper.

Library of Congress Cataloging-in-Publication Data

Das, Braja M., 1941–
    Principles of geotechnical engineering/Braja M. Das. — 3rd ed.
      p.   cm.
    Includes bibliographical references and index.
    ISBN 0-534-93375-0
    1. Soil mechanics.   I. Title.
TA710.D264 1993   93-5738
624.1'5136—dc20   CIP

Sponsoring Editor: Jonathan Plant
Production Coordinator: Robine Andrau
Manufacturing Coordinator: Ruth Graham
Assistant Editor: Mary Thomas
Editorial Assistant: Cynthia Harris
Marketing Manager: Nathan Wilbur
Production: Cecile Joyner/The Cooper Company
Interior Designer: Cynthia Bogue
Interior Illustrator: Network Graphics
Cover Printer: John P. Pow Company, Inc.
Typesetter: Santype International Limited
Printer and Binder: Arcata Graphics/Martinsburg

Printed and bound in the United States of America

96 97 98 — 10 9 8 7 6 5 4

To Janice and Valerie

PREFACE

Principles of Geotechnical Engineering, Third Edition is intended for introductory courses in soils and geotechnical engineering taken by virtually all civil engineering majors. It is also useful for professionals and other readers wanting a general introduction to this important branch of engineering. As in the first and second editions of the book (1985 and 1990, respectively), the new third edition offers an overview of soil properties and mechanics, together with coverage of field practices and basic engineering procedures. *Principles of Foundation Engineering*, Second Edition (1990), by the same author, goes on to apply these general concepts and procedures to earth, earth-supported, and earth-retaining structures, with an emphasis on design. *Principles of Geotechnical Engineering*, Third Edition provides the background information needed to support study in later design-oriented courses or in professional practice.

Changes in the Third Edition

The new edition, consisting of fourteen chapters, includes a number of new features that were incorporated in response to suggestions made by professors, students, and professionals familiar with the earlier versions of the book. The major changes are the following:

- ▶ Chapter 13, which provides an overview of environmental geotechnology, has been added. Disposal of solid and hazardous waste material has become a crucial issue worldwide; new materials and design procedures are rapidly being developed to protect our soils and groundwater. This new chapter introduces the major activities in this dynamic field and outlines their relationship to basic geotechnical principles.
- ▶ Soil compaction has been moved to an earlier part of the book (Chapter 4) and now directly follows coverage of soil classification.
- ▶ Reference tables previously included within the text have been moved to a new appendix, thus providing readers with a more convenient reference tool.
- ▶ Chapter 11 includes new material on slope stability, including slope stability in earthquake zones.
- ▶ More than 70 percent of the problems and 60 percent of the examples provided are either new or have been changed in this third edition.

▶ In response to student, professor, and professional feedback, the illustrations have been dramatically improved in quality. Two-color treatment has been added to make illustrations more effective and accessible to students and other readers.

▶ Besides complete worked solutions, the *Solutions Manual* for the book now includes a set of transparency masters to aid instructors in their classroom presentation.

Acknowledgments

I am grateful to my wife, Janice, for typing the manuscript and preparing some of the figures and tables. She has been the driving force in the completion of this edition of the text.

For their helpful reviews during the first and second editions of the text, the core of which is still intact, I would like to thank the following reviewers:

Robert D'Andrea
Worcester Polytechnic Institute

J. K. Jaypalan
Formerly of the University of Wisconsin—Madison

James L. Jorgenson
North Dakota State University

Robert Koerner
Drexel University

Shiou-san Kuo
University of Central Florida

Thomas J. Siller
Colorado State University

M. C. Wang
Pennsylvania State University

T. Leslie Youd
Brigham Young University

Thomas F. Zimmie
Rensselaer Polytechnic Institute

For their reviews and helpful suggestions for the development of the third edition of the text, I would like to thank

Thomas L. Brandon
Virginia Polytechnic Institute and State University

Linford L. Harley
Pennsylvania State University—Harrisburg

Norman L. Jones
Brigham Young University

Derek V. Morris
Texas A & M University

Mysore S. Nataraj
University of New Orleans

Thomas F. Wolff
Michigan State University

And, finally, many thanks are due to the production staff at PWS Publishing Company for the final development and production of the book.

Braja M. Das
Carbondale, Illinois

CONTENTS

CHAPTER THREE

CLASSIFICATION OF SOIL   67

CHAPTER FOUR

SOIL COMPACTION   88

CHAPTER FIVE

FLOW OF WATER IN SOIL: PERMEABILITY AND SEEPAGE   129

CHAPTER SIX

EFFECTIVE STRESS CONCEPTS   182

CHAPTER SEVEN

STRESSES IN A SOIL MASS   212

CHAPTER EIGHT

COMPRESSIBILITY OF SOIL 253

CHAPTER NINE

SHEAR STRENGTHS OF SOIL 316

CHAPTER TEN

LATERAL EARTH PRESSURE 380

SOIL-BEARING CAPACITY FOR SHALLOW FOUNDATIONS 465

SLOPE STABILITY 520

ENVIRONMENTAL GEOTECHNOLOGY   581

SUBSOIL EXPLORATION   603

APPENDICES

SOILS AND ROCKS

For engineering purposes, *soil* is defined as an uncemented aggregate of mineral grains and decayed organic matter (solid particles) with liquid and gas in the empty spaces between the solid particles. Soil is used as a construction material in various civil engineering projects, and it supports structural foundations. Thus, civil engineers must study the properties of soil, such as its origin, grain-size distribution, ability to drain water, compressibility, shear strength, and load-bearing capacity. *Soil mechanics* is the branch of science that deals with the study of the physical properties of soil and the behavior of soil masses subjected to various types of forces. *Soils engineering* is the application of the principles of soil mechanics to practical problems.

The record of a person's first use of soil as a construction material is lost in antiquity. For years, the art of soils engineering was based on only past experience. With the growth of science and technology, the need for better and more economical structural design and construction became critical. This led to a detailed study of the nature and properties of soil as they related to engineering during the early part of the twentieth century. The publication of *Erdbaumechanik* by Karl Terzaghi in 1925 gave birth to modern soil mechanics. This book presents the fundamental principles of soil mechanics on which more advanced studies are based.

Geotechnical engineering is defined as the subdiscipline of civil engineering that involves natural materials found close to the surface of the earth. It includes the application of the principles of soil mechanics and rock mechanics to the design of foundations, retaining structures, and earth structures.

1.1 ROCK CYCLE AND THE ORIGIN OF SOIL

The mineral grains that form the solid phase of a soil aggregate are the product of rock weathering. The size of the individual grains varies over a wide range. Many of the physical properties of soil are dictated by the size, shape, and chemical composition of the grains. To get a better understanding of these factors, one must be familiar with the basic types of rock that form the earth's crust, the rock-forming minerals, and the weathering process.

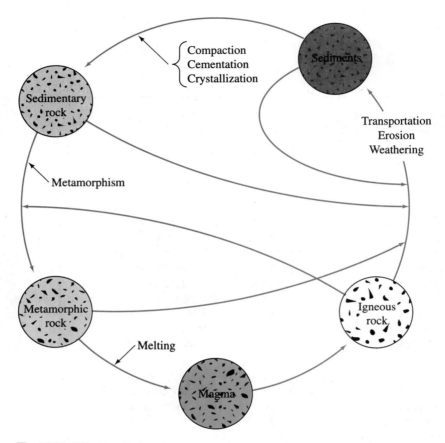

Compaction
Cementation
Crystallization

Transportation
Erosion
Weathering

Metamorphism

Melting

▼   **FIGURE 1.1**   Rock cycle

Based on their mode of origin, rocks can be divided into three basic types: *igneous, sedimentary,* and *metamorphic.* Figure 1.1 shows a diagram of the formation cycle of different types of rock and the processes associated with them. This is called the *rock cycle.* Brief discussions of each element of the rock cycle follow.

Igneous Rock

Igneous rocks are formed by the solidification of molten *magma* ejected from deep in the earth's mantle. After ejection by either *fissure eruption* or *volcanic eruption,* some of the molten magma cools on the surface of the earth. Sometimes magma ceases its mobility below the earth's surface and cools to form intrusive igneous rocks that are called *plutons.* Intrusive rocks formed in the past may be exposed at the surface as a result of the continuous process of erosion of the materials that covered them at one time.

The types of igneous rock formed by the cooling of magma depend on factors such as the composition of the magma and the rate of cooling associated with it. After conducting several laboratory tests, Bowen (1922) was able to explain the relation of the

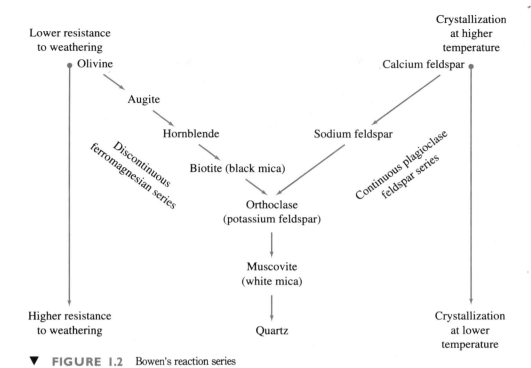

▼ **FIGURE 1.2**   Bowen's reaction series

rate of magma cooling to the formation of different types of rock. This explanation—known as *Bowen's reaction principle*—describes the sequence by which new minerals are formed as magma cools. The mineral crystals grow larger and some of them settle. The crystals that remain suspended in the liquid react with the remaining melt to form a new mineral at a lower temperature. This process continues until the entire body of melt is solidified. Bowen classified these reactions into two groups: (1) *discontinuous ferromagnesian reaction series*, in which the minerals formed are different in their chemical composition and crystalline structure, and (2) *continuous plagioclase feldspar reaction series*, in which the minerals formed have different chemical compositions with similar crystalline structures. Figure 1.2 shows Bowen's reaction series. The chemical compositions of the minerals are given in Table 1.1.

Thus, depending on the proportions of minerals available, different types of igneous rock are formed. Granite, gabbro, and basalt are some of the common types of igneous rock generally encountered in the field. Table 1.2 shows the general composition of some igneous rocks.

Weathering

Weathering is the process of breaking down rocks by mechanical and chemical processes into smaller pieces. Mechanical weathering may be caused by the expansion and contraction of rocks from the continuous gain and loss of heat, resulting in ultimate disintegration. Frequently, water seeps into the pores and existing cracks in rocks. As

▼ **TABLE 1.1**   **Composition of Minerals Shown in Bowen's Reaction Series**

Mineral	Composition
Olivine	$(Mg, Fe)_2SiO_4$
Augite	$Ca, Na(Mg, Fe, Al)(Al, Si_2O_6)$
Hornblende	Complex ferromagnesian silicate of Ca, Na, Mg, Ti, and Al
Biotite (black mica)	$K(Mg, Fe)_3AlSi_3O_{10}(OH)_2$
Plagioclase { calcium feldspar	$Ca(Al_2Si_2O_8)$
Plagioclase { sodium feldspar	$Na(AlSi_3O_8)$
Orthoclase (potassium feldspar)	$K(AlSi_3O_8)$
Muscovite (white mica)	$KAl_3Si_3O_{10}(OH)_2$
Quartz	SiO_2

the temperature drops, the water freezes and expands. The pressure exerted by ice because of volume expansion is large enough to break down even large rocks. Other physical agents that help disintegrate rocks are glacier ice, wind, the running water of streams and rivers, and ocean waves. It is important to realize that in mechanical weathering, large rocks are broken down into smaller pieces without any change in the chemical composition.

In chemical weathering, the original rock minerals are transformed into new minerals by chemical reaction. Water and carbon dioxide from the atmosphere form carbonic acid, which reacts with the existing rock minerals to form new minerals and soluble

▼ **TABLE 1.2**   **Composition of Some Igneous Rocks**

Name of rock	Mode of occurrence	Texture	Abundant minerals	Less abundant minerals
Granite	Intrusive	Coarse	Quartz, sodium feldspar, potassium feldspar	Biotite, muscovite, hornblende
Rhyolite	Extrusive	Fine	Quartz, sodium feldspar, potassium feldspar	Biotite, muscovite, hornblende
Gabbro	Intrusive	Coarse	Plagioclase, pyroxines, olivine	Hornblende, biotite, magnetite
Basalt	Extrusive	Fine	Plagioclase, pyroxines, olivine	Hornblende, biotite, magnetite
Diorite	Intrusive	Coarse	Plagioclase, hornblende	Biotite, pyroxenes (quartz usually absent)
Andesite	Extrusive	Fine	Plagioclase, hornblende	Biotite, pyroxenes (quartz usually absent)
Syenite	Intrusive	Coarse	Potassium feldspar	Sodium feldspar, biotite, hornblende
Trachyte	Extrusive	Fine	Potassium feldspar	Sodium feldspar, biotite, hornblende
Peridotite	Intrusive	Coarse	Olivine, pyroxenes	Oxides of iron

salts. Soluble salts present in the ground water and organic acids formed from decayed organic matter also cause chemical weathering. An example of the chemical weathering of orthoclase to form clay minerals, silica, and soluble potassium carbonate follows:

$$H_2O + CO_2 \rightarrow H_2CO_3 \rightarrow H^+ + (HCO_3)^-$$
$$\text{Carbonic acid}$$

$$\underset{\text{Orthoclase}}{2K(AlSi_3O_8)} + 2H^+ + H_2O \rightarrow 2K^+ + \underset{\text{Silica}}{4SiO_2} + \underset{\substack{\text{Kaolinite} \\ \text{(clay mineral)}}}{Al_2Si_2O_5(OH)_4}$$

Most of the potassium ions released are carried away in solution as potassium carbonate is taken up by plants.

The chemical weathering of plagioclase feldspars is similar to that of orthoclase in that it produces clay minerals, silica, and different soluble salts. Ferromagnesian minerals also form the decomposition products of clay minerals, silica, and soluble salts. Additionally, the iron and magnesium in ferromagnesian minerals result in other end products such as hematite and limonite. Quartz is highly resistant to weathering and only slightly soluble in water. Figure 1.2 shows the susceptibility of rock-forming minerals to weathering. The minerals formed at higher temperatures in Bowen's reaction series are less resistant to weathering than the ones formed at lower temperatures (Figure 1.2).

The weathering process is not limited to igneous rocks. As shown in the rock cycle (Figure 1.1), sedimentary and metamorphic rocks also weather in a similar manner.

Thus, from the preceding brief discussion, we can see how the weathering process changes solid rock masses into smaller fragments of various sizes that can range from large boulders to very small clay particles. Uncemented aggregates of these small grains in various proportions form different types of soil. The clay minerals, which are a product of chemical weathering of feldspars, ferromagnesians, and micas, are the minerals whose presence gives the plastic property to soils. There are three important clay minerals: (1) *kaolinite*, (2) *illite*, and (3) *montmorillonite*. We discuss these clay minerals later in this chapter.

Transportation of Weathering Products

The products of weathering may stay in the same place or may be moved to other places by ice, water, wind, and gravity.

The soils formed by the weathered products at their place of origin are called *residual soils*. An important characteristic of residual soil is the gradation of particle size. Fine-grained soil is found at the surface, and the grain size increases with depth. At greater depths, angular rock fragments may also be found.

The transported soils may be classified into several groups, depending on their mode of transportation and deposition:

1. *Glacial soils*—formed by transportation and deposition of glaciers
2. *Alluvial soils*—transported by running water and deposited along streams
3. *Lacustrine soils*—formed by deposition in quiet lakes

4. *Marine soils*—formed by deposition in the seas
5. *Aeolian soils*—transported and deposited by wind
6. *Colluvial soils*—formed by movement of soil from its original place by gravity, such as during landslides

Sedimentary Rock

The deposits of gravel, sand, silt, and clay formed by weathering may become compacted by overburden pressure and cemented by agents like iron oxide, calcite, dolomite, and quartz. Cementing agents are generally carried in solution by ground water. They fill the spaces between particles and form sedimentary rock. Rocks formed in this way are called *detrital sedimentary rocks*. Conglomerate, breccia, sandstone, mudstone, and shale are some examples of the detrital type.

Sedimentary rock can also be formed by chemical process. Rocks of this type are classified as *chemical sedimentary rock*. Limestone, chalk, dolomite, gypsum, anhydrite, and others belong to this category. Limestone is formed mostly of calcium carbonate that originates from calcite deposited either by organisms or by an inorganic process. Dolomite is calcium magnesium carbonate $[CaMg(CO_3)_2]$. It is formed either by the chemical deposition of mixed carbonates or by the reaction of magnesium in water with limestone. Gypsum and anhydrite results from the precipitation of soluble $CaSO_4$ because of evaporation of ocean water. They belong to a class of rocks generally referred to as *evaporites*. Rock salt (NaCl) is another example of an evaporite that originates from the salt deposits of sea water.

Sedimentary rock may undergo weathering to form sediments or may be subjected to the process of *metamorphism* to become metamorphic rock.

Metamorphic Rock

Metamorphism is the process of changing the composition and texture of rocks, without melting, by heat and pressure. During metamorphism, new minerals are formed and mineral grains are sheared to give a foliated texture to metamorphic rocks. Granite, diorite, and gabbro become gneisses by high-grade metamorphism. Shales and mudstones are transformed into slates and phyllites by low-grade metamorphism. Schists are a type of metamorphic rock with well-foliated texture and visible flakes of platy and micaceous minerals.

Marble is formed from calcite and dolomite by recrystallization. The mineral grains in marble are larger than those present in the original rock.

Quartzite is a metamorphic rock formed from quartz-rich sandstones. Silica enters into the void spaces between the quartz and sand grains and acts as a cementing agent. Quartzite is one of the hardest rocks.

Under extreme heat and pressure, metamorphic rocks may melt to form magma and the cycle is repeated.

1.2 SOIL-PARTICLE SIZE

As discussed in the preceding section, the sizes of particles that make up soil vary over a wide range. Soils are generally called *gravel*, *sand*, *silt*, or *clay*, depending on the predominant size of particles within the soil. To describe soils by their particle size, several organizations have developed particle-size classifications. Table 1.3 shows the particle-size classifications developed by the Massachusetts Institute of Technology, the U.S. Department of Agriculture, the American Association of State Highway and Transportation Officials, and the U.S. Army Corps of Engineers and U.S. Bureau of Reclamation. In this table, the MIT system is presented for illustration purposes only. This system is important in the history of the development of the size limits of particles present in soils; however, the Unified system is now almost universally accepted. The Unified Soil Classification System has been adopted by the American Society for Testing and Materials (ASTM). Figure 1.3 shows the particle-size classification systems in a graphic form.

▼ **TABLE 1.3** **Particle-Size Classifications**

Name of organization	Grain size (mm)			
	Gravel	Sand	Silt	Clay
Massachusetts Institute of Technology (MIT)	>2	2 to 0.06	0.06 to 0.002	<0.002
U.S. Department of Agriculture (USDA)	>2	2 to 0.05	0.05 to 0.002	<0.002
American Association of State Highway and Transportation Officials (AASHTO)	76.2 to 2	2 to 0.075	0.075 to 0.002	<0.002
Unified Soil Classification System (U.S. Army Corps of Engineers, U.S. Bureau of Reclamation, and American Society for Testing and Materials)	76.2 to 4.75	4.75 to 0.075	Fines (i.e., silts and clays) <0.075	

Gravels are pieces of rocks with occasional particles of quartz, feldspar, and other minerals.

Sand particles are made of mostly quartz and feldspar. Other mineral grains may also be present at times.

Silts are the microscopic soil fractions that consist of very fine quartz grains and some flake-shaped particles that are fragments of micaceous minerals.

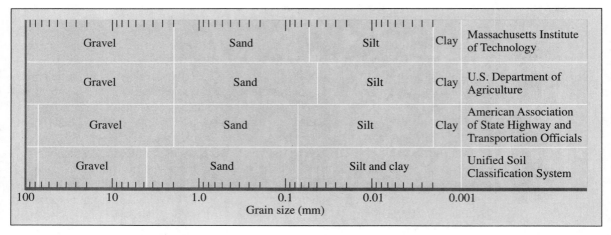

▼   **FIGURE 1.3**   Particle-size classification by various systems

Clays are mostly flake-shaped microscopic and submicroscopic particles of mica, clay minerals, and other minerals. As shown in Table 1.3, clays are generally defined as particles smaller than 0.002 mm. However, in some cases, particles between 0.002 mm and 0.005 mm in size are also referred to as clay (see ASTM D-653). Particles are classified as *clay* on the basis of their size; they do not necessarily contain clay minerals. Clays have been defined as those particles "which develop plasticity when mixed with a limited amount of water" (Grim, 1953). (Plasticity is the putty-like property of clays that contain a certain amount of water.) Nonclay soils can contain particles of quartz, feldspar, or mica that are small enough to be within the clay classification. Hence, it is appropriate for soil particles smaller than 2 microns (2 μm), or 5 microns (5 μm) as defined under different systems, to be called clay-sized particles rather than clay. Clay particles are mostly in the colloidal size range (< 1 μm), and 2 μm appears to be the upper limit.

1.3 CLAY MINERALS

Clay minerals are complex aluminum silicates composed of two basic units: (1) *silica tetrahedron* and (2) *alumina octahedron*. Each tetrahedron unit consists of four oxygen atoms surrounding a silicon atom (Figure 1.4a). The combination of tetrahedral silica units gives a *silica sheet* (Figure 1.4b). Three oxygen atoms at the base of each tetrahedron are shared by neighboring tetrahedra. The octahedral units consist of six hydroxyls surrounding an aluminum atom (Figure 1.4c), and the combination of the octahedral aluminum hydroxyl units gives an *octahedral sheet*. (This is also called a *gibbsite sheet*—Figure 1.4d.) Sometimes magnesium replaces the aluminum atoms in the octahedral units; in that case, the octahedral sheet is called a *brucite sheet*.

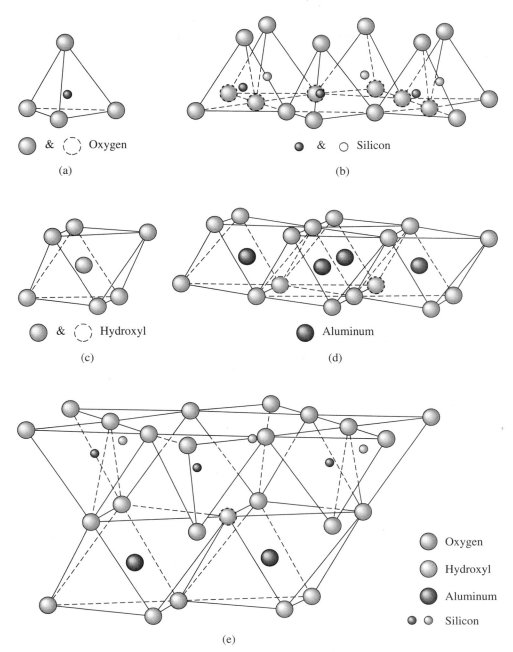

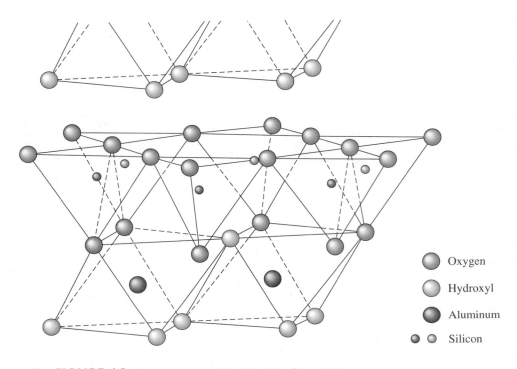

▼ **FIGURE 1.5**   Atomic structure of kaolinite (after Grim, 1959)

Oxygen

Hydroxyl

Aluminum

Silicon

In a silica sheet, each silicon atom with a positive charge of four is linked to four oxygen atoms with a total negative charge of eight. But each oxygen atom at the base of the tetrahedron is linked to two silicon atoms. This means that the top oxygen atom of each tetrahedral unit has a negative charge of one to be counterbalanced. When the silica sheet is stacked over the octahedral sheet as shown in Figure 1.4e, these oxygen atoms replace the hydroxyls to balance their charges.

Kaolinite consists of repeating layers of elemental silica-gibbsite sheets in a 1 : 1 lattice as shown in Figures 1.5 and 1.8a. Each layer is about 7.2 Å thick. The layers are held together by hydrogen bonding. Kaolinite occurs as platelets, each with a lateral dimension of 1000 to 20,000 Å and a thickness of 100 to 1000 Å. The surface area of the kaolinite particles per unit mass is about 15 m^2/g. The surface area per unit mass is defined as *specific surface*. Figure 1.6 shows scanning electron micrographs of a kaolinite specimen.

Illite consists of a gibbsite sheet bonded to two silica sheets—one at the top and another at the bottom (Figure 1.8b). It is sometimes called *clay mica*. The illite layers are bonded together by potassium ions. The negative charge to balance the potassium ions comes from the substitution of aluminum for some silicon in the tetrahedral sheets. Substitution of one element for another with no change in the crystalline form is known as *isomorphous substitution*. Illite particles generally have lateral dimensions ranging

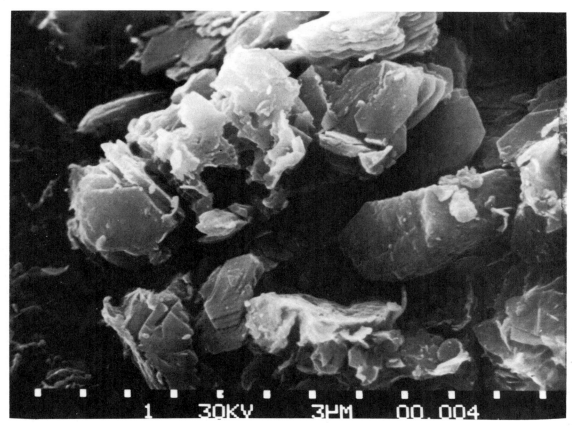

▼ **FIGURE 1.6** Scanning electron micrograph of a kaolinite specimen (courtesy of U.S. Geological Survey)

from 1000 to 5000 Å and thicknesses from 50 to 500 Å. The specific surface of the particles is about 80 m^2/g.

Montmorillonite has a structure similar to that of illite—that is, one gibbsite sheet sandwiched between two silica sheets (Figures 1.7 and 1.8c). In montmorillonite there is isomorphous substitution of magnesium and iron for aluminum in the octahedral sheets. Potassium ions are not present here as in illite, and a large amount of water is attracted into the space between the layers. Particles of montmorillonite have lateral dimensions of 1000 to 5000 Å and thicknesses of 10 to 50 Å. The specific surface is about 800 m^2/g.

Besides kaolinite, illite, and montmorillonite, other common clay minerals generally found are chlorite, halloysite, vermiculite, and attapulgite.

The clay particles carry a net negative charge on their surfaces. This is the result both of isomorphous substitution and of a break in continuity of the structure at its edges. Larger negative charges are derived from larger specific surfaces. Some positively charged sites also occur at the edges of the particles. A list of the reciprocal of the

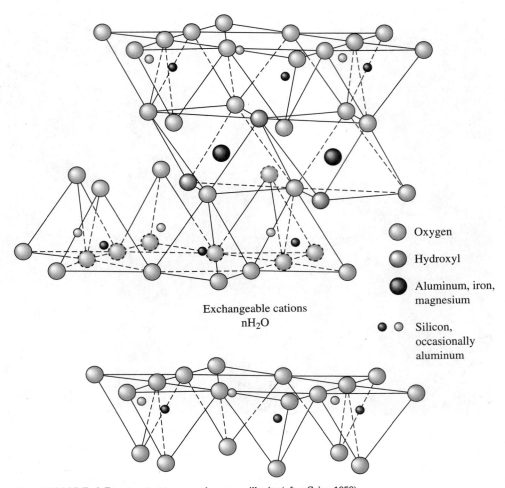

▼ FIGURE 1.7   Atomic structure of montmorillonite (after Grim, 1959)

average surface densities of the negative charges on the surfaces of some clay minerals follows (Yong and Warkentin, 1966):

Clay mineral	Reciprocal of average surface density of charge (Å^2/electronic charge)
Kaolinite	25
Clay mica and chlorite	50
Montmorillonite	100
Vermiculite	75

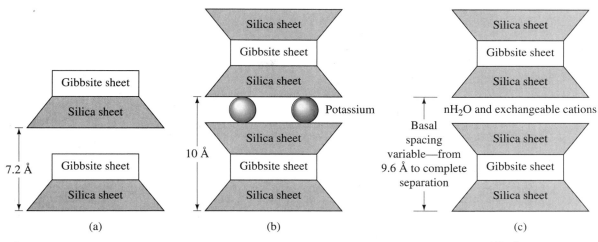

▼ **FIGURE 1.8**   Diagram of the structures of (a) kaolinite; (b) illite; (c) montmorillonite

In dry clay, the negative charge is balanced by exchangeable cations like Ca^{2+}, Mg^{2+}, Na^{+}, and K^{+} surrounding the particles being held by electrostatic attraction. When water is added to clay, these cations and a small number of anions float around the clay particles. This is referred to as a *diffuse double layer* (Figure 1.9a). The cation concentration decreases with the distance from the surface of the particle (Figure 1.9b).

Water molecules are polar. Hydrogen atoms are not axisymmetric around an oxygen atom; instead, they occur at a bonded angle of 105° (Figure 1.10). As a result, a

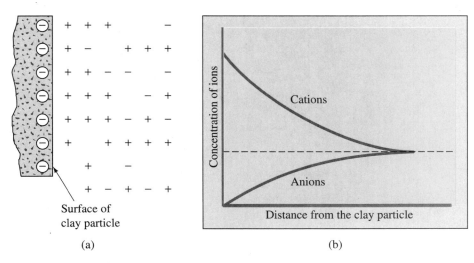

▼ **FIGURE 1.9**   Diffuse double layer

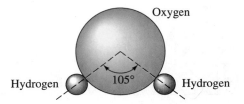

▼   **FIGURE 1.10**   Dipolar character of water

water molecule has a positive charge at one side and a negative charge at the other side. It is known as a *dipole*.

Dipolar water is attracted both by the negatively charged surface of the clay particles and by the cations in the double layer. The cations, in turn, are attracted to the soil particles. A third mechanism by which water is attracted to clay particles is *hydrogen bonding*, where hydrogen atoms in the water molecules are shared with oxygen atoms on the surface of the clay. Some partially hydrated cations in the pore water are also attracted to the surface of clay particles. These cations attract dipolar water molecules. All of these possible mechanics of attraction of water to clay are shown in

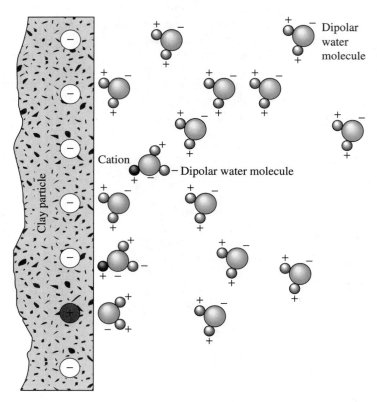

▼   **FIGURE 1.11**   Attraction of dipolar molecules in diffuse double layer

Figure 1.11. The force of attraction between water and clay decreases with distance from the surface of the particles. All of the water held to clay particles by force of attraction is known as *double-layer water*. The innermost layer of double-layer water, which is held very strongly by clay, is known as *adsorbed water*. This water is more viscous than is free water.

Figure 1.12 shows the adsorbed and double-layer water for typical montmorillonite and kaolinite particles. This orientation of water around the clay particles gives clay soils their plastic properties.

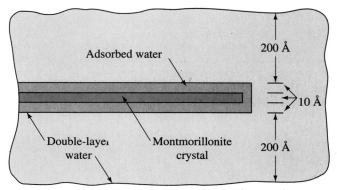

Typical montmorillonite particle, 1000 Å by 10 Å

(a)

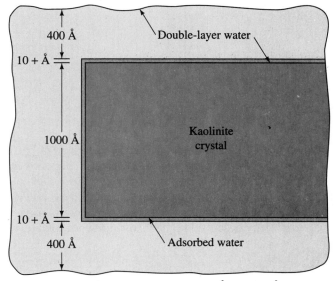

Typical kaolinite particle, 10,000 Å by 1000 Å

(b)

▼ **FIGURE 1.12** Clay water (redrawn after Lambe, 1958)

1.4 SPECIFIC GRAVITY (G_s)

Specific gravity is defined as the ratio of the unit weight of a given material to the unit weight of water. The specific gravity of soil solids is often needed for various calculations in soil mechanics. It can be determined accurately in the laboratory. Table 1.4 shows the specific gravity of some common minerals found in soils. Most of the values fall within a range of 2.6 to 2.9. The specific gravity of solids of light colored sand, which is mostly made of quartz, may be estimated to be about 2.65; for clayey and silty soils, it may vary from 2.6 to 2.9.

▼ **TABLE 1.4   Specific Gravity of Common Minerals**

Mineral	Specific gravity, G_s
Quartz	2.65
Kaolinite	2.6
Illite	2.8
Montmorillonite	2.65–2.80
Halloysite	2.0–2.55
Potassium feldspar	2.57
Sodium and calcium feldspar	2.62–2.76
Chlorite	2.6–2.9
Biotite	2.8–3.2
Muscovite	2.76–3.1
Hornblende	3.0–3.47
Limonite	3.6–4.0
Olivine	3.27–3.7

1.5 MECHANICAL ANALYSIS OF SOIL

Mechanical analysis is the determination of the size range of particles present in a soil, expressed as a percentage of the total dry weight. There are two methods generally used to find the particle-size distribution of soil: (1) *sieve analysis*—for particle sizes larger than 0.075 mm in diameter, and (2) *hydrometer analysis*—for particle sizes smaller than 0.075 mm in diameter. The basic principles of sieve analysis and hydrometer analysis are briefly described in the following two sections.

Sieve Analysis

Sieve analysis consists of shaking the soil sample through a set of sieves that have progressively smaller openings. U.S. standard sieve numbers and the sizes of openings are given in Table 1.5.

▼ **TABLE 1.5** U.S. Standard Sieve Sizes

Sieve no.	Opening (mm)
4	4.750
6	3.350
8	2.360
10	2.000
16	1.180
20	0.850
30	0.600
40	0.425
50	0.300
60	0.250
80	0.180
100	0.150
140	0.106
170	0.088
200	0.075
270	0.053

First the soil is oven dried and then all lumps are broken into small particles before they are passed through the sieves. Figure 1.13 shows a set of sieves in a sieve shaker used for conducting the test in the laboratory. After the completion of the shaking period, the mass of soil retained on each sieve is determined. When cohesive soils are analyzed, it may be difficult to break lumps into individual particles. In that case, the soil may be mixed with water to make a slurry and then washed through the sieves. Portions retained on each sieve are collected separately and oven dried before the mass retained on each sieve is measured.

The results of sieve analysis are generally expressed in terms of the percentage of the total weight of soil that passed through different sieves. Table 1.6 shows an example of the calculations required in a sieve analysis.

Hydrometer Analysis

Hydrometer analysis is based on the principle of sedimentation of soil grains in water. When a soil specimen is dispersed in water, the particles settle at different velocities, depending on their shape, size, weight, and the viscosity of the water. For simplicity, it is assumed that all the soil particles are spheres and the velocity of soil particles can be

▼ **FIGURE 1.13**   Set of sieves in a sieve shaker (courtesy of Soiltest, Inc., Lake Bluff, Illinois)

▼ **TABLE 1.6**   **Sieve Analysis (Mass of Dry Soil Sample = 450 g)**

Sieve no. (1)	Diameter (mm) (2)	Mass of soil retained on each sieve (g) (3)	Percent of soil retained on each sieve[a] (4)	Percent passing[b] (5)
10	2.000	0	0	100.00
16	1.180	9.90	2.20	97.80
30	0.600	24.66	5.48	92.32
40	0.425	17.60	3.91	88.41
60	0.250	23.90	5.31	83.10
100	0.150	35.10	7.80	75.30
200	0.075	59.85	13.30	62.00
Pan	—	278.99	62.00	0

[a] Column 4 = [(column 3)/(total mass of soil)] × 100
[b] This is also referred to as *percent finer*.

expressed by *Stokes' law*, according to which

$$v = \frac{\gamma_s - \gamma_w}{18\eta} D^2 \tag{1.1}$$

where v = velocity
  γ_s = unit weight of soil particles
  γ_w = unit weight of water
  η = viscosity of water
  D = diameter of soil particles

Thus, from Eq. (1.1),

$$D = \sqrt{\frac{18\eta v}{\gamma_s - \gamma_w}} = \sqrt{\frac{18\eta}{\gamma_s - \gamma_w}} \sqrt{\frac{L}{t}} \tag{1.2}$$

where $v = \dfrac{\text{distance}}{\text{time}} = \dfrac{L}{t}$.

Note that

$$\gamma_s = G_s \gamma_w \tag{1.3}$$

Thus, combining Eqs. (1.2) and (1.3) gives

$$D = \sqrt{\frac{18\eta}{(G_s - 1)\gamma_w}} \sqrt{\frac{L}{t}} \tag{1.4}$$

If the units of η are $(g \cdot sec)/cm^2$, γ_w is in g/cm^3, L is in cm, t is in min, and D is in mm, then

$$\frac{D \text{ (mm)}}{10} = \sqrt{\frac{18\eta \ [(g \cdot sec)/cm^2]}{(G_s - 1)\gamma_w \ (g/cm^3)}} \sqrt{\frac{L \text{ (cm)}}{t \text{ (min)} \times 60}}$$

or

$$D = \sqrt{\frac{30\eta}{(G_s - 1)\gamma_w}} \sqrt{\frac{L}{t}}$$

Assume γ_w to be approximately equal to $1 \ g/cm^3$, so that

$$D \text{ (mm)} = K \sqrt{\frac{L \text{ (cm)}}{t \text{ (min)}}} \tag{1.5}$$

where

$$K = \sqrt{\frac{30\eta}{(G_s - 1)}} \qquad (1.6)$$

Note that the value of K is a function of G_s and η, which are dependent on the temperature of the test. Table 1.7 gives the variation of K with the test temperature and the specific gravity of soil solids.

In the laboratory, the hydrometer test is conducted in a sedimentation cylinder with 50 g of oven-dried sample. The sedimentation cylinder is 18 in. (457.2 mm) high and 2.5 in. (63.5 mm) in diameter. It is marked for a volume of 1000 ml. Sodium hexametaphosphate is generally used as the *dispersing agent*. The volume of the dispersed soil suspension is brought up to 1000 ml by adding distilled water. Figure 1.14 shows an ASTM 152H type of hydrometer.

When a hydrometer is placed in the soil suspension at a time t, measured from the start of sedimentation, it measures the specific gravity in the vicinity of its bulb at a depth L (Figure 1.15). The specific gravity is a function of the amount of soil particles present per unit volume of suspension at that depth. Also, at a time t, the soil particles

▼ **TABLE 1.7**   **Values of K from Eq. (1.6)[a]**

Temperature (°C)	G_s							
	2.45	2.50	2.55	2.60	2.65	2.70	2.75	2.80
16	0.01510	0.01505	0.01481	0.01457	0.01435	0.01414	0.01394	0.01374
17	0.01511	0.01486	0.01462	0.01439	0.01417	0.01396	0.01376	0.01356
18	0.01492	0.01467	0.01443	0.01421	0.01399	0.01378	0.01359	0.01339
19	0.01474	0.01449	0.01425	0.01403	0.01382	0.01361	0.01342	0.01323
20	0.01456	0.01431	0.01408	0.01386	0.01365	0.01344	0.01325	0.01307
21	0.01438	0.01414	0.01391	0.01369	0.01348	0.01328	0.01309	0.01291
22	0.01421	0.01397	0.01374	0.01353	0.01332	0.01312	0.01294	0.01276
23	0.01404	0.01381	0.01358	0.01337	0.01317	0.01297	0.01279	0.01261
24	0.01388	0.01365	0.01342	0.01321	0.01301	0.01282	0.01264	0.01246
25	0.01372	0.01349	0.01327	0.01306	0.01286	0.01267	0.01249	0.01232
26	0.01357	0.01334	0.01312	0.01291	0.01272	0.01253	0.01235	0.01218
27	0.01342	0.01319	0.01297	0.01277	0.01258	0.01239	0.01221	0.01204
28	0.01327	0.01304	0.01283	0.01264	0.01244	0.01225	0.01208	0.01191
29	0.01312	0.01290	0.01269	0.01249	0.01230	0.01212	0.01195	0.01178
30	0.01298	0.01276	0.01256	0.01236	0.01217	0.01199	0.01182	0.01169

[a] After ASTM (1991)

▼ **FIGURE 1.14** ASTM 152H
hydrometer (courtesy of Soiltest, Inc.,
Lake Bluff, Illinois)

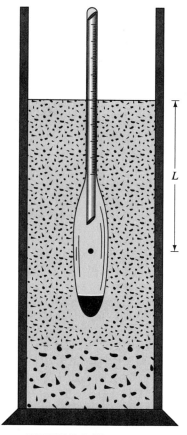

▼ **FIGURE 1.15** Definition of L in
hydrometer test

in suspension at a depth L will have a diameter smaller than D as calculated in Eq. (1.5).
The larger particles would have settled beyond the zone of measurement. Hydrometers
are designed to give the amount of soil, in grams, that is still in suspension. Hydrometers are calibrated for soils that have a specific gravity, G_s, of 2.65; for soils of other
specific gravity, it is necessary to make a correction.

By knowing the amount of soil in suspension, L, and t, we can calculate the
percentage of soil by weight finer than a given diameter. Note that L is the depth
measured from the surface of the water to the center of gravity of the hydrometer bulb
at which the density of the suspension is measured. The value of L will change with
time t; its variation with the hydrometer readings is given in the Annual Book of
ASTM Standards (1991—see Test Designation D-422, Table 2). Hydrometer analysis is
effective for separating soil fractions down to a size of about 0.5 μ.

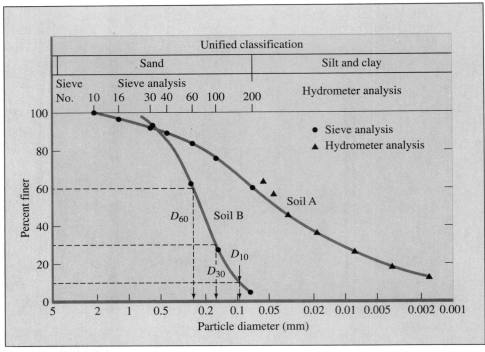

▼   **FIGURE 1.16**   Particle-size distribution curves

Particle-Size Distribution Curve

The results of mechanical analysis (sieve and hydrometer analyses) are generally presented by semilogarithmic plots known as *particle-size distribution curves*. The particle diameters are plotted in log scale, and the corresponding percent finer in arithmetic scale. As an example, the particle-size distribution curves for two soils are shown in Figure 1.16. The particle-size distribution curve for soil A is the combination of the sieve analysis results presented in Table 1.6 and the results of the hydrometer analysis for the finer fraction. When the results of sieve analysis and hydrometer analysis are combined, a discontinuity generally occurs in the range where they overlap. This is because soil particles are generally irregular in shape. Sieve analysis gives the intermediate dimension of a particle; hydrometer analysis gives the diameter of a sphere that would settle at the same rate as the soil particle.

The percentages of gravel, sand, silt, and clay-size particles present in a soil can be obtained from the particle-size distribution curve. According to the Unified soil classification, soil A in Figure 1.16 has:

▶   *Gravel* (size limits—greater than 4.75 mm) = 0%
▶   *Sand* (size limits—4.75 to 0.075 mm) = percent finer than 4.75 mm diameter − percent finer than 0.075 mm diameter = 100 − 62 = 38%
▶   *Silt and clay* (size limits—less than 0.075 mm) = 62%

1.6 EFFECTIVE SIZE, UNIFORMITY COEFFICIENT, AND COEFFICIENT OF GRADATION

The particle-size distribution curves can be used for comparing different soils. Also, three basic soil parameters can be determined from these curves, and they can be used to classify granular soils. These parameters are:

1. Effective size
2. Uniformity coefficient
3. Coefficient of gradation

The diameter in the particle-size distribution curve corresponding to 10% finer is defined as the *effective size*, or D_{10}. The *uniformity coefficient* is given by the relation

$$C_u = \frac{D_{60}}{D_{10}} \qquad (1.7)$$

where C_u = uniformity coefficient
D_{60} = the diameter corresponding to 60% finer in the particle-size distribution curve

The *coefficient of gradation* may be expressed as

$$C_c = \frac{D_{30}^2}{D_{60} \times D_{10}} \qquad (1.8)$$

where C_c = coefficient of gradation
D_{30} = diameter corresponding to 30% finer

For the particle-size distribution curve of soil B shown in Figure 1.16, the values of D_{10}, D_{30}, and D_{60} are 0.096 mm, 0.16 mm, and 0.24 mm, respectively. The uniformity coefficient and coefficient of gradation are

$$C_u = \frac{D_{60}}{D_{10}} = \frac{0.24}{0.096} = 2.5$$

$$C_c = \frac{D_{30}^2}{D_{60} \times D_{10}} = \frac{(0.16)^2}{0.24 \times 0.096} = 1.11$$

The particle-size distribution curve shows not only the range of particle sizes present in a soil but also the type of distribution of various size particles. This is demonstrated in Figure 1.17. Curve I represents a type of soil in which most of the soil grains are the same size. This is called *poorly graded* soil. Curve II represents a soil in

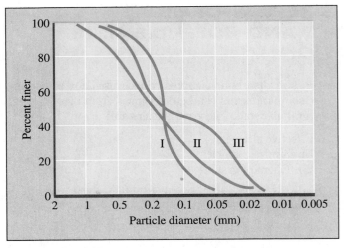

FIGURE 1.17   Different types of particle-size distribution curves

which the particle sizes are distributed over a wide range, termed *well graded*. A well-graded soil will have a uniformity coefficient greater than about 4 for gravels and 6 for sands, and a coefficient of gradation between 1 and 3 (for gravels and sands). A soil might have a combination of two or more uniformly graded fractions. Curve III represents such a soil. This type of soil is termed *gap graded*.

▼ **EXAMPLE 1.1**

Following are the results of a sieve analysis:

U.S. sieve no.	Mass of soil retained (g)
4	28
10	42
20	48
40	128
60	221
100	86
200	40
Pan	24

   a. Determine the percent finer than each sieve and plot a grain-size distribution curve.
   b. Determine D_{10}, D_{30}, and D_{60} from the grain-size distribution curve.
   c. Calculate the uniformity coefficient, C_u.
   d. Calculate the coefficient of gradation, C_c.

Solution

a. The following table can be prepared for obtaining the percent finer:

Sieve no.	Mass of soil retained on each sieve (g)	Percent retained on each sieve	Percent finer
4	28	4.54	95.46
10	42	6.81	88.65
20	48	7.78	80.87
40	128	20.75	60.12
60	221	35.82	24.3
100	86	13.93	10.37
200	40	6.48	3.89
Pan	24	3.89	0
	617 g		

The grain-size distribution is shown in Figure 1.18.

b. From Figure 1.18,

$$D_{10} = \textbf{0.14 mm}$$

$$D_{30} = \textbf{0.27 mm}$$

$$D_{60} = \textbf{0.42 mm}$$

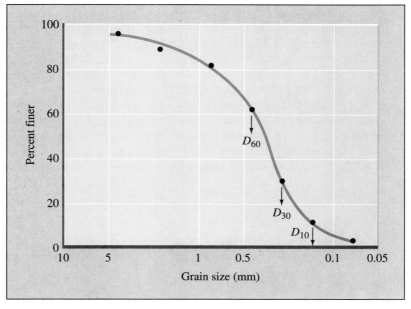

▼ **FIGURE 1.18** Grain-size distribution

c. From Eq. (1.7),

$$C_u = \frac{D_{60}}{D_{10}} = \frac{0.42}{0.14} = \mathbf{3.0}$$

d. From Eq. (1.8),

$$C_c = \frac{D_{30}^2}{D_{60} \times D_{10}} = \frac{(0.27)^2}{(0.42)(0.14)} = \mathbf{1.24} \qquad \blacktriangledown$$

1.7 GENERAL COMMENTS

This chapter was made up of primarily three descriptions. The first was about the formation of soil by various natural processes. In many instances, knowing the process of formation of a given soil deposit helps a geotechnical engineer to anticipate problem areas in the design and construction of a structure's foundation. It helps an engineer to watch for possible areas of high clay content and also to locate borrow material for various types of construction work.

The second description was about clay and clay minerals. It needs to be well recognized that the presence of clay minerals in a soil aggregate has a great influence on the engineering properties of the soil as a whole. When moisture is present, the engineering behavior of a soil will change greatly as the percentage of clay mineral content increases. For all practical purposes, when the clay content is about 50% or more, the sand and silt particles float in a clay matrix, and the clay minerals primarily dictate the engineering properties of the soil.

The third topic was the particle-size distribution in a given soil. The range of particle sizes present in a given soil can be very large. The effective size, D_{10}, of a granular soil is a good measure to estimate the permeability and drainage through the soil (see Chapter 5). The uniformity coefficient, C_u [Eq. (1.7)], of granular soils can vary widely. For well-graded sands, it can be as large as 10 to 15 or more. The coefficient of gradation, C_c, is defined in Eq. (1.8). In some instances, it is also referred to as the *coefficient of curvature* or the *coefficient of concavity*.

Although the grain-size distribution is generally used for the engineering classification of granular soils (see Chapter 3), the shape of particles present in the soil has a great influence in controlling engineering properties such as compressibility (Chapter 8) and shear strength (Chapter 9). Geologists use such terms as *angular*, *subangular*, *rounded*, and *subrounded* (Figure 1.19) to describe the shape of bulky particles. Particles

Angular

Subangular

Subrounded

Rounded

▼ **FIGURE 1.19**   Shapes of bulky particles

derived from artificially crushed rocks have highly angular shapes, whereas mica and clay minerals have flaky shapes.

PROBLEMS **1.1** For a soil with $D_{60} = 0.38$ mm, $D_{30} = 0.19$ mm, and $D_{10} = 0.11$ mm, calculate the uniformity coefficient and the coefficient of gradation.

1.2 Repeat Problem 1.1 with the following values: $D_{10} = 0.18$ mm, $D_{30} = 0.32$ mm, and $D_{60} = 0.78$ mm.

1.3 Following are the results of a sieve analysis:

U.S. sieve no.	Mass retained on each sieve (g)
4	0
10	40
20	60
40	89
60	140
80	122
100	210
200	56
Pan	12

a. Determine the percent finer than each sieve size and plot a grain-size distribution curve.
b. Determine D_{10}, D_{30}, and D_{60} from the grain-size distribution curve.
c. Calculate the uniformity coefficient, C_u.
d. Calculate the coefficient of gradation, C_c.

1.4 Repeat Problem 1.3 with the following results of a sieve analysis:

U.S. sieve no.	Mass of soil retained on each sieve (g)
4	0
10	18.5
20	53.2
40	90.5
60	81.8
100	92.2
200	58.5
Pan	26.5

1.5   Repeat Problem 1.3 with the following results of a sieve analysis:

U.S. sieve no.	Mass of soil retained on each seive (g)
4	0
10	41.2
20	55.1
40	80.0
60	91.6
100	60.5
200	35.6
Pan	21.5

1.6   Repeat Problem 1.3 with the following results of a sieve analysis:

U.S. sieve no.	Mass of soil retained on each seive (g)
4	0
6	0
10	20.1
20	19.5
40	210.5
60	85.6
100	22.7
200	15.5
Pan	23.5

1.7   The particle-size characteristics of a soil are given in this table. Draw the particle-size distribution curve.

Size (mm)	Percent finer
0.425	100
0.033	90
0.018	80
0.01	70
0.0062	60
0.0035	50
0.0018	40
0.001	35

Determine the percentages of gravel, sand, silt, and clay:
a.   according to the USDA system.
b.   according to the AASHTO system.

1.8 Repeat Problem 1.7 with the following data:

Size (mm)	Percent finer
0.425	100
0.1	92
0.052	84
0.02	62
0.01	46
0.004	32
0.001	22

1.9 Repeat Problem 1.7 with the following values:

Size (mm)	Percent finer
0.425	100
0.1	79
0.04	57
0.02	48
0.01	40
0.002	35
0.001	33

1.10 Repeat Problem 1.7 with the following data:

Size (mm)	Percent finer
0.425	100
0.07	90
0.046	80
0.034	70
0.026	60
0.019	50
0.014	40
0.009	30
0.0054	20
0.0019	10

1.11 In a hydrometer test, the results are: $G_s = 2.60$, temperature of water $= 24°C$, and $L = 9.2$ cm at 60 min after the start of sedimentation (see Figure 1.15). What is the diameter, D, of the smallest size particles that have settled beyond the zone of measurement at that time (that is, $t = 60$ min)?

1.12 Repeat Problem 1.11 with the following values: $G_s = 2.70$, temperature $= 23°C$, $t = 120$ min, and $L = 12.8$ cm.

REFERENCES

American Society for Testing and Materials (1991). *ASTM Book of Standards*, Sec. 4, Vol. 04.08, Philadelphia. Pa.

Bowen, N. L. (1922). "The Reaction Principles in Petrogenesis," *Journal of Geology*, Vol. 30, 177–198.

Grim, R. E. (1953). *Clay Mineralogy*, McGraw-Hill, New York.

Grim, R. E. (1959). "Physico-Chemical Properties of Soils: Clay Minerals," *Journal of the Soil Mechanics and Foundations Division*, ASCE, Vol. 85, No. SM2, 1–17.

Lambe, T. W. (1958). "The Structure of Compacted Clay," *Journal of the Soil Mechanics and Foundations Division*, ASCE, Vol. 84, No. SM2, 1655-1 to 1655-35.

Terzaghi, K. (1925). *Erdbaumechanik auf Bodenphysikalischer Grundlage*, Deuticke, Vienna.

Yong, R. N., and Warkentin, B. P. (1966). *Introduction of Soil Behavior*, Macmillan, New York.

Supplementary References for Further Study

Mitchell, J. K. (1993). *Fundamentals of Soil Behavior*, 2nd ed., Wiley, New York.

Van Olphen, H. (1963). *An Introduction to Clay Colloid Chemistry*, Wiley Interscience, New York.

SOIL COMPOSITION

The preceding chapter presented the geological processes by which soils are formed, the description of limits on the sizes of soil particles, and the mechanical analysis of soils. In natural occurrence, soils are three-phase systems consisting of soil solids, water, and air. This chapter discusses the weight-volume relationships of soil aggregates, along with their structures and plasticity.

2.1 WEIGHT-VOLUME RELATIONSHIPS

Figure 2.1a shows an element of soil of volume V and weight W as it would exist in a natural state. To develop the weight-volume relationship, the three phases (that is, solid, water, and air) are separated as shown in Figure 2.1b. Thus, the total volume of a given soil sample can be expressed as

$$V = V_s + V_v = V_s + V_w + V_a \tag{2.1}$$

where V_s = volume of soil solids
V_v = volume of voids
V_w = volume of water in the voids
V_a = volume of air in the voids

Assuming the weight of the air is negligible, we can give the total weight of the sample as

$$W = W_s + W_w \tag{2.2}$$

where W_s = weight of soil solids
W_w = weight of water

The *volume relationships* commonly used for the three phases in a soil element are *void ratio*, *porosity*, and *degree of saturation*. Void ratio is defined as the ratio of the volume of voids to the volume of solids. Thus,

$$e = \frac{V_v}{V_s} \tag{2.3}$$

where e = void ratio.

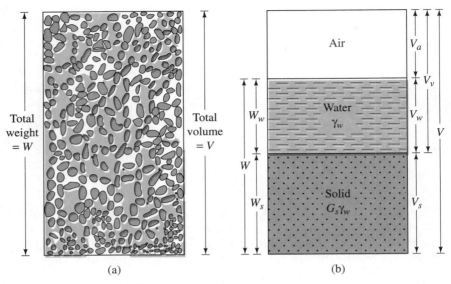

▼ **FIGURE 2.1**   (a) Soil element in natural state; (b) three phases of the soil element

Porosity is defined as the ratio of the volume of voids to the total volume, or

$$n = \frac{V_v}{V} \qquad (2.4)$$

where n = porosity.

The degree of saturation is defined as the ratio of the volume of water to the volume of voids, or

$$S = \frac{V_w}{V_v} \qquad (2.5)$$

where S = degree of saturation. It is commonly expressed as a percentage.

The relationship between void ratio and porosity can be derived from Eqs. (2.1), (2.3), and (2.4) as follows:

$$e = \frac{V_v}{V_s} = \frac{V_v}{V - V_v} = \frac{\left(\dfrac{V_v}{V}\right)}{1 - \left(\dfrac{V_v}{V}\right)} = \frac{n}{1 - n} \qquad (2.6)$$

Also, from Eq. (2.6),

$$n = \frac{e}{1 + e} \tag{2.7}$$

The common terms used for *weight relationships* are *moisture content* and *unit weight*. The definitions of these terms follow.

Moisture content (*w*) is also referred to as *water content* and is defined as the ratio of the weight of water to the weight of solids in a given volume of soil:

$$w = \frac{W_w}{W_s} \tag{2.8}$$

Unit weight (γ) is the weight of soil per unit volume. Thus,

$$\gamma = \frac{W}{V} \tag{2.9}$$

The unit weight can also be expressed in terms of the weight of soil solids, the moisture content, and the total volume. From Eqs. (2.2), (2.8), and (2.9),

$$\gamma = \frac{W}{V} = \frac{W_s + W_w}{V} = \frac{W_s\left[1 + \left(\dfrac{W_w}{W_s}\right)\right]}{V} = \frac{W_s(1 + w)}{V} \tag{2.10}$$

Soils engineers sometimes refer to the unit weight defined by Eq. (2.9) as the *moist unit weight*.

Often in earthwork problems it is necessary to know the weight per unit volume of soil, excluding water. This is referred to as the *dry unit weight*, γ_d. Thus,

$$\gamma_d = \frac{W_s}{V} \tag{2.11}$$

From Eqs. (2.10) and (2.11), the relationship of unit weight, dry unit weight, and moisture content can be given as

$$\gamma_d = \frac{\gamma}{1 + w} \tag{2.12}$$

Unit weight is expressed in English units (a gravitational system of measurement) as pounds per cubic foot (lb/ft^3). In SI (Système Internationale), the unit used is Newtons per cubic meter (N/m^3).

Since the Newton is a derived unit, it may sometimes be convenient to work with mass densities (ρ) of soil. The SI unit of mass density is kilograms per cubic meter (kg/m^3). We can write the density equations [similar to Eqs. (2.9) and (2.11)] as

$$\rho = \frac{m}{V} \tag{2.13a}$$

and

$$\rho_d = \frac{m_s}{V} \tag{2.13b}$$

where ρ = density of soil (kg/m^3)
ρ_d = dry density of soil (kg/m^3)
m = total mass of the soil sample (kg)
m_s = mass of soil solids in the sample (kg)

The unit of total volume, V, is m^3.

The unit weights of soil in N/m^3 can be obtained from densities in kg/m^3 as

$$\gamma = \rho g = 9.81\rho \tag{2.14a}$$

and

$$\gamma_d = \rho_d g = 9.81\rho_d \tag{2.14b}$$

where g = acceleration due to gravity = 9.81 m/sec^2.

Conversion factors for unit weight from English to SI units are as follows:

$$1 \text{ lb/ft}^3 = 0.1572 \text{ kN/m}^3$$
$$1 \text{ lb/ft}^3 = 16.0256 \text{ kgf/m}^3$$
$$\text{Unit weight of water, } \gamma_W = \begin{cases} 62.4 \text{ lb/ft}^3 \\ 9.81 \text{ kN/m}^3 \\ 1000 \text{ kgf/m}^3 \end{cases}$$

2.2 RELATIONSHIP AMONG UNIT WEIGHT, VOID RATIO, MOISTURE CONTENT, AND SPECIFIC GRAVITY

To obtain a relationship among unit weight (or density), void ratio, and moisture content, consider a volume of soil in which the volume of the soil solids is 1, as shown in Figure 2.2. If the volume of the soil solids is 1, then the volume of voids is numerically equal to the void ratio, e [from Eq. (2.3)]. The weights of soil solids and water can

Weight   Volume

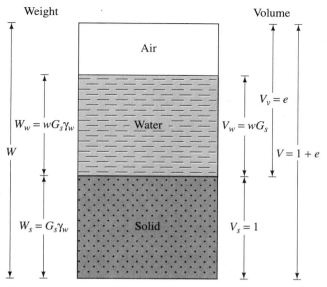

▼ **FIGURE 2.2**   Three separate phases of a soil element with volume of soil solids equal to 1

be given as

$$W_s = G_s \gamma_w$$

$$W_w = w W_s = w G_s \gamma_w$$

where G_s = specific gravity of soil solids
$\quad w$ = moisture content
$\quad \gamma_w$ = unit weight of water

Now, using the definitions of unit weight and dry unit weight [Eqs. (2.9) and (2.11)], we can write

$$\gamma = \frac{W}{V} = \frac{W_s + W_w}{V} = \frac{G_s \gamma_w + w G_s \gamma_w}{1 + e} = \frac{(1 + w) G_s \gamma_w}{1 + e} \tag{2.15}$$

and

$$\gamma_d = \frac{W_s}{V} = \frac{G_s \gamma_w}{1 + e} \tag{2.16}$$

or

$$e = \frac{G_s \gamma_w}{\gamma_d} - 1 \tag{2.17}$$

Since the weight of water for the soil element under consideration is $wG_s\gamma_w$, the volume occupied by it is

$$V_w = \frac{W_w}{\gamma_w} = \frac{wG_s\gamma_w}{\gamma_w} = wG_s$$

Hence, from the definition of degree of saturation [Eq. (2.5)],

$$S = \frac{V_w}{V_v} = \frac{wG_s}{e}$$

or

$$Se = wG_s \tag{2.18}$$

This is a very useful equation for solving problems involving three-phase relationships.

If the soil sample is *saturated*—that is, the void spaces are completely filled with water (Figure 2.3)—the relationship for saturated unit weight can be derived in a

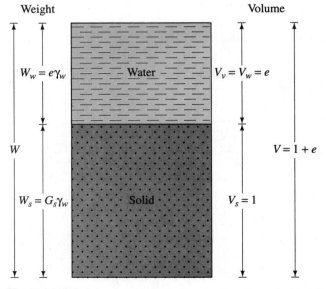

▼ **FIGURE 2.3**   Saturated soil element with volume of soil solids equal to 1

similar manner:

$$\gamma_{sat} = \frac{W}{V} = \frac{W_s + W_w}{V} = \frac{G_s \gamma_w + e\gamma_w}{1 + e} = \frac{(G_s + e)\gamma_w}{1 + e} \tag{2.19}$$

where γ_{sat} = saturated unit weight of soil.
Also, from Eq. (2.18) with $S = 1$,

$$e = wG_s \tag{2.20}$$

▼ **EXAMPLE 2.1**

For a soil, show that

$$\gamma_{sat} = \left(\frac{e}{w}\right)\left(\frac{1 + w}{1 + e}\right)\gamma_w$$

Solution   From Eqs. (2.19) and (2.20),

$$\gamma_{sat} = \frac{(G_s + e)\gamma_w}{1 + e} \tag{a}$$

and

$$e = wG_s$$

or

$$G_s = \frac{e}{w} \tag{b}$$

Combining Eqs. (a) and (b) gives

$$\gamma_{sat} = \frac{\left(\frac{e}{w} + e\right)\gamma_w}{1 + e} = \left(\frac{e}{w}\right)\left(\frac{1 + w}{1 + e}\right)\gamma_w \qquad ▼$$

▼ **EXAMPLE 2.2**

A moist soil has these values: $V = 0.25$ ft^3, $W = 30.75$ lb, $w = 9.8\%$, and $G_s = 2.66$.
Determine the following:

    a. γ (lb/ft^3)
    b. γ_d (lb/ft^3)
    c. e
    d. n
    e. S (%)
    f. Volume occupied by water (lb)

Solution

a. From Eq. (2.9),

$$\gamma = \frac{W}{V} = \frac{30.75}{0.25} = \textbf{123 lb/ft}^3$$

b. From Eq. (2.12),

$$\gamma_d = \frac{\gamma}{1 + w} = \frac{123}{1 + \left(\dfrac{9.8}{100}\right)} = \textbf{112 lb/ft}^3$$

c. From Eq. (2.17),

$$e = \frac{G_s \gamma_w}{\gamma_d} - 1$$

$$e = \frac{(2.66)(62.4)}{112} - 1 = \textbf{0.48}$$

d. From Eq. (2.7),

$$n = \frac{e}{1 + e} = \frac{0.48}{1 + 0.48} = \textbf{0.324}$$

e. From Eq. (2.18),

$$S(\%) = \left(\frac{wG_s}{e}\right)(100) = \frac{(0.098)(2.66)}{0.48}(100) = \textbf{54.3\%}$$

f. Weight of soil solids is

$$W_s = \frac{W}{1 + w} = \frac{30.75}{1 + 0.098} = 28 \text{ lb}$$

Thus, weight of water is

$$W_w = W - W_s = 30.75 - 28 = 2.75 \text{ lb}$$

Volume of water is

$$V_w = \frac{W_w}{\gamma_w} = \frac{2.75}{62.4} = \textbf{0.044 ft}^3$$

Alternate Solution   Dry weight of soil is

$$W_s = \frac{W}{1 + w} = \frac{30.75}{1 + \left(\dfrac{9.8}{100}\right)} = 28 \text{ lb}$$

Thus, weight of water is

$$W_w = W - W_s = 30.75 - 28 = 2.75 \text{ lb}$$

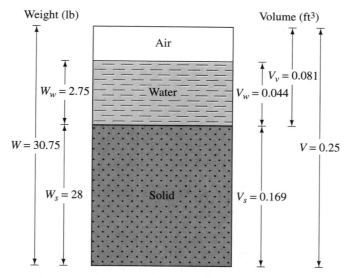

▼ FIGURE 2.4

Volume of soil solids is

$$V_s = \frac{W_s}{G_s \gamma_w} = \frac{28}{(2.66)(62.4)} = 0.169 \text{ ft}^3$$

Thus,

$$V_v = V - V_s = 0.25 - 0.169 = 0.081 \text{ ft}^3$$

Volume of water is

$$V_w = \frac{W_w}{\gamma_w} = \frac{2.75}{62.4} = 0.044 \text{ ft}^3$$

Now refer to Figure 2.4.

    a. From Eq. (2.9),

$$\gamma = \frac{W}{V} = \frac{30.75}{0.25} = \textbf{123 lb/ft}^3$$

    b. From Eq. (2.11),

$$\gamma_d = \frac{W_s}{V} = \frac{28}{0.25} = \textbf{112 lb/ft}^3$$

    c. From Eq. (2.3),

$$e = \frac{V_v}{V_s} = \frac{0.081}{0.169} = \textbf{0.48}$$

d. From Eq. (2.4),

$$n = \frac{V_v}{V} = \frac{0.081}{0.25} = \mathbf{0.324}$$

e. From Eq. (2.5),

$$S = \frac{V_w}{V_v} = \frac{0.044}{0.081} = 0.54 = \mathbf{54\%}$$

f. Weight of water is

$$V_w = \mathbf{0.044 \ ft^3} \qquad \blacktriangledown$$

▼ **EXAMPLE 2.3**

The dry density of a sand with a porosity of 0.387 is 1600 kg/m³. Find the void ratio of the soil and the specific gravity of the soil solids.

Solution   Void ratio

From $n = 0.387$ and Eq. (2.6),

$$e = \frac{n}{1-n} = \frac{0.387}{1 - 0.387} = \mathbf{0.631}$$

Specific gravity of soil solids
      From Eqs. (2.14b) and (2.16),

$$\rho_d = \frac{\gamma_d}{g} = \frac{G_s\left(\dfrac{\gamma_w}{g}\right)}{1 + e}$$

or

$$\rho_d = \frac{G\rho_w}{1 + e}$$

where ρ_d = dry density of soil
        ρ_w = density of water = 1000 kg/m³

Thus,

$$1600 = \frac{G_s(1000)}{1 + 0.631}$$

$$G_s = \mathbf{2.61} \qquad \blacktriangledown$$

▼ **EXAMPLE 2.4**

For a saturated soil, given $w = 40\%$ and $G_s = 2.71$, determine the saturated and dry unit weights in lb/ft³ and kN/m³.

Solution   For saturated soil, from Eq. (2.20),

$$e = wG_s = (0.4)(2.71) = 1.084$$

From Eq. (2.19),

$$\gamma_{sat} = \frac{(G_s + e)\gamma_w}{1 + e} = \frac{(2.71 + 1.084)62.4}{1 + 1.084} = \textbf{113.6 lb/ft}^3$$

Also

$$\gamma_{sat} = (113.6)\left(\frac{9.81}{62.4}\right) = \textbf{17.86 kN/m}^3$$

From Eq. (2.16),

$$\gamma_d = \frac{G_s\gamma_w}{1 + e} = \frac{(2.71)(62.4)}{1 + 1.084} = \textbf{81.2 lb/ft}^3$$

Also

$$\gamma_d = (81.2)\left(\frac{9.81}{62.4}\right) = \textbf{12.76 kN/m}^3 \quad \blacktriangledown$$

▼ **EXAMPLE 2.5**

The mass of a moist soil sample collected from the field is 465 g, and its own dry mass is 405.76 g. The specific gravity of the soil solids was determined in the laboratory to be 2.68. If the void ratio of the soil in the natural state is 0.83, find the following:

a. The moist unit weight of the soil in the field (lb/ft³)
b. The dry unit weight of the soil in the field (lb/ft³)
c. The weight of water (in lb) to be added per cubic foot of soil in the field for saturation

Solution

a. From Eq. (2.8),

$$w = \frac{W_w}{W_s} = \frac{465 - 405.76}{405.76} = \frac{59.24}{405.76} = 14.6\%$$

From Eq. (2.15),

$$\gamma = \frac{(1 + w)G_s\gamma_w}{1 + e} = \frac{(1 + 0.146)(2.68)(62.4)}{1 + 0.83} = \textbf{104.73 lb/ft}^3$$

b. From Eq. (2.16),

$$\gamma_d = \frac{G_s\gamma_w}{1 + e} = \frac{(2.68)(62.4)}{1 + 0.83} = \textbf{91.38 lb/ft}^3$$

c.   From Eq. (2.19),

$$\gamma_{\text{sat}} = \frac{(G_s + e)\gamma_w}{1 + e} = \frac{(2.68 + 0.83)(62.4)}{1 + 0.83} = 119.69 \ \text{lb/ft}^3$$

So, the weight of water to be added is

$$\gamma_{\text{sat}} - \gamma = 119.69 - 104.73 = \textbf{14.96 lb/ft}^3 \quad \blacktriangledown$$

2.3   RELATIONSHIP AMONG UNIT WEIGHT, POROSITY, AND MOISTURE CONTENT

The relationship among *unit weight*, *porosity*, and *moisture content* can be developed in a manner similar to that presented in the preceding section. Consider a soil that has a total volume equal to 1, as shown in Figure 2.5. From Eq. (2.4),

$$n = \frac{V_v}{V}$$

If V is equal to 1, then V_v is equal to n, so $V_s = 1 - n$, The weight of soil solids (W_s) and the weight of water (W_w) can then be expressed as follows:

$$W_s = G_s \gamma_w (1 - n) \tag{2.21}$$

$$W_w = w W_s = w G_s \gamma_w (1 - n) \tag{2.22}$$

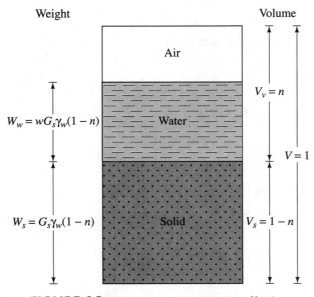

▼   **FIGURE 2.5**   Soil element with total volume $V = 1$

Weight   Volume

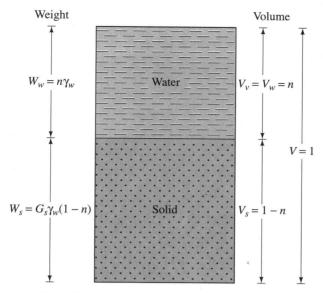

$W_w = n\gamma_w$ Water $V_v = V_w = n$

$V = 1$

$W_s = G_s\gamma_w(1 - n)$ Solid $V_s = 1 - n$

▼ **FIGURE 2.6**   Saturated soil element with total volume $V = 1$

So the dry unit weight equals

$$\gamma_d = \frac{W_s}{V} = \frac{G_s\gamma_w(1 - n)}{1} = G_s\gamma_w(1 - n) \tag{2.23}$$

The moist unit weight equals

$$\gamma = \frac{W_s + W_w}{V} = G_s\gamma_w(1 - n)(1 + w) \tag{2.24}$$

Figure 2.6 shows a soil sample that is saturated and has volume $V = 1$. According to this figure,

$$\gamma_{\text{sat}} = \frac{W_s + W_w}{V} = \frac{(1 - n)G_s\gamma_w + n\gamma_w}{1} = [(1 - n)G_s + n]\gamma_w \tag{2.25}$$

The moisture content of a saturated soil sample can be expressed as

$$w = \frac{W_w}{W_s} = \frac{n\gamma_w}{(1 - n)\gamma_w G_s} = \frac{n}{(1 - n)G_s} \tag{2.26}$$

In Sections 2.2 and 2.3, we derived the fundamental relationships for the moist unit weight, dry unit weight, and saturated unit weight of soil. Various forms of relationships that can be obtained for γ, γ_d, and γ_{sat} are given in Appendix B. Some typical values of void ratio, moisture content in a saturated condition, and dry unit weight encountered in a natural state are given in Table 2.1.

▼   **TABLE 2.1**   **Void Ratio, Moisture Content, and Dry Unit Weight for Some Typical Soils in a Natural State**

Type of soil	Void ratio, e	Natural moisture content in a saturated state (%)	Dry unit weight, γ_d	
			lb/ft³	kN/m³
Loose uniform sand	0.8	30	92	14.5
Dense uniform sand	0.45	16	115	18
Loose angular-grained silty sand	0.65	25	102	16
Dense angular-grained silty sand	0.4	15	121	19
Stiff clay	0.6	21	108	17
Soft clay	0.9–1.4	30–50	73–93	11.5–14.5
Loess	0.9	25	86	13.5
Soft organic clay	2.5–3.2	90–120	38–51	6–8
Glacial till	0.3	10	134	21

2.4   RELATIVE DENSITY

The term *relative density* is commonly used to indicate the *in situ* denseness or looseness of granular soil. It is defined as

$$D_r = \frac{e_{max} - e}{e_{max} - e_{min}} \qquad (2.27)$$

where D_r = relative density, usually given as a percent
  e = *in situ* void ratio of the soil
  e_{max} = void ratio of the soil in the loosest condition
  e_{min} = void ratio of the soil in the densest condition

The values of D_r may vary from a minimum of 0 for very loose soil to a maximum of 1 for very dense soil. Soils engineers qualitatively describe the granular soil deposits according to their relative densities, as shown in Table 2.2. In-place soils seldom have relative densities less than 20% to 30%. Compacting a granular soil to a relative density greater than about 85% is difficult.

By using the definition of dry unit weight given in Eq. (2.16), we can express relative density in terms of maximum and minimum possible dry unit weights. Thus,

$$D_r = \frac{\left[\dfrac{1}{\gamma_{d(min)}}\right] - \left[\dfrac{1}{\gamma_d}\right]}{\left[\dfrac{1}{\gamma_{d(min)}}\right] - \left[\dfrac{1}{\gamma_{d(max)}}\right]} = \left[\frac{\gamma_d - \gamma_{d(min)}}{\gamma_{d(max)} - \gamma_{d(min)}}\right]\left[\frac{\gamma_{d(max)}}{\gamma_d}\right] \qquad (2.28)$$

▼   **TABLE 2.2**   **Qualitative Description of Granular Soil Deposits**

Relative density (%)	Description of soil deposit
0–15	Very loose
15–50	Loose
50–70	Medium
70–85	Dense
85–100	Very dense

where $\gamma_{d(min)}$ = dry unit weight in the loosest condition (at a void ratio of e_{max})

γ_d = *in situ* dry unit weight (at a void ratio of e)

$\gamma_{d(max)}$ = dry unit weight in the densest condition (at a void ratio of e_{min})

ASTM Test Designation D-2049 provides a procedure for the determination of the minimum and maximum dry unit weights of granular soils so that they can be used in Eq. (2.28) to measure the relative density of compaction in the field. For sands, this is done by using a mold with a volume of 0.1 ft³ (2830 cm³). For a determination of the *minimum dry unit weight*, sand is loosely poured into the mold from a funnel with a $\frac{1}{2}$-in. (12.7-mm) diameter spout. The average height of the fall of sand into the mold is kept at about 1 in. (25.4 min). The value of $\gamma_{d(min)}$ can then be determined as

$$\gamma_{d(min)} = \frac{W_s}{V_m} \qquad (2.29)$$

where W_s = weight of sand required to fill the mold

V_m = volume of the mold ($=0.1$ ft³)

The *maximum dry unit weight* is determined by vibrating sand in the mold for 8 min. A surcharge of 2 lb/in.² (13.8 kN/m²) is added to the top of the sand in the mold. The mold is placed on a table that vibrates at a frequency of 3600 cycles/min and that has an amplitude of vibration of 0.025 in. (0.635 mm). The value of $\gamma_{d(max)}$ can be determined at the end of the vibrating period with knowledge of the weight and volume of the sand. Several factors control the magnitude of $\gamma_{d(max)}$: the magnitude of acceleration, the surcharge load, and the geometry of acceleration. Hence, it is possible to obtain a larger value $\gamma_{d(max)}$ than that obtained by the ASTM standard method.

2.5   CONSISTENCY OF SOIL

When clay minerals are present in fine-grained soil, that soil can be remolded in the presence of some moisture without crumbling. This cohesive nature is caused by the adsorbed water surrounding the clay particles. In the early 1900s, a Swedish scientist named Atterberg developed a method to describe the consistency of fine-grained soils

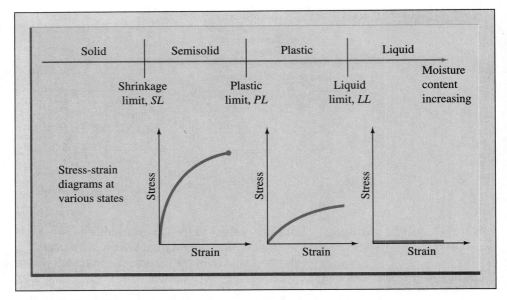

▼   **FIGURE 2.7**   Atterberg limits

with varying moisture contents. At a very low moisture content, soil behaves more like a solid. When the moisture content is very high, the soil and water may flow like a liquid. Hence, on an arbitrary basis, depending on the moisture content, the behavior of soil can be divided into four basic states. They are *solid, semisolid, plastic,* and *liquid,* as shown in Figure 2.7.

The moisture content, in percent, at which the transition from solid to semisolid state takes place is defined as the *shrinkage limit.* The moisture content at the point of transition from semisolid to plastic state is the *plastic limit,* and from plastic to liquid state is the *liquid limit.* These are also known as *Atterberg limits.* We describe the procedures for laboratory determination of Atterberg limits.

Liquid Limit (LL)

A schematic diagram (side view) of a liquid limit device is shown in Figure 2.8a. This device consists of a brass cup and hard rubber base. The brass cup can be dropped onto the base by a cam operated by a crank. In order to perform the liquid limit test, a soil paste is placed in the cup. A groove is cut at the center of the soil pat with the standard grooving tool (Figure 2.8b). Then, by the use of the crank-operated cam, the cup is lifted and dropped from a height of 0.3937 in. (10 mm). The moisture content, in percent, required to close a distance of 0.5 in. (12.7 mm) along the bottom of the groove (see Figures 2.8c and d) after 25 blows is defined as the *liquid limit.*

It is difficult to adjust the moisture content in the soil to meet the required 0.5-in. (12.7-mm) closure of the groove in the soil pat at 25 blows. Hence, at least four tests for the same soil are made at varying moisture contents, with the number of blows, N, required to achieve closure varying between 15 and 35. Figure 2.9 shows a photograph

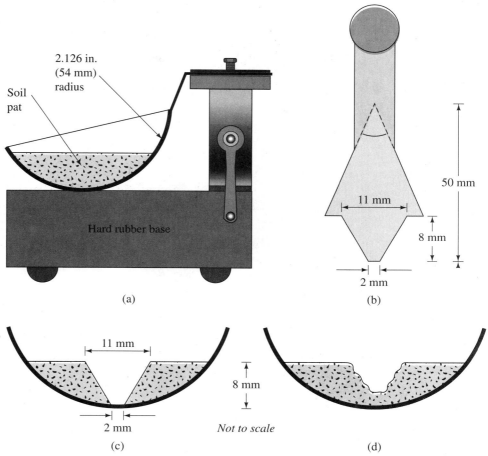

▼ **FIGURE 2.8**   Liquid limit test: (a) liquid limit device; (b) grooving tool; (c) soil pat before test; (d) soil pat after test

of a liquid limit test device and grooving tools. The moisture content of the soil, in percent, and the corresponding number of blows are plotted on semilogarithmic graph paper (Figure 2.10). The relationship between moisture content and log N is approximated as a straight line. This is referred to as the *flow curve*. The moisture content corresponding to $N = 25$, determined from the flow curve, gives the liquid limit of the soil. The slope of the flow line is defined as the *flow index* and may be written as

$$I_F = \frac{w_1 - w_2}{\log\left(\dfrac{N_2}{N_1}\right)} \tag{2.30}$$

where I_F = flow index

w_1 = moisture content of soil, in percent, corresponding to N_1 blows

w_2 = moisture content corresponding to N_2 blows

▼ **FIGURE 2.9**   Liquid limit test device and grooving tools (courtesy of Soiltest, Inc., Lake Bluff, Illinois)

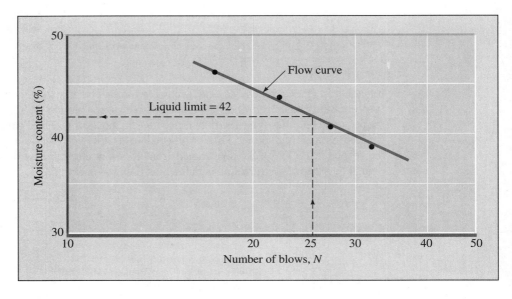

▼ **FIGURE 2.10**   Flow curve for liquid limit determination of a silty clay

Note that w_2 and w_1 are exchanged to get a positive value even though the slope of the flow line is actually negative. Thus, the equation of the flow line can be written in a general form as

$$w = -I_F \log N + C \tag{2.31}$$

From the analysis of hundreds of liquid limit tests, the Waterways Experiment Station in Vicksburg, Mississippi (1949), proposed an empirical equation of the form

$$LL = w_N \left(\frac{N}{25}\right)^{\tan \beta} \tag{2.32}$$

where N = number of blows in the liquid limit device for a 0.5-in. groove closure
w_N = corresponding moisture content
$\tan \beta = 0.121$ (but note that $\tan \beta$ is not equal to 0.121 for all soils)

Equation (2.32) generally yields good results for the number of blows between 20 and 30. For routine laboratory tests, it may be used to determine the liquid limit when only one test is run for a soil. It is generally referred to as the *one point method*. This has also been adopted by ASTM under designation D-4318. The reason that the one point method yields fairly good results is that a small range of moisture content is involved where $N = 20$ to $N = 30$. Table 2.3 shows the values of the term $(N/25)^{0.121}$ given in Eq. (2.32) for $N = 20$ to $N = 30$. Atterberg's limit values for several clay minerals are given in Table 2.4.

Casagrande (1932) concluded that each blow in a standard liquid limit device corresponds to a soil shear strength of about 1 g/cm^2 ($\simeq 0.1$ kN/m^2). Hence, the liquid limit of a fine-grained soil gives the moisture content at which the shear strength of the soil is approximately 25 g/cm^2 ($\simeq 2.5$ kN/m^2).

Plastic Limit (PL)

The *plastic limit* is defined as the moisture content, in percent, at which the soil crumbles, when rolled into threads of $\frac{1}{8}$ in. (3.2 mm) in diameter. The plastic limit is the lower

▼ **TABLE 2.3** **Values of**
$(N/25)^{0.121}$

N	$\left(\dfrac{N}{25}\right)^{0.121}$	N	$\left(\dfrac{N}{25}\right)^{0.121}$
20	0.973	26	1.005
21	0.979	27	1.009
22	0.985	28	1.014
23	0.990	29	1.018
24	0.995	30	1.022
25	1.000		

▼ **TABLE 2.4**   **Atterberg Limits Values for Clay Minerals***

Mineral	Liquid limit	Plastic limit	Shrinkage limit
Montmorillonite	100–900	50–100	8.5–15
Nontronite	37–72	19–27	
Illite	60–120	35–60	15–17
Kaolinite	30–110	25–40	25–29
Hydrated halloysite	50–70	47–60	
Dehydrated halloysite	35–55	30–45	
Attapulgite	160–230	100–120	
Chlorite	44–47	36–40	
Allophane	200–250	130–140	

* After Mitchell (1976)

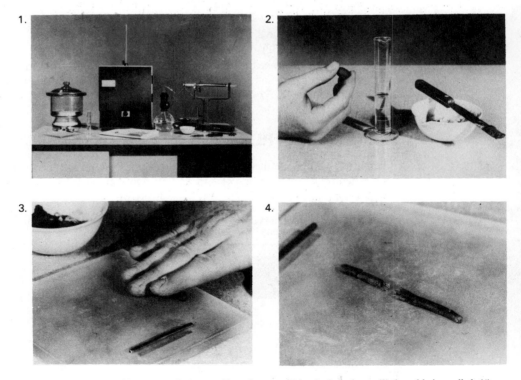

▼ **FIGURE 2.11**   Plastic limit test: (1) equipment; (2) beginning of test; (3) thread being rolled; (4) crumbled soil (courtesy of Soiltest, Inc., Lake Bluff, Illinois)

limit of the plastic stage of soil. The test is simple and is performed by repeated rollings of an ellipsoidal size soil mass by hand on a ground glass plate (Figure 2.11).

The *plasticity index* (*PI*) is the difference between the liquid limit and the plastic limit of a soil, or

$$PI = LL - PL$$

(2.33)

The procedure for the plastic limit test is given by ASTM in Test Designation D-4318.

Shrinkage Limit (SL)

Soil shrinks as moisture is gradually lost from it. With continuing loss of moisture, a stage of equilibrium is reached at which more loss of moisture will result in no further volume change (Figure 2.12). The moisture content, in percent, at which the volume of the soil mass ceases to change is defined as the *shrinkage limit*.

Shrinkage limit tests (ASTM Test Designation D-427) are performed in the laboratory with a porcelain dish about $1\frac{3}{4}$ in. (44.4 mm) in diameter and about $\frac{1}{2}$ in. (12.7 mm) in height. The inside of the dish is coated with petroleum jelly and is then filled completely with wet soil. Excess soil standing above the edge of the dish is struck off with a straightedge. The mass of the wet soil inside the dish is recorded. The soil pat in the dish is then oven dried. The volume of the oven-dried soil pat is determined by the displacement of mercury.

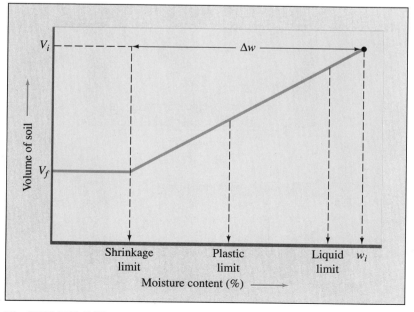

▼ **FIGURE 2.12**   Definition of shrinkage limit

By reference to Figure 2.12, the shrinkage limit can be determined in the following manner:

$$SL = w_i(\%) - \Delta w(\%) \tag{2.34}$$

where w_i = initial moisture content when the soil is placed in the shrinkage limit dish
  Δw = change in moisture content (that is, between the initial moisture content and the moisture content at the shrinkage limit)

However,

$$w_i(\%) = \frac{m_1 - m_2}{m_2} \times 100 \tag{2.35}$$

where m_1 = mass of the wet soil pat in the dish at the beginning of the test (g)
  m_2 = mass of the dry soil pat (g) (see Figure 2.13)

Also

$$\Delta w(\%) = \frac{(V_i - V_f)\rho_w}{m_2} \times 100 \tag{2.36}$$

where V_i = initial volume of the wet soil pat (that is, inside volume of the dish, cm³)
  V_f = volume of the oven-dried soil pat (cm³)
  ρ_w = density of water (g/cm³)

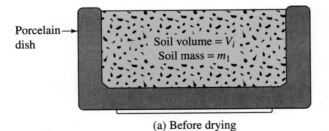

Porcelain dish →

Soil volume = V_i
Soil mass = m_1

(a) Before drying

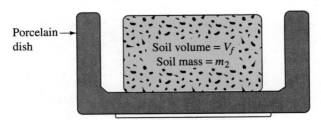

Porcelain dish →

Soil volume = V_f
Soil mass = m_2

(b) After drying

▼   FIGURE 2.13   Shrinkage limit test

Now, combining Eqs. (2.34), (2.35), and (2.36) gives

$$SL = \left(\frac{m_1 - m_2}{m_2}\right)(100) - \left(\frac{V_i - V_f}{m_2}\right)(\rho_w)(100) \tag{2.37}$$

▼ **EXAMPLE 2.6**

Following are the results of a shrinkage limit test:

- ▶ Initial volume of soil in a saturated state = 24.6 cm³
- ▶ Final volume of soil in a dry state = 15.9 cm³
- ▶ Initial mass in a saturated state = 44 g
- ▶ Final mass in a dry state = 30.1 g

Determine the shrinkage limit of the soil.

Solution   From Eq. (2.37),

$$SL = \left(\frac{m_1 - m_2}{m_2}\right)(100) - \left(\frac{V_i - V_f}{m_2}\right)(\rho_w)(100)$$

$$m_1 = 44 \text{ g} \qquad V_i = 24.6 \text{ cm}^3 \qquad \rho_w = 1 \text{ g/cm}^3$$

$$m_2 = 30.1 \text{ g} \qquad V_f = 15.9 \text{ cm}^3$$

$$SL = \left(\frac{44 - 30.1}{30.1}\right)(100) - \left(\frac{24.6 - 15.9}{30.1}\right)(1)(100)$$

$$= 46.18 - 28.9 = \textbf{17.28\%} \qquad ▼$$

2.6 LIQUIDITY INDEX

The relative consistency of a cohesive soil in the natural state can be defined by a ratio called the *liquidity index* (*LI*):

$$LI = \frac{w - PL}{LL - PL} \tag{2.38}$$

where w = *in situ* moisture content of soil.

The *in situ* moisture content for a sensitive clay may be greater than the liquid limit. In that case (Figure 2.14),

$$LI > 1$$

These soils, when remolded, can be transformed into a viscous form to flow like a liquid.

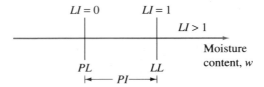

▼ **FIGURE 2.14**   Liquidity index

Soil deposits that are heavily overconsolidated may have a natural moisture content less than the plastic limit. In that case (Figure 2.14),

$$LI < +\varnothing$$

The values of the liquidity index for some of these soils may be negative.

2.7 ACTIVITY

Since the plasticity of soil is caused by the adsorbed water that surrounds the clay particles, we can expect that the type of clay minerals and their proportional amounts in a soil will affect the liquid and plastic limits. Skempton (1953) observed that the plasticity index of a soil linearly increases with the percent of clay-size fraction (% finer than 2 μm by weight) present in it. The correlations of PI with the clay-size fractions for different clays plot separate lines. This is because of the type of clay minerals in each soil. On the basis of these results, Skempton defined a quantity called *activity*, which is the slope of the line correlating PI and % finer than 2 μm. This activity may be expressed as

$$A = \frac{PI}{(\% \text{ of clay-size fraction, by weight})} \qquad (2.39)$$

where A = activity.

Activity is used as an index for identifying the swelling potential of clay soils. Typical values of activities for various clay minerals are given in Table 2.5.

Seed, Woodward, and Lundgren (1964a) studied the plastic property of several artificially prepared mixtures of sand and clay. They concluded that, although the relationship of the plasticity index to the percent of clay-size fraction is linear, as observed by Skempton, it may not always pass through the origin. This is shown in Figure 2.15. Thus, the activity can be redefined as in Eq. (2.40):

▼ **TABLE 2.5** **Activities of Clay Minerals***

Mineral	Activity, A
Smectites	1–7
Illite	0.5–1
Kaolinite	0.5
Halloysite ($2H_2O$)	0.5
Holloysite ($4H_2O$)	0.1
Attapulgite	0.5–1.2
Allophane	0.5–1.2

* After Mitchell (1976)
Note: Smectites are montmorillonite type of clay.

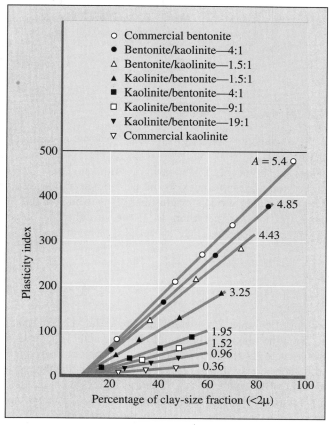

▼ **FIGURE 2.15**   Relationship between plasticity index and clay-size fraction by weight for kaolinite/bentonite clay mixtures (after Seed, Woodward, and Lundgren, 1964a)

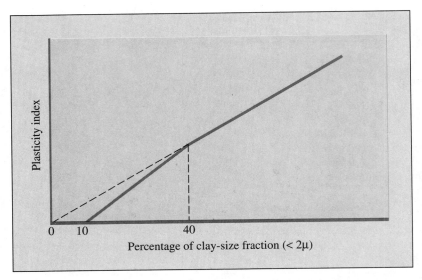

▼   **FIGURE 2.16**   Simplified relationship between plasticity index and percentage of clay-size fraction by weight (after Seed, Woodward, and Lundgren, 1964b)

$$A = \frac{PI}{\% \text{ of clay-size fraction} - C'} \tag{2.40}$$

where C' is a constant for a given soil. For the experimental results shown in Figure 2.15, $C' = 9$.

Further works of Seed, Woodward, and Lundgren (1964b) showed that the relationship of the plasticity index to the percentage of clay-size fractions present in a soil can be represented by two straight lines. This is shown qualitatively in Figure 2.16. For clay-size fractions greater than 40%, the straight line passes through the origin when it is projected back.

2.8 PLASTICITY CHART

Liquid and plastic limits are determined by relatively simple laboratory tests that provide information about the nature of cohesive soils. The tests have been used extensively by engineers for the correlation of several physical soil parameters as well as for soil identification. Casagrande (1932) studied the relationship of the plasticity index to the liquid limit of a wide variety of natural soils. On the basis of the test results, he proposed a plasticity chart as shown in Figure 2.17. The important feature of this chart is the empirical A-line that is given by the equation $PI = 0.73(LL - 20)$. The A-line separates the inorganic clays from the inorganic silts. Plots of plasticity indices against liquid limits for inorganic clays lie above the A-line, and those for inorganic silts lie

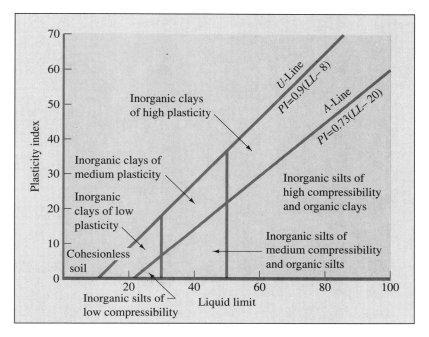

▼ **FIGURE 2.17**   Plasticity chart

below the A-line. Organic silts plot in the same region (below the A-line and with LL ranging from 30 to 50) as the inorganic silts of medium compressibility. Organic clays plot in the same region as inorganic silts of high compressibility (below the A-line and LL greater than 50). The information provided in the plasticity chart is of great value, and this is the basis for the classification of fine-grained soils in the Unified Soil Classification System (see Chapter 3).

Note that there is a line called the U-line above the A-line. The U-line is approximately the upper limit of the relationship of the plasticity index to the liquid limit for any soil found so far. The equation for the U-line can be given as

$$PI = 0.9(LL - 8) \tag{2.41}$$

There is another use for the A-line and the U-line. Casagrande suggested that the shrinkage limit of a soil can be approximately determined if its plasticity index and liquid limit are known (see Holtz and Kovacs, 1981). This can be done in the following manner with reference to Figure 2.18:

1. Plot the plasticity index against the liquid limit of a given soil, such as point A in Figure 2.18.
2. Project the A-line and the U-line downward to meet at point B. Point B will have the coordinates $LL = -43.5$ and $PI = -46.5$.
3. Join points B and A with a straight line. This line will intersect the liquid limit axis at point C. The abscissa of point C is the estimated shrinkage limit.

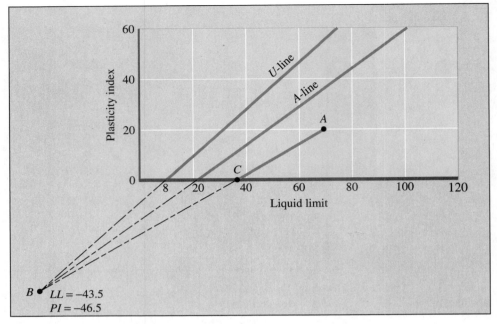

▼   **FIGURE 2.18**   Estimation of shrinkage from plasticity chart

2.9   SOIL STRUCTURE

Soil structure is defined as the geometric arrangement of soil particles with respect to one another. Among the many factors that affect the structure are the shape, size, and mineralogical composition of soil particles, and the nature and composition of soil water. For purposes of general discussion, soils can be placed into two groups: cohesionless and cohesive. The structures developed in soils under each group are described.

Structures in Cohesionless Soil

The structures generally encountered in cohesionless soils can be divided into two major categories: *single-grained* and *honeycombed*. In single-grained structures, soil particles are in stable positions, with each particle in contact with the surrounding ones. The shape and size distribution of the soil particles and their relative positions influence the denseness of packing (Figure 2.19), thus giving a wide range of void ratios. To get an idea of the variation of void ratios caused by the relative positions of the particles, consider the mode of packing of equal spheres shown in Figure 2.20. For a very loose state of packing, the void ratio is 0.91. However, the void ratio is reduced to 0.35 when the same spheres are rearranged into a very dense state of packing. Real soil differs from the equal-spheres model in that soil particles are neither equal-sized nor spherical. The smaller-sized particles may occupy the void spaces between the larger ones, thus

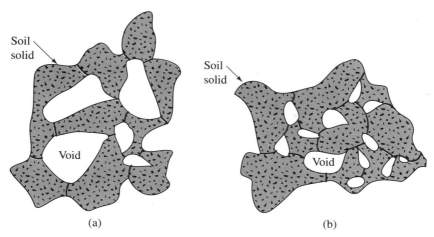

▼ FIGURE 2.19   Single-grained structures: (a) loose; (b) dense

tending to show a decrease in the void ratio of soils, as compared to that for equal spheres. However, the irregularity in the particle shapes generally tends to show an increase in the void ratio of soils. As a result of these two factors, the void ratios encountered in real soils have approximately the same range as those obtained in equal spheres.

In the honeycombed structure (Figure 2.21), relatively fine sand and silt form small arches with chains of particles. Soils that exhibit honeycombed structure have large void ratios, and they can carry an ordinary static load. However, under a heavy load or when subjected to shock loading, the structure breaks down, resulting in large settlement.

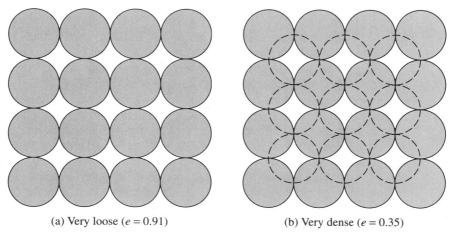

(a) Very loose ($e = 0.91$)   (b) Very dense ($e = 0.35$)

▼ FIGURE 2.20   Mode of packing of equal spheres (plan views): (a) very loose packing ($e = 0.91$); (b) very dense packing ($e = 0.35$)

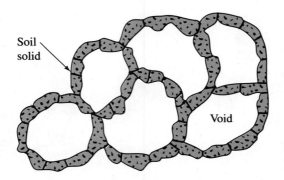

Soil solid

Void

▼ **FIGURE 2.21**   Honeycombed structure

Structures in Cohesive Soils

To understand the basic structures in cohesive soils, we need to know the types of forces that act between clay particles suspended in water. In the previous chapter, we discussed the negative charge on the surface of the clay particles and the diffuse double layer surrounding each particle. When two clay particles in suspension come close to each other, the tendency for interpenetration of the diffuse double layers results in repulsion between the particles. At the same time, there exists an attractive force between the clay particles that is caused by Van der Waal's forces and is independent of the characteristics of water. Both repulsive and attractive forces increase with decreasing distance between the particles, but at different rates. When the spacing between the particles is very small, the force of attraction is greater than the force of repulsion. These are the forces treated by colloidal theories.

The fact that local concentrations of positive charges occur at the edges of clay particles was discussed in Chapter 1. If the clay particles are very close to each other, the positively charged edges can be attracted to the negatively charged faces of the particles.

Let us consider the behavior of clay in the form of a dilute suspension. When the clay is initially dispersed in water, the particles repel one another. This is because with larger interparticle spacing, the forces of repulsion between the particles are greater than the forces of attraction (Van der Waal's forces). The force of gravity on each particle is negligible. Thus, the individual particles may settle very slowly or remain in suspension, undergoing *Brownian motion* (a random zigzag motion of colloidal particles in suspension). The sediment formed by the settling of the individual particles has a dispersed structure, and all particles will be oriented more or less parallel to each other (Figure 2.22a).

If the clay particles initially dispersed in water come close to each other during random motion in suspension, they might tend to aggregate into visible flocs with edge-to-face contact. In this instance, the particles are being held together by electrostatic attraction of positively charged edges to negatively charged faces. This is known as *flocculation*. When the flocs become large, they will settle under the force of gravity. The sediment formed in this manner will have a flocculent structure (Figure 2.22b).

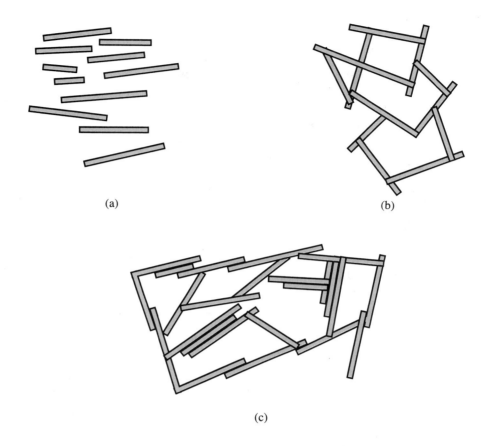

(a)

(b)

(c)

▼ **FIGURE 2.22**   Sediment structures: (a) dispersion; (b) non-salt flocculation; (c) salt flocculation (adapted from Lambe, 1958)

When salt is added to a clay-water suspension that has been initially dispersed, the ions tend to depress the double layer around the particles. This reduces the inter-particle repulsion. The clay particles are attracted to each other to form flocs and settle. The flocculent structure of the sediments formed is shown in Figure 2.22c. In the salt-type flocculent structure of sediment, the particle orientation approaches a large degree of parallelism. This is due to Van der Waal's forces.

Clays that have flocculent structures are lightweight and possess high void ratios. Clay deposits formed in the sea are highly flocculent. Most of the sediment deposits formed from fresh water possess an intermediate structure between dispersed and flocculent.

A deposit of pure clay minerals is rare in nature. When a soil has 50% or more particles with sizes of 0.002 mm or less, it is generally termed clay. Studies with scanning electron microscopes (Collins and McGown, 1974; Pusch, 1978; Yong and Sheeran, 1973) have shown that individual clay particles tend to be aggregated or flocculated in

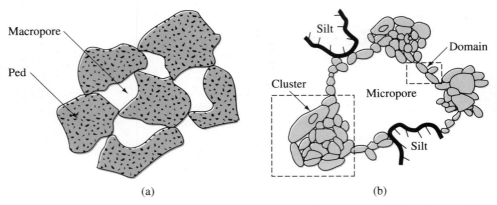

▼ FIGURE 2.23   Soil structure: (a) arrangement of peds and macropore spaces; (b) arrangement of domains and clusters with silt-size particles

submicroscopic units. These units are referred to as *domains*. The domains then group together, and these groups are called *clusters*. Clusters can be seen under a light microscope. This grouping to form clusters is caused by primarily interparticle forces. The clusters, in turn, group together to form *peds*. Peds can be seen without a microscope. Groups of peds are macrostructural features along with joints and fissures.

Figure 2.23a shows the arrangement of the peds and macropore spaces. The arrangement of domains and clusters with silt-size particles is shown in Figure 2.23b.

From the preceding discussion, it can be seen that the structure of cohesive soils is highly complex. Macrostructures have an important influence on the behavior of soils from an engineering point of view. The microstructure is more important from a fundamental point of view. Table 2.6 summarizes the macrostructures of clay soils.

▼ TABLE 2.6   Structure of Clay Soils

Item	Remarks
Dispersed structure	Formed by settlement of individual clay particles. More or less parallel orientation (Figure 2.22a).
Flocculent structure	Formed by settlement of flocs of clay particles (Figures 2.22b and c).
Domain	Aggregated or flocculated submicroscopic units of clay particles.
Cluster	Domains group to form clusters. Can be seen under light microscope.
Peds	Clusters group to form peds. Can be seen without microscope.

2.10 GENERAL COMMENTS

In this chapter, we covered the basic concepts related to (1) weight-volume relationships and (2) the plasticity of cohesive soils. Both of these topics are of fundamental importance to geotechnical engineers. The grain-size distribution (see Chapter 1) and the plasticity of a given soil are taken into consideration when soil is classified. Classification of soils is the topic of discussion in Chapter 3. Casagrande (1932) suggested that the liquid limit of a soil corresponds to a moisture content at which the shear strength of a soil is about 2.5 kN/m^2. The liquid limit is used as a parameter for correlation of the swell potential of expansive soils and the compression index of saturated cohesive soils (see Chapter 8). Expansive soils, found in many parts of the world, are always a potential problem in housing construction. The potential swell of a given soil can be generally assessed from Atterberg limits as given here (O'Neill and Poormoayed, 1980):

Liquid limit	Plasticity index	Potential swell classification
< 50	< 25	Low
50–60	25–35	Medium
> 60	> 35	High

PROBLEMS

2.1 In its natural state, a moist soil has a volume of 0.33 ft^3 and weighs 39.93 lb. The oven-dry weight of the soil is 34.54 lb. If $G_s = 2.67$, calculate the moisture content, moist unit weight, dry unit weight, void ratio, porosity, and degree of saturation.

2.2 Repeat Problem 2.1 with the following values:
▶ Volume of moist soil = 0.25 ft^3
▶ Weight of moist soil = 30 lb
▶ Weight of dry soil = 26.1 lb
▶ $G_s = 2.63$

2.3 The moist weight of 0.2 ft^3 of a soil is 23 lb. The moisture content and the specific gravity of the soil solids are determined in the laboratory to be 11% and 2.7, respectively. Calculate the following:
a. Moist unit weight (lb/ft^3)
b. Dry unit weight (lb/ft^3)
c. Void ratio
d. Porosity
e. Degree of saturation (%)
f. Volume occupied by water (ft^3)

2.4 For a given soil, show that

$$\gamma_{sat} = \gamma_d + n\gamma_w$$

2.5 For a given soil, show that

$$\gamma_{sat} = \gamma_d + \left(\frac{e}{1+e}\right)\gamma_w$$

2.6 For a given soil, show that

$$\gamma_d = \frac{eS\gamma_w}{(1 + e)w}$$

2.7 The saturated unit weight of a soil is 126 lb/ft³. The moisture content of the soil is 18.2%. Determine the following:

a. Dry unit weight
b. Void ratio
c. Specific gravity of soil solids

2.8 The unit weight of a soil is 14.9 kN/m³. The moisture content of this soil is 17% when the degree of saturation is 60%. Determine:

a. Saturated unit weight
b. Void ratio
c. Specific gravity of soil solids

2.9 For a given soil, the following are given: $G_s = 2.67$; moist unit weight, $\gamma = 112$ lb/ft³; and moisture content, $w = 10.8\%$. Determine:

a. Dry unit weight
b. Void ratio
c. Porosity
d. Degree of saturation

2.10 Refer to Problem 2.9. Determine the weight of water, in pounds, to be added per cubic foot of soil for:

a. 80% degree of saturation
b. 100% degree of saturation

2.11 The moist unit weight of a soil is 16.5 kN/m³. Given $w = 15\%$ and $G_s = 2.7$, determine:

a. Dry unit weight
b. Porosity
c. Degree of saturation
d. Mass of water, in kg/m³, to be added to reach full saturation

2.12 The dry density of a soil is 1750 kg/m³. Given $G_s = 2.66$, what would be the moisture content of the soil when saturated?

2.13 The porosity of a soil is 0.35. Given $G_s = 2.69$, calculate:

a. saturated unit weight (kN/m³)
b. Moisture content when moist unit weight = 17.5 kN/m³

2.14 A saturated soil has $w = 28\%$ and $G_s = 2.66$. Determine its saturated and dry unit weights.

2.15 A soil has $e = 0.75$, $w = 21.5\%$, and $G_s = 2.71$. Determine:

a. Moist unit weight (lb/ft³)
b. Dry unit weight (lb/ft³)
c. Degree of saturation (%)

2.16 Repeat Problem 2.15 with the following: $e = 0.6$, $w = 16\%$, and $G_s = 2.65$.

2.17 A soil has $w = 18.2\%$, $G_s = 2.67$, and $S = 80\%$. Determine the moist and dry unit weights of the soil, in lb/ft³.

2.18 The moist unit weight of a soil is 112.32 lb/ft³ at a moisture content of 10%. Given $G_s = 2.7$, determine:

a. e
b. Saturated unit weight

2.19 The moist unit weights and degrees of saturation of a soil are given in the table.

γ (lb/ft³)	S (%)
105.73	50
112.67	75

Determine:
a. G_s
b. e

2.20 Refer to Problem 2.19. Determine the weight of water, in lb, that will be in 2.5 ft³ of the soil when it is saturated.

2.21 For a given sand, the maximum and minimum void ratios are 0.78 and 0.43, respectively. Given $G_s = 2.67$, determine the dry unit weight of the soil when the relative density is 65%.

2.22 For a given sandy soil, $e_{max} = 0.75$, $e_{min} = 0.46$, and $G_s = 2.68$. What will be the moist unit weight of compaction (kN/m³) in the field if $D_r = 78\%$ and $w = 9\%$?

2.23 For a given sandy soil, the maximum and minimum dry unit weights are 108 lb/ft³ and 92 lb/ft³, respectively. Given $G_s = 2.65$, determine the moist unit weight of this soil when the relative density is 60% and the moisture content is 8%.

2.24 Following are the results from the liquid and plastic limit tests for a soil:

Liquid limit test:

Number of blows, N	Moisture content (%)
15	42
20	40.8
28	39.1

Plastic limit = 18.7%
a. Draw the flow curve and obtain the liquid limit.
b. What is the plasticity index of the soil?

2.25 Refer to Problem 2.24. Determine the liquidity index of the soil when the *in situ* moisture content is 30%.

2.26 Repeat Problem 2.24 with the following values:

Number of blows, N	Moisture content (%)
13	33
18	27
29	22

Plastic limit = 16.5%

2.27 Determine the liquidity index of the soil referred to in Problem 2.26 when the *in situ* moisture content of the soil is 14.8%.

2.28 A saturated soil used in the determination of the shrinkage limit has initial volume, $V_i = 19.65$ cm^3; final volume, $V_f = 13.5$ cm^3; mass of wet soil, $m_1 = 36$ g; and mass of dry soil, $m_2 = 25$ g. Determine the shrinkage limit.

REFERENCES

American Society for Testing and Materials (1986). *ASTM Book of Standards*, Sec. 4, Vol. 04.08, Philadelphia, Pa.

Casagrande, A. (1932). "Research of Atterberg Limits of Soils," *Public Roads*, Vol. 13, No. 8, 121–136.

Collins, K., and McGown, A. (1974). "The Form and Function of Microfabric Features in a Variety of Natural Soils," *Geotechnique*, Vol. 24, No. 2, 223–254.

Holtz, R. D., and Kovacs, W. D. (1981). *An Introduction to Geotechnical Engineering*, Prentice-Hall, Englewood Cliffs, N.J.

Lambe, T. W. (1958). "The Structure of Compacted Clay," *Journal of the Soil Mechanics and Foundations Division*, ASCE, Vol. 85, No. SM2, 1654-1 to 1654-35.

Mitchell, J. K. (1976). *Fundamentals of Soil Behavior*, Wiley, New York.

O'Neill, M. W., and Poormoayed, N. (1980). "Methodology for Foundations on Expansive Clays," *Journal of the Geotechnical Engineering Division*, American Society of Civil Engineers, Vol. 106, No. GT12, pp. 1345–1367.

Pusch, R. (1978). "General Report on Physico-Chemical Processes Which Affect Soil Structure and Vice Versa," *Proceedings*, International Symposium on Soil Structure, Gothenburg, Sweden, Appendix, 33.

Seed, H. B., Woodward, R. J., and Lundgren, R. (1964a). "Clay Mineralogical Aspects of Atterberg Limits," *Journal of the Soil Mechanics and Foundations Division*, ASCE, Vol. 90, No. SM4, 107–131.

Seed, H. B., Woodward, R. J., and Lundgren, R. (1964b). "Fundamental Aspects of the Atterberg Limits," *Journal of the Soil Mechanics and Foundations Division*, ASCE, Vol. 90, No. SM6, 75–105.

Skempton, A. W. (1953). "The Colloidal Activity of Clays," *Proceedings*, 3rd International Conference on Soil Mechanics and Foundation Engineering, London, Vol. 1, 57–61.

Waterways Experiment Station (1949). U.S. Corps of Engineers, *Technical Memo 3-286*, Vicksburg, Mississippi.

Yong, R. N., and Sheeran, D. E. (1973). "Fabric Unit Interaction and Soil Behavior," *Proceedings*, International Symposium on Soil Structure, Gothenburg, Sweden, 176–183.

CLASSIFICATION OF SOIL

Different soils with similar properties may be classified into groups and subgroups based on their engineering behavior. Classification systems provide a common language to concisely express the general characteristics of soils, which are infinitely varied, without detailed descriptions. Most of the soil classification systems that have been developed for engineering purposes are based on simple index properties such as particle-size distribution and plasticity. Although several classification systems are now in use, none is totally definitive of any soil for all possible applications because of the wide diversity of soil properties.

3.1 TEXTURAL CLASSIFICATION

In a general sense, *texture* of soil refers to its surface appearance. Soil texture is influenced by the size of the individual particles present in it. Table 1.3 divided soils into gravel, sand, silt, and clay categories on the basis of particle size. In most cases, natural soils are mixtures of particles from several size groups. In the textural classification system, the soils are named after their principal components, such as sandy clay, silty clay, and so forth.

A number of textural classification systems were developed in the past by different organizations to serve their needs, and several of those are in use today. Figure 3.1 shows the textural classification system developed by the U.S. Department of Agriculture. It is based on the particle-size limits as described under the USDA system in Table 1.3; that is:

▶ *Sand-size:* 2.0 to 0.05 mm in diameter
▶ *Silt-size:* 0.05 to 0.002 mm in diameter
▶ *Clay-size:* smaller than 0.002 mm in diameter

The use of this chart can best be demonstrated by an example. If the particle-size distribution of soil A shows 30% sand, 40% silt, and 30% clay-size particles, its textural classification can be determined by proceeding in the manner indicated by the arrows in Figure 3.1. This soil falls into the zone of *clay loam*. Note that this

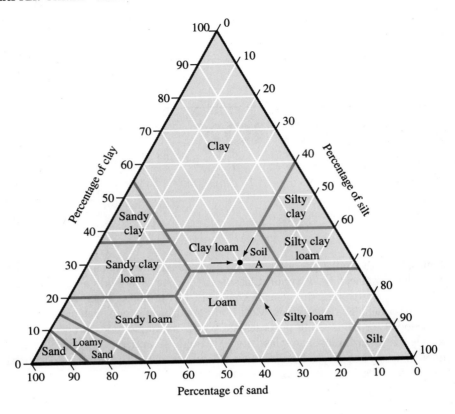

▼ **FIGURE 3.1**   U.S. Department of Agriculture textural classification

chart is based on only the fraction of soil that passes through the No. 10 sieve. Hence, if the particle-size distribution of a soil is such that a certain percentage of the soil particles are larger than 2 mm in diameter, then a correction will be necessary. For example, if soil B has a particle-size distribution of 20% gravel, 10% sand, 30% silt, and 40% clay, the modified textural compositions are

$$Sand\text{-}size: \frac{10 \times 100}{100 - 20} = 12.5\%$$

$$Silt\text{-}size: \frac{30 \times 100}{100 - 20} = 37.5\%$$

$$Clay\text{-}size: \frac{40 \times 100}{100 - 20} = 50.0\%$$

Based on the preceding modified percentages, the USDA textural classification is clay. However, due to the large percentage of gravel, it may be called gravelly clay.

Several other textural classification systems are also used, but they are no longer useful for civil engineering purposes.

▼ **EXAMPLE 3.1**

Classify the following soils by using the USDA textural classification system.

Soil no.	Gravel (%)	Sand (%)	Silt (%)	Clay (%)
1	18	51	22	9
2	0	30	30	40
3	15	20	35	30
4	0	16	37	47

Solution   Modified textural composition:

Soil no.	Modified % sand $\left(\dfrac{\% \text{ sand}}{100 - \% \text{ gravel}}\right)(100)$	Modified % silt $\left(\dfrac{\% \text{ silt}}{100 - \% \text{ gravel}}\right)(100)$	Modified % clay $\left(\dfrac{\% \text{ clay}}{100 - \% \text{ gravel}}\right)(100)$
1	62.2	26.83	10.96
2	30	30	40
3	23.53	41.18	35.29
4	16	37	47

Classification:

With the modified compositions calculated above, we can refer to Figure 3.1 to determine the zones in which the soils fall. They are as follows:

Soil no.	Classification
1	Gravelly sandy loam[a]
2	Clay and clay loam (borderline)
3	Gravelly clay loam[a]
4	Clay

[a] The word *gravelly* has been added because of the large percentage of gravel present in the soil.

▼

3.2 CLASSIFICATION BY ENGINEERING BEHAVIOR

Although the textural classification of soil is relatively simple, it is based entirely on the particle-size distribution. The amount and type of clay minerals present in fine-grained soils dictate to a great extent their physical properties. Hence, it is necessary to consider *plasticity*, which results from the presence of clay minerals, in order to interpret soil characteristics. Since textural classification systems do not take plasticity into account and are not totally indicative of many important soil properties, they are inadequate for most engineering purposes. At the present time, two more elaborate classification systems are commonly used by soils engineers. Both systems take into consideration the particle-size distribution and Atterberg limits. They are the AASHTO classification system and the Unified classification system. The AASHTO classification system is used mostly by state and county highway departments. Geotechnical engineers generally prefer the Unified system.

AASHTO Classification System

This system of soil classification was developed in 1929 as the Public Road Administration Classification System. It has undergone several revisions, with the present version proposed by the Committee on Classification of Materials for Subgrades and Granular Type Roads of the Highway Research Board in 1945 (ASTM designation D-3282; AASHTO method M145).

The AASHTO classification in present use is given in Table 3.1. According to this system, soil is classified into seven major groups: A-1 through A-7. Soils classified under groups A-1, A-2, and A-3 are granular materials where 35% or less of the particles pass through the No. 200 sieve. Soils where more than 35% pass through the No. 200 sieve are classified under groups A-4, A-5, A-6, and A-7. These are mostly silt and clay-type materials. The classification system is based on the following criteria:

1. *Grain size*
   Gravel: fraction passing the 75 mm (3 in.) sieve and retained on the No. 10 (2 mm) U.S. sieve
   Sand: fraction passing the No. 10 (2 mm) U.S. sieve and retained on the No. 200 (0.075 mm) U.S. sieve
   Silt and clay: fraction passing the No. 200 U.S. sieve
2. *Plasticity:* The term *silty* is applied when the fine fractions of the soil have a plasticity index of 10 or less. The term *clayey* is applied when the fine fractions have a plasticity index of 11 or more.
3. If cobbles and *boulders* (size larger than 75 mm) are encountered, they are excluded from the portion of the soil sample on which classification is made. However, the percentage of such material is recorded.

To classify a soil according to Table 3.1, the test data are applied from left to right. By process of elimination, the first group from the left into which the test data will fit is the correct classification.

▼ **TABLE 3.1**   **Classification of Highway Subgrade Materials**

General classification	Granular materials (35% or less of total sample passing No. 200)						
	A-1		A-3	A-2			
Group classification	A-1-a	A-1-b		A-2-4	A-2-5	A-2-6	A-2-7
Sieve analysis (percent passing)							
No. 10	50 max.						
No. 40	30 max.	50 max.	51 min.				
No. 200	15 max.	25 max.	10 max.	35 max.	35 max.	35 max.	35 max.
Characteristics of fraction passing No. 40							
Liquid limit				40 max.	41 min.	40 max.	41 min.
Plasticity index	6 max.		NP	10 max.	10 max.	11 min.	11 min.
Usual types of significant constituent materials	Stone fragments, gravel, and sand		Fine sand	Silty or clayey gravel and sand			
General subgrade rating	Excellent to good						

General classification	Silt-clay materials (more than 35% of total sample passing No. 200)			
Group classification	A-4	A-5	A-6	A-7 A-7-5[a] A-7-6[b]
Sieve analysis (percent passing)				
No. 10				
No. 40				
No. 200	36 min.	36 min.	36 min.	36 min.
Characteristics of fraction passing No. 40				
Liquid limit	40 max.	41 min.	40 max.	41 min.
Plasticity index	10 max.	10 max.	11 min.	11 min.
Usual types of significant constituent materials	Silty soils		Clayey soils	
General subgrade rating	Fair to poor			

[a] For A-7-5, $PI \leq LL - 30$
[b] For A-7-6, $PI > LL - 30$

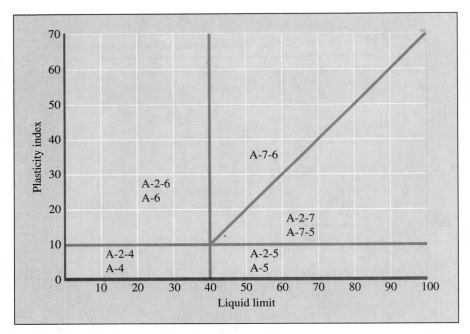

▼ **FIGURE 3.2**   Range of liquid limit and plasticity index for soils in groups A-2, A-4, A-5, A-6, and A-7

Figure 3.2 shows a plot of the range of the liquid limit and the plasticity index for soils that fall into groups A-2, A-4, A-5, A-6, and A-7.

To evaluate the quality of a soil as a highway subgrade material, a number called the *group index* (*GI*) is also incorporated with the groups and subgroups of the soil. This is written in parentheses after the group or subgroup designation. The group index is given by the equation

$$GI = (F - 35)[0.2 + 0.005(LL - 40)] + 0.01(F - 15)(PI - 10) \qquad (3.1)$$

where F = percent passing the No. 200 sieve
LL = liquid limit
PI = plasticity index

The first term of Eq. (3.1)—that is, $(F - 35)[0.2 + 0.005(LL - 40)]$—is the partial group index determined from the liquid limit. The second term—that is, $0.01(F - 15)(PI - 10)$—is the partial group index determined from the plasticity index. Following are some rules for determining group index:

1. If Eq. (3.1) yields a negative value for *GI*, it is taken as 0.
2. The group index calculated from Eq. (3.1) is rounded off to the nearest whole number (for example, *GI* = 3.4 is rounded off to 3; *GI* = 3.5 is rounded off to 4).

3. There is no upper limit for the group index.
4. The group index of soils belonging to groups A-1-a, A-1-b, A-2-4, A-2-5, and A-3 will always be 0.
5. When calculating the group index for soils that belong to groups A-2-6 and A-2-7, use the partial group index for PI, or

$$GI = 0.01(F - 15)(PI - 10) \qquad (3.2)$$

In general, the quality of performance of a soil as a subgrade material is inversely proportional to the group index.

▼ **EXAMPLE 3.2**

Classify the following soils by the AASHTO classification system.

Soil no.	Sieve analysis; % finer			Plasticity for the minus No. 40 fraction	
	No. 10 sieve	No. 40 sieve	No. 200 sieve	Liquid limit	Plasticity index
1	100	82	38	42	23
2	48	29	8	—	2
3	100	80	64	47	29
4	90	76	34	37	12

Solution Soil 1: Percent passing the No. 200 sieve is 38%, which is greater than 35%, so it is a silty clay material. Proceeding from left to right in Table 3.1, we see that it falls under A-7. For this case, $PI = 23 > LL - 30$, so it is A-7-6. From Eq. (3.1),

$$GI = (F - 35)[0.2 + 0.005(LL - 40)] + 0.01(F - 15)(PI - 10)$$

For this soil, $F = 38$, $LL = 42$, and $PI = 23$, so

$$GI = (38 - 35)[0.2 + 0.005(42–40)] + 0.01(38 - 15)(23 - 10) = 3.88 \approx 4$$

Hence, the soil is **A-7-6(4)**.

Soil 2: Percent passing the No. 200 sieve is less than 35%, so it is a granular material. Proceeding from left to right in Table 3.1, we find that it is A-1-a. The group index is 0, so the soil is **A-1-a(0)**.

Soil 3: Percent passing the No. 200 sieve is greater than 35%, so this is a silty clay material. Proceeding from left to right in Table 3.1, we find it to be A-7-6.

$$GI = (F - 35)[0.2 + 0.005(LL - 40)] + 0.01(F - 15)(PI - 10)$$

Given $F = 64$, $LL = 47$, and $PI = 29$, we have

$$GI = (64 - 35)[0.2 + 0.005(47 - 40)] + 0.01(64 - 15)(29 - 10) = 16.1 \approx 16$$

Hence, the soil is **A-7-6(16)**.

Soil 4: Percent passing the No. 200 sieve is less than 35%, so it is a granular material. From Table 3.1, it is A-2-6.

$$GI = 0.01(F - 15)(PI - 10)$$

Now, $F = 34$ and $PI = 12$, so

$$GI = 0.01(34 - 15)(12 - 10) = 0.38 \approx 0$$

Thus, the soil is **A-2-6(0)**.   ▼

Unified Classification System

The original form of this system was proposed by Casagrande in 1942 for use in the air field construction works undertaken by the Army Corps of Engineers during World War II. In cooperation with the U.S. Bureau of Reclamation, this system was revised in 1952. At present, it is widely used by engineers (ASTM designation D-2487). The

▼   **TABLE 3.2**   **Unified Classification System—Group Symbols for Gravelly Soil**

Group symbol	Criteria
GW	Less than 5% passing No. 200 sieve; $C_u = D_{60}/D_{10}$ greater than or equal to 4; $C_c = (D_{30})^2/(D_{10} \times D_{60})$ between 1 and 3
GP	Less than 5% passing No. 200 sieve; not meeting both criteria for GW
GM	More than 12% passing No. 200 sieve; Atterberg's limits plot below A-line (Figure 3.3) or plasticity index less than 4
GC	More than 12% passing No. 200 sieve; Atterberg's limits plot above A-line (Figure 3.3); plasticity index greater than 7
GC-GM	More than 12% passing No. 200 sieve; Atterberg's limits fall in hatched area marked CL-ML in Figure 3.3
GW-GM	Percent passing No. 200 sieve is 5 to 12; meets the criteria for GW and GM
GW-GC	Percent passing No. 200 sieve is 5 to 12; meets the criteria for GW and GC
GP-GM	Percent passing No. 200 sieve is 5 to 12; meets the criteria for GP and GM
GP-GC	Percent passing No. 200 sieve is 5 to 12; meets the criteria for GP and GC

Unified classification system is presented in Tables 3.2, 3.3, and 3.4. This system classifies soils into two broad categories:

1. Coarse-grained soils that are gravelly and sandy in nature with less than 50% passing through the No. 200 sieve. The group symbols start with prefixes of either G or S. G stands for gravel or gravelly soil, and S for sand or sandy soil.
2. Fine-grained soils with 50% or more passing through the No. 200 sieve. The group symbols start with prefixes of M, which stands for inorganic silt, C for inorganic clay, and O for organic silts and clays. The symbol Pt is used for peat, muck, and other highly organic soils.

Other symbols used for the classification are

▶ W—well graded
▶ P—poorly graded
▶ L—low plasticity (liquid limit less than 50)
▶ H—high plasticity (liquid limit more than 50)

▼ **TABLE 3.3** **Unified Classification System—Group Symbols for Sandy Soil**

Group symbol	Criteria
SW	Less than 5% passing No. 200 sieve; $C_u = D_{60}/D_{10}$ greater than or equal to 6; $C_c = (D_{30})^2/(D_{10} \times D_{60})$ between 1 and 3
SP	Less than 5% passing No. 200 sieve; not meeting both criteria for SW
SM	More than 12% passing No. 200 sieve; Atterberg's limits plot below A-line (Figure 3.3) or plasticity index less than 4
SC	More than 12% passing No. 200 sieve; Atterberg's limits plot above A-line (Figure 3.3); plasticity index greater than 7
SC-SM	More than 12% passing No. 200 sieve; Atterberg's limits fall in hatched area marked CL-ML in Figure 3.3
SW-SM	Percent passing No. 200 sieve is 5 to 12; meets the criteria for SW and SM
SW-SC	Percent passing No. 200 sieve is 5 to 12; meets the criteria for SW and SC
SP-SM	Percent passing No. 200 sieve is 5 to 12; meets the criteria for SP and SM
SP-SC	Percent passing No. 200 sieve is 5 to 12; meets the criteria for SP and SC

▼ **TABLE 3.4**   **Unified Classification System—Group Symbols for Silty and Clayey Soils**

Group symbol	Criteria
CL	Inorganic; $LL < 50$; $PI > 7$; plots on or above A-line (see CL zone in Figure 3.3)
ML	Inorganic; $LL < 50$; $PI < 4$ or plots below A-line (see ML zone in Figure 3.3)
OL	Organic; (LL—oven-dried)/(LL—not dried) < 0.75; $LL < 50$ (see OL zone in Figure 3.3)
CH	Inorganic; $LL \geq 50$; PI plots on or above A-line (see CH zone in Figure 3.3)
MH	Inorganic; $LL \geq 50$; PI plots below A-line (see MH zone in Figure 3.3)
OH	Organic; (LL—oven-dried)/(LL—not dried) < 0.75; $LL \geq 50$ (see OH zone in Figure 3.3)
CL-ML	Inorganic; plot in the hatched zone in Figure 3.3
Pt	Peat, muck, and other highly organic soils

For proper classification according to this system, some or all of the following need to be known:

1. Percent of gravel—that is, the fraction passing the 3-in. sieve (76.2-mm opening) and retained on the No. 4 sieve (4.75-mm opening)
2. Percent of sand—that is, the fraction passing the No. 4 sieve (4.75-mm opening) and retained on the No. 200 sieve (0.075-mm opening)
3. Percent of silt and clay—that is, the fraction finer than the No. 200 sieve (0.075-mm opening)
4. Uniformity coefficient (C_u) and the coefficient of gradation (C_c)
5. Liquid limit and plasticity index of the portion of soil passing the No. 40 sieve

The group symbols for coarse-grained gravelly soils are GW, GP, GM, GC, GC-GM, GW-GM, GW-GC, GP-GM, and GP-GC. Similarly, the group symbols for fine-grained soils are CL, ML, OL, CH, MH, OH, CL-ML, and Pt. Following is a step-by-step procedure for the classification of soils:

Step 1: Determine the percent of soil passing through the No. 200 sieve (F). If $F < 50\%$, it is a coarse-grained soil—that is, gravelly or sandy soil (where F = percent finer than No. 200 sieve). Go to Step 2. If $F \geq 50\%$, it is a fine-grained soil. Go to Step 3.

Step 2: For a coarse-grained soil, $100 - F$ is the coarse fraction in percent. Determine the percent of soil passing the No. 4 sieve and retained on the No. 200 sieve, F_1. If

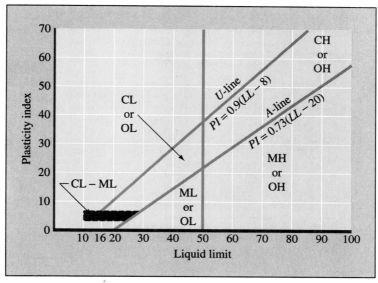

▼ **FIGURE 3.3** Plasticity chart

$F_1 < (100 - F)/2$, then it has more gravel than sand, so it is a gravelly soil. Go to Table 3.2 and Figure 3.3 for determination of the *group symbol*, and then proceed to Figure 3.4 for the proper *group name* of the soil. If $F_1 \geq (100 - F)/2$, then it is a sandy soil. Go to Table 3.3 and Figure 3.3 to determine the group symbol, and to Figure 3.4 for the group name of the soil.

Step 3: For a fine-grained soil, go to Table 3.4 and Figure 3.3 to obtain the *group symbol*. If it is an inorganic soil, go to Figure 3.5 to obtain the *group name*. If it is an organic soil, go to Figure 3.6 to get the group name.

Note that Figure 3.3 is the plasticity chart originally developed by Casagrande (1948) and modified to some extent here.

▼ **EXAMPLE 3.3**

A given soil has the following values:

▶ Gravel fraction (retained on No. 4 sieve) = 30%
▶ Sand fraction (passing No. 4 sieve but retained on No. 200 sieve) = 40%
▶ Silt and clay (passing No. 200 sieve) = 30%
▶ Liquid limit = 33
▶ Plasticity index = 12

Classify the soil by the Unified classification system, giving the group symbol and the group name.

Group Symbol   **Group Name**

GW → <15% sand → Well-graded gravel
→ ≥15% sand → Well-graded gravel with sand
GP → <15% sand → Poorly graded gravel
→ ≥15% sand → Poorly graded gravel with sand

GW-GM → <15% sand → Well-graded sand with silt
→ ≥15% sand → Well-graded gravel with silt and sand
GW-GC → <15% sand → Well-graded gravel with clay (or silty clay)
→ ≥15% sand → Well-graded gravel with clay and sand (or silty clay and sand)

GP-GM → <15% sand → Poorly graded gravel with silt
→ ≥15% sand → Poorly graded gravel with silt and sand
GP-GC → <15% sand → Poorly graded gravel with clay (or silty clay)
→ ≥15% sand → Poorly graded gravel with clay and sand (or silty clay and sand)

GM → <15% sand → Silty gravel
→ ≥15% sand → Silty gravel with sand
GC → <15% sand → Clayey gravel
→ ≥15% sand → Clayey gravel with sand
GC-GM → <15% sand → Silty clayey gravel
→ ≥15% sand → Silty clayey gravel with sand

SW → <15% gravel → Well-graded sand
→ ≥15% gravel → Well-graded sand with gravel
SP → <15% gravel → Poorly graded sand
→ ≥15% gravel → Poorly graded sand with gravel

SW-SM → <15% gravel → Well-graded sand with silt
→ ≥15% gravel → Well-graded sand with silt and gravel
SW-SC → <15% gravel → Well-graded sand with clay (or silty clay)
→ ≥15% gravel → Well-graded sand with clay and gravel (or silty clay and gravel)

SP-SM → <15% gravel → Poorly graded sand with silt
→ ≥15% gravel → Poorly graded sand with silt and gravel
SP-SC → <15% gravel → Poorly graded sand with clay (or silty clay)
→ ≥15% gravel → Poorly graded sand with clay and gravel (or silty clay and gravel)

SM → <15% gravel → Silty sand
→ ≥15% gravel → Silty sand with gravel
SC → <15% gravel → Clayey sand
→ ≥15% gravel → Clayey sand with gravel
SC-SM → <15% gravel → Silty clayey sand
→ ≥15% gravel → Silty clayey sand with gravel

▼ **FIGURE 3.4**   Flowchart group names for gravelly and sandy soil (after ASTM, 1991)

Solution   We are given $F = 30$ (i.e., $< 50\%$); hence, it is a coarse-grained soil. Also $F_1 = 40$, so

$$F_1 = 40 > \frac{100 - F}{2} = \frac{100 - 30}{2} = 35$$

and it is a sandy soil. From Table 3.3 and Figure 3.3, it can be seen that the soil is **SC**. Since the soil has more than 15% gravel (Figure 3.4), its group name is **clayey sand with gravel**.   ▼

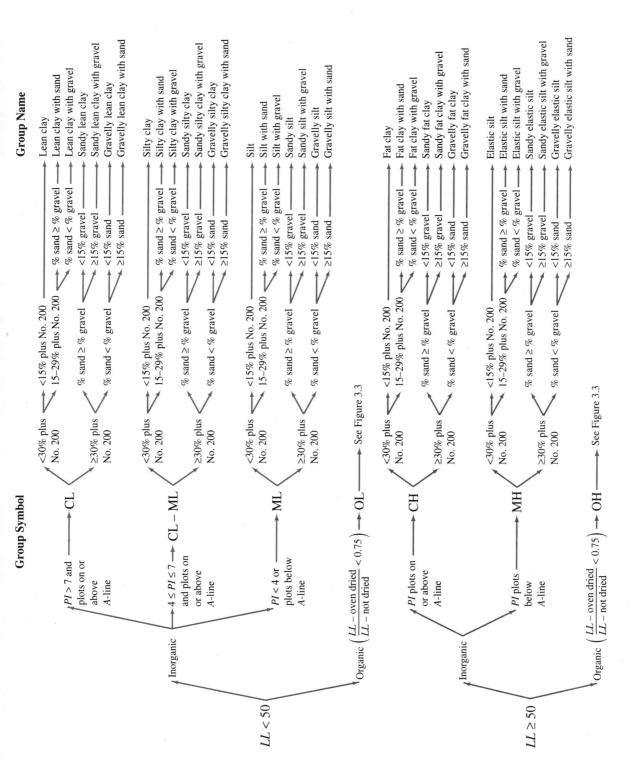

▼ **FIGURE 3.5** Flowchart group names for inorganic silty and clayey soils (after ASTM, 1991)

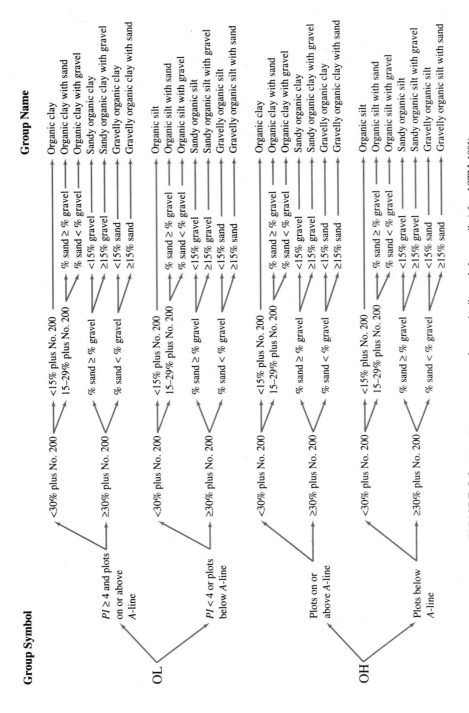

Group Symbol ... **Group Name**

▶ **FIGURE 3.6** Flowchart group names for organic silty and clayey soils (after ASTM, 1991)

▼ **EXAMPLE 3.4**

A soil has these values:

▶ Gravel fraction (retained on No. 4 sieve) = 10%
▶ Sand fraction (passing No. 4 sieve but retained on No. 200 sieve) = 82%
▶ Silt and clay (passing No. 200 sieve) = 8%
▶ Liquid limit = 39
▶ Plasticity index = 8
▶ C_u = 3.9
▶ C_c = 2.1

Classify the soil by the Unified classification system, giving the group symbol and the group name.

Solution   We are given $F = 8$, so it is a coarse-grained soil.

$$F_1 = 82 > \frac{100 - F}{2} = 46$$

hence, it is a sandy soil. Since F is between 5 and 12, dual symbols are necessary. From Table 3.3 and Figure 3.3, since C_u is less than 6, the soil is **SP-SM**. Now, from Figure 3.4, since the soil contains less than 15% gravel, its group name is **poorly graded sand with silt**.   ▼

▼ **EXAMPLE 3.5**

For a given soil:

▶ Percent passing No. 4 sieve = 100
▶ Percent passing No. 200 sieve = 86
▶ Liquid limit = 55
▶ Plasticity index = 28

Classify the soil using the Unified classification system, giving the group symbol and the group name.

Solution   We are given that the percent passing the No. 200 sieve is $F = 86$ (i.e., >50%), so it is a fine-grained soil. From Table 3.4 and Figure 3.3, the group symbol is **CH**. From Figure 3.5, the group name is **fat clay**.   ▼

3.3 COMPARISON BETWEEN THE AASHTO AND UNIFIED SYSTEMS

Both of the soil classification systems, AASHTO and Unified, are based on the texture and plasticity of soil. Also, both systems divide the soils into two major categories, coarse-grained and fine-grained, as separated by the No. 200 sieve. According to the

AASHTO system, a soil is considered fine-grained when more than 35% passes through the No. 200 sieve. According to the Unified system, a soil is considered fine-grained when more than 50% passes through the No. 200 sieve. A coarse-grained soil that has about 35% fine grains will behave like a fine-grained material. This is because there are enough fine grains to fill the voids between the coarse grains and hold them apart. In that respect, the AASHTO system appears to be more appropriate. In the AASHTO system, the No. 10 sieve is used to separate gravel from sand; in the Unified system, the No. 4 sieve is used. From the point of view of soil-separate size limits, the No. 10 sieve is the more accepted upper limit for sand. It is used in concrete and highway base-course technology.

In the Unified system, the gravelly and sandy soils are clearly separated; in the AASHTO system, they are not. The A-2 group, in particular, contains a large variety of soils. Symbols like GW, SM, CH, and others that are used in the Unified system are more descriptive of the soil properties than the A symbols used in the AASHTO system.

MIDDLETON PEAT   WAUPACA PEAT

PORTAGE PEAT   FOND DU LAC PEAT

100 μ

▼ FIGURE 3.7   Scanning electron micrographs for four peat samples (after Dhowian and Edil, 1980)

▼ **TABLE 3.5** **Properties of the Peats Shown in Figure 3.7**

Source of peat	Moisture content (%)	Unit weight		Specific gravity, G_s	Ash content (%)
		kN/m³	lb/ft³		
Middleton	510	9.1	57.9	1.41	12.0
Waupaca County	460	9.6	61.1	1.68	15.0
Portage	600	9.6	61.1	1.72	19.5
Fond du Lac County	240	10.2	64.9	1.94	39.8

The classification of organic soils as OL, OH, and Pt is provided in the Unified system. Under the AASHTO system, there is no place for organic soils. Peats usually have a high moisture content, low specific gravity of soil solids, and low unit weight. Figure 3.7 shows the scanning electron micrographs of four peat samples collected in Wisconsin. Some of the properties of the peats are given in Table 3.5.

▼ **TABLE 3.6** **Comparison of the AASHTO System with the Unified System***

Soil group in AASHTO system	Comparable soil groups in Unified system		
	Most probable	Possible	Possible but improbable
A-1-a	GW, GP	SW, SP	GM, SM
A-1-b	SW, SP, GM, SM	GP	—
A-3	SP	—	SW, GP
A-2-4	GM, SM	GC, SC	GW, GP, SW, SP
A-2-5	GM, SM	—	GW, GP, SW, SP
A-2-6	GC, SC	GM, SM	GW, GP, SW, SP
A-2-7	GM, GC, SM, SC	—	GW, GP, SW, SP
A-4	ML, OL	CL, SM, SC	GM, GC
A-5	OH, MH, ML, OL	—	SM, GM
A-6	CL	ML, OL, SC	GC, GM, SM
A-7-5	OH, MH	ML, OL, CH	GM, SM, GC, SC
A-7-6	CH, CL	ML, OL, SC	OH, MH, GC, GM, SM

* After Liu (1967)

Liu (1967) compared the AASHTO and Unified systems. The results of his study are presented in Tables 3.6 and 3.7.

▼   **TABLE 3.7   Comparison of the Unified System with the AASHTO System***

| Soil group in Unified system | Comparable soil groups in AASHTO system | | |
	Most probable	Possible	Possible but improbable
GW	A-1-a	—	A-2-4, A-2-5, A-2-6, A-2-7
GP	A-1-a	A-1-b	A-3, A-2-4, A-2-5, A-2-6, A-2-7
GM	A-1-b, A-2-4, A-2-5, A-2-7	A-2-6	A-4, A-5, A-6, A-7-5, A-7-6, A-1-a
GC	A-2-6, A-2-7	A-2-4, A-6	A-4, A-7-6, A-7-5
SW	A-1-b	A-1-a	A-3, A-2-4, A-2-5, A-2-6, A-2-7
SP	A-3, A-1-b	A-1-a	A-2-4, A-2-5, A-2-6, A-2-7
SM	A-1-b, A-2-4, A-2-5, A-2-7	A-2-6, A-4, A-5	A-6, A-7-5, A-7-6, A-1-a
SC	A-2-6, A-2-7	A-2-4, A-6 A-4, A-7-6	A-7-5
ML	A-4, A-5	A-6, A-7-5	—
CL	A-6, A-7-6	A-4	—
OL	A-4, A-5	A-6, A-7-5, A-7-6	—
MH	A-7-5, A-5	—	A-7-6
CH	A-7-6	A-7-5	—
OH	A-7-5, A-5	—	A-7-6
Pt	—	—	—
* After Liu (1967)			

PROBLEMS **3.1** Classify the following soils using the USDA textural classification chart:

Soil	Sand	Silt	Clay
	Particle-size distribution (%)		
A	20	20	60
B	55	5	40
C	45	35	20
D	50	15	35
E	70	15	15
F	30	58	12
G	40	25	35
H	30	25	45
I	5	45	50
J	45	45	10

3.2 Classify the following soils using the USDA textural classification system:

Soil	Gravel	Sand	Silt	Clay
	Particle-size distribution (%)			
A	18	51	22	9
B	10	20	41	29
C	21	12	35	32
D	0	18	24	58
E	12	22	26	40

3.3 The sieve analysis of 10 soils and the liquid and plastic limits of the fraction passing through the No. 40 sieve are given in the table. Classify the soils by the AASHTO classification system and give the group indices.

Soil	No. 10	No. 40	No. 200	Liquid limit	Plastic limit
	Sieve analysis, % finer				
1	98	80	50	38	29
2	100	92	80	56	23
3	100	88	65	37	22
4	85	55	45	28	20
5	92	75	62	43	28
6	97	60	30	25	16
7	100	55	8	—	NP
8	94	80	63	40	21
9	83	48	20	20	15
10	100	92	86	70	38

3.4 Classify the following soils using the AASHTO classification system and give the group indices:

Soil	No. 10	No. 40	No. 200	Liquid limit	Plasticity index
		Sieve analysis, % finer			
A	48	28	6	—	NP
B	87	62	30	32	8
C	90	76	34	37	12
D	100	78	8	—	NP
E	92	74	32	44	9

3.5 Classify the following soils using the AASHTO classification system and give the group indices also:

Sieve size	A	B	C	D	E
			Percent passing		
No. 4	94	98	100	100	100
No. 10	63	86	100	100	100
No. 20	21	50	98	100	100
No. 40	10	28	93	99	94
No. 60	7	18	88	95	82
No. 100	5	14	83	90	66
No. 200	3	10	77	86	45
0.01 mm	—	—	65	42	26
0.002 mm	—	—	60	47	21
Liquid limit	—	—	63	55	36
Plasticity index	NP	NP	25	28	22

3.6 Classify soils 1–6 given in Problem 3.3 using the Unified classification system. Give the group symbols and the group names.

3.7 Classify the soils given in Problem 3.5 using the Unified classification system. Give the group symbols and the group names.

3.8 Classify the following soils using the Unified classification system. Give the group symbols and the group names.

Soil	Sieve analysis, % finer		Liquid limit	Plasticity index
	No. 4	No. 200		
A	92	48	30	8
B	60	40	26	4
C	99	76	60	32
D	90	60	41	12
E	80	35	24	2

REFERENCES

American Association of State Highway and Transportation Officials (1982). *AASHTO Materials, Part I, Specifications*, Washington, D.C.

American Society for Testing and Materials (1991). *ASTM Book of Standards*, Sec. 4, Vol. 04.08, Philadelphia, Pa.

Casagrande, A. (1948). "Classification and Identification of Soils," *Transactions*, ASCE, Vol. 113, 901–930.

Dhowian, A. W., and Edil, T. B. (1980). "Consolidation Behavior of Peats," *Geotechnical Testing Journal*, ASTM, Vol. 3, No. 3, 105–114.

Liu, T. K. (1967). "A Review of Engineering Soil Classification Systems," *Highway Research Record No. 156*, National Academy of Sciences, Washington, D.C., 1–22.

CHAPTER FOUR

SOIL COMPACTION

In the construction of highway embankments, earth dams, and many other engineering structures, loose soils must be compacted to increase their unit weights. Compaction increases the strength characteristics of soils, thereby increasing the bearing capacity of foundations constructed over them. Compaction also decreases the amount of undesirable settlement of structures and increases the stability of slopes of embankments. Smooth-wheel rollers, sheepsfoot rollers, rubber-tired rollers, and vibratory rollers are generally used in the field for soil compaction. Vibratory rollers are used mostly for the densification of granular soils. Vibroflots are also used for compacting granular soil deposits to a considerable depth. Compaction of soil in this manner is known as *vibroflotation*. This chapter discusses in some detail the principles of soil compaction in the laboratory and in the field.

4.1 COMPACTION—GENERAL PRINCIPLES

Compaction, in general, is the densification of soil by removal of air, which requires mechanical energy. The degree of compaction of a soil is measured in terms of its dry unit weight. When water is added to the soil during compaction, it acts as a softening agent on the soil particles. The soil particles slip over each other and move into a densely packed position. The dry unit weight after compaction first increases as the moisture content increases (Figure 4.1). Note that at a moisture content $w = 0$, the moist unit weight (γ) is equal to the dry unit weight (γ_d), or

$$\gamma = \gamma_{d(w=0)} = \gamma_1$$

When the moisture content is gradually increased and the same compactive effort is used for compaction, the weight of the soil solids in a unit volume gradually increases. For example, at $w = w_1$, the moist unit weight is equal to

$$\gamma = \gamma_2$$

However, the dry unit weight at this moisture content is given by

$$\gamma_{d(w=w_1)} = \gamma_{d(w=0)} + \Delta\gamma_d$$

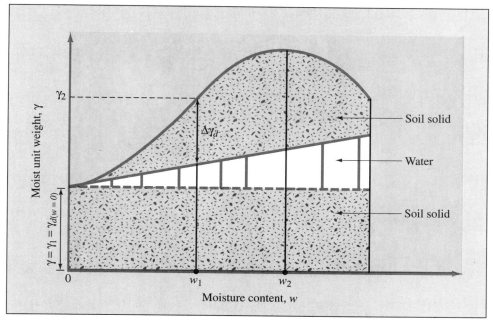

▼ **FIGURE 4.1** Principles of compaction

Beyond a certain moisture content $w = w_2$ (Figure 4.1), any increase in the moisture content tends to reduce the dry unit weight. This is because the water takes up the spaces that would have been occupied by the solid particles. The moisture content at which the maximum dry unit weight is attained is generally referred to as the *optimum moisture content.*

The laboratory test generally used to obtain the maximum dry unit weight of compaction and the optimum moisture content is called the *Proctor compaction test* (Proctor, 1933). The procedure for conducting this type of test is described in the following section.

4.2 STANDARD PROCTOR TEST

In the Proctor test, the soil is compacted in a mold that has a volume of $\frac{1}{30}$ ft^3 (943.3 cm^3). The diameter of the mold is 4 in. (101.6 mm). During the laboratory test, the mold is attached to a base plate at the bottom and to an extension at the top (Figure 4.2a). The soil is mixed with varying amounts of water and then compacted (Figure 4.3) in three equal layers by a hammer (Figure 4.2b) that delivers 25 blows to each layer. The hammer weighs 5.5 lb (mass = 2.5 kg) and has a drop of 12 in. (304.8 mm). For each test, the moist unit weight of compaction γ can be calculated as

$$\gamma = \frac{W}{V_{(m)}}$$

(4.1)

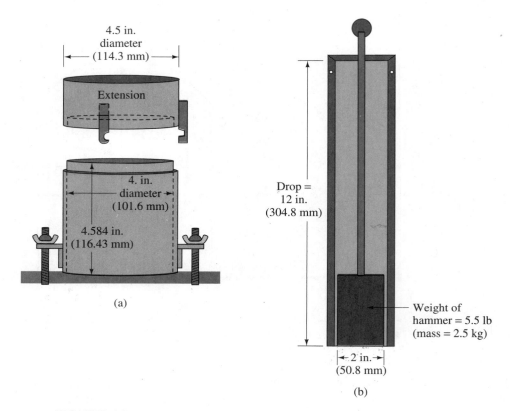

4.5 in.
diameter
(114.3 mm)

Extension

4. in.
diameter
(101.6 mm)

4.584 in.
(116.43 mm)

(a)

Drop =
12 in.
(304.8 mm)

Weight of
hammer = 5.5 lb
(mass = 2.5 kg)

2 in.
(50.8 mm)

(b)

▼ **FIGURE 4.2**   Standard Proctor test equipment: (a) mold; (b) hammer

where W = weight of the compacted soil in the mold

$V_{(m)}$ = volume of the mold ($= \frac{1}{30}$ ft^3)

For each test, the moisture content of the compacted soil is determined in the laboratory. With known moisture content, the dry unit weight γ_d can be calculated as

$$\gamma_d = \frac{\gamma}{1 + \frac{w(\%)}{100}}$$

(4.2)

where $w(\%)$ = percentage of moisture content.

The values of γ_d determined from Eq. (4.2) can be plotted against the corresponding moisture contents to obtain the maximum dry unit weight and the optimum moisture content for the soil. Figure 4.4 shows such a compaction for a silty-clay soil.

The procedure for the standard Proctor test is elaborated in ASTM Test Designation D-698 and AASHTO Test Designation T-99.

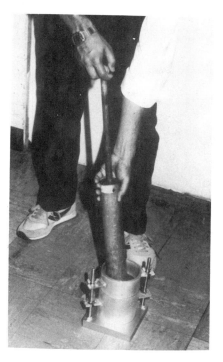

▼ **FIGURE 4.3** Compaction of soil using standard Proctor hammer (courtesy of John Hester, Carterville, Illinois)

For a given moisture content, the theoretical maximum dry unit weight is obtained when there is no air in the void spaces—that is, when the degree of saturation equals 100%. Thus, the maximum dry unit weight at a given moisture content with zero air voids can be given by

$$\gamma_{zav} = \frac{G_s \gamma_w}{1 + e}$$

where γ_{zav} = zero-air-void unit weight

γ_ω = unit weight of water

e = void ratio

G_s = specific gravity of soil solids

For 100% saturation, $e = wG_s$, so

$$\gamma_{zav} = \frac{G_s \gamma_w}{1 + \omega G_s} = \frac{\gamma_w}{w + \dfrac{1}{G_s}} \tag{4.3}$$

where w = moisture content.

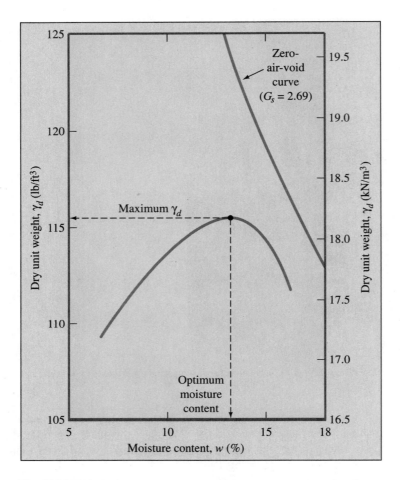

▼   **FIGURE 4.4**   Standard Proctor compaction test results for a silty clay

To obtain the variation of γ_{zav} with moisture content, use the following procedure:

1. Determine the specific gravity of soil solids.
2. Know the unit weight of water (γ_w).
3. Assume several values of w, such as 5%, 10%, 15%, and so on.
4. Use Eq. (4.3) to calculate γ_{zav} for various values of w.

Figure 4.4 also shows the variation of γ_{zav} with moisture content and its relative location with respect to the compaction curve. Under no circumstances should any part of the compaction curve lie to the right of the zero-air-void curve.

4.3 FACTORS AFFECTING COMPACTION

The preceding section showed that moisture content has a great influence on the degree of compaction achieved by a given soil. Besides moisture content, other important factors that affect compaction are soil type and compaction effort (energy per unit volume). The importance of each of the preceding two factors is described in more detail in the following two sections.

Effect of Soil Type

The soil type—that is, grain-size distribution, shape of the soil grains, specific gravity of soil solids, and amount and type of clay minerals present—has a great influence on the maximum dry unit weight and optimum moisture content. Figure 4.5 shows typical

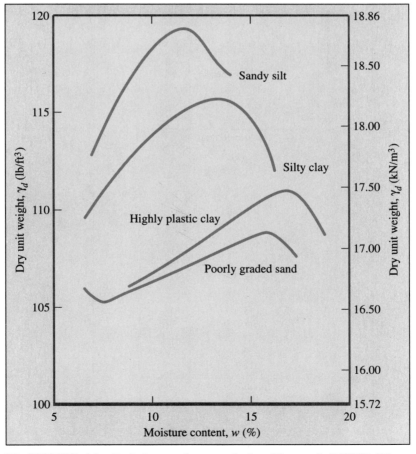

▼ **FIGURE 4.5** Typical compaction curves for four different soils (ASTM D-698)

compaction curves obtained from four different soils. The laboratory tests were conducted in accordance with ASTM Test Designation D-698.

Note also that the bell-shaped compaction curve shown in Figure 4.4 is typical of most clayey soils. Figure 4.5 shows that for sands, the dry unit weight has a general tendency first to decrease as moisture content increases, and then to increase to a maximum value with further increase of moisture. The initial decrease of dry unit weight with increase of moisture content can be attributed to the capillary tension effect.

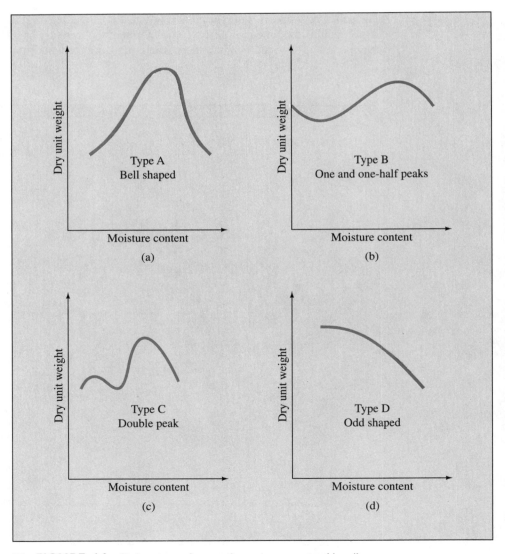

▼ **FIGURE 4.6**   Various types of compaction curves encountered in soils

At lower moisture contents, the capillary tension in the pore water inhibits the tendency of the soil particles to move around and be densely compacted.

Lee and Suedkamp (1972) studied compaction curves for 35 different soil samples. They observed that four different types of compaction curves can be found. These are shown in Figure 4.6. Type A compaction curves are the ones that have a single peak. This type of curve is generally found in soils that have a liquid limit between 30 and 70. Curve type B is a one and one-half peak curve, and curve type C is a double peak curve. Compaction curves of types B and C can be found in soils that have a liquid limit less than about 30. Compaction curves of type D are ones that do not have a definite peak. They are termed odd-shaped. Soils with a liquid limit greater than about 70 may exhibit compaction curves of type C or D. Soils that produce C- and D-type curves are not very common.

Effect of Compaction Effort

The compaction energy per unit volume (E) used for the standard Proctor test described in Section 4.2 can be given as

$$E = \frac{\left(\begin{array}{c}\text{number}\\\text{of blows}\\\text{per layer}\end{array}\right) \times \left(\begin{array}{c}\text{number}\\\text{of}\\\text{layers}\end{array}\right) \times \left(\begin{array}{c}\text{weight}\\\text{of}\\\text{hammer}\end{array}\right) \times \left(\begin{array}{c}\text{height of}\\\text{drop of}\\\text{hammer}\end{array}\right)}{\text{volume of mold}} \tag{4.4}$$

or

$$E = \frac{(25)(3)(5.5)(1)}{(1/30)} = 12{,}375 \text{ ft-lb/ft}^3 \ (\simeq 592.5 \text{ kJ/m}^3)$$

If the compaction effort per unit volume of soil is changed, the moisture-unit weight curve will also change. This can be demonstrated with the aid of Figure 4.7. This figure shows four compaction curves for a sandy clay. The standard Proctor mold and hammer were used to obtain compaction curves. The number of layers of soil used for compaction was kept at three for all cases. However, the number of hammer blows per each layer varied from 20 to 50. The compaction energy used per unit volume of soil for each curve can be easily calculated by using Eq. (4.4). These are tabulated in the table.

Curve number in Figure 4.7	Number of blows/layer	Compaction energy (ft-lb/ft^3)[a]
1	20	9,900
2	25	12,375
3	30	14,850
4	50	24,750

[a] 1 ft-lb/ft^3 = 47.88 J/m^3

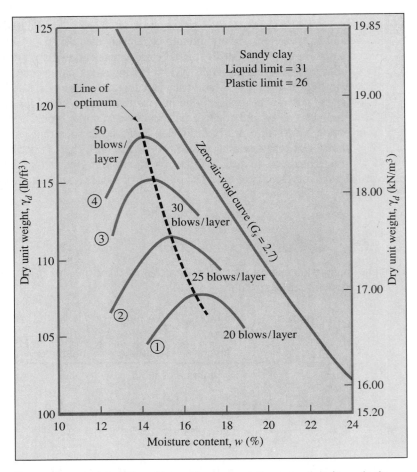

▼ **FIGURE 4.7** Effect of compaction energy on the compaction of a sandy clay

From the preceding tabulation and Figure 4.7, it can be seen that:

1. As the compaction effort is increased, the maximum dry unit weight of compaction is also increased.
2. As the compaction effort is increased, the optimum moisture content is decreased to some extent.

The preceding statements are true for all soils. Note, however, that the degree of compaction is not directly proportional to the compaction effort.

4.4 MODIFIED PROCTOR TEST

With the development of heavy rollers and their use in field compaction, the standard Proctor test was modified to better represent field conditions. This is sometimes referred to as the *modified Proctor test* (ASTM Test Designation D-1557 and AASHTO Test

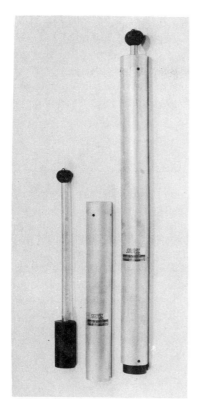

▼ **FIGURE 4.8** Comparison of hammers used in standard Proctor test and modified Proctor test. *Note:* At the left and middle are the standard Proctor hammer and the cylinder disassembled; at the right is the modified Proctor hammer (courtesy of Soiltest, Inc., Lake Bluff, Illinois)

Designation T-180). For conducting the modified Proctor test, the same mold is used with a volume of $\frac{1}{30}$ ft^3 (944 cm^3) as in the case of the standard Proctor test. However, the soil is compacted in five layers by a hammer that weighs 10 lb (mass = 4.54 kg). The drop of the hammer is 18 in. (457.2 mm). The number of hammer blows for each layer is kept at 25 as in the case of the standard Proctor test. Figure 4.8 shows a comparison between the hammers used for the standard Proctor test and the modified Proctor test. The compaction energy for unit volume of soil in the modified test can be calculated as

$$E = \frac{(5 \text{ layers})(25 \text{ blows/layer})(10 \text{ lb})(1.5 \text{ ft/drop})}{(1/30 \text{ ft}^3)}$$

$$= 56,250 \text{ ft-lb/ft}^3 \ (2693.3 \text{ kJ/m}^3)$$

Because it increases the compactive effort, the modified Proctor test results in an increase of the maximum dry unit weight of the soil. The increase of the maximum dry unit weight is accompanied by a decrease of the optimum moisture content.

In the preceding discussions, the specifications given for Proctor tests adopted by ASTM and AASHTO regarding the volume of the mold ($\frac{1}{30}$ ft^3) and the number of blows (25 blows/layer) are generally the ones adopted for fine-grained soils that pass the U.S. No. 4 sieve. However, under each test designation, there are four different suggested methods that reflect the size of the mold, the number of blows per layer, and the maximum particle-size in a soil aggregate used for testing. A summary of the test methods is given in Appendix C.

▼ **EXAMPLE 4.1**

The laboratory test data for a standard Proctor test are given here. Find the maximum dry unit weight and the optimum moisture content.

Volume of Proctor mold (ft^3)	Weight of wet soil in the mold (lb)	Moisture content (%)
1/30	3.88	12
1/30	4.09	14
1/30	4.23	16
1/30	4.28	18
1/30	4.24	20
1/30	4.19	22

Solution   The following table can now be prepared:

Volume, V (ft^3)	Weight of wet soil, W (lb)	Moist unit weight, γ[a] (lb/ft^3)	Moisture content, w (%)	Dry unit weight, γ_d[b] (lb/ft^3)
1/30	3.88	116.4	12	103.9
1/30	4.09	122.7	14	107.6
1/30	4.23	126.9	16	109.4
1/30	4.28	128.4	18	108.8
1/30	4.24	127.2	20	106.0
1/30	4.19	125.7	22	103.0

[a] $\gamma = \dfrac{W}{V}$

[b] $\gamma_d = \dfrac{\gamma}{1 + \dfrac{w\%}{100}}$

The plot of γ_d against w is shown in Figure 4.9. From the graph,

▶   Maximum dry unit weight = **109.5 lb/ft^3**
▶   Optimum moisture content = **16.3%**

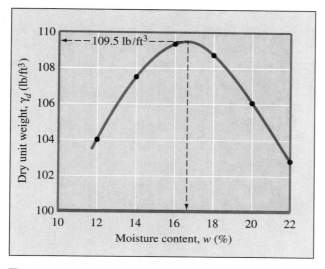

▼   **FIGURE 4.9**   ▼

4.5   STRUCTURE OF COMPACTED COHESIVE SOIL

Lambe (1958a) studied the effect of compaction on the structure of clay soils, and the results of his study are illustrated in Figure 4.10. If clay is compacted with a moisture content on the dry side of the optimum, as represented by point A, it will possess a flocculent structure. This is because, at low moisture content, the diffuse double layers of ions surrounding the clay particles cannot be fully developed; hence, the interparticle repulsion is reduced. This results in a more random particle orientation and a lower dry unit weight. When the moisture content of compaction is increased, as shown by point B, the diffuse double layers around the particles expand, thus increasing the repulsion between the clay particles and giving a lower degree of flocculation and a higher dry unit weight. A continued increase of moisture content from B to C will expand the double layers more, and this will result in a continued increase of repulsion between the particles. This will give a still greater degree of particle orientation and a more or less dispersed structure. However, the dry unit weight will decrease because the added water will dilute the concentration of soil solids per unit volume.

At a given moisture content, higher compactive effort tends to give a more parallel orientation to the clay particles, thereby giving a more dispersed structure. The particles are closer and the soil has a higher unit weight of compaction. This can be seen by comparing point A with point E in Figure 4.10.

Figure 4.11 shows the variation in the degree of particle orientation with molding water content on compacted Boston blue clay. Works of Seed and Chan (1959) have also shown similar results on compacted kaolin clay.

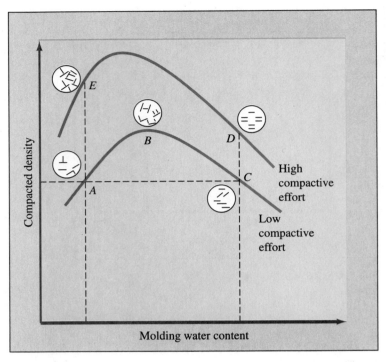

▼ **FIGURE 4.10**   Effect of compaction on structure of clay soils (redrawn after Lambe, 1958a)

* 4.6   EFFECT OF COMPACTION ON COHESIVE SOIL PROPERTIES

Compaction induces variations in the structure of cohesive soils. Results of these structural variations include changes in permeability, compressibility, and strength. Figure 4.12 shows the results of permeability tests (see Chapter 5) on Jamaica sandy clay. The samples used for the tests were compacted at various moisture contents by the same compactive effort. The coefficient of permeability, which is a measure of how easily water flows through soil, decreases with the increase in moisture content. It reaches a minimum value at approximately the optimum moisture content. Beyond the optimum moisture content, the permeability coefficient increases slightly. The high value of the coefficient of permeability on the dry side of the optimum moisture content is caused by the random orientation of clay particles that results in larger pore spaces.

One-dimensional compressibility characteristics (see Chapter 8) of clay soils compacted on the dry side of the optimum and compacted on the wet side of the optimum are shown in Figure 4.13. Under lower pressure, a soil that is compacted on the wet side

---

* This section may be combined with Chapters 5, 8, and 9.

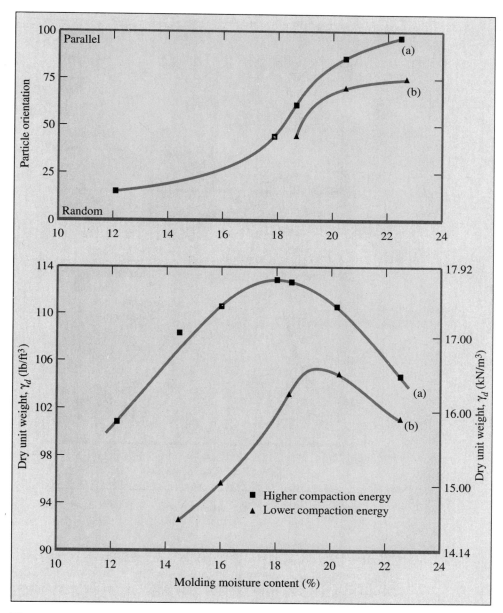

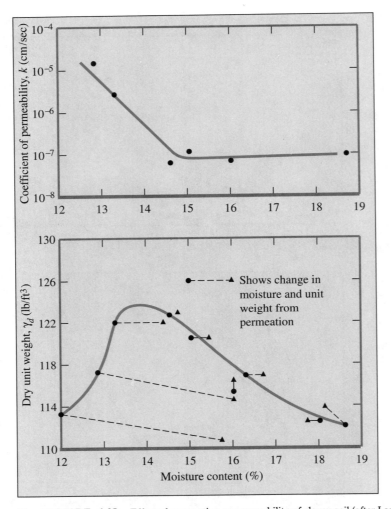

▼ **FIGURE 4.12**   Effect of compaction on permeability of clayey soil (after Lambe, 1958a)

of the optimum is more compressible than a soil that is compacted on the dry side of the optimum. This is shown in Figure 4.13a. Under high pressure, the trend is exactly the opposite, and this is shown in Figure 4.13b. For samples compacted on the dry side of the optimum, the pressure tends to orient the particles normal to its direction of application. The space between the clay particles is reduced at the same time. However, for samples compacted on the wet side of the optimum, pressure merely reduces the space between the clay particles. At very high pressure, it is possible to have identical structures for samples compacted on the dry and wet sides of optimum.

The strength of compacted clayey soils (see Chapter 9) generally decreases with the molding moisture content. This is shown in Figure 4.14. Note that at approximately optimum moisture content there is a great loss of strength. This means that, if two

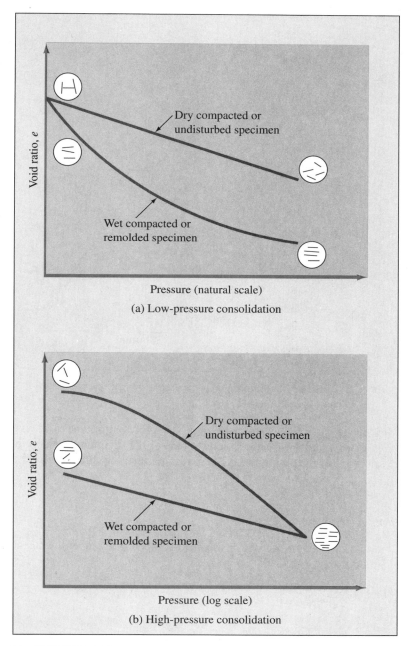

FIGURE 4.13   Effect of compaction on one-dimensional compressibility of clayey soil (redrawn after Lambe, 1958b)

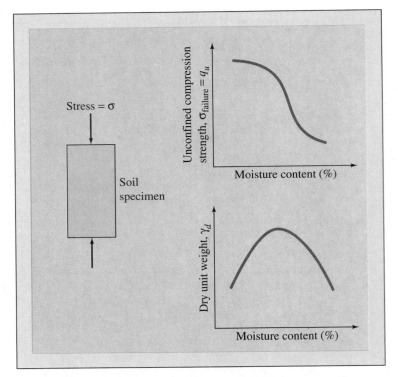

▼ **FIGURE 4.14**   Effect of compaction on the strength of clayey soils

samples are compacted to the same dry unit weight, one of them on the dry side of the optimum and the other on the wet side of the optimum, the specimen compacted on the dry side of the optimum (that is, with flocculent structure) will exhibit greater strength.

4.7 FIELD COMPACTION

Most of the compaction in the field is done with rollers. The most common types of rollers are:

1. Smooth-wheel roller (or smooth-drum roller)
2. Pneumatic rubber-tired roller
3. Sheepsfoot roller
4. Vibratory roller

Smooth-wheel rollers (Figure 4.15) are suitable for proofrolling subgrades and for finishing operation of fills with sandy and clayey soils. They provide 100% coverage under the wheels with ground contact pressures as high as 45 to 55 lb/in.^{2} (310 to 380 kN/m^2). They are not suitable for producing high unit weights of compaction when used on thicker layers.

▼ **FIGURE 4.15** Smooth-wheel roller (courtesy of David A. Carroll, Austin, Texas)

Pneumatic rubber-tired rollers (Figure 4.16) are better in many respects than the smooth-wheel rollers. The former are heavily loaded wagons with several rows of tires. These tires are closely spaced—four to six in a row. The contact pressure under the tires can range from 85 to 100 lb/in.^{2} (585 to 690 kN/m^2), and they produce about 70% to 80% coverage. Pneumatic rollers can be used for sandy and clayey soil compaction. Compaction is achieved by a combination of pressure and kneading action.

Sheepsfoot rollers (Figure 4.17) are drums with a large number of projections. The area of each of these projections may range from 4 to 13 in.^{2} ($\simeq 25$ to 85 cm^2). They are most effective in compacting clayey soils. The contact pressure under these projections can range from 200 to 1000 lb/in.^{2} (1380 to 6900 kN/m^2). During compaction in the field, the initial passes compact the lower portion of a lift. Compaction at the top and middle of a lift is done at a later stage.

Vibratory rollers are very efficient in compacting granular soils. Vibrators can be attached to smooth-wheel, pneumatic rubber-tired, or sheepsfoot rollers to provide vibratory effects to the soil. Figure 4.18 demonstrates the principles of vibratory rollers. The vibration is produced by rotating off-center weights.

Hand-held vibrating plates can be used for effective compaction of granular soils over a limited area. Vibrating plates are also gang-mounted on machines. These can be used in less restricted areas.

In addition to soil type and moisture content, other factors must be considered to achieve the desired unit weight of compaction in the field. These factors include the thickness of lift, the intensity of pressure applied by the compacting equipment, and the area over which the pressure is applied. This is because the pressure applied at the

▼ **FIGURE 4.16**   Pneumatic rubber-tired roller (courtesy of David A. Carroll, Austin, Texas)

▼ **FIGURE 4.17**   Sheepsfoot roller (courtesy of David A. Carroll, Austin, Texas)

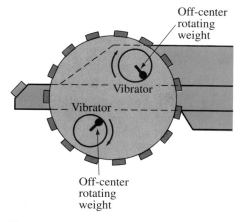

Off-center
rotating
weight

▼ **FIGURE 4.18** Principles of vibratory rollers

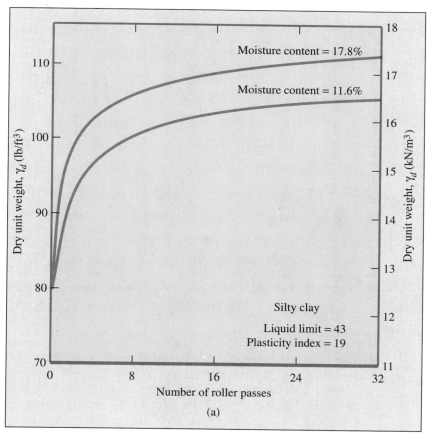

▼ **FIGURE 4.19** (a) Growth curve for a silty clay—relationship between dry unit weight and number of passes of 9.5 ton (84.5 kN) three-wheel roller when compacted in 9 in. (228.6 mm) loose layers at different moisture contents (redrawn after Johnson and Sallberg, 1960); see p. 108 for Figure 4.19(b).

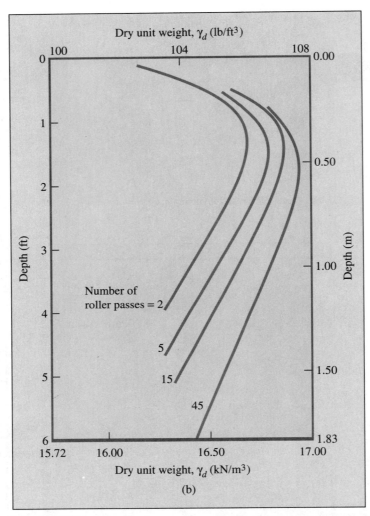

▼ **FIGURE 4.19**   (b) vibratory compaction of a sand—variation of dry unit weight with number of roller passes; thickness of lift = 8 ft (2.44 m) (redrawn after D'Appolonia, Whitman, and D'Appolonia, 1969)

surface decreases with depth, resulting in a decrease in the degree of compaction of soil. During compaction, the dry unit weight of soil is also affected by the number of roller passes. Figure 4.19a shows the growth curves for a silty clay soil. The dry unit weight of a soil at a given moisture content will increase up to a certain point with the number of passes of the roller. Beyond this point it will remain approximately constant. In most cases, about 10–15 roller passes yield the maximum dry unit weight economically attainable.

Figure 4.19b shows the variation in the unit weight of compaction with depth for a poorly graded dune sand for which compaction was achieved by a vibratory drum roller. Vibration was produced by mounting an eccentric weight on a single rotating

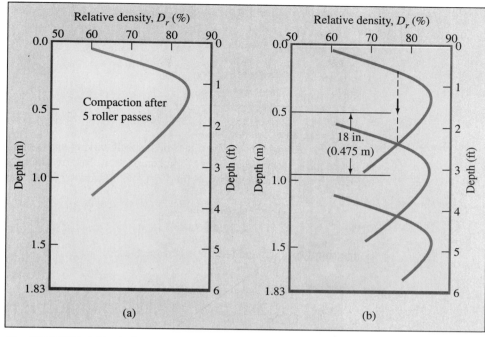

▼ **FIGURE 4.20**   Estimation of compaction lift thickness for minimum required relative density of 75% with five roller passes (after D'Appolonia, Whitman, and D'Appolonia, 1969)

shaft within the drum cylinder. The weight of the roller used for this compaction was 12.5 kip (55.6 kN) and the drum diameter was 47 in. (1.19 m). The lifts were kept at 8 ft (2.44 m). Note that, at any given depth, the dry unit weight of compaction increases with the number of roller passes. However, the rate of increase of unit weight gradually decreases after about 15 passes. Another fact to note from Figure 4.19b is the variation of dry unit weight with depth for any given number of roller passes. The dry unit weight and hence the relative density, D_r, reach maximum values at a depth of about 1.5 ft ($\approx$ 0.5 m) and gradually decrease at lesser depths. This is because of the lack of confining pressure toward the surface. Once the relationship between depth and relative density (or dry unit weight) for a given soil with a given number of roller passes is determined, it is easy to estimate the approximate thickness of each lift. This procedure is shown in Figure 4.20 (D'Appolonia, Whitman, and D'Appolonia, 1969).

4.8 SPECIFICATIONS FOR FIELD COMPACTION

In most specifications for earth work, the contractor is specified to achieve a compacted field dry unit weight of 90% to 95% of the maximum dry unit weight determined in the laboratory by either the standard or modified Proctor test. This is a specification for relative compaction R, which can be expressed as

$$R(\%) = \frac{\gamma_{d(\text{field})}}{\gamma_{d(\text{max}-\text{lab})}} \times 100 \tag{4.5}$$

where R = relative compaction.

In the compaction of granular soils, specifications are sometimes written in terms of the required relative density D_r or compaction. Relative density should not be confused with relative compaction. From Eq. (2.28) in Chapter 2, we can write

$$D_r = \left[\frac{\gamma_{d(\text{field})} - \gamma_{d(\text{min})}}{\gamma_{d(\text{max})} - \gamma_{d(\text{min})}} \right] \left[\frac{\gamma_{d(\text{max})}}{\gamma_{d(\text{field})}} \right] \tag{4.6}$$

Comparing Eqs. (4.5) and (4.6), one can see that

$$R = \frac{R_0}{1 - D_r(1 - R_0)} \qquad \leftarrow \text{Do not use (error)} \tag{4.7}$$

where

$$R_0 = \frac{\gamma_{d(\text{min})}}{\gamma_{d(\text{max})}} . \qquad \leftarrow \text{Do not use (error)} \tag{4.8}$$

Based on the observation of 47 soil samples, Lee and Singh (1971) gave a correlation between R and D_r for granular soils:

$$R = 80 + 0.2D_r \tag{4.9}$$

The specification for field compaction based on relative compaction or on relative density is an end-product specification. The contractor is expected to achieve a minimum dry unit weight regardless of the field procedure adopted. The most economical compaction condition can be explained with the aid of Figure 4.21. The compaction curves A, B, and C are for the same soil with varying compactive effort. Let curve A represent the conditions of maximum compactive effort that can be obtained from the existing equipment. Let it be required to achieve a minimum dry unit weight of $\gamma_{d(\text{field})} = R\gamma_{d(\text{max})}$. To achieve this, the moisture content w needs to be between w_1 and w_2. However, as can be seen from compaction curve C, the required $\gamma_{d(\text{field})}$ can be achieved with a lower compactive effort at a moisture content $w = w_3$. However, for most practical conditions, a compacted field unit weight of $\gamma_{d(\text{field})} = R\gamma_{d(\text{max})}$ cannot be achieved by the minimum compactive effort. Hence, equipment with slightly more than the minimum compactive effort should be used. The compaction curve B represents this condition. Now it can be seen from Figure 4.21 that the most economical moisture

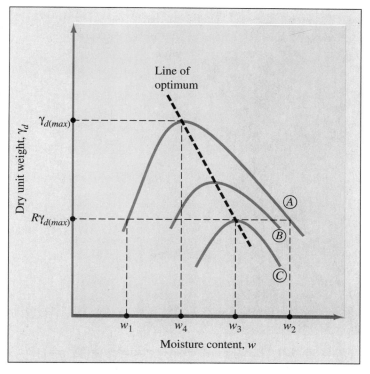

▼   **FIGURE 4.21**   Most economical compaction condition

content is between w_3 and w_4. Note that $w = w_4$ is the optimum moisture content for curve A, which is for the maximum compactive effort.

The concept described in the preceding paragraph, along with Figure 4.21, is historically attributed to Seed (1964), who was a giant in modern geotechnical engineering. It is elaborated on in more detail in Holtz and Kovacs (1981).

4.9   DETERMINATION OF FIELD UNIT WEIGHT OF COMPACTION

When the compaction work is progressing in the field, it is useful to know whether or not the unit weight specified is achieved. The standard procedures for determining the field unit weight of compaction include:

1.   Sand cone method
2.   Rubber balloon method
3.   Nuclear method

Following is a brief description of each of these methods.

▼ **FIGURE 4.22**   Glass jar with sand cone attached (*Note:* The jar is filled with Ottawa sand)

Sand Cone Method (ASTM Designation D-1556)

The sand cone device consists of a glass or plastic jar with a metal cone attached at its top (Figure 4.22). The jar is filled with very uniform dry Ottawa sand. The weight of the jar, the cone, and the sand filling the jar is determined (W_1). In the field, a small hole is excavated in the area where the soil has been compacted. If the weight of the moist soil excavated from the hole (W_2) is determined and the moisture content of the excavated soil is known, the dry weight of the soil W_3 can be obtained as

$$W_3 = \frac{W_2}{1 + \dfrac{w(\%)}{100}} \tag{4.10}$$

where w = moisture content.

After excavation of the hole, the cone with the sand-filled jar attached to it is inverted and placed over the hole (Figure 4.23). Sand is allowed to flow out of the jar to fill the hole and the cone. After that, the weight of the jar, the cone, and the remaining sand in the jar is determined (W_4), so

$$W_5 = W_1 - W_4 \tag{4.11}$$

where W_5 = weight of sand to fill the hole and cone.

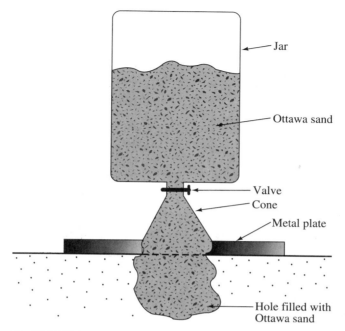

▼ **FIGURE 4.23**   Field unit weight by sand cone method

The volume of the hole excavated can now be determined as

$$V = \frac{W_5 - W_c}{\gamma_{d(sand)}}$$

(4.12)

where W_c = weight of sand to fill the cone only
$\gamma_{d(sand)}$ = dry unit weight of Ottawa sand used

The values of W_c and $\gamma_{d(sand)}$ are determined from the calibration done in the laboratory. The dry unit weight of compaction made in the field can now be determined as

$$\gamma_d = \frac{\text{dry weight of the soil excavated from the hole}}{\text{volume of the hole}} = \frac{W_3}{V}$$

(4.13)

Rubber Balloon Method (ASTM Designation D-2167)

The procedure for the rubber balloon method is similar to that for the sand cone method; a test hole is made and the moist weight of soil removed from the hole and its moisture content are determined. However, the volume of the hole is determined by introducing into it a rubber balloon filled with water from a calibrated vessel, from

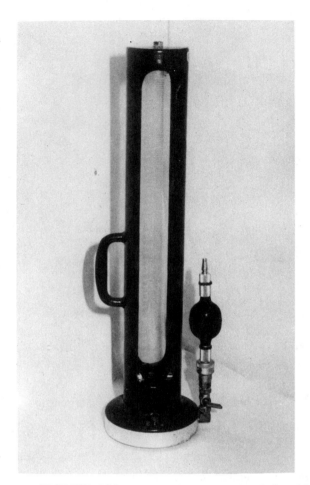

▼ **FIGURE 4.24**   Calibrated vessel with rubber balloon (courtesy of John Hester, Carterville, Illinois)

which the volume can be read directly. The dry unit weight of the compacted soil can be determined by using Eq. (4.13). Figure 4.24 shows a calibrated vessel with a rubber balloon.

Nuclear Method

Nuclear density meters are now often used for determining the compacted dry unit weight of soil. The density meters operate either in drilled holes or from the ground surface. The instrument measures the weight of wet soil per unit volume and also the weight of water present in a unit volume of soil. The dry unit weight of compacted soil can be determined by subtracting the weight of water from the moist unit weight of soil. Figure 4.25 shows a photograph of a nuclear density meter.

▼ **FIGURE 4.25**   Nuclear density meter (courtesy of David A. Carroll, Austin, Texas)

▼   **EXAMPLE 4.2**

Following are the results of a field unit weight determination test using the sand cone method:

- ▶ Calibrated dry unit weight of Ottawa sand $= 104$ lb/ft^3
- ▶ Weight of Ottawa sand to fill the cone $= 0.258$ lb
- ▶ Weight of jar + cone + sand (before use) $= 13.21$ lb
- ▶ Weight of jar + cone + sand (after use) $= 6.2$ lb
- ▶ Weight of moist soil from hole $= 7.3$ lb
- ▶ Moisture content of moist soil $= 11.6\%$

Determine the dry unit weight of compaction in the field.

Solution   The weight of the sand needed to fill the hole and cone is

$$13.21 - 6.2 = 7.01 \text{ lb}$$

The weight of the sand used to fill the hole is

$$7.01 - 0.258 = 6.752 \text{ lb}$$

So the volume of the hole is

$$V = \frac{6.752}{\text{dry unit weight of Ottawa sand}} = \frac{6.752}{104} = 0.0649 \text{ ft}^3$$

From Eq. (4.10), the dry weight of soil from the field is

$$W_3 = \frac{W_2}{1 + \dfrac{w(\%)}{100}} = \frac{7.3}{1 + \dfrac{11.6}{100}} = 6.54 \text{ lb}$$

Hence, the dry unit weight of compaction is

$$\gamma_d = \frac{W_3}{V} = \frac{6.54}{0.0649} = \textbf{100.77 lb/ft}^3 \qquad \blacktriangledown$$

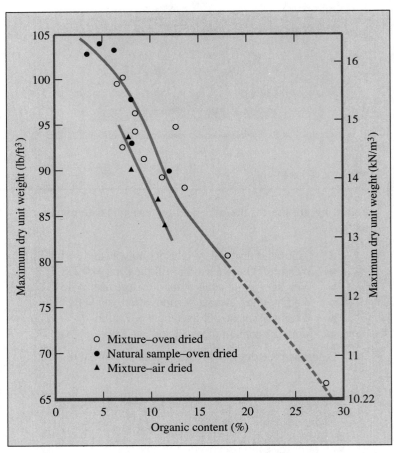

▼   **FIGURE 4.26**   Variation of maximum dry unit weight with organic content (after Franklin, Orozco, and Semrau, 1973)

4.10   COMPACTION OF ORGANIC SOILS

The presence of organic materials in a given soil tends to reduce the soil strength. In most cases, soils with high organic content are generally discarded as fill material. However, in certain economic circumstances, slightly organic soils are used for compaction. The organic content, *OC*, of a given soil may be defined as (Franklin, Orozco, and Semrau, 1973)

$$OC = \frac{\text{loss of dry weight due to heating in oven from } 105°C \text{ to } 400°C}{\text{dry weight of soil at } 105°C} \tag{4.14}$$

Franklin, Orozco, and Semrau conducted several laboratory tests to observe the effect of organic content on the compaction characteristics of soil. In that test program, a number of natural soils and soil mixtures were tested. Figure 4.26 shows the effect of organic content on the maximum dry unit weight. When the organic content exceeds 8% to 10%, the maximum dry unit weight of compaction decreases very rapidly. The optimum moisture content for a given compactive effort increases with the increase of organic content. This trend is shown in Figure 4.27. The maximum unconfined compression strength (see Chapter 9) obtained from a compacted soil (with a given compactive effort) also decreases with the increase of organic content of a soil (Figure 4.28). From these facts, it can be seen that soils with organic contents higher than about 10% are very undesirable for compaction work.

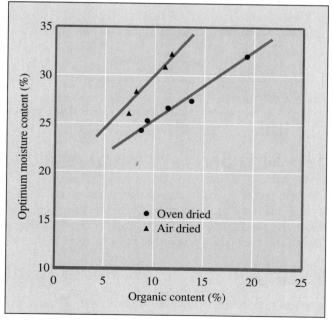

▼ FIGURE 4.27   Variation of optimum moisture content with organic content (after Franklin, Orozco, and Semrau, 1973)

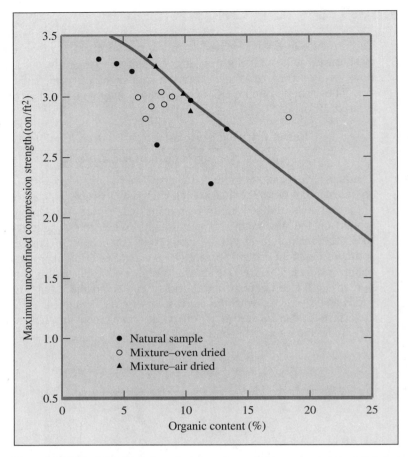

▼   **FIGURE 4.28**   Variation of maximum unconfined compression strength (of compacted soil samples) with organic content (after Franklin, Orozco, and Semrau, 1973)

4.11 SPECIAL COMPACTION TECHNIQUES

Several special types of compaction techniques have been developed for deep compaction of in-place soils, and they are used in the field for large-scale compaction works. Among these, the popular methods are vibroflotation, dynamic compaction, and blasting. Details of these methods are described in more detail in the following sections.

Vibroflotation

Vibroflotation is a technique for *in situ* densification of thick layers of loose granular soil deposits. It was developed in Germany in the 1930s. The first vibroflotation device was used in the United States about 10 years later. The process involves the use of a *vibroflot*, as shown in Figure 4.29 (also called the *vibrating unit*), which is about 7 ft

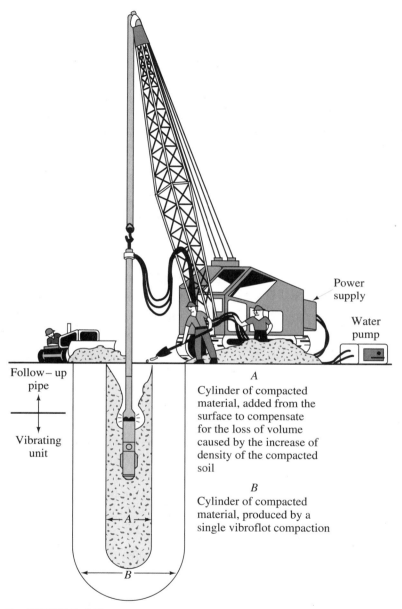

Power
supply

Water
pump

Follow – up
pipe

Vibrating
unit

A
Cylinder of compacted
material, added from the
surface to compensate
for the loss of volume
caused by the increase of
density of the compacted
soil

B
Cylinder of compacted
material, produced by a
single vibroflot compaction

←*A*→

←*B*→

▼ **FIGURE 4.29**   Vibroflotation unit (after Brown, 1977)

($\simeq 2.13$ m) in length. This vibrating unit has an eccentric weight inside it and can develop a centrifugal force. This enables the vibrating unit to vibrate horizontally. There are openings at the bottom and top of the vibrating unit for water jets. The vibrating unit is attached to a follow-up pipe. Figure 4.29 shows the entire assembly of equipment necessary for conducting the field compaction.

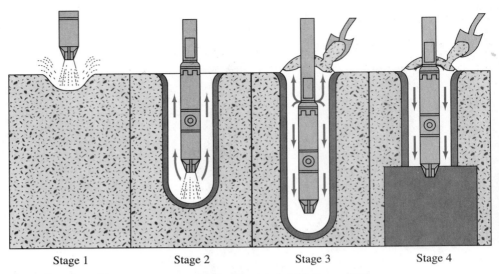

Stage 1   Stage 2   Stage 3   Stage 4

▼   **FIGURE 4.30**   Compaction by vibroflotation process (after Brown, 1977)

The entire compaction process in the field can be divided into four stages (Figure 4.30):

Stage 1: The jet at the bottom of the vibroflot is turned on and lowered into the ground.

Stage 2: The water jet creates a quick condition in the soil. It allows the vibrating unit to sink into the ground.

Stage 3: Granular material is poured from the top of the hole. The water from the lower jet is transferred to the jet located at the top of the vibrating unit. This water carries the granular material down the hole.

Stage 4: The vibrating unit is gradually raised in about 1 ft ($\simeq 0.3$ m) lifts and held vibrating for about 30 seconds at each lift. This process compacts the soil to the desired unit weight.

The details of various types of vibroflot units used in the United States are given in Table 4.1. Note that 30-hp electric units have been used since the latter part of the 1940s. The 100-hp units were introduced in the early 1970s.

The zone of compaction around a single probe will vary with the type of vibroflot used. The cylindrical zone of compaction will have a radius of about 6 to 7 ft ($\simeq 2$ m) for a 30-hp unit. This radius can extend to about 10 ft ($\simeq 3$ m) for a 100-hp unit.

Compaction by vibroflotation is done in various probe spacings, depending on the zone of compaction. This is shown in Figure 4.31. The capacity of successful densification of *in situ* soil will depend on several factors, the most important of which is the grain-size distribution of the soil itself and the type of backfill used to fill the holes during the withdrawal period of the vibroflot. The range of the grain-size distribution of

▼ **TABLE 4.1** **Types of Vibroflot Units***

Motor type	100-hp electric and hydraulic	30-hp electric
a. Vibrating tip		
Length (ft)	7.0	6.11
Diameter (in.)	16	15
Weight (lb)	4000	4000
Maximum movement when full (in.)	0.49	0.3
Centrifugal force (ton)	18	10
b. Eccentric		
Weight (lb)	260	170
Offset (in.)	1.5	1.25
Length (in.)	24	15.25
Speed (rpm)	1800	1800
c. Pump		
Operating flow rate (gal/min)	0–400	0–150
Pressure (lb/in.²)	100–150	100–150
d. Lower follow-up pipe and extensions		
Diameter (in.)	12	12
Weight (lb/ft)	250	250

* After Brown (1977)
Note: 1 ft = 0.305 m; 1 in. = 25.4 mm; 1 lb = 4.448 N;
1 ton = 8.9 kN; 1 gal/min = 0.004 m³/min;
1 lb/in.² = 6.9 kN/m²; 1 lb/ft = 14.6 N/m

in situ soil marked Zone 1 in Figure 4.32 is most suitable for compaction by vibroflotation. Soils that contain excessive amounts of fine sand and silt-size particles are difficult
to compact, and considerable effort is needed to reach the proper relative density of
compaction. Zone 2 in Figure 4.32 is the approximate lower limit of grain-size distribution for which compaction by vibroflotation is effective. Soil deposits whose grain-size

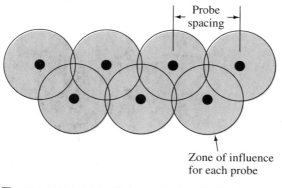

Probe
spacing

Zone of influence
for each probe

▼ **FIGURE 4.31** Probe spacing for vibroflotation

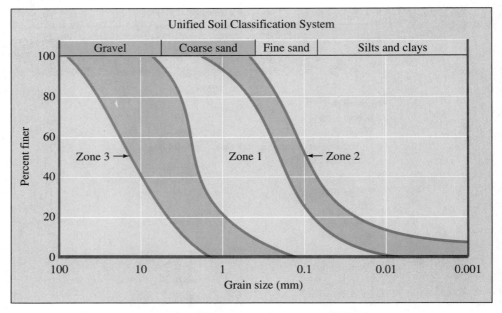

▼ **FIGURE 4.32**   Effective range of grain-size distribution of soil for vibroflotation

distribution fall in Zone 3 contain appreciable amounts of gravel. For these soils, the rate of probe penetration may be rather slow and might prove to be uneconomical in the long run.

The grain-size distribution of the backfill material is an important factor that controls the rate of densification. Brown (1977) has defined a quantity called the suitability number (S_N) for rating backfill material:

$$S_N = 1.7 \sqrt{\frac{3}{(D_{50})^2} + \frac{1}{(D_{20})^2} + \frac{1}{(D_{10})^2}} \qquad (4.15)$$

where D_{50}, D_{20}, and D_{10} are the diameters (in mm) through which, respectively, 50%, 20%, and 10% of the material pass.

The smaller the value of S_N, the more desirable is the backfill material. Following is a backfill rating system proposed by Brown:

Range of S_N	Rating as backfill
0–10	Excellent
10–20	Good
20–30	Fair
30–50	Poor
> 50	Unsuitable

Dynamic Compaction

Dynamic compaction is a technique that has gained popularity in the United States for the densification of granular soil deposits. This process primarily consists of dropping a heavy weight repeatedly on the ground at regular intervals. The weight of the hammer used varies over a range of 18 to 80 kip (80 to 356 kN), and the height of the hammer drop varies between 25 and 100 ft ($\simeq 7.5$ and 30.5 m). The stress waves generated by the hammer drops help in the densification. The degree of compaction achieved at a given site depends on the following factors:

1. Weight of the hammer
2. Height of hammer drop
3. Spacing of the locations at which the hammer is dropped

Leonards, Cutter, and Holtz (1980) suggested that the significant depth of influence for compaction can be given approximately by the equation

$$D \simeq \left(\frac{1}{2}\right)\sqrt{W_H h} \tag{4.16}$$

where D = significant depth of densification (m)
W_H = dropping weight (metric ton)
h = height of drop (in.)

In English units, the preceding equation takes the following form:

$$D = 0.61\sqrt{W_H h} \tag{4.17}$$

where the units of D and h are ft, and the unit of W_H is kip.

Blasting

Blasting is a technique that has been used successfully in many projects (Mitchell, 1970) for the densification of granular soils. The general soil grain sizes suitable for compaction by blasting are the same as those for compaction by vibroflotation. The process involves the detonation of explosive charges such as 60% dynamite at a certain depth below the ground surface in saturated soil. The lateral spacing of the charges varies from about 10 to 30 ft (3 to 10 m). Three to five successful detonations are usually necessary to achieve the desired compaction. Compaction (up to a relative density of about 80%) up to a depth of about 60 ft (20 m) over a large area can be easily achieved by this process. Usually the explosive charges are placed at a depth of about two-thirds of the thickness of the soil layer desired to be compacted. The sphere of influence of compaction by a 60% dynamite charge can be given as (Mitchell, 1970)

$$r = \sqrt{\frac{W_{EX}}{0.0025}} \tag{4.18}$$

where r = sphere of influence (ft)
W_{EX} = weight of explosive—60% dynamite (lb)

▼   **EXAMPLE 4.3**

Following are the details for the backfill material used in a vibroflotation project:

▶ $D_{10} = 0.36$ mm

▶ $D_{20} = 0.52$ mm

▶ $D_{50} = 1.42$ mm

Determine the suitability number, S_N. What would be its rating as a backfill?

Solution   From Eq. (4.15),

$$S_N = 1.7 \sqrt{\frac{3}{(D_{50})^2} + \frac{1}{(D_{20})^2} + \frac{1}{(D_{10})^2}}$$

$$= 1.7 \sqrt{\frac{3}{(1.42)^2} + \frac{1}{(0.52)^2} + \frac{1}{(0.36)^2}}$$

$$= 6.1$$

Rating: **Excellent**   ▼

▼   **EXAMPLE 4.4**

For a dynamic compaction test we are given: weight of hammer = 15 metric tons and height of drop = 12 m. Determine the significant depth, D, of influence for compaction in meters.

Solution   From Eq. (4.16),

$$D = \left(\frac{1}{2}\right)\sqrt{W_H h} = \left(\frac{1}{2}\right)\sqrt{(15)(12)} = \textbf{6.71 m}   ▼$$

4.12 GENERAL COMMENTS

In this chapter, we discussed the following topics:

1.   Laboratory compaction tests and related standards
2.   Field compaction equipment and special field compaction techniques
3.   Procedures used for determination of field unit weight compaction

Laboratory standard and modified Proctor compaction tests described in this chapter are essentially *impact* or *dynamic* compaction of soil; however, in the laboratory, *static compaction* and *kneading compaction* can also be used. It is important to realize that the compaction of clayey soils achieved by rollers in the field is essentially the kneading type. The relationships of dry unit weight (γ_d) and moisture content (w) obtained by dynamic and kneading compaction are not the same. Proctor compaction test results obtained in the laboratory are used primarily to see whether the roller compaction in the field is sufficient. The structures of compacted cohesive soil at a similar dry unit weight obtained by dynamic and kneading compaction may be different. This in turn affects physical properties such as permeability, compressibility, and strength.

For most fill operations, the final selection of the burrow site will depend on such factors as the type of soil and the cost of excavation and hauling. Table 4.2 provides tentative requirements for the relative compaction, R, of various types of soil in the field.

Fill materials for compaction are generally brought to the site by trucks and wagons. The fill material may be *end-dumped*, *side-dumped*, or *bottom-dumped* at the site in piles. If the material is too wet, it may be cut and turned to aerate and dry before

▼ **TABLE 4.2** **Tentative Requirements for Relative Compaction,**
$R \mid = \gamma_{d(\text{field})}/\gamma_{d(\text{max})\text{-standard Proctor}} \mid$*

Soil class (Unified classification system)	Required relative compaction, $R(\%)$		
	Class 1	Class 2	Class 3
GW	97	94	90
GP	97	94	90
GM	98	94	90
GC	98	94	90
SW	97	95	91
SP	98	95	91
SM	98	95	91
SC	99	96	92
ML	100	96	92
CL	100	96	92
OL	—	96	93
MH	—	97	93
CH	—	—	93
OH	—	97	93

Class 1	Upper 9 ft (3 m) of fills supporting one- or two-story buildings
	Upper 3 ft (1 m) of subgrade under pavements
	Upper 1 ft (0.3 m) of subgrade under floors
Class 2	Deeper parts of fills under buildings
	Deeper parts [30 ft (10 m)] of fills under pavements, floors
	Earth dams
Class 3	All other fills requiring some degree of strength or compressibility

* After Sowers (1979)

being spread in lifts for compaction. If it is too dry, the desired amount of water is added by sprinkling irrigation.

In summary, proper selection and careful compaction of soil can prevent problems in the future.

PROBLEMS

4.1 Calculate the zero-air-void unit weight for a soil (in lb/ft³) at $w = 5\%$, 8%, 10%, 12%, and 15%, given $G_s = 2.68$.

4.2 For a slightly organic soil, $G_s = 2.54$. For this soil, calculate and plot the variation of γ_{zav} (in lb/ft³) against w (in percent) (with w varying from 5% to 20%).

4.3 **a.** Derive an equation for obtaining the theroetical dry unit weight for different degrees of saturation, S (i.e., γ_d as a function of G_s, γ_w, S, and w), for a soil.
   b. For a given soil, if $G_s = 2.6$, calculate the theoretical variation of γ_d with w for 90% saturation.

4.4 For a compacted soil, given $G_s = 2.72$, $w = 18\%$, and $\gamma_d = 0.9\gamma_{zav}$, determine the dry unit weight of the compacted soil.

4.5 The results of a standard Proctor test are given here. Determine the maximum dry unit weight of compaction and the optimum moisture content. Also determine the moisture contents required to achieve 95% of $\gamma_{d(max)}$.

Volume of Proctor mold (ft³)	Weight of wet soil in the mold (lb)	Moisture content (%)
1/30	3.63	10
1/30	3.86	12
1/30	4.02	14
1/30	3.98	16
1/30	3.88	18
1/30	3.73	20

4.6 Repeat Problem 4.5 with the following values:

Weight of wet soil in standard Proctor mold (lb)	Moisture content (%)
3.26	8.4
4.15	10.2
4.67	12.3
4.02	14.6
3.36	16.8

Volume of mold $= \frac{1}{30}$ ft³.

4.7 A field unit weight determination test for the soil described in Problem 4.5 gave the following data: moisture content = 15% and moist unit weight = 107 lb/ft³.
   a. Determine the relative compaction.
   b. If G_s is 2.68, what was the degree of saturation in the field?

4.8 The maximum and minimum dry unit weights of a sand were determined in the laboratory to be 104 lb/ft³ and 93 lb/ft³, respectively. What would be the relative compaction in the field if the relative density is 78%?

4.9 The maximum and minimum dry unit weights of a sand were determined in the laboratory to be 16.5 kN/m³ and 14.5 kN/m³, respectively. In the field, if the relative density of compaction of the same sand is 70%, what are its relative compaction (%) and dry unit weight (kN/m³)?

4.10 The relative compaction of a sand in the field is 94%. The maximum and minimum dry unit weights of the sand are 16.2 kN/m³ and 14.9 kN/m³, respectively. For the field condition, determine:
   a. Dry unit weight
   b. Relative density of compaction
   c. Moist unit weight at a moisture content of 8%

4.11 Laboratory compaction test results on a clayey silt are given here.

Moisture content (%)	Dry unit weight (kN/m³)
6	14.80
8	17.45
9	18.52
11	18.9
12	18.5
14	16.9

Following are the results of a field unit weight determination test on the same soil with the sand cone method:

   ▶ Calibrated dry density of Ottawa sand = 1570 kg/m³
   ▶ Calibrated mass of Ottawa sand to fill the cone = 0.545 kg
   ▶ Mass of jar + cone + sand (before use) = 7.59 kg
   ▶ Mass of jar + cone + sand (after use) = 4.78 kg
   ▶ Mass of moist soil from hole = 3.007 kg
   ▶ Moisture content of moist soil = 10.2%

Determine:
   a. Dry unit weight of compaction in the field
   b. Relative compaction in the field

4.12 The backfill material for a vibroflotation project has the following grain sizes:
   ▶ $D_{10} = 0.11$ mm
   ▶ $D_{20} = 0.19$ mm
   ▶ $D_{50} = 1.3$ mm
Determine the suitability number, S_N.

4.13   Repeat Problem 4.12 with the following values:

$$D_{10} = 0.09 \text{ mm}$$
$$D_{20} = 0.25 \text{ mm}$$
$$D_{50} = 0.61 \text{ mm}$$

REFERENCES

American Association of State Highway and Transportation Officials (1982). *AASHTO Materials, Part II*, Washington, D.C.

American Society for Testing and Materials (1982). *ASTM Standards, Part 19*, Philadelphia, Pa.

Brown, E. (1977). "Vibroflotation Compaction of Cohesionless Soils," *Journal of the Geotechnical Engineering Division*, ASCE, Vol. 103, No. GT12, 1437–1451.

D'Appolonia, D. J., Whitman, R. V., and D'Appolonia, E. D. (1969). "Sand Compaction with Vibratory Rollers," *Journal of the Soil Mechanics and Foundations Division*, ASCE, Vol. 95, No. SM1, 263–284.

Franklin, A. F., Orozco, L. F., and Semrau, R. (1973). "Compaction of Slightly Organic Soils," *Journal of the Soil Mechanics and Foundations Division*, ASCE, Vol. 99, No. SM7, 541–557.

Holtz, R. D., and Kovacs, W. D. (1981). *An Introduction to Geotechnical Engineering*, Prentice Hall, Englewood Cliffs, New Jersey.

Johnson, A. W., and Sallberg, J. R. (1960). "Factors That Influence Field Compaction of Soil," Highway Research Board, *Bulletin No. 272*.

Lambe, T. W. (1958a). "The Structure of Compacted Clay," *Journal of the Soil Mechanics and Foundations Division*, ASCE, Vol. 84, No. SM2, 1654-1–1654-34.

Lambe, T. W. (1958b). "The Engineering Behavior of Compacted Clay," *Journal of the Soil Mechanics and Foundations Division*, ASCE, Vol. 84, No. SM2, 1655-1 to 1655-35.

Lee, K. W., and Singh, A. (1971). "Relative Density and Relative Compaction," *Journal of the Soil Mechanics and Foundations Division*, ASCE, Vol. 97, No. SM7, 1049–1052.

Lee, P. Y., and Suedkamp, R. J. (1972). "Characteristics of Irregularly Shaped Compaction Curves of Soils," *Highway Research Record No. 381*, National Academy of Sciences, Washington, D.C., 1–9.

Leonards, G. A., Cutter, W. A., and Holtz, R. D. (1980). "Dynamic Compaction of Granular Soils," *Journal of the Geotechnical Engineering Division*, ASCE, Vol. 106, No. GT1, 35–44.

Mitchell, J. K. (1970). "In-Place Treatment of Foundation Soils," *Journal of the Soil Mechanics and Foundations Division*, ASCE, Vol. 96, No. SM1, 73–110.

Proctor, R. R. (1933). "Design and Construction of Rolled Earth Dams," *Engineering News Record*, Vol. 3, 245–248, 286–289, 348–351, 372–376.

Seed, H. B. (1964). Lecture Notes, CE 271, Seepage and Earth Dam Design, University of California, Berkeley.

Seed, H. B., and Chan, C. K. (1959). "Structure and Strength Characteristics of Compacted Clays," *Journal of the Soil Mechanics and Foundations Division*, ASCE, Vol. 85, No. SM5, 87–128.

Sowers, G. F. (1979). *Introductory Soil Mechanics and Foundations: Geotechnical Engineering*, 4th ed., Macmillan, New York.

FLOW OF WATER IN SOIL: PERMEABILITY AND SEEPAGE

Soils have interconnected voids through which water can flow from points of high energy to points of low energy. The study of the flow of water through porous soil media is important in soil mechanics. It is necessary for estimating the quantity of underground seepage under various hydraulic conditions, for investigating problems involving the pumping of water for underground construction, and for making stability analyses of earth dams and earth-retaining structures that are subject to seepage forces.

5.1 HYDRAULIC GRADIENT

From fluid mechanics we know that according to Bernoulli's equation, the total head at a point in water under motion can be given by the sum of the pressure, velocity, and elevation heads, or

$$h = \underset{\underset{\text{head}}{\underset{\uparrow}{\text{Pressure}}}}{\frac{p}{\gamma_w}} + \underset{\underset{\text{head}}{\underset{\uparrow}{\text{Velocity}}}}{\frac{v^2}{2g}} + \underset{\underset{\text{head}}{\underset{\uparrow}{\text{Elevation}}}}{Z} \qquad (5.1)$$

where h = total head

p = pressure

v = velocity

g = acceleration due to gravity

γ_w = unit weight of water

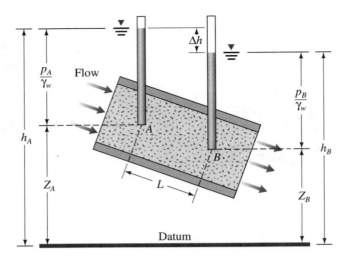

▼ **FIGURE 5.1**   Pressure, elevation, and total head for flow through soil

Note that the elevation head, Z, is the vertical distance of a given point above or below a datum plane. The pressure head is the water pressure, p, at that point divided by the unit weight of water, γ_w.

If Bernoulli's equation is applied to the flow of water through a porous soil medium, the term containing the velociy head can be neglected, since the seepage velocity is small and the total head at any point can be adequately represented by

$$h = \frac{p}{\gamma_w} + Z \qquad\qquad (5.2)$$

Figure 5.1 shows the relationship among pressure, elevation, and total heads for the flow of water through soil. Open standpipes called piezometers are installed at points A and B. The levels to which water rises in the peizometer tubes situated at points A and B are known as the *piezometric levels* of points A and B, respectively. The pressure head at a point is the height of the vertical column of water in the piezometer installed at that point.

The loss of head between two points, A and B, can be given by

$$\Delta h = h_A - h_B = \left(\frac{p_A}{\gamma_w} + Z_A\right) - \left(\frac{p_B}{\gamma_w} + Z_B\right) \qquad\qquad (5.3)$$

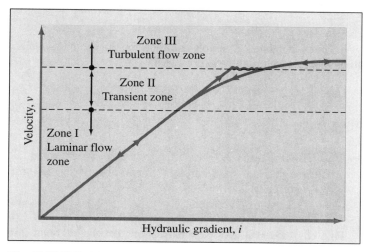

▼ FIGURE 5.2 Nature of variation v with hydraulic gradient i

The head loss, Δh, can be expressed in a nondimensional form as

$$i = \frac{\Delta h}{L} \qquad (5.4)$$

where $i =$ hydraulic gradient
       $L =$ distance between points A and B—that is, the length of flow over which the loss of head occurred

    In general, the variation of the velocity v with the hydraulic gradient i is as shown in Figure 5.2. This figure has been divided into three zones:

1. Laminar flow zone (Zone I)
2. Transition zone (Zone II)
3. Turbulent flow zone (Zone III)

When the hydraulic gradient is gradually increased, the flow remains laminar in Zones I and II, and the velocity v bears a linear relationship to the hydraulic gradient. At a higher hydraulic gradient, the flow becomes turbulent (Zone III). When the hydraulic gradient is decreased, laminar flow conditions will exist only in Zone I.

    In most soils, the flow of water through the void spaces can be considered laminar and thus

$$v \propto i \qquad (5.5)$$

In fractured rock, stones, gravels, and very coarse sands, turbulent flow conditions may exist, and then, Eq. (5.5) may not be valid.

5.2 DARCY'S LAW

In 1856, Darcy published a simple equation for the discharge velocity of water through saturated soils, which may be expressed as

$$v = ki$$

(5.6)

where v = *discharge velocity*, which is the quantity of water flowing in unit time through a unit gross cross-sectional area of soil at right angles to the direction of flow

k = coefficient of permeability

This equation was based primarily on the observations made by Darcy on the flow of water through clean sands. Note that Eq. (5.6) is similar to Eq. (5.5); both are valid for laminar flow conditions and applicable for a wide range of soils.

In Eq. 5.6, v is the discharge velocity of water based on the gross cross-sectional area of the soil. However, the actual velocity of water (that is, the seepage velocity) through the void spaces is greater than v. A relationship between the discharge velocity and the seepage velocity can be derived by referring to Figure 5.3, which shows a soil of length L with a gross cross-sectional area A. If the quantity of water flowing through the soil in unit time is q, then

$$q = vA = A_v v_s$$

(5.7)

where v_s = seepage velocity

A_v = area of void in the cross-section of the specimen

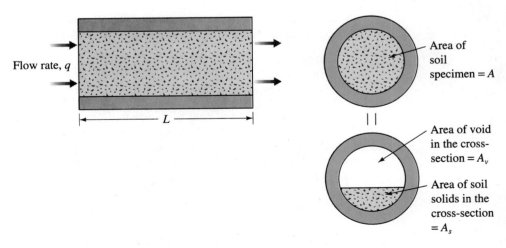

Flow rate, q

L

Area of soil specimen = A

Area of void in the cross-section = A_v

Area of soil solids in the cross-section = A_s

▼ **FIGURE 5.3**   Derivation of Eq. (5.10)

However,

$$A = A_v + A_s \tag{5.8}$$

where A_s = area of solid in the cross-section of the specimen.

Combining Eqs. (5.7) and (5.8) gives

$$q = v(A_v + A_s) = A_v v_s$$

or

$$v_s = \frac{v(A_v + A_s)}{A_v} = \frac{v(A_v + A_s)L}{A_v L} = \frac{v(V_v + V_s)}{V_v} \tag{5.9}$$

where V_v = volume of voids in the specimen
V_s = volume of soil solids in the specimen

Equation (5.9) can be rewritten as

$$v_s = v\left[\frac{1 + \left(\dfrac{V_v}{V_s}\right)}{\dfrac{V_v}{V_s}}\right] = v\left(\frac{1 + e}{e}\right) = \frac{v}{n} \tag{5.10}$$

where e = void ratio
n = porosity

Darcy's law as defined by Eq. (5.6) implies that the discharge velocity v bears a linear relationship to the hydraulic gradient i and will pass through the origin as shown in Figure 5.4. Hansbo (1960) reported the test results on four undisturbed natural clays.

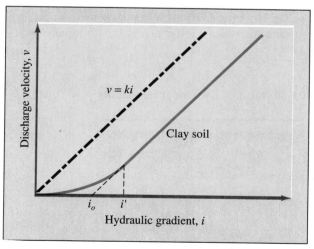

▼ **FIGURE 5.4** Variation of discharge velocity with hydraulic gradient in clay

Based on his results, it appears that there is a hydraulic gradient i' (Figure 5.4) at which

$$v = k(i - i_0) \qquad (\text{for } i \geq i') \tag{5.11}$$

and

$$v = ki^m \qquad (\text{for } i < i') \tag{5.12}$$

The preceding equation implies that, for very low hydraulic gradients, the relationship between v and i is nonlinear. The value of m in Eq. (5.12) for four Swedish clays was about 1.5. However, there are several other studies to refute the preceding findings. These have been discussed in detail by Mitchell (1976). Taking all points into consideration, Mitchell concluded that Darcy's law is valid.

5.3 COEFFICIENT OF PERMEABILITY

The *coefficient of permeability* has the same unit as velocity. The term *coefficient of permeability* is mostly used by geotechnical engineers; geologists express the same idea as *hydraulic conductivity*. In English units, the coefficient of permeability is generally expressed in ft/min or ft/day, and discharge in ft^3. In SI units, the coefficient of permeability is expressed as m/sec, and discharge in m^3.

The coefficient of permeability of soils is dependent on several factors: fluid viscosity, pore-size distribution, grain-size distribution, void ratio, roughness of mineral particles, and degree of soil saturation. In clayey soils, structure plays an important role in the coefficient of permeability. Other major factors that affect the permeability of clays are the ionic concentration and the thickness of layers of water held to the clay particles.

The value of the coefficient of permeability (k) varies widely for different soils. Some typical values of permeability coefficients are given in Table 5.1. The permeability coefficient of unsaturated soils is lower and increases rapidly with the degree of saturation.

▼ **TABLE 5.1** **Typical Values of Permeability Coefficients**

Soil type	k	
	cm/sec	ft/min
Clean gravel	1.0–100	2.0–200
Coarse sand	1.0–0.01	2.0–0.02
Fine sand	0.01–0.001	0.02–0.002
Silty clay	0.001–0.00001	0.002–0.00002
Clay	Less than 0.000001	Less than 0.000002

The coefficient of permeability is also related to the properties of the fluid flowing through it by the following equation:

$$k = \frac{\gamma_w}{\eta} \bar{K}$$

(5.13)

where γ_w = unit weight of water
η = viscosity of water
$\bar{K}$ = absolute permeability

The *absolute permeability*, $\bar{K}$, has a unit of L^2 (that is, cm^2, ft^2, and so forth).

5.4 LABORATORY DETERMINATION OF COEFFICIENT OF PERMEABILITY

There are two standard laboratory test procedures for determining the coefficient of permeability of soil: the constant head test and the falling head test. A brief description of each test follows.

Constant Head Test

A typical arrangement of the constant head permeability test is shown in Figure 5.5. In this type of laboratory setup, the water supply at the inlet is adjusted in such a way that the difference of head between the inlet and the outlet remains constant during the period of the test. After a constant rate of flow is established, water is collected in a graduated flask for a known duration.

The total volume of water collected may be expressed as

$$Q = Avt = A(ki)t$$

(5.14)

where Q = volume of water collected
A = area of cross-section of the soil specimen
t = duration of collection of water

But

$$i = \frac{h}{L}$$

(5.15)

where L = length of the specimen.

Substitution of Eq. (5.15) in Eq. (5.14) gives

$$Q = A\left(k\frac{h}{L}\right)t$$

(5.16)

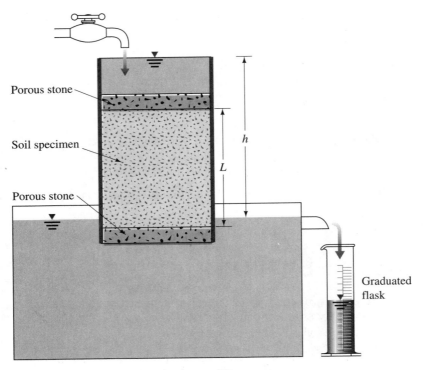

Porous stone

Soil specimen

Porous stone

Graduated flask

▼ **FIGURE 5.5**   Constant head permeability test

or

$$k = \frac{QL}{Aht}$$ (5.17)

Constant head tests are more suitable for coarse-grained soils that have higher coefficients of permeability.

Falling Head Test

A typical arrangement of the falling head permeability test is shown in Figure 5.6. Water from a standpipe flows through the soil. Initial head difference h_1 at time $t = 0$ is recorded, and water is allowed to flow through the soil specimen such that the final head difference at time $t = t_F$ is h_2.

The rate of flow of the water through the specimen at any time t can be given by

$$q = k \frac{h}{L} A = - a \frac{dh}{dt}$$ (5.18)

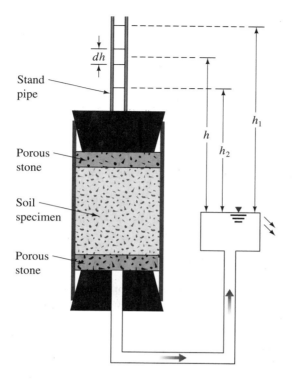

▼ **FIGURE 5.6**   Falling head permeability test

where q = rate of flow
        a = cross-sectional area of the standpipe
        A = cross-sectional area of the soil specimen

Rearrangement of the above equation gives

$$dt = \frac{aL}{Ak}\left(-\frac{dh}{h}\right)$$

(5.19)

Integration of the left side of Eq. (5.19) with limits of time from 0 to t and the right side with limits of head difference from h_1 to h_2 gives

$$t = \frac{aL}{Ak}\log_e\frac{h_1}{h_2}$$

or

$$k = 2.303\frac{aL}{At}\log_{10}\frac{h_1}{h_2}$$

(5.20)

The falling head test is more appropriate for fine-grained soils with a low coefficient of permeability.

5.5 EFFECT OF TEMPERATURE OF WATER ON k

Equation (5.13) showed that the coefficient of permeability is a function of the unit weight and the viscosity of water, which is in turn a function of the temperature at which the test is conducted. So, from Eq. (5.13),

$$\frac{k_{T_1}}{k_{T_2}} = \frac{\eta_{T_2}}{\eta_{T_1}} \frac{\gamma_{w(T_1)}}{\gamma_{w(T_2)}} \tag{5.21}$$

where
$$k_{T_1}, k_{T_2} = \text{coefficient of permeability at temperatures } T_1 \text{ and } T_2, \text{ respectively}$$
$$\eta_{T_1}, \eta_{T_2} = \text{viscosity of water at temperatures } T_1 \text{ and } T_2, \text{ respectively}$$
$$\gamma_{w(T_1)}, \gamma_{w(T_2)} = \text{unit weight of water at temperatures } T_1 \text{ and } T_2, \text{ respectively}$$

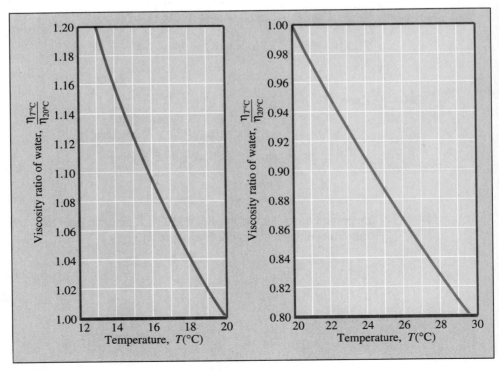

▼ **FIGURE 5.7**   Variation of $\eta_{T°c}/\eta_{20°c}$ with test temperature, $T°C$

It is a conventional practice to express the value of k at a temperature of 20°C. Within the range of test temperatures, we can assume that $\gamma_{w(T_1)} \simeq \gamma_{w(T_2)}$. So, from Eq. (5.21),

$$k_{20°C} = \left(\frac{\eta_{T°C}}{\eta_{20°C}}\right) k_{T°C}$$

(5.22)

The variation of $\eta_{T°C}/\eta_{20°C}$ as the test temperature T varies from 13°C to 30°C is given in Figure 5.7.

▼   **EXAMPLE 5.1**

Refer to the constant head permeability test arrangement shown in Figure 5.5. A test gives these values:

► $L = 18$ in.
► $A = $ area of the specimen $= 3.5$ in.^{2}
► Constant head difference, $h = 28$ in.
► Water collected in a period of 3 min $= 21.58$ in.^{3}

Calculate the coefficient of permeability in in./sec.

Solution   From Eq. (5.17),

$$k = \frac{QL}{Aht}$$

Given $Q = 21.58$ in.^{3}, $L = 18$ in., $A = 3.5$ in.^{2}, $h = 28$ in., and $t = 3$ min, we have

$$k = \frac{(21.58)(18)}{(3.5)(28)(3)(60)} = \textbf{0.022 in./sec}$$ ▼

▼   **EXAMPLE 5.2**

For a variable head permeability test, the following values are given:

► Length of specimen $= 200$ mm
► Area of soil specimen $= 1000$ mm^2
► Area of standpipe $= 40$ mm^2
► Head difference at time $t = 0 = 500$ mm
► Head difference at time $t = 180$ sec $= 300$ mm

Determine the coefficient of permeability of the soil in cm/sec.

Solution   From Eq. (5.20),

$$k = 2.303 \frac{aL}{At} \log_{10}\left(\frac{h_1}{h_2}\right)$$

We are given $a = 40$ mm^2, $L = 200$ mm, $A = 1000$ mm^2, $t = 180$ sec, $h_1 = 500$ mm, and $h_2 = 300$ mm,

$$k = 2.303 \frac{(40)(200)}{(1000)(180)} \log_{10}\left(\frac{500}{300}\right) = 0.0227 \text{ mm/sec}$$

$$= \mathbf{2.27 \times 10^{-2} \ cm/sec} \quad \blacktriangledown$$

▼ **EXAMPLE 5.3**

The coefficient of permeability of a clayey soil is 3×10^{-7} cm/sec. The viscosity of water at 25°C is 0.0911×10^{-4} g · sec/cm^2. Calculate the absolute permeability $\bar{K}$ of the soil.

Solution   From Eq. (5.13),

$$k = \frac{\gamma_w}{\eta} \bar{K} = 3 \times 10^{-7} \text{ cm/sec}$$

so

$$3 \times 10^{-7} = \left(\frac{1 \text{ g/cm}^3}{0.0911 \times 10^{-4}}\right)\bar{K}$$

$$\bar{K} = \mathbf{0.2733 \times 10^{-11} \ cm^2} \quad \blacktriangledown$$

▼ **EXAMPLE 5.4**

A permeable soil layer is underlain by an impervious layer, as shown in Figure 5.8a. With $k = 4.8 \times 10^{-3}$ cm/sec for the permeable layer, calculate the rate of seepage through it in ft^3/hr/ft width if $H = 10$ ft and $\alpha = 5°$.

Solution   From Figure 5.8b,

$$i = \frac{\text{head loss}}{\text{length}} = \frac{L \tan \alpha}{\left(\dfrac{L}{\cos \alpha}\right)} = \sin \alpha$$

$$q = kiA = (k)(\sin \alpha)(10 \cos \alpha)(1)$$

$$k = 4.8 \times 10^{-3} \text{ cm/sec} = 0.000158 \text{ ft/sec}$$

$$q = (0.000158)(\sin 5°)(10 \underset{\substack{\cos}}{\cancel{\sin}} 5°)(3600) = \mathbf{0.493 \ ft^3/hr/ft}$$

$$\uparrow$$
$$\text{To change to}$$
$$\text{ft/hr}$$

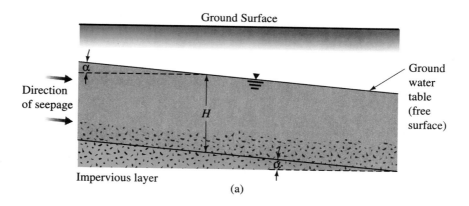

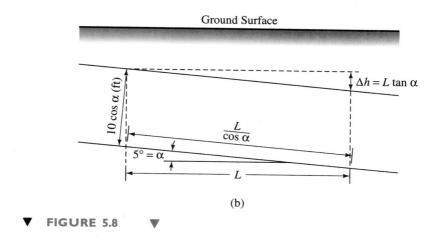

▼   FIGURE 5.8   ▼

5.6 EMPIRICAL RELATIONS FOR COEFFICIENT OF PERMEABILITY

Several empirical equations for estimating the coefficient of permeability have been proposed in the past. Some of these are briefly discussed in this section.

For fairly uniform sand (that is, small uniformity coefficient), Hazen (1930) proposed an empirical relationship for the coefficient of permeability in the form

$$k \text{ (cm/sec)} = cD_{10}^2$$

(5.23)

where c = a constant that varies from 1.0 to 1.5
D_{10} = the effective size, in mm

Equation (5.23) is based primarily on observations made by Hazen on loose, clean filter sands. A small quantity of silts and clays, when present in a sandy soil, may change the coefficient of permeability substantially.

Casagrande proposed a simple relationship for the coefficient of permeability for fine- to medium-clean sand in the following form:

$$k = 1.4e^2 k_{0.85} \qquad (5.24)$$

where k = coefficient of permeability at a void ratio e
$k_{0.85}$ = the corresponding value at a void ratio of 0.85

Another form of equation that gives fairly good results in estimating the coefficient of permeability of sandy soils is based on the Kozeny-Carman equation. The derivation of this equation will not be presented here. Interested readers are referred to any advanced soil mechanics book (for example, Das, 1983). An application of the Kozeny-Carman equation yields

$$k \propto \frac{e^3}{1+e} \qquad (5.25)$$

where k = coefficient of permeability at a void ratio of e.

Equation (5.25) can be written as

$$k = C_1 \frac{e^3}{1+e} \qquad (5.26)$$

where C_1 = a constant.

It was mentioned at the end of Section 5.1 that in very coarse sands and gravels, turbulent flow conditions may exist and Darcy's law may not be valid. However, under a low hydraulic gradient, laminar flow conditions usually exist. Kenny, Lau, and Ofoegbu (1984) conducted laboratory tests on granular soils in which the particle sizes in various specimens ranged from 0.074 to 25.4 mm. The uniformity coefficients, C_u, of these specimens ranged from 1.04 to 12. All permeability tests were conducted at a relative density of 80% or more. These tests showed that for laminar flow conditions,

$$k \text{ (mm}^2) = (0.05 \text{ to } 1)D_5^2 \qquad (5.27)$$

where D_5 = diameter through which 5% of soil passes.

Figures 5.9a and b show the results on which Eq. (5.27) is based.

According to their experimental observations, Samarasinghe, Huang, and Drnevich (1982) suggested that the coefficient of permeability of normally consolidated clays (see Chapter 8 for definition) can be given by the equation

$$k = C_3 \frac{e^n}{1+e} \qquad (5.28)$$

where C_3 and n are constants to be determined experimentally.

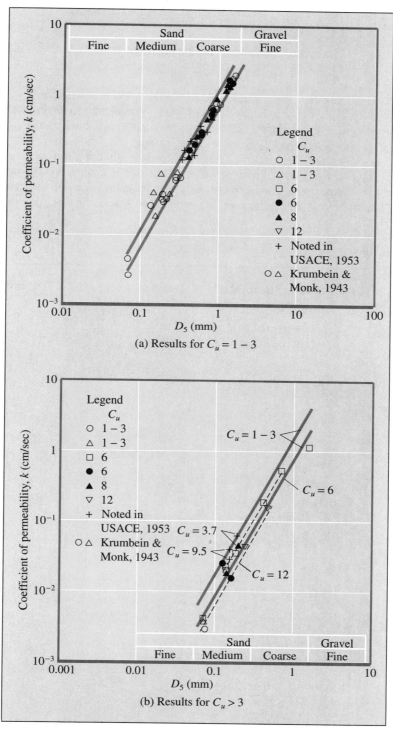

▼ **FIGURE 5.9** Results of permeability tests on which Eq. (5.27) is based (after Kenny, Lau, and Ofoegbu, 1984)

The preceding equation can be rewritten as

$$\log [k(1 + e)] = \log C_3 + n \log e \tag{5.29}$$

Hence, for any given clayey soil, if the variation of k with the void ratio is known, then a log-log graph can be plotted with $k(1 + e)$ against e to determine the values of C_3 and n (Figure 5.10). Figure 5.11 shows the variation of k with $e^n/(1 + e)$ for New Liskeard clay as determined by Samarasinghe, Huang, and Drnevich.

Some other empirical relationships for estimating the coefficient of permeability in sand and clayey soils are given in Table 5.2. It should be kept in mind, however, that any empirical relationship of this type is for estimation only, since the magnitude of k is a highly variable parameter and depends on several factors. (See also Figure 5.12.)

▼ **EXAMPLE 5.5**

The coefficient of permeability of a sand at a void ratio of 0.62 is 0.03 cm/sec. Estimate its coefficient of permeability at a void ratio of 0.48.

    a. Use Eq. (5.24).
    b. Use Eq. (5.25).

Solution

    a. From Eq. (5.24),

$$k \propto e^2$$

so

$$\frac{k_1}{k_2} = \frac{e_1^2}{e_2^2}$$

$$\frac{0.03}{k_2} = \frac{(0.62)^2}{(0.48)^2}$$

$$k_2 = \textbf{0.018 cm/sec}$$

    b. From Eq. (5.25),

$$\frac{k_1}{k_2} = \frac{\dfrac{e_1^3}{1 + e_1}}{\dfrac{e_2^3}{1 + e_2}}$$

$$\frac{0.03}{k_2} = \frac{\dfrac{(0.62)^3}{1 + 0.62}}{\dfrac{(0.48)^3}{1 + 0.48}}$$

$$k_2 = \textbf{0.015 cm/sec} \qquad ▼$$

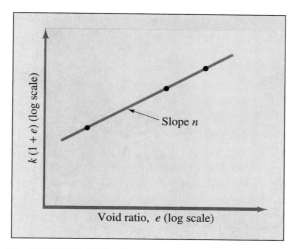

▼ **FIGURE 5.10** Nature of variation of $\log [k(1 + e)]$ with $\log e$ for normally consolidated clay

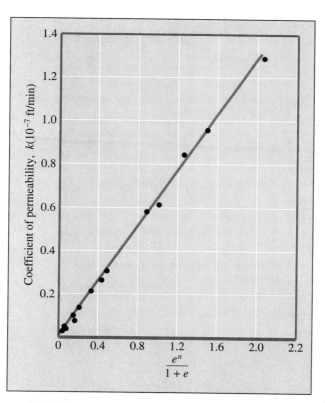

▼ **FIGURE 5.11** Variation of k with $e^n/(1 + e)$ for normally consolidated New Liskeard clay (after Samarasinghe, Huang, and Drnevich, 1982)

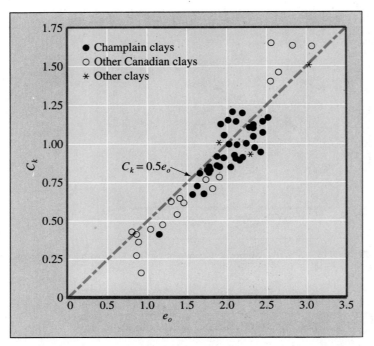

▼ **FIGURE 5.12**   Relationship between C_k and e_0; see Table 5.2 (after Tavenas et al., 1983)

▼ **TABLE 5.2   Empirical Relationships for Estimating the Coefficient of Permeability**

Type of soil	Source	Relationship	Comments
Sand	Amer and Awad (1974)	$k = C_2 D_{10}^{2.32} C_u^{0.6} \dfrac{e^3}{1+e}$	
	Shahabi, Das, and Tarquin (1984)	$k = 1.2 C_2^{0.735} D_{10}^{0.89} \dfrac{e^3}{1+e}$	Medium to fine sand
Clay	Mesri and Olson (1971)	$\log k = A' \log e + B'$	
	Taylor (1948)	$\log k = \log k_0 - \dfrac{e_0 - e}{C_k}$	For $e < 2.5$, $C_k \approx 0.5 e_0$ (see Fig. 5.12)

D_{10} = effective size
C_u = uniformity coefficient
C_2 = a constant
k_0 = *in situ* coefficient of permeability at void ratio e_0
k = coefficient of permeability at void ratio e
C_k = permeability change index

▼ **EXAMPLE 5.6**

For a normally consolidated clay soil, the following values are given:

Void ratio	k (cm/sec)
1.1	0.302×10^{-7}
0.9	0.12×10^{-7}

Estimate the coefficient of permeability of the clay at a void ratio of 0.75. Use Eq. (5.28).

Solution   From Eq. (5.28),

$$k = C_3 \left(\frac{e^n}{1 + e} \right)$$

$$\frac{k_1}{k_2} = \frac{\left(\dfrac{e_1^n}{1 + e_1} \right)}{\left(\dfrac{e_2^n}{1 + e_2} \right)}$$

$$\frac{0.302 \times 10^{-7}}{0.12 \times 10^{-7}} = \frac{\dfrac{(1.1)^n}{1 + 1.1}}{\dfrac{(0.9)^n}{1 + 0.9}}$$

$$2.517 = \left(\frac{1.9}{2.1} \right) \left(\frac{1.1}{0.9} \right)^n$$

$$2.782 = (1.222)^n$$

$$n = \frac{\log (2.782)}{\log (1.222)} = \frac{0.444}{0.087} = 5.1$$

so

$$k = C_3 \left(\frac{e^{5.1}}{1 + e} \right)$$

To find C_3,

$$0.302 \times 10^{-7} = C_3 \left[\frac{(1.1)^{5.1}}{1 + 1.1} \right] = \left(\frac{1.626}{2.1} \right) C_3$$

$$C_3 = \frac{(0.302 \times 10^{-7})(2.1)}{1.626} = 0.39 \times 10^{-7}$$

Hence,

$$k = (0.39 \times 10^{-7} \text{ cm/sec})\left(\frac{e^n}{1 + e}\right)$$

At a void ratio of 0.75,

$$k = (0.39 \times 10^{-7})\left(\frac{0.75^{5.1}}{1 + 0.75}\right) = \mathbf{0.514 \times 10^{-8} \text{ cm/sec}} \quad \blacktriangledown$$

5.7 EQUIVALENT PERMEABILITY IN STRATIFIED SOIL

Depending on the nature of a soil deposit, the coefficient of permeability of a given layer of soil may vary with the direction of flow. In a stratified soil deposit where the permeability coefficient for flow in a given direction changes from layer to layer, an equivalent permeability determination becomes necessary to simplify calculations (see also Terzaghi and Peck, 1967). The following derivations relate to the equivalent permeabilities for flow in vertical and horizontal directions through multilayered soils with horizontal stratification.

Figure 5.13 shows n layers of soil with flow in the *horizontal direction*. Let us consider a cross-section of unit length passing through the n layer and perpendicular to

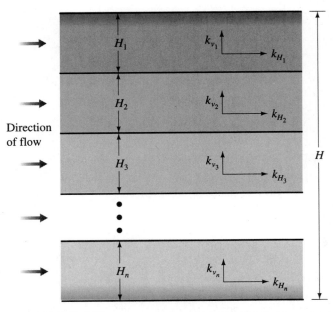

▼ **FIGURE 5.13** Equivalent coefficient of permeability determination—horizontal flow in stratified soil

the direction of flow. The total flow through the cross-section in unit time can be written as

$$q = v \cdot 1 \cdot H$$
$$= v_1 \cdot 1 \cdot H_1 + v_2 \cdot 1 \cdot H_2 + v_3 \cdot 1 \cdot H_3 + \cdots + v_n \cdot 1 \cdot H_n \qquad (5.30)$$

where v = average discharge velocity

$v_1, v_2, v_3, \ldots, v_n$ = discharge velocities of flow in layers denoted by the subscripts

If $k_{H_1}, k_{H_2}, k_{H_3}, \ldots, k_{H_n}$ are the coefficients of permeability of the individual layers in the horizontal direction, and $k_{H(eq)}$ is the equivalent coefficient of permeability in the horizontal direction, then from Darcy's law,

$$v = k_{H(eq)} i_{eq}; \quad v_1 = k_{H_1} i_1; \quad v_2 = k_{H_2} i_2; \quad v_3 = k_{H_3} i_3; \ldots;$$
$$v_n = k_{H_n} i_n$$

Substituting the preceding relations for velocities in Eq. (5.30) and noting $i_{eq} = i_1 = i_2 = i_3 = \cdots = i_n$ result in

$$k_{H(eq)} = \frac{1}{H} (k_{H_1} H_1 + k_{H_2} H_2 + k_{H_3} H_3 + \cdots + k_{H_n} H_n) \qquad (5.31)$$

Figure 5.14 shows n layers of soil with flow in the vertical direction. In this case, the velocity of flow through all the layers is the same. However, the total head loss, h, is equal to the sum of the head losses in all layers. Thus,

$$v = v_1 = v_2 = v_3 = \cdots = v_n \qquad (5.32)$$

and

$$h = h_1 + h_2 + h_3 + \cdots + h_n \qquad (5.33)$$

Using Darcy's law, we can rewrite Eq. (5.32) as

$$k_{V(eq)} \frac{h}{H} = k_{V_1} i_1 = k_{V_2} i_2 = k_{V_3} i_3 = \cdots = k_{V_n} i_n \qquad (5.34)$$

where $k_{V_1}, k_{V_2}, k_{V_3}, \ldots, k_{V_n}$ are the coefficients of permeability of the individual layers in the vertical direction and $k_{V(eq)}$ is the equivalent coefficient of permeability.

Again, from Eq. (5.33),

$$h = H_1 i_1 + H_2 i_2 + H_3 i_3 + \cdots + H_n i_n \qquad (5.35)$$

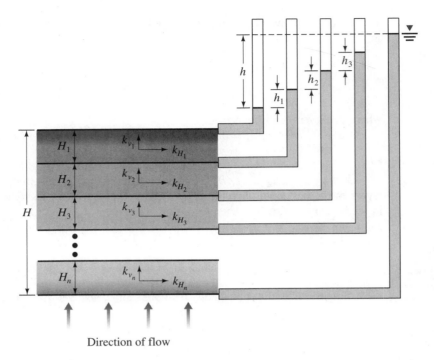

FIGURE 5.14   Equivalent permeability determination—vertical flow in stratified soil

Solving of Eqs. (5.34) and (5.35) gives

$$k_{V(eq)} = \frac{H}{\left(\dfrac{H_1}{k_{V_1}}\right) + \left(\dfrac{H_2}{k_{V_2}}\right) + \left(\dfrac{H_3}{k_{V_3}}\right) + \cdots + \left(\dfrac{H_n}{k_{V_n}}\right)} \tag{5.36}$$

▼   **EXAMPLE 5.7**

A layered soil is shown in Figure 5.15. Given:

- ▶ $H_1 = 3$ ft $k_1 = 10^{-4}$ cm/sec
- ▶ $H_2 = 4$ ft $k_2 = 3.2 \times 10^{-2}$ cm/sec
- ▶ $H_3 = 6$ ft $k_3 = 4.1 \times 10^{-5}$ cm/sec

Estimate the ratio of equivalent permeability

$$\frac{k_{H(eq)}}{k_{V(eq)}}$$

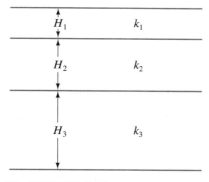

▼ **FIGURE 5.15**

Solution   From Eq. (5.31),

$$k_{H(eq)} = \frac{1}{H}(k_{H_1}H_1 + k_{H_2}H_2 + k_{H_3}H_3)$$

$$= \frac{1}{(3 + 4 + 6)}[(10^{-4})(3) + (3.2 \times 10^{-2})(4) + (4.1 \times 10^{-5})(6)]$$

$$= 98.88 \times 10^{-4} \text{ cm/sec}$$

Again, from Eq. (5.36),

$$k_{V(eq)} = \frac{H}{\left(\dfrac{H_1}{k_{V_1}}\right) + \left(\dfrac{H_2}{k_{V_2}}\right) + \left(\dfrac{H_3}{k_{V_3}}\right)}$$

$$= \frac{3 + 4 + 6}{\left(\dfrac{3}{10^{-4}}\right) + \left(\dfrac{4}{3.2 \times 10^{-2}}\right) + \left(\dfrac{6}{4.1 \times 10^{-5}}\right)}$$

$$= 0.737 \times 10^{-4} \text{ cm/sec}$$

Hence,

$$\frac{k_{H(eq)}}{k_{V(eq)}} = \frac{98.88 \times 10^{-4}}{0.737 \times 10^{-4}} = \textbf{134.17} \qquad ▼$$

5.8 PERMEABILITY TEST IN THE FIELD BY PUMPING FROM WELLS

In the field, the average coefficient of permeability of a soil deposit in the direction of flow can be determined by performing pumping tests from wells. Figure 5.16 shows a case where the top permeable layer, whose coefficient of permeability has to be determined, is unconfined and underlain by an impermeable layer. During the test, water is

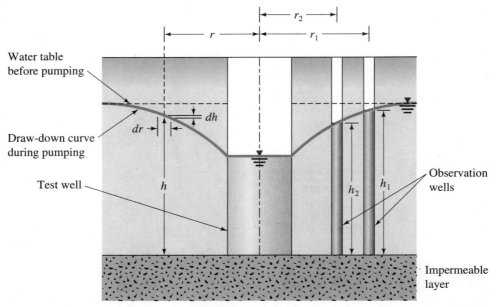

FIGURE 5.16   Pumping test from a well in an unconfined permeable layer underlain by an impermeable stratum

pumped out at a constant rate from a test well that has a perforated casing. Several
observation wells at various radial distances are made around the test well. Continuous
observations of the water level in the test well and the observation wells are made after
the start of pumping, until a steady state is reached. The steady state is established
when the water level in the test well and observation wells becomes constant. The
expression for the rate of flow of ground water, q, into the well, which is equal to the
rate of discharge from pumping, can be written as

$$q = k\left(\frac{dh}{dr}\right)2\pi rh \tag{5.37}$$

or

$$\int_{\gamma_2}^{\gamma_1} \frac{dr}{r} = \left(\frac{2\pi k}{q}\right)\int_{h_2}^{h_1} h\, dh$$

Thus,

$$k = \frac{2.303q\,\log_{10}\left(\dfrac{r_1}{r_2}\right)}{\pi(h_1^2 - h_2^2)} \tag{5.38}$$

From field measurements, if q, r_1, r_2, h_1, and h_2 are known, then the coefficient of
permeability can be calculated from the simple relationship presented in Eq. (5.38).

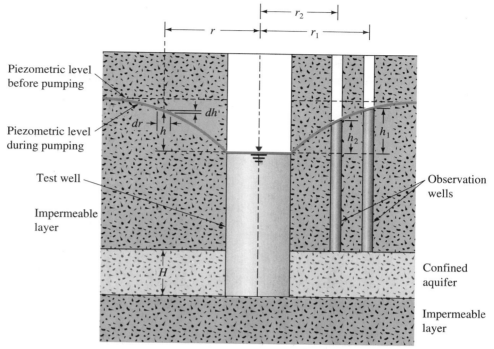

▼ **FIGURE 5.17**   Pumping test from a well penetrating the full depth in a confined aquifer

The average coefficient of permeability for a confined aquifer can also be determined by conducting a pumping test from a well with a perforated casing that penetrates the full depth of the aquifer and by observing the piezometric level in a number of observation wells at various radial distances (Figure 5.17). Pumping is continued at a uniform rate q until a steady state is reached.

Since water can enter the test well only from the aquifer of thickness H, the steady state of discharge is

$$q = k \left(\frac{dh}{dr} \right) 2\pi r H \tag{5.39}$$

or

$$\int_{r_2}^{r_1} \frac{dr}{r} = \int_{h_2}^{h_1} \frac{2\pi k H}{q} \, dh$$

This gives the coefficient of permeability in the direction of flow as

$$k = \frac{q \log_{10} \left(\dfrac{r_1}{r_2} \right)}{2.727 H (h_1 - h_2)} \tag{5.40}$$

▼   **EXAMPLE 5.8**

A pumping test from a confined aquifer yielded the following results: $q = 80$ gal/min, $h_1 = 8$ ft, $h_2 = 5$ ft, $r_1 = 60$ ft, $r_2 = 30$ ft, and $H = 10$ ft. Refer to Figure 5.17 and determine the magnitude of k of the permeable layer.

Solution   From Eq. (5.40),

$$k = \frac{q \log_{10}\left(\dfrac{r_1}{r_2}\right)}{2.727H(h_1 - h_2)}$$

We have $q = 80$ gal/min $= 10.7$ ft^3/min, so

$$k = \frac{(10.7)\log_{10}\left(\dfrac{60}{30}\right)}{(2.727)(10)(8 - 5)} = \mathbf{0.039 \ ft/min} \qquad ▼$$

5.9   COEFFICIENT OF PERMEABILITY FROM AUGER HOLES

The coefficient of permeability can also be estimated in the field from single auger holes (Figure 5.18). These types of tests are often called *slug tests*. Holes are made in the field that extend to a depth L below the ground water table. Water is first bailed out of the

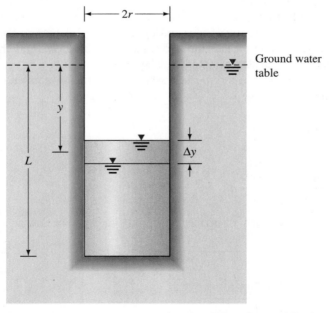

▼   **FIGURE 5.18**   Determination of coefficient of permeability from an auger hole

hole. This creates a flow of ground water into the auger hole through its perimeter and from the bottom. The rise of the water level in the auger hole with time is recorded. The coefficient of permeability can be calculated from these readings (Ernst, 1950; also see Dunn, Anderson, and Kiefer, 1980) as,

$$k = \frac{40}{\left(20 + \dfrac{L}{r}\right)\left(2 - \dfrac{y}{L}\right)} \frac{r}{y} \frac{\Delta y}{\Delta t} \tag{5.41}$$

where r = radius of the auger hole (m)

$\quad\quad y$ = average value of the distance of the water level in the auger hole measured from the ground water table during a time interval of Δt (m)

Note that in the preceding equation, L and Δy are in meters and k is in m/sec or m/min, depending on the unit of time Δt.

Determination of the coefficient of permeability from auger holes does not yield very accurate results; however, it does provide an estimate of k.

5.10   EQUATION OF CONTINUITY

In the preceding sections of this chapter, we considered some simple cases for which direct application of Darcy's law was required to calculate the flow of water through soil. In many instances, the flow of water through soil is not in one direction only, nor is it uniform over the entire area perpendicular to the flow. In such cases, the ground water flow is generally calculated by the use of graphs referred to as flow nets. The concept of the flow net is based on *Laplace's equation of continuity*, which governs the steady flow condition for a given point in the soil mass.

To derive the Laplace differential equation of continuity, let us take a single row of sheet piles that have been driven into a permeable soil layer, as shown in Figure 5.19a. The row of sheet piles is assumed to be impervious. The steady state flow of water from the upstream to the downstream side through the permeable layer is a two-dimensional flow. For flow at a point A, we consider an elemental soil block. The block has dimensions dx, dy, and dz (length dy is perpendicular to the plane of the paper); it is shown in an enlarged scale in Figure 5.19b. Let v_x and v_z be the components of the discharge velocity in the horizontal and vertical directions, respectively. The rate of flow of water into the elemental block in the horizontal direction is equal to $v_x\, dz\, dy$, and in the vertical direction it is $v_z\, dx\, dy$. Again, the rates of outflow from the block in the horizontal and vertical directions are

$$\left(v_x + \frac{\partial v_x}{\partial x}\, dx\right) dz\, dy$$

and

$$\left(v_z + \frac{\partial v_z}{\partial z}\, dz\right) dx\, dy$$

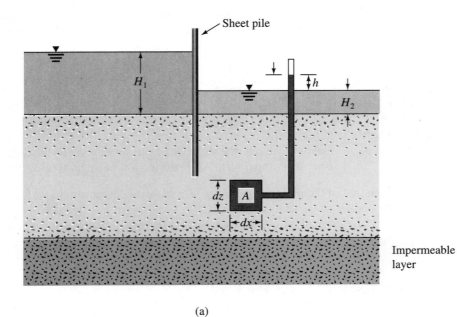

(a)

(b)

▼ **FIGURE 5.19** (a) Single-row sheet piles driven into permeable layer; (b) flow at A

Assuming water is incompressible and no volume change in the soil mass occurs, we know the total rate of inflow should be equal to the total rate of outflow. Thus,

$$\left[\left(v_x + \frac{\partial v_x}{\partial x}\,dx\right)dz\,dy + \left(v_z + \frac{\partial v_z}{\partial z}\,dz\right)dx\,dy\right] - [v_x\,dz\,dy + v_z\,dx\,dy] = 0$$

or

$$\frac{\partial v_x}{\partial x} + \frac{\partial v_z}{\partial z} = 0 \tag{5.42}$$

with Darcy's law, the discharge velocities can be expressed as

$$v_x = k_x i_x = k_x \frac{\partial h}{\partial x} \tag{5.43}$$

and

$$v_z = k_z i_z = k_z \frac{\partial h}{\partial z} \tag{5.44}$$

where k_x and k_z are the coefficients of permeability in the vertical and horizontal directions, respectively.

From Eqs. (5.42), (5.43), and (5.44), we can write

$$k_x \frac{\partial^2 h}{\partial x^2} + k_z \frac{\partial^2 h}{\partial z^2} = 0 \tag{5.45}$$

If the soil is isotropic with respect to the permeability coefficients—that is, $k_x = k_z$—the preceding continuity equation for two-dimensional flow simplifies to

$$\frac{\partial^2 h}{\partial x^2} + \frac{\partial^2 h}{\partial z^2} = 0 \tag{5.46}$$

5.11 CONTINUITY EQUATION FOR SOLUTION OF SIMPLE FLOW PROBLEMS

The continuity equation given in Eq. (5.46) can be used in solving some simple flow problems. In order to illustrate this, let us consider a one-dimensional flow problem as shown in Figure 5.20, in which a constant head is maintained across a two-layered soil for the flow of water. The head difference between the top of the soil layer no. 1 and the

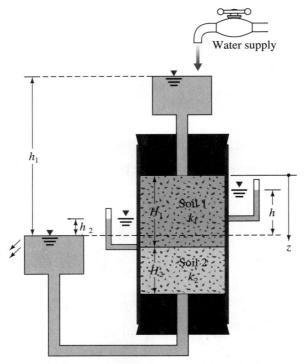

▼   **FIGURE 5.20**   Flow through a two-layered soil

bottom of the soil layer no. 2 is h_1. Since the flow is in only the z direction, the continuity equation [Eq. 5.46] is simplified to the form

$$\frac{\partial^2 h}{\partial z^2} = 0 \tag{5.47}$$

or

$$h = A_1 z + A_2 \tag{5.48}$$

where A_1 and A_2 are constants.

In order to obtain A_1 and A_2 for flow through soil layer no. 1, the known boundary conditions are as follows:

▶   *Condition 1:* At $z = 0$, $h = h_1$.
▶   *Condition 2:* At $z = H_1$, $h = h_2$.

Combining Eq. (5.48) and condition 1 gives

$$A_2 = h_1 \tag{5.49}$$

Similarly, combining Eq. (5.48) and condition 2 gives

$$h_2 = A_1 H_1 + h_1$$

or

$$A_1 = -\left(\frac{h_1 - h_2}{H_1}\right) \tag{5.50}$$

Combining Eqs. (5.48), (5.49), and (5.50), we obtain

$$h = -\left(\frac{h_1 - h_2}{H_1}\right)z + h_1 \quad \text{(for } 0 \leq H_1) \tag{5.51}$$

Again, for flow through soil layer no. 2, the boundary conditions are

▶ *Condition 1:* At $z = H_1$, $h = h_2$.
▶ *Condition 2:* At $z = H_1 + H_2$, $h = 0$.

From condition 1 and Eq. (5.48),

$$A_2 = h_2 - A_1 H_1 \tag{5.52}$$

Also, from condition 2 and Eq. (5.48),

$$0 = A_1(H_1 + H_2) + (h_2 - A_1 H_1)$$
$$A_1 H_1 + A_1 H_2 + h_2 - A_1 H_1 = 0$$

or

$$A_1 = -\frac{h_2}{H_2} \tag{5.53}$$

So, from Eqs. (5.48), (5.52), and (5.53),

$$h = -\left(\frac{h_2}{H_2}\right)z + h_2\left(1 + \frac{H_1}{H_2}\right) \quad \text{(for } H_1 \leq z \leq H_1 + H_2) \tag{5.54}$$

At any given time, flow through soil layer no. 1 = flow through soil layer no. 2, so

$$q = k_1\left(\frac{h_1 - h_2}{H_1}\right)A = k_2\left(\frac{h_2 - 0}{H_2}\right)A$$

where A = area of cross-section of the soil
  k_1 = coefficient of permeability of soil layer no. 1
  k_2 = coefficient of permeability of soil layer no. 2

or

$$h_2 = \frac{h_1 k_1}{H_1\left(\dfrac{k_1}{H_1} + \dfrac{k_2}{H_2}\right)} \tag{5.55}$$

Substituting Eq. (5.55) into Eq. (5.51), we obtain

$$h = h_1 \left(1 - \frac{k_2 z}{k_1 H_2 + k_2 H_1} \right) \qquad \text{(for } 0 \leq z \leq H_1)$$

(5.56)

Similarly, combining Eqs. (5.54) and (5.55) gives

$$h = h_1 \left[\left(\frac{k_1}{k_1 H_2 + k_2 H_1} \right)(H_1 + H_2 - z) \right]$$

$$\text{(for } H_1 \leq z \leq H_1 + H_2)$$

(5.57)

▼　**EXAMPLE 5.9**

Refer to Figure 5.20. Given $H_1 = 300$ mm; $H_2 = 500$ mm; $h_1 = 600$ mm; and at $z = 200$ mm, $h = 500$ mm, determine h at $z = 600$ mm.

Solution　We know $z = 200$ mm is located in soil layer no. 1, so Eq. (5.56) is valid:

$$h = h_1 \left(1 - \frac{k_2 z}{k_1 H_2 + k_2 H_1} \right)$$

$$500 = 600 \left[1 - \frac{k_2(200)}{k_1(500) + k_2(300)} \right]$$

or

$$\frac{k_1}{k_2} = \mathbf{1.795}$$

Since $z = 600$ mm is located in soil layer no. 2, Eq. (5.57) is valid:

$$h = h_1 \left[\frac{1}{H_2 + \left(\dfrac{k_2}{k_1} \right) H_1} (H_1 + H_2 - z) \right]$$

or

$$h = 600 \left[\left(\frac{1}{500 + \dfrac{300}{1.795}} \right)(300 + 500 - 600) \right] = \mathbf{178.9 \ mm} \qquad ▼$$

5.12 FLOW NETS

The continuity equation [Eq. (5.46)] in an isotropic medium represents two orthogonal families of curves—that is, the flow lines and the equipotential lines. A *flow line* is a line along which a water particle will travel from upstream to the downstream side in the permeable soil medium. An *equipotential line* is a line along which the potential head at all points is equal. Thus, if piezometers are placed at different points along an equipotential line, the level of water will rise to the same elevation in all of them. Figure 5.21a demonstrates the definition of flow and equipotential lines for flow in the permeable soil layer around the row of sheet piles shown in Figure 5.19 (for $k_x = k_z = k$).

A combination of a number of flow lines and equipotential lines is called a *flow net*. As mentioned in the preceding section, flow nets are constructed for the calculation of ground water flow and the evaluation of heads in the media. To complete the graphical construction of a flow net, the flow and equipotential lines are drawn in such a way that:

1. The equipotential lines intersect the flow lines at right angles.
2. The flow elements formed are approximate squares.

Figure 5.21b is an example of a completed flow net. Another example of a flow net in an isotropic permeable layer is given in Figure 5.22.

Drawing a flow net takes several trials. While constructing the flow net, keep the boundary conditions in mind. For the flow net shown in Figure 5.21b, the following boundary conditions apply:

1. The upstream and downstream surfaces of the permeable layer (lines *ab* and *de*) are equipotential lines.
2. Because *ab* and *de* are equipotential lines, all of the flow lines intersect them at right angles.
3. The boundary of the impervious layer—that is, line *fg*—is a flow line, and so is the surface of the impervious sheet pile, line *acd*.
4. The equipotential lines intersect *acd* and *fg* at right angles.

Seepage Calculation from a Flow Net

In any flow net, the strip between any two adjacent flow lines is called a *flow channel*. Figure 5.23 shows a flow channel with the equipotential lines forming square elements. Let h_1, h_2, h_3, h_4, ..., h_n be the piezometric levels corresponding to the equipotential lines. The rate of seepage through the flow channel per unit length (perpendicular to the vertical section through the permeable layer) can be calculated as follows. Since there is no flow across the flow lines,

$$\Delta q_1 = \Delta q_2 = \Delta q_3 = \cdots = \Delta q \qquad (5.58)$$

From Darcy's law, the rate of flow is equal to kiA. Thus, Eq. (5.58) can be written as

$$\Delta q = k\left(\frac{h_1 - h_2}{l_1}\right)l_1 = k\left(\frac{h_2 - h_3}{l_2}\right)l_2 = k\left(\frac{h_3 - h_4}{l_3}\right)l_3 = \cdots \qquad (5.59)$$

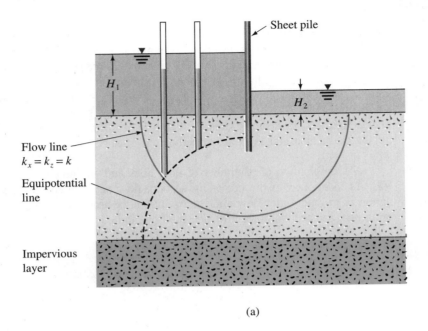

(a)

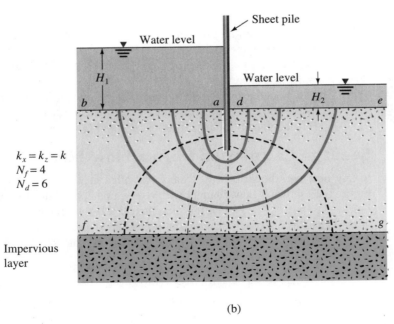

(b)

▼ **FIGURE 5.21**   (a) Definition of flow lines and equipotential lines; (b) completed flow net

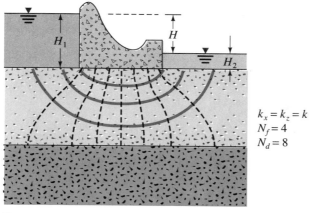

▼ FIGURE 5.22 Flow net under a dam

Eq. (5.59) shows that, if the flow elements are drawn as approximate squares, the drop in the piezometric level between any two adjacent equipotential lines is the same. This is called the *potential drop*. Thus,

$$h_1 - h_2 = h_2 - h_3 = h_3 - h_4 = \cdots = \frac{H}{N_d} \tag{5.60}$$

and

$$\Delta q = k \frac{H}{N_d} \tag{5.61}$$

where H = difference of head between the upstream and downstream sides
N_d = number of potential drops

In Figure 5.21a, for any flow channel, $H = H_1 - H_2$ and $N_d = 6$.

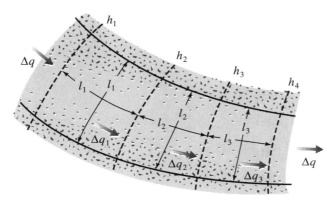

▼ FIGURE 5.23 Seepage through a flow channel with square elements

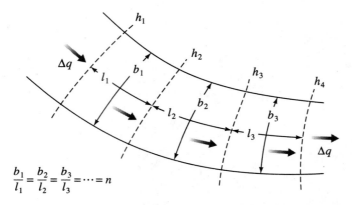

$$\frac{b_1}{l_1} = \frac{b_2}{l_2} = \frac{b_3}{l_3} = \cdots = n$$

▼ **FIGURE 5.24**   Seepage through a flow channel with rectangular elements

If the number of flow channels in a flow net is equal to N_f, then the total rate of flow through all the channels per unit length can be given by

$$q = k\,\frac{HN_f}{N_d} \tag{5.62}$$

Although convenient, it is not always necessary to draw square elements for a flow net. It is also possible to draw a rectangular mesh for a flow channel as shown in Figure 5.24, provided the width-to-length ratios for all the rectangular elements in the flow net are the same. In that case, Eq. (5.59) for rate of flow through the channel can be modified to

$$k\left(\frac{h_1 - h_2}{l_1}\right)b_1 = k\left(\frac{h_2 - h_3}{l_2}\right)b_2 = k\left(\frac{h_3 - h_4}{l_3}\right)b_3 = \cdots \tag{5.63}$$

If $b_1/l_1 = b_2/l_2 = b_3/l_3 = \cdots = n$ (i.e., the elements are not square), then Eqs. (5.61) and (5.62) can be modified

$$\Delta q = kH\left(\frac{n}{N_d}\right) \tag{5.64}$$

or

$$q = kH\left(\frac{N_f}{N_d}\right)n \tag{5.65}$$

Figure 5.25 shows a flow net for seepage around a single row of sheet piles. Note that flow channels No. 1 and 2 have square elements. Hence, the rate of flow through

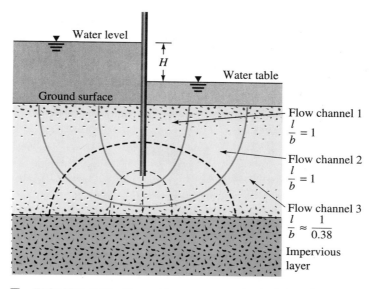

▼ FIGURE 5.25 Flow net for seepage around a single row of sheet piles

these two channels can be obtained from Eq. (5.61):

$$\Delta q_1 + \Delta q_2 = \frac{k}{N_d} H + \frac{k}{N_d} H = \frac{2kH}{N_d}$$

However, flow channel 3 has rectangular elements. These elements have a width-to-length ratio of about 0.38; hence, from Eq. (5.64),

$$\Delta q_3 = \frac{k}{N_d} H(0.38)$$

So, the total rate of seepage can be given as

$$q = \Delta q_1 + \Delta q_2 + \Delta q_3 = 2.38 \frac{kH}{N_d}$$

5.13 FLOW NET IN ANISOTROPIC SOIL

The flow net construction described above and the derived Eqs. (5.62) and (5.65) for seepage calculation have been based on the assumption that the soil is isotropic. However, in nature, most soils exhibit some degree of anisotropy. In order to account for soil anisotropy with respect to permeability, some modification of the flow net construction is necessary.

The differential equation of continuity for a two-dimensional flow [Eq. (5.45)] is

$$k_x \frac{\partial^2 h}{\partial x^2} + k_z \frac{\partial^2 h}{\partial z^2} = 0$$

For anisotropic soils, $k_x \neq k_z$. In that case, the equation represents two families of curves that do not meet at 90°. However, we can rewrite the preceding equation as

$$\frac{\partial^2 h}{(k_z/k_x)\, \partial x^2} + \frac{\partial^2 h}{\partial z^2} = 0 \qquad (5.66)$$

Substituting $x' = \sqrt{k_z/k_x}\, x$, we can express Eq. (5.66) as

$$\frac{\partial^2 h}{\partial x'^2} + \frac{\partial^2 h}{\partial z^2} = 0 \qquad (5.67)$$

Now Eq. (5.67) is in a form similar to Eq. (5.46), with x replaced by x', which is the new transformed coordinate. To construct the flow net, use the following procedure:

1. Adopt a vertical scale (that is, z-axis) for drawing the cross-section.
2. Adopt a horizontal scale (that is, x-axis) such that horizontal scale $= \sqrt{k_z/k_x}$ · (vertical scale).
3. With scales adopted as in Steps 1 and 2, plot the vertical section through the permeable layer parallel to the direction of flow.
4. Draw the flow net for the permeable layer on the section obtained from Step 3, with flow lines intersecting equipotential lines at right angles and the elements as approximate squares.

The rate of seepage per unit length can be calculated by modifying Eq. (5.62) to

$$\boxed{q = \sqrt{k_x k_z}\, \frac{H N_f}{N_d}} \qquad (5.68)$$

where H = total head loss

N_f and N_d = number of flow channels and potential drops, respectively (from flow net drawn in Step 4)

Note that when flow nets are drawn in transformed sections (in anisotropic soils), the flow lines and the equipotential lines are orthogonal. However, when they are redrawn in a true section, the flow lines and the equipotential lines will not be at right angles to each other. This is shown in Figure 5.26. In this figure it is assumed that $k_x = 6k_z$. Figure 5.26a shows a flow element in a transformed section. The flow element has been redrawn in a true section in Figure 5.26b.

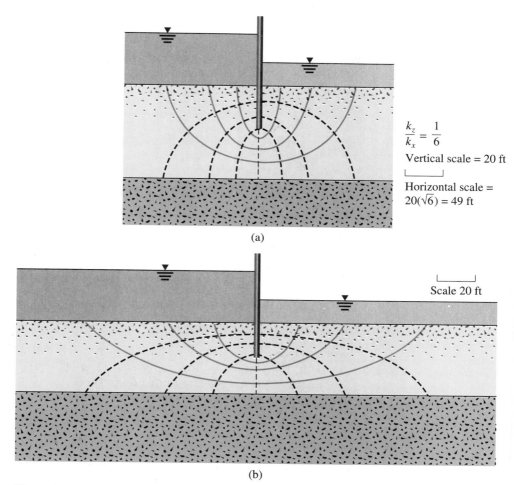

$$\frac{k_z}{k_x} = \frac{1}{6}$$

Vertical scale = 20 ft

Horizontal scale = $20(\sqrt{6}) = 49$ ft

(a)

Scale 20 ft

(b)

▼ **FIGURE 5.26**   A flow element in anisotropic soil: (a) in transformed section; (b) in true section

▼ **EXAMPLE 5.10**

A flow net for flow around a single row of sheet piles in a permeable soil layer is shown in Figure 5.27. Given $k_x = k_z = k = 5 \times 10^{-3}$ cm/sec, determine:

a. How high (above the ground surface) the water will rise if piezometers are placed at points a, b, c, and d

b. The rate of seepage through the flow channel II per unit length (perpendicular to the section shown)

Solution

a. From Figure 5.27, $N_f = 3$ and $N_d = 6$. The difference of head between the upstream and downstream sides is 10 ft, so the loss of head for each drop is $10/6 = 1.667$ ft.

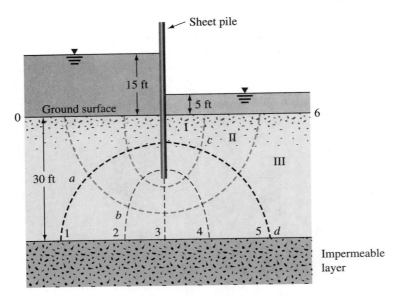

Point a is located on equipotential line 1, which means that the potential drop at a is 1×1.667 ft. The water in the piezometer at a will rise to an elevation of $(15 - 1.667) = $ **13.333 ft above the ground surface.**

Similarly, the piezometric levels for

$$b = (15 - 2 \times 1.667) = \textbf{11.67 ft above the ground surface}$$

$$c = (15 - 5 \times 1.667) = \textbf{6.67 ft above the ground surface}$$

$$d = (15 - 5 \times 1.667) = \textbf{6.67 ft above the ground surface}$$

b. From Eq. (5.61),

$$\Delta q = k \frac{H}{N_d}$$

$$k = 5 \times 10^{-3} \text{ cm/sec} = 5 \times 10^{-3} \times 0.03281 \text{ ft/sec} = \textbf{1.64} \times \textbf{10}^{-4} \textbf{ ft/sec}$$

$$\Delta q = (1.64 \times 10^{-4})(1.667) = \textbf{2.73} \times \textbf{10}^{-4} \textbf{ ft}^3\textbf{/sec/ft} \qquad ▼$$

5.14 MATHEMATICAL SOLUTION FOR SEEPAGE

The seepage under several simple hydraulic structures can be solved mathematically. Harr (1962) has analyzed many such conditions. Figure 5.28 shows a nondimensional plot for the rate of seepage around a single row of sheet piles. In a similar manner, Figure 5.29 is a nondimensional plot for the rate of seepage under a dam.

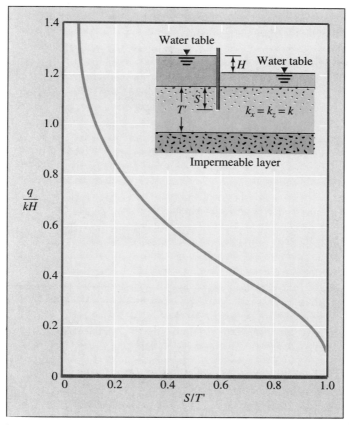

▼ **FIGURE 5.28**   Plot of q/kH against S/T' for flow around a single row of sheet piles (after Harr, 1962)

5.15 UPLIFT PRESSURE UNDER HYDRAULIC STRUCTURES

Flow nets can be used to determine the uplift pressure at the base of a hydraulic structure. This general concept can be demonstrated by a simple example. Figure 5.30a shows a weir, the base of which is 6 ft below the ground surface. The necessary flow net has also been drawn (assuming $k_x = k_z = k$). The pressure distribution diagram at the base of the weir can be obtained from the equipotential lines as follows.

There are seven equipotential drops (N_d) in the flow net, and the difference in the water levels between the upstream and downstream sides is $H = 21$ ft. The loss of head for each potential drop is $H/7 = 21/7 = 3$ ft. The uplift pressure at

a (left corner of the base) = (pressure head at a) × (γ_w)

$$= [(21 + 6) - 3]\gamma_w = 24\gamma_w$$

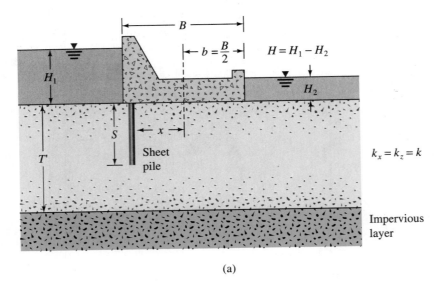

(a)

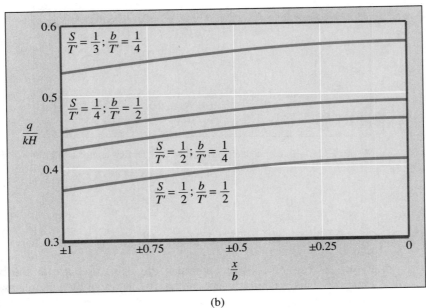

(b)

▼ **FIGURE 5.29**   Seepage under a dam (after Harr, 1962)

Similarly, at

$$b = [27 - (2)(3)]\gamma_w = 21\gamma_w$$

and at

$$f = [27 - (6)(3)]\gamma_w = 9\gamma_w$$

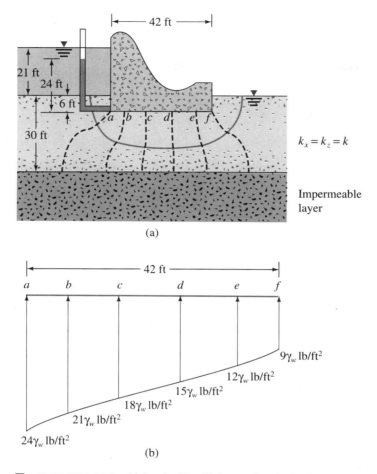

42 ft

$k_x = k_z = k$

Impermeable
layer

(a)

42 ft

$9\gamma_w$ lb/ft^2

$12\gamma_w$ lb/ft^2

$15\gamma_w$ lb/ft^2

$18\gamma_w$ lb/ft^2

$21\gamma_w$ lb/ft^2

$24\gamma_w$ lb/ft^2

(b)

▼ **FIGURE 5.30**   (a) A weir; (b) uplift force under a hydraulic structure

The uplift pressures have been plotted in Figure 5.30b. The uplift force per unit length measured along the axis of the weir can be calculated by finding the area of the pressure diagram.

5.16 SEEPAGE THROUGH AN EARTH DAM ON AN IMPERVIOUS BASE

Figure 5.31 shows a homogeneous earth dam resting on an impervious base. Let the coefficient of permeability of the compacted material of which the earth dam is made be equal to k. The free surface of the water passing through the dam is given by *abcd*. It is

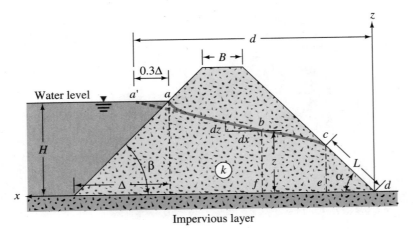

Flow through an earth dam constructed over an impervious base

assumed that $a'bc$ is parabolic. The slope of the free surface can be assumed to be equal to the hydraulic gradient. It is also assumed that this hydraulic gradient is constant with depth (Dupuit, 1863), so

$$i \simeq \frac{dz}{dx} \tag{5.69}$$

Considering the triangle cde, we can give the rate of seepage per unit length of the dam (at right angles to the cross-section shown in Figure 5.31) as

$$q = kiA$$

$$i = \frac{dz}{dx} = \tan \alpha$$

$$A = (\overline{ce})(1) = L \sin \alpha$$

So

$$q = k(\tan \alpha)(L \sin \alpha) = kL \tan \alpha \sin \alpha \tag{5.70}$$

Again, the rate of seepage (per unit length of the dam) through the section bf is

$$q = kiA = k\left(\frac{dz}{dx}\right)(z \times 1) = kz \frac{dz}{dx} \tag{5.71}$$

For continuous flow

$$q_{\text{Eq. (5.70)}} = q_{\text{Eq. (5.71)}}$$

or

$$kz \frac{dz}{dx} = kL \tan \alpha \sin \alpha$$

or

$$\int_{z=L \sin \alpha}^{z=H} kz \, dz = \int_{x=L \cos \alpha}^{x=d} (kL \tan \alpha \sin \alpha) \, dx$$

$$\frac{1}{2}(H^2 - L^2 \sin^2 \alpha) = L \tan \alpha \sin \alpha (d - L \cos \alpha)$$

$$\frac{H^2}{2} - \frac{L^2 \sin^2 \alpha}{2} = Ld\left(\frac{\sin^2 \alpha}{\cos \alpha}\right) - L^2 \sin^2 \alpha$$

$$\frac{H^2 \cos \alpha}{2 \sin^2 \alpha} - \frac{L^2 \cos \alpha}{2} = Ld - L^2 \cos \alpha$$

or

$$L^2 \cos \alpha - 2Ld + \frac{H^2 \cos \alpha}{\sin^2 \alpha} = 0$$

$$L = \frac{d}{\cos \alpha} - \sqrt{\frac{d^2}{\cos^2 \alpha} - \frac{H^2}{\sin^2 \alpha}} \qquad (5.72)$$

Following is a step-by-step procedure to obtain the seepage rate q (per unit length of the dam).

1. Obtain α.
2. Calculate Δ (Figure 5.31) and then 0.3Δ.
3. Calculate d.
4. With known values of α and d, calculate L from Eq. (5.72).
5. With known values of L, calculate q from Eq. (5.70).

▼ **EXAMPLE 5.11**

Refer to Figure 5.31. Given height of embankment $= 40$ ft, $H = 30$ ft, width of top of embankment $B = 15$ ft, upstream and downstream slope of embankment $= 1V:1H$, and $k = 4.2 \times 10^{-4}$ ft/min, determine the seepage rate, q, in ft^3/day/ft.

Solution   From Figure 5.31 we are given $\beta = 45°$ and $\alpha = 45°$. Thus,

$$\Delta = \frac{H}{\tan \beta} = \frac{30}{\tan 45°} = 30 \text{ ft}$$

$$d = 0.3\Delta + \frac{(40 - 30)}{\tan \beta} + B + \frac{40}{\tan \alpha}$$

$$= (0.3)(30) + \frac{(40 - 30)}{\tan 45°} + 15 + \frac{40}{\tan 45°} = 74 \text{ ft}$$

From Eq. (5.72),

$$L = \frac{d}{\cos \alpha} - \sqrt{\frac{d^2}{\cos^2\alpha} - \frac{H^2}{\sin^2\alpha}}$$

$$= \frac{74}{\cos 45°} - \sqrt{\left(\frac{74}{\cos 45°}\right)^2 - \left(\frac{30}{\sin 45°}\right)^2} = 8.98 \text{ ft}$$

$$q = kL \tan \alpha \sin \alpha = (4.2 \times 10^{-4})(8.98)(\tan 45°)(\sin 45°)$$

$$= 26.67 \times 10^{-4} \text{ ft}^3/\text{min/ft} = \textbf{3.84 ft}^3\textbf{/day/ft} \quad \blacktriangledown$$

5.17 GENERAL COMMENTS

This chapter covered Darcy's law and the definition of the coefficient of permeability, k, of soil; laboratory and field procedures for the determination of k; and calculations of seepage under hydraulic structures and seepage through earth dams located on an impermeable layer.

Among other factors, seepage is a function of the coefficient of permeability of the soil layer(s) involved, which is highly variable. The accuracy of the values of k determined in the laboratory depends on several factors:

1. Temperature of the fluid
2. Viscosity of the fluid
3. Trapped air bubbles present in the soil specimen
4. Degree of saturation of the soil specimen
5. Migration of fines during testing
6. Duplication of field conditions in the laboratory

The coefficient of permeability of saturated cohesive soils can also be determined by laboratory consolidation tests (see Example 8.11 in Chapter 8). The actual value of the coefficient of permeability in the field may also be somewhat different than that obtained in the laboratory because of the nonhomogeneity of the soil. Hence, proper care should be taken in assessing the order of the magnitude of k.

PROBLEMS **5.1** Find the rate of flow in $ft^3/hr/ft$ length (at right angles to the cross-section shown) through the permeable soil layer shown in Figure 5.32 given $H = 10$ ft, $H_1 = 3.5$ ft, $h = 4.6$ ft, $L = 120$ ft, $\alpha = 14°$, and $k = 0.16 \times 10^{-2}$ ft/sec.

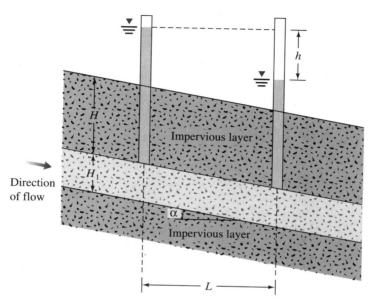

▼ **FIGURE 5.32**

5.2 Repeat Problem 5.1 with the following values: $H = 3.5$ m, $H_1 = 1.5$ m, $h = 2.5$ m, $\alpha = 20°$, $L = 8$ m, and $k = 0.062$ cm/sec. Flow rate is required in $m^3/hr/m$ length (at right angles to the cross-section shown).

5.3 For a constant head laboratory permeability test on a fine sand, the following values are given (refer to Figure 5.5):

  ▶ Length of specimen = 10 in.
  ▶ Diameter of specimen = 2.5 in.
  ▶ Head difference = 18 in.
  ▶ Water collected in a period of 2 min = 0.031 in.^{3}

Determine:
  a. The coefficient of permeability, k, of the soil (in in./min)
  b. Discharge velocity (in in./min)
  c. Seepage velocity (in in./min)
The void ratio of the soil specimen is 0.46.

5.4 Repeat Problem 5.3 with the following values:

  ▶ Length of specimen = 300 mm
  ▶ Diameter of specimen = 200 mm
  ▶ Head difference = 400 mm
  ▶ Volume of water collected in 4 min = 420 cc
  ▶ Void ratio of the soil specimen = 0.55

5.5 For a constant head permeability test, the following values are given:

▶ $L = 300$ mm
▶ $A =$ specimen area $= 32$ cm^2
▶ $k = 0.0244$ cm/sec

The head difference was slowly changed in steps to 800 mm, 700 mm, 600 mm, 500 mm, and 400 mm. Calculate and plot the rate of flow, q, through the specimen in cm^3/sec against the head difference.

5.6 For a variable head permeability test, the following values are given:

▶ Area of the soil specimen $= 4.9$ in.^{2}
▶ Length of the soil specimen $= 18$ in.
▶ Area of the standpipe $= 0.2$ in.^{2}
▶ At time $t = 0$, head difference $= 30$ in.
▶ At time $t = 2$ min, head difference $= 20$ in.

Find the coefficient of permeability of the soil in in./min.

5.7 For the permeability test described in Problem 5.6, what was the head difference at time $t = 1$ min?

5.8 Repeat Problem 5.6 with the following values:

▶ Area of the soil specimen $= 1200$ mm^2
▶ Length of the soil specimen $= 150$ mm
▶ Area of the standpipe $= 50$ mm^2
▶ At time $t = 0$, head difference $= 400$ mm
▶ At time $t = 2$ min, head difference $= 200$ mm

Find the coefficient of permeability of the soil in cm/sec.

5.9 For a variable head permeability test, the following are given: length of specimen $= 15$ in., area of specimen $= 3$ in.^{2}, and $k = 0.0688$ in./min. What should be the area of the standpipe for the head to drop from 25 in. to 12 in. in 8 min?

5.10 The coefficient of permeability, k, of a soil is 10^{-5} cm/sec at a temperature of 28°C. Determine its absolute permeability at 20°C given that at 20°C, $\gamma_w = 9.789$ kN/m^3 and $\eta = 1.005 \times 10^{-3}$ N.s/m^2 [Newton second per meter squared].

5.11 The coefficient of permeability of a sand at a temperature of 25°C and at a void ratio of 0.55 is 0.1 ft/min. Estimate its permeability coefficient at a temperature of 20°C and a void ratio of 0.7. Use Eq. (5.24).

5.12 Redo Problem 5.11 by using Eq. (5.25).

5.13 For a sand, the coefficient of permeability is 0.05 cm/sec at a porosity of 0.42. Determine the value of k at a porosity of 0.3:
a. By using Eq. (5.24)
b. By using Eq. (5.25)

5.14 The maximum dry unit weight determined in the laboratory for a quartz sand is 102 lb/ft^3. In the field, if the relative compaction is 90%, determine the coefficient of permeability of the sand in the field compaction condition (given k for the sand at the maximum dry unit weight condition to be 0.03 cm/sec and $G_s = 2.7$). Use Eq. (5.25).

5.15 For a sandy soil, we are given:

▶ Maximum void ratio $= 0.68$
▶ Minimum void ratio $= 0.42$
▶ Coefficient of permeability of sand at a relative density of 70% $= 0.005$ cm/sec

Determine the coefficient of permeability of the sand at a relative density of 32%. Use Eq. (5.24).

5.16 Repeat Problem 5.15 using Eq. (5.25).

5.17 For a normally consolidated clay, the following values are given:

Void ratio, e	k (cm/sec)
0.75	1.2×10^{-6}
1.2	2.8×10^{-6}

Estimate the coefficient of permeability of the clay at void ratio $e = 0.6$. Use Eq. (5.28).

5.18 For a normally consolidated clay, the following values are given:

Void ratio, e	k (cm/sec)
0.95	0.2×10^{-6}
1.6	0.91×10^{-6}

Determine the magnitude of k at a void ratio of 1.1. Use Eq. (5.28).

5.19 Solve Problem 5.17 using the Mesri and Olson equation given in Table 5.2.

5.20 Solve Problem 5.18 using the Mesri and Olson equation given in Table 5.2.

5.21 The *in situ* void ratio of a soft clay deposit is 2.1, and the coefficient of permeability of the clay at that void ratio is 0.86×10^{-6} cm/sec. What would be the coefficient of permeability if the soil is compressed to have a void ratio of 1.3? Use Taylor's equation given in Table 5.2.

5.22 Figure 5.33 shows the layers of soil in a tube that is 100 mm × 100 mm in cross-section.

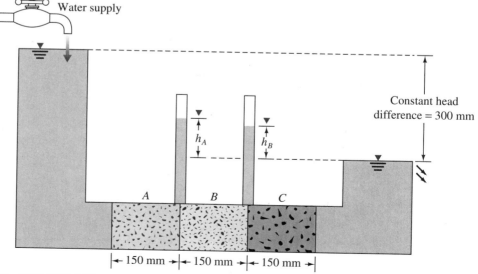

▼ **FIGURE 5.33**

Water is supplied to maintain a constant head difference of 300 mm across the sample. The coefficients of permeability of the soils in the direction of flow through them are as follows:

Soil	k (cm/sec)
A	10^{-2}
B	3×10^{-3}
C	4.9×10^{-4}

Find the rate of water supply in cm^3/hr.

5.23 Refer to Problem 5.22. Using the rate of flow determined, calculate the head differences h_A and h_B.

5.24 A layered soil is shown in Figure 5.34. Estimate the equivalent permeabilities $k_{V(eq)}$ and $k_{H(eq)}$ in cm/sec. Also calculate the ratio of $k_{H(eq)}/k_{V(eq)}$.

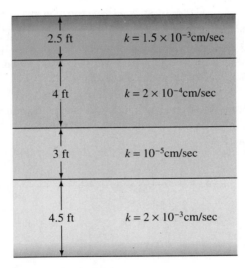

2.5 ft	$k = 1.5 \times 10^{-3}$ cm/sec
4 ft	$k = 2 \times 10^{-4}$ cm/sec
3 ft	$k = 10^{-5}$ cm/sec
4.5 ft	$k = 2 \times 10^{-3}$ cm/sec

▼ **FIGURE 5.34**

5.25 Refer to Figure 5.16 for a field pumping test from a well. For a steady state condition, we are given:

▶ $q = 26$ ft^3/min
▶ $h_1 = 18.0$ ft at $r_1 = 200$ ft
▶ $h_2 = 15.7$ ft at $r_2 = 100$ ft

Calculate the coefficient of permeability (in ft/min) of the permeable layer.

5.26   Refer to Figure 5.35. Given:

▶ $H_1 = 20$ ft $D = 10$ ft
▶ $H_2 = 5$ ft $D_1 = 20$ ft

draw a flow net. Calculate the seepage loss per foot length of the sheet pile (at right angle to the cross-section shown).

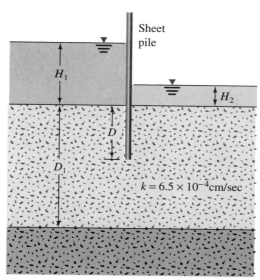

▼   **FIGURE 5.35**

5.27   Draw a flow net for the single row of sheet piles driven into a permeable layer as shown in Figure 5.35 given:

▶ $H_1 = 5$ m $D = 4$ m
▶ $H_2 = 0.7$ m $D_1 = 10$ m

Calculate the seepage loss per meter length of the sheet pile (at right angle to the cross-section shown).

5.28   Draw a flow net for the weir shown in Figure 5.36. Calculate the rate of seepage under the weir.

5.29   For the flow net drawn in Problem 5.28, calculate the uplift force at the base of the weir per foot length (measured along the axis) of the structure.

5.30   Refer to the constant head permeability test arrangement in a two-layered soil as shown in Figure 5.20. During the test it was seen that when a constant head $h_1 = 180$ mm was maintained, the magnitude of h_2 was 60 mm. If k_1 is 0.002 cm/sec, determine the value of k_2 given $H_1 = 70$ mm and $H_2 = 100$ mm.

5.31   Refer to the earth dam shown in Figure 5.31. Given $\beta = 45°$, $\alpha = 30°$, $B = 10$ ft, $H = 20$ ft, height of dam $= 25$ ft, and $k = 2 \times 10^{-4}$ ft/min, calculate the seepage rate, q, in ft³/day/ft length.

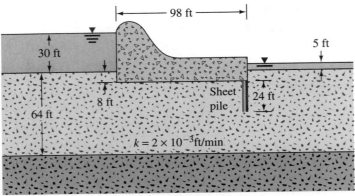

|←————— 98 ft —————→|

30 ft

5 ft

Sheet pile

8 ft

64 ft

24 ft

$k = 2 \times 10^{-3}$ ft/min

Impermeable layer

▼   **FIGURE 5.36**

REFERENCES

Amer, A. M., and Awad, A. A. (1974). "Permeability of Cohesionless Soils," *Journal of the Geotechnical Engineering Division*, ASCE, Vol. 100, No. GT12, 1309–1316.

Darcy, H. (1856). *Les Fontaines Publiques de la Ville de Dijon*, Dalmont, Paris.

Das, B. M. (1983). *Advanced Soil Mechanics*, McGraw-Hill, New York.

Dunn, I. S., Anderson, L. R., and Kiefer, F. W. (1980). *Fundamentals of Geotechnical Analysis*, Wiley, New York.

Dupuit, J. (1863). *Etudes Theoriques et Practiques sur le Mouvement des Eaux dans les Canaux Decouverts et a Travers les Terrains Permeables*, Dunod, Paris.

Ernst, L. F. (1950). "Een nieuwe formule voor de berekening van de doorlaatfactor met de boorgatenmethode," Rap. Landbouw-proefsta. en Bodemkundig Inst. T.N.O., Groningen, The Netherlands.

Hansbo, S. (1960). "Consolidation of Clay with Special Reference to Influence of Vertical Sand Drains," Swedish Geotechnical Institute, *Proc. No. 18*, 41–61.

Harr, M. E. (1962). *Ground Water and Seepage*, McGraw-Hill, New York.

Hazen, A. (1930). "Water Supply," in *American Civil Engineers Handbook*, Wiley, New York.

Kenny, T. C., Lau, D., and Ofoegbu, G. I. (1984). "Permeability of Compacted Granular Materials," *Canadian Geotechnical Journal*, Vol. 21, No. 4, 726–729.

Krumbein, W. C., and Monk, G. D. (1943). "Permeability as a Function of the Size Parameters of Unconsolidated Sand," *Transactions*, AIMME (Petroleum Division), Vol. 151, pp. 153–163.

Mesri, G., and Olson, R. E. (1971). "Mechanism Controlling the Permeability of Clays," *Clay and Clay Minerals*, Vol. 19, 151–158.

Mitchell, J. K. (1976). *Fundamentals of Soil Behavior*, Wiley, New York.

Samarasinghe, A. M., Huang, Y. H., and Drnevich, V. P. (1982). "Permeability and Consolidation of Normally Consolidated Soils," *Journal of Geotechnical Engineering Division*, ASCE, Vol. 108, No. GT6, 835–850.

Shahabi, A. A., Das, B. M., and Tarquin, A. J. (1984). "An Empirical Relation for Coefficient of Permeability of Sand," *Proceedings*, Fourth Australia–New Zealand Conference on Geomechanics, Vol. 1, 54–57.

Tavenas, F., Jean, P., Leblond, F. T. P., and Leroueil, S. (1983). "The Permeability of Natural Soft Clays. Part II: Permeability Characteristics," *Canadian Geotechnical Journal*, Vol. 20, No. 4, 645–660.

Taylor, D. W. (1948). *Fundamentals of Soil Mechanics*, Wiley, New York.

Terzaghi, K., and Peck, R. B. (1967). *Soil Mechanics in Engineering Practice*, 2nd ed., Wiley, New York.

USACE (1953). *Filter Experiments and Design Criteria*, Technical Memorandum No. 3–360, U.S. Army Waterways Experiment Station, Vicksburg, MS.

Supplementary References for Further Study

Chan, H. T., and Kenney, T. C. (1973). "Laboratory Investigation of Permeability Ratio of New Liskeard Varved Soil," *Canadian Geotechnical Journal*, Vol. 10, No. 3, 453–472.

Leblond, F. T. P., Jean, P., and Leroueil, S. (1983). "The Permeability of Natural Soft Clays. Part I: Methods and Laboratory Measurement," *Canadian Geotechnical Journal*, Vol. 20, No. 4, 629–644.

Olsen, H. W. (1962). "Hydraulic Flow Through Saturated Clays," *Proceedings*, 9th National Conference on Clay and Clay Minerals, Vol. 9, Pergamon Press, New York, 131–161.

Olsen, H. W. (1965). "Deviations from Darcy's Law in Saturated Clays," *Proceedings*, Soil Science Society of America, Vol. 29, No. 2, 135–140.

Olson, R. E., and Daniel, D. E. (1981). "Measurement of the Hydraulic Conductivity of Fine-Grained Soils," *Special Technical Publication No. 746*, ASTM, 18–64.

CHAPTER
SIX

EFFECTIVE STRESS CONCEPTS

As described in Chapter 2, soils are multiphase systems. In a given volume of soil, the solid particles are distributed randomly with void spaces in between. The void spaces are continuous and are occupied by water and/or air. To analyze problems such as compressibility of soils, bearing capacity of foundations, stability of embankments, and lateral pressure on earth retaining structures, we need to know the nature of the distribution of stress along a given cross-section of the soil profile. We can begin the analysis by considering a saturated soil with no seepage.

6.1 STRESSES IN SATURATED SOIL WITHOUT SEEPAGE

Figure 6.1a shows a column of saturated soil mass with no seepage of water in any direction. The total stress at the elevation of point A can be obtained from the saturated unit weight of the soil and the unit weight of water above it. Thus,

$$\sigma = H\gamma_w + (H_A - H)\gamma_{\text{sat}} \tag{6.1}$$

where σ = total stress at the elevation of point A
     γ_w = unit weight of water
     γ_{sat} = saturated unit weight of the soil
     H = height of water table from the top of the soil column
     H_A = distance between point A and the water table

The total stress, σ, given by Eq. (6.1) can be divided into two parts:

1. A portion is carried by water in the continuous void spaces. This portion acts with equal intensity in all directions.
2. The rest of the total stress is carried by the soil solids at their points of contact. The sum of the vertical components of the forces developed at the points of contact of the solid particles per unit cross-sectional area of the soil mass is called the *effective stress*.

This can be seen by drawing a wavy line, a-a, through the point A that passes only through the points of contacts of the solid particles. Let $P_1, P_2, P_3, \ldots, P_n$ be the forces that act at the points of contact of the soil particles (Figure 6.1b). The sum of the vertical components of all such forces over the unit cross-sectional area is equal to the effective stress, σ', or

$$\sigma' = \frac{P_{1(v)} + P_{2(v)} + P_{3(v)} + \cdots + P_{n(v)}}{\bar{A}} \qquad (6.2)$$

where $P_{1(v)}, P_{2(v)}, P_{3(v)}, \ldots, P_{n(v)}$ are the vertical components of $P_1, P_2, P_3, \ldots, P_n$, respectively, and $\bar{A}$ is the cross-sectional area of the soil mass under consideration.

Again, if a_s is the cross-sectional area occupied by solid-to-solid contacts (that is, $a_s = a_1 + a_2 + a_3 + \cdots + a_n$), then the space occupied by water equals $(\bar{A} - a_s)$. So we can write

$$\sigma = \sigma' + \frac{u(\bar{A} - a_s)}{\bar{A}} = \sigma' + u(1 - a'_s) \qquad (6.3)$$

where $u = H_A \gamma_w$ = pore water pressure (that is, the hydrostatic pressure at A)
$a'_s = a_s/\bar{A}$ = fraction of unit cross-sectional area of the soil mass occupied by solid-to-solid contacts

The value of a'_s is very small and can be neglected for pressure ranges generally encountered in practical problems. Thus, Eq. (6.3) can be approximated by

$$\boxed{\sigma = \sigma' + u} \qquad (6.4)$$

where u is also referred to as *neutral stress*. Substitution of Eq. (6.1) for σ in Eq. (6.4) gives

$$\sigma' = [H\gamma_w + (H_A - H)\gamma_{\text{sat}}] - H_A \gamma_w$$
$$= (H_A - H)(\gamma_{\text{sat}} - \gamma_w)$$
$$= (\text{height of the soil column}) \times \gamma' \qquad (6.5)$$

where $\gamma' = \gamma_{\text{sat}} - \gamma_w$ equals the submerged unit weight of soil. Thus, it can be seen that the effective stress at any point A is independent of the depth of water, H, above the submerged soil.

Figure 6.2a shows a layer of submerged soil in a tank where there is no seepage. Figures 6.2b, c, and d show plots of the variations of the total stress, pore water pressure, and effective stress, respectively, with depth for a submerged layer of soil placed in a tank with no seepage.

The principle of effective stress [Eq. (6.4)] was first developed by Terzaghi (1925, 1936). Skempton (1960) extended the work of Terzaghi and proposed the relationship between total and effective stress in the form of Eq. (6.3).

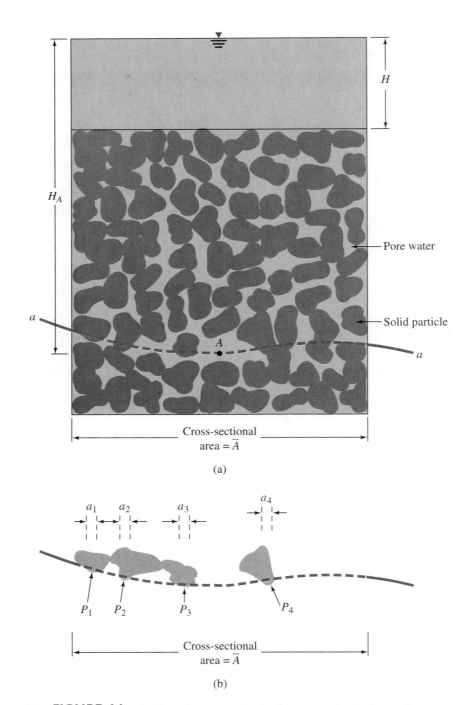

(a)

(b)

▼ **FIGURE 6.1** (a) Effective stress consideration for a saturated soil column without seepage; (b) forces acting at the points of contact of soil particles at the level of point A

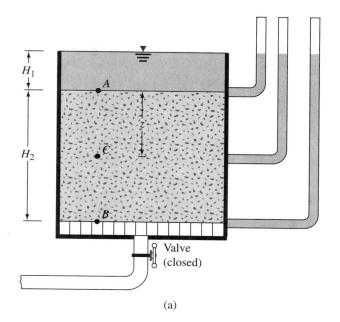

(a)

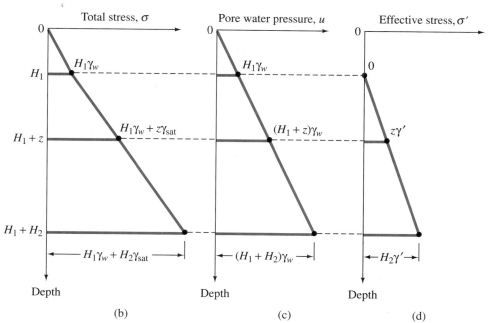

(b)   (c)   (d)

▼ **FIGURE 6.2** (a) Layer of soil in a tank where there is no seepage; variation of (b) total stress; (c) pore water pressure; (d) effective stress with depth in a submerged soil layer without seepage

In summary, effective stress is approximately the force per unit area carried by the soil skeleton. It is the effective stress in a soil mass that controls its volume change and strength. Increasing the effective stress will induce soil to move into denser state packing.

6.2 STRESSES IN SATURATED SOIL WITH SEEPAGE

If water is seeping, the effective stress at any point in a soil mass will be different from the static case. It will increase or decrease, depending on the direction of seepage.

Upward Seepage

Figure 6.3a shows a layer of granular soil in a tank where upward seepage is caused by adding water through the valve at the bottom of the tank. The rate of water supply is kept constant. The loss of head caused by upward seepage between the levels of A and B is h. Keeping in mind that the total stress at any point in the soil mass is due solely to the weight of soil and water above it, we find the effective stress calculations at points A and B are as follows:

At A

▶ Total stress: $\sigma_A = H_1 \gamma_w$
▶ Pore water pressure: $u_A = H_1 \gamma_w$
▶ Effective stress: $\sigma'_A = \sigma_A - u_A = 0$

At B

▶ Total stress: $\sigma_B = H_1 \gamma_w + H_2 \gamma_{sat}$
▶ Pore water pressure: $u_B = (H_1 + H_2 + h)\gamma_w$
▶ Effective stress: $\sigma'_B = \sigma_B - u_B$
$$= H_2(\gamma_{sat} - \gamma_w) - h\gamma_w$$
$$= H_2 \gamma' - h\gamma_w$$

Similarly, the effective stress at a point C located at a depth z below the top of the soil surface can be calculated:

At C

▶ Total stress: $\sigma_C = H_1 \gamma_w + z\gamma_{sat}$

▶ Pore water pressure: $u_C = \left(H_1 + z + \dfrac{h}{H_2} z\right)\gamma_w$

▶ Effective stress: $\sigma'_C = \sigma_C - u_C$

$$= z(\gamma_{sat} - \gamma_w) - \frac{h}{H_2} z\gamma_w$$

$$= z\gamma' - \frac{h}{H_2} z\gamma_w$$

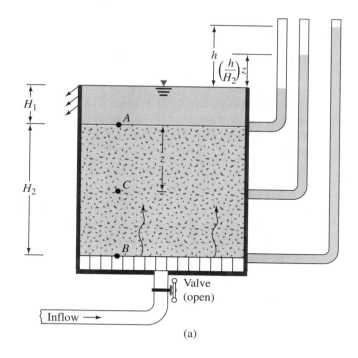

(a)

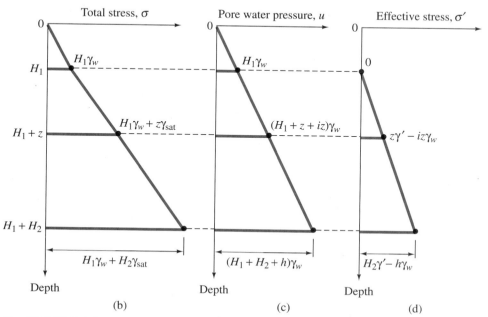

(b)   (c)   (d)

▼ **FIGURE 6.3**   (a) Layer of soil in a tank with upward seepage; variation of (b) total stress; (c) pore water pressure; (d) effective stress with depth in a soil layer with upward seepage

Note that h/H_2 is the hydraulic gradient i caused by the flow, and so

$$\sigma'_C = z\gamma' - iz\gamma_w \qquad (6.6)$$

The variations of total stress, pore water pressure, and effective stress with depth are plotted in Figures 6.3b, c, and d, respectively. A comparison of Figures 6.2d and 6.3d shows that the effective stress at a point located at a depth z measured from the surface of a soil layer is reduced by an amount $iz\gamma_w$ because of upward seepage of water. If the rate of seepage and thereby the hydraulic gradient are gradually increased, a limiting condition will be reached, at which point

$$\sigma'_C = z\gamma' - i_{cr} z\gamma_w = 0 \qquad (6.7)$$

where i_{cr} = critical hydraulic gradient (for zero effective stress).

Under such a situation, the stability of the soil will be lost. This is generally referred to as *boiling*, or *quick condition*.

From Eq. (6.7),

$$\boxed{i_{cr} = \frac{\gamma'}{\gamma_w}} \qquad (6.8)$$

For most soils, the value of i_{cr} varies from 0.9 to 1.1, with an average of 1.

Downward Seepage

The condition of downward seepage is shown in Figure 6.4a. The level of water in the soil tank is held constant by adjusting the supply from the top and the outflow at the bottom.

The hydraulic gradient caused by the downward seepage equals $i = h/H_2$. The total stress, pore water pressure, and effective stress at any point C are, respectively,

$$\sigma_C = H_1\gamma_w + z\gamma_{sat}$$

$$u_C = (H_1 + z - iz)\gamma_w$$

$$\sigma'_C = (H_1\gamma_w + z\gamma_{sat}) - (H_1 + z - iz)\gamma_w$$

$$= z\gamma' + iz\gamma_w \qquad (6.9)$$

The variations of total stress, pore water pressure, and effective stress with depth are also shown graphically in Figure 6.4b, c, and d.

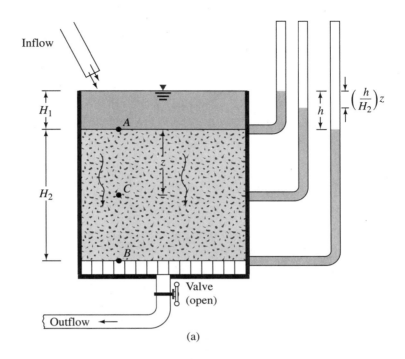

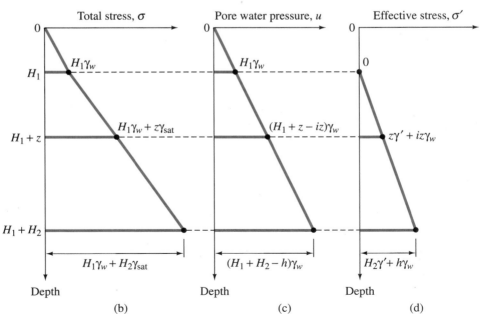

FIGURE 6.4 (a) Layer of soil in a tank with downward seepage; variation of (b) total stress; (c) pore water pressure; (d) effective stress with depth in a soil layer with downward seepage

▼   **EXAMPLE 6.1**

A soil profile is shown in Figure 6.5. Calculate the total stress, pore water pressure, and effective stress at A, B, C, and D.

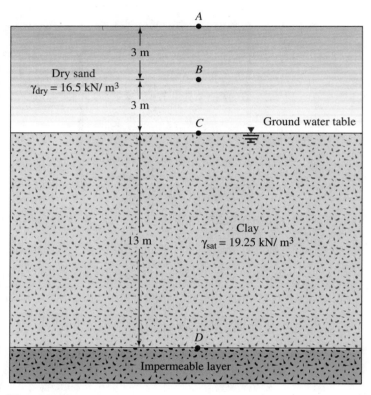

▼   **FIGURE 6.5**

Solution   At A: Total stress: $\sigma_A = 0$
Pore water pressure: $u_A = 0$
Effective stress: $\sigma'_A = 0$

At B: $\sigma_B = 3\gamma_{dry(sand)} = 3 \times 16.5 = $ **49.5 kN/m²**
$u_B = $ **0 kN/m²**
$\sigma'_B = 49.5 - 0 = $ **49.5 kN/m²**

At C: $\sigma_C = 6\gamma_{dry(sand)} = 6 \times 16.5 = $ **99 kN/m²**
$u_C = $ **0 kN/m²**
$\sigma'_C = 99 - 0 = $ **99 kN/m²**

At D: $\sigma_C = 6\gamma_{dry(sand)} + 13\gamma_{sat(clay)}$
$= 6 \times 16.5 + 13 \times 19.25$
$= 99 + 250.25 = \mathbf{349.25 \ kN/m^2}$

$u_D = 13\gamma_w = 13 \times 9.81 = \mathbf{127.53 \ kN/m^2}$

$\sigma'_D = 349.25 - 127.53 = \mathbf{221.72 \ kN/m^2}$ ▼

▼ **EXAMPLE 6.2**

An exploratory drill hole was made in a saturated stiff clay (Figure 6.6). It was observed that the sand layer underlying the clay was under artesian pressure. Water in the drill hole rose to a height of H_1 above the top of the sand layer. If an open excavation is to be made in the clay, how deep can the excavation proceed before the bottom heaves? We are given $H = 8$ m, $H_1 = 4$ m, and $w = 32\%$.

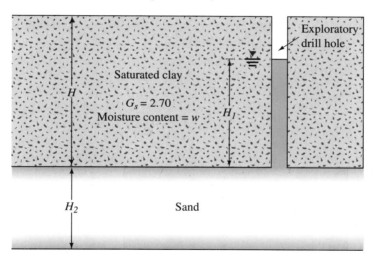

▼ **FIGURE 6.6**

Solution   Consider a point at the sand-clay interface. For heaving, $\sigma' = 0$, so

$$(H - H_{exc})\gamma_{sat(clay)} - H_1\gamma_w = 0$$

$$\gamma_{sat(clay)} = \frac{G_s\gamma_w + wG_s\gamma_w}{1 + e} = \frac{[2.7 + (0.32)(2.7)](9.81)}{1 + (0.32)(2.7)}$$

$$= 18.76 \ kN/m^3$$

Thus,

$$(8 - H_{exc})(18.76) - \overset{4}{(3)}(9.81) = 0$$

$$H_{exc} = 8 - \frac{\overset{4}{(3)}(9.81)}{18.76} \overset{5.91}{=} \mathbf{6.43 \ m}$$ ▼

▼   **EXAMPLE 6.3**

Refer to Figure 6.3a, in which there is an upward seepage of the soil. We are given $H_1 = 1.5$ ft, $H_2 = 4.5$ ft, $h = 1.75$ ft, and $\gamma_{sat} = 122$ lb/ft^3.

  a.   Calculate the total stress, pore water pressure, and effective stress at C. (*Note:* $z = 2$ ft.)

  b.   What is the upward seepage per unit volume of soil?

Solution

  a.   At C: $\sigma = 1.5\gamma_w + 2\gamma_{sat} = (1.5)(62.4) + (2)(122)$

$$= \mathbf{337.6\ lb/ft^2}$$

$$u = \left[1.5 + 2 + \left(\frac{h}{H_2}\right)(2)\right]\gamma_w$$

$$= \left[1.5 + 2 + \left(\frac{1.75}{4.5}\right)(2)\right](62.4)$$

$$= \mathbf{266.93\ lb/ft^2}$$

$$\sigma' = \sigma - u = 337.6 - 266.93 = \mathbf{70.67\ lb/ft^2}$$

  b.   Upward seepage per unit volume is

$$i\gamma_w = \left(\frac{1.75}{4.5}\right)(62.4) = \mathbf{24.27\ lb/ft^3}\qquad▼$$

6.3 SEEPAGE FORCE

The preceding section showed that the effect of seepage is to increase or decrease the effective stress at a point in a layer of soil. It is often convenient to express the seepage force per unit volume of soil.

In Figure 6.2 it was shown that, with no seepage, the effective stress at a depth z measured from the surface of the soil layer in the tank is equal to $z\gamma'$. Thus, the effective force on an area A is

$$P'_1 = z\gamma'A$$

(The direction of the force P'_1 is shown in Figure 6.7a.)

Again, if there is an upward seepage of water in the vertical direction through the same soil layer (Figure 6.3), the effective force on an area A at a depth z can be given by

$$P'_2 = (z\gamma' - iz\gamma_w)A$$

Hence, the decrease in the total force because of seepage is

$$P'_1 - P'_2 = iz\gamma_w A \qquad\qquad (6.10)$$

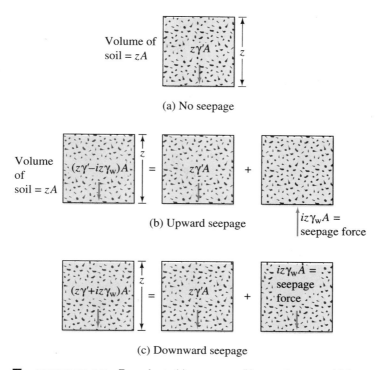

(a) No seepage

(b) Upward seepage

(c) Downward seepage

▼ **FIGURE 6.7**   Force due to (a) no seepage; (b) upward seepage; (c) downward seepage on a volume of soil

The volume of the soil contributing to the effective force equals zA, so the seepage force per unit volume of soil is

$$\frac{P'_1 - P'_2}{\text{(volume of soil)}} = \frac{iz\gamma_w A}{zA} = i\gamma_w \tag{6.11}$$

The force per unit volume, $i\gamma_w$, for this case acts in the upward direction—that is, in the direction of flow. This is demonstrated in Figure 6.7b. Similarly, for downward seepage, it can be shown that the seepage force in the downward direction per unit volume of soil is $i\gamma_w$ (Figure 6.7c).

From the preceding discussions, we can conclude that the seepage force per unit volume of soil is equal to $i\gamma_w$, and in isotropic soils the force acts in the same direction as the direction of flow. This statement is true for flow in any direction. Flow nets can be used to find the hydraulic gradient at any point and, thus, the seepage force per unit volume of soil.

This concept of seepage force can be effectively used to obtain the factor of safety against heave on the downstream side of a hydraulic structure. This is discussed in the following section.

6.4 HEAVING IN SOIL CAUSED BY FLOW AROUND SHEET PILES

The seepage force per unit volume of soil can be calculated for checking the possible failure of sheet-pile structures where underground seepage may cause heaving of soil on the downstream side (Figure 6.8a). After conducting several model tests, Terzaghi (1922) concluded that heaving generally occurs within a distance of $D/2$ from the sheet piles (when D equals the depth of embedment of sheet piles into the permeable layer). Therefore, we need to investigate the stability of soil in a zone measuring D by $D/2$ in cross-section, as shown in Figure 6.8a.

The factor of safety against heaving can be given by (Figure 6.8b)

$$FS = \frac{W'}{U} \tag{6.12}$$

where FS = factor of safety
W' = submerged weight of soil in the heave zone per unit length of sheet pile = $D(D/2)(\gamma_{sat} - \gamma_w) = (1/2)D^2\gamma'$
U = uplifting force caused by seepage on the same volume of soil

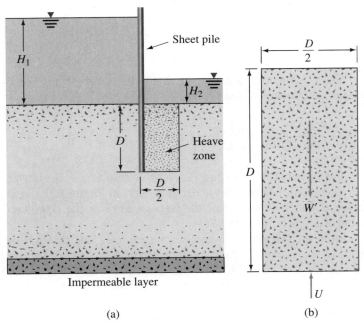

(a)   (b)

▼ **FIGURE 6.8**   (a) Check for heaving on the downstream side for a row of sheet piles driven into a permeable layer; (b) enlargement of heave zone

From Eq. (6.11),

$$U = (\text{soil volume}) \times (i_{av} \gamma_w) = \frac{1}{2} D^2 i_{av} \gamma_w$$

where i_{av} = average hydraulic gradient at the bottom of the block of soil.
Substituting the values of W' and U in Eq. (6.12), we can write

$$FS = \frac{\gamma'}{i_{av} \gamma_w} \qquad (6.13)$$

▼ **EXAMPLE 6.4**

Figure 6.9 shows the flow net for seepage of water around a single row of sheet piles driven into a permeable layer. Calculate the factor of safety against downstream heave given γ_{sat} for the permeable layer = 112.32 lb/ft³.

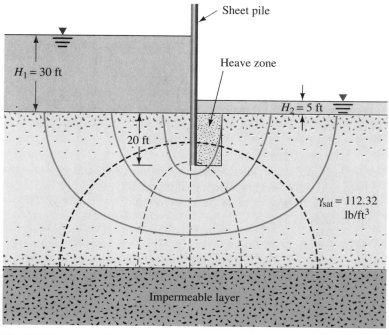

▼ **FIGURE 6.9** Flow net for seepage of water around sheet piles driven into permeable layer

Solution   From the dimensions given in Figure 6.9, the soil prism to be considered is 20 ft × 10 ft in cross-section.

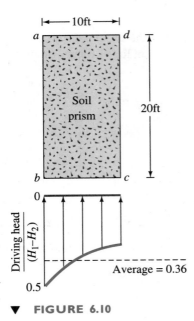

▼ **FIGURE 6.10**

The soil prism is drawn to an enlarged scale in Figure 6.10. By use of the flow net, we can calculate the head loss through the prism as follows:

▶ At b, the driving head $= \dfrac{3}{6}(H_1 - H_2)$

▶ At c, the driving head $\approx \dfrac{1.6}{6}(H_1 - H_2)$

Similarly, for other intermediate points along bc, the approximate driving heads have been calculated and are shown in Figure 6.10.

The average value of the head loss in the prism is $0.36(H_1 - H_2)$, and the average hydraulic gradient is

$$i_{\text{av}} = \frac{0.36(H_1 - H_2)}{D}$$

Thus, the factor of safety [Eq. (6.13)] is

$$FS = \frac{\gamma'}{i_{\text{av}}\gamma_w} = \frac{\gamma'D}{0.36(H_1 - H_2)\gamma_w} = \frac{(112.32 - 62.4)20}{0.36(30 - 5) \times 62.4} = \mathbf{1.78} \qquad \blacktriangledown$$

6.5 USE OF FILTERS TO INCREASE THE FACTOR OF SAFETY AGAINST HEAVE

The factor of safety against heave as calculated in Example 6.4 is rather low. In actual practice, a minimum factor of safety of about 4 to 5 is required for the safety of the structure. Such a high factor of safety is recommended primarily because of the inaccuracies inherent in the analysis. One way to increase the factor of safety against heave is to use a *filter* in the downstream side of the sheet-pile structure (Figure 6.11a). A filter is a granular material with openings small enough to prevent the movement of the soil particles upon which it is placed and which, at the same time, is pervious enough to offer little resistance to seepage through it. In Figure 6.11a, the thickness of the filter material is D_1. In that case, the factor of safety against heave can be calculated as follows (Figure 6.11b):

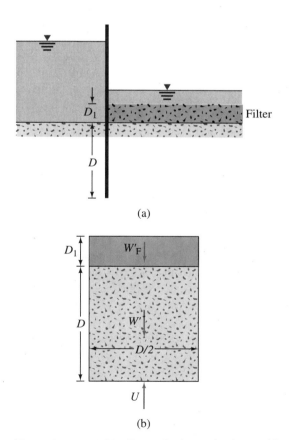

(a)

(b)

▼ **FIGURE 6.11**   Factor of safety against heave with a filter

Submerged weight of the soil and the filter in the heave zone per unit length of sheet pile $= W' + W'_F$

$$W' = (D)\left(\frac{D}{2}\right)(\gamma_{sat} - \gamma_w) = \frac{1}{2} D^2 \gamma'$$

$$W'_F = (D_1)\left(\frac{D}{2}\right)(\gamma'_F) = \frac{1}{2} D_1 D \gamma'_F$$

where $\gamma'_F =$ effective unit weight of the filter.

The uplifting force caused by seepage on the same volume of soil is equal to U:

$$U = \frac{1}{2} D^2 i_{av} \gamma_w$$

The preceding relationship is derived in Section 6.4.

The factor of safety against heave is thus

$$FS = \frac{W' + W'_F}{U} = \frac{\frac{1}{2} D^2 \gamma' + \frac{1}{2} D_1 D \gamma'_F}{\frac{1}{2} D^2 i_{av} \gamma_w} = \frac{\gamma' + \left(\frac{D_1}{D}\right)\gamma'_F}{i_{av} \gamma_w} \qquad (6.14)$$

The application of Eq. (6.14) is given in Example 6.5.

▼ **EXAMPLE 6.5**

Refer to Example 6.4. Assume that a filter is placed on the downstream side. The thickness of the filter has to be such that a factor of safety of 3 is maintained. For the filter, $G_s = 2.68$ and $e = 0.46$. Determine the thickness of the filter.

Solution   We have

$$\gamma'_F = \gamma_{sat(F)} - \gamma_w = \frac{(G_s + e)\gamma_w}{1 + e} - \gamma_w$$

or

$$\gamma'_F = \frac{(G_s - 1)\gamma_w}{1 + e} = \frac{(2.68 - 1)(62.4)}{1 + 0.46} = 71.8 \text{ lb/ft}^3$$

From Eq. (6.14),

$$FS = \frac{\gamma' + \left(\dfrac{D_1}{D}\right)\gamma'_F}{i_{av}\,\gamma_w}$$

$$3.0 = \frac{(112.32 - 62.4) + \left(\dfrac{D_1}{20}\right)(71.8)}{\left[(0.36)\left(\dfrac{30 - 5}{20}\right)\right](62.4)}$$

$$D_1 = \textbf{9.56 ft} \qquad \blacktriangledown$$

6.6 SELECTION OF FILTER MATERIAL

It is extremely important that the filter material mentioned in Section 6.5 be chosen carefully, taking into consideration that the soil is to be protected. To describe the selection criteria of a filter, refer to Figure 6.12. Note that, in this figure, the soil to be protected is referred to as the *base material*. Terzaghi and Peck (1948) suggested the following criteria for selection of the filter material:

1. $\dfrac{D_{15(F)}}{D_{85(B)}} < 4$

2. $\dfrac{D_{15(F)}}{D_{15(B)}} > 4$

where $D_{15(F)}$, $D_{15(B)}$ = diameters through which 15% of the filter and base material, respectively, will pass

$D_{85(B)}$ = diameter through which 85% of the base material will pass

The first criterion is for prevention of the movement of the soil particles of the base material (that is, the soil to be protected) through the filter.

The application of the above-stated filter selection criteria can be explained by using Figure 6.13, in which curve *a* is the grain-size distribution curve of the base

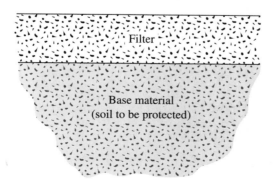

▼ **FIGURE 6.12** Definition of base material and filter material

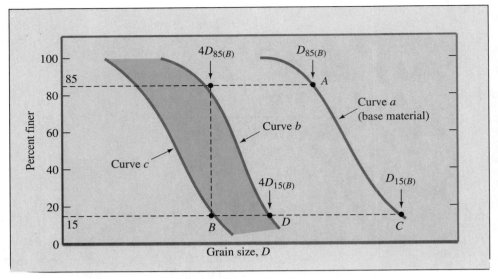

▼ **FIGURE 6.13**   Filter selection criteria

material. From criterion 1, $D_{15(F)} < 4D_{85(B)}$. The abscissa of point A is $D_{85(B)}$, so the magnitude of $4D_{85(B)}$ can be calculated and point B, whose abscissa is $4D_{85(B)}$, can be plotted. Similarly, from criterion 2, $D_{15(F)} > 4D_{15(B)}$. The abscissas of points C and D are $D_{15(B)}$ and $4D_{15(B)}$, respectively. The curves b and c are drawn, which are geometrically similar to curve a and are within the limits of points B and D. A soil whose grain-size curve falls within the bounds of curves b and c is a good filter material.

Several other filter design criteria have been suggested in the past based on the results of laboratory experiments. These are summarized in Appendix D.

6.7   EFFECTIVE STRESS IN PARTIALLY SATURATED SOIL

In partially saturated soil, water in the void spaces is not continuous, and it is a three-phase system—that is, solid, pore water, and pore air (Figure 6.14). Hence, the total stress at any point in a soil profile consists of intergranular, pore air, and pore water pressures. From laboratory test results, Bishop and colleagues (1960) gave the following equation for effective stress in partially saturated soils:

$$\sigma' = \sigma - u_a + \chi(u_a - u_w) \tag{6.15}$$

where σ' = effective stress
      σ = total stress
      u_a = pore air pressure
      u_w = pore water pressure

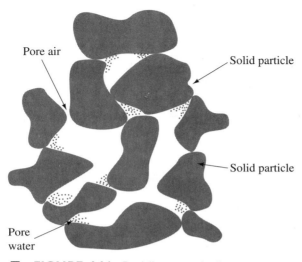

▼   **FIGURE 6.14**   Partially saturated soil

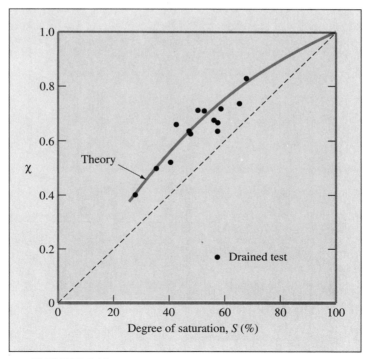

▼   **FIGURE 6.15**   Relationship between the parameter χ and
the degree of saturation for Bearhead silt
(after Bishop et al., 1960)

In Eq. (6.15), χ represents the fraction of a unit cross-sectional area of the soil occupied by water. For dry soil $\chi = 0$, and for saturated soil $\chi = 1$.

Bishop et al. pointed out that the intermediate values of χ will depend primarily on the degree of saturation S. However, these values will also be influenced by factors such as soil structure. The nature of variation of χ with the degree of saturation for a silt is shown in Figure 6.15.

6.8 CAPILLARY RISE IN SOILS

The continuous void spaces in soil can behave as bundles of capillary tubes of variable cross-section. Because of surface tension force, water may rise above the phreatic surface.

Figure 6.16 shows the fundamental concept of the height of rise in a capillary tube. The height of rise of water in the capillary tube can be given by summing the forces in the vertical direction, or

$$\left(\frac{\pi}{4} d^2\right) h_c \gamma_w = \pi d T \cos \alpha$$

$$h_c = \frac{4T \cos \alpha}{d\gamma_w} \tag{6.16}$$

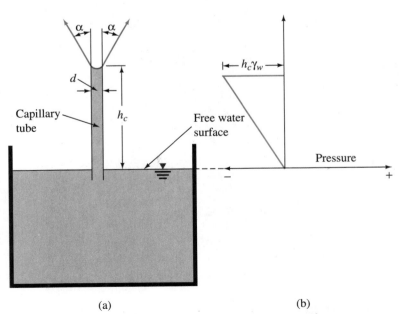

(a)   (b)

▼ **FIGURE 6.16**   (a) Rise of water in the capillary tube; (b) pressure within the height of rise in the capillary tube (atmospheric pressure taken as datum)

where T = surface tension
      α = angle of contact
      d = diameter of capillary tube
      γ_w = unit weight of water

From Eq. (6.16), it may be seen that with T, α, and γ_w remaining constant,

$$h_c \propto \frac{1}{d} \tag{6.17}$$

The pressure at any point in the capillary tube above the free water surface will be negative with respect to the atmospheric pressure and the magnitude may be given by $h\gamma_w$ (where h = height above the free water surface).

Although the concept of capillary rise as demonstrated for an ideal capillary tube can be applied to soils, it must be realized that the capillary tubes formed in soils because of the continuity of voids have variable cross-sections. The results of the non-uniformity on capillary rise can be seen when a dry column of sandy soil is placed in contact with water (Figure 6.17). After the lapse of a given amount of time, the variation of the degree of saturation with the height of the soil column caused by capillary rise will be roughly as shown in Figure 6.17b. The degree of saturation is about 100% up to a height of h_2 and this corresponds to the largest voids. Beyond the height h_2, water can occupy only the smaller voids; hence, the degree of saturation will be less than 100%. The maximum height of capillary rise corresponds to the smallest voids. Hazen (1930) gave a formula for the approximate determination of the height of capillary rise

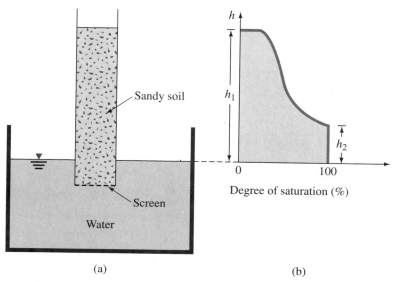

(a)   (b)

▼ **FIGURE 6.17** Capillary effect in sandy soil: (a) a soil column in contact with water; (b) variation of degree of saturation in the soil column

▼ **TABLE 6.1**   **Approximate Range of Capillary Rise in Soils**

Soil type	Range of capillary rise	
	ft	m
Coarse sand	0.4–0.6	0.12–0.18
Fine sand	1–4	0.3–1.2
Silt	2.5–25	0.76–7.6
Clay	25–75	7.6–23

in the form

$$h_1 \text{ (mm)} = \frac{C}{eD_{10}} \qquad (6.18)$$

where D_{10} = effective size (mm)

e = void ratio

C = a constant that varies from 10 to 50 mm^2

Equation (6.18) has an approach similar to Eq. (6.17). With the decrease of D_{10}, the pore size in soil will decrease, causing higher capillary rise. Table 6.1 shows the approximate range of capillary rise that is encountered in various types of soils.

Capillary rise is important in the formation of some types of soils such as *caliche*, which can be found in the desert Southwest of the United States. Caliche is a mixture of sand, silt, and gravel bonded together by calcareous deposits. The calcareous deposits are brought to the surface by a net upward migration of water by capillary action. The water evaporates in the high local temperature. Because of sparse rainfall, the carbonates are not washed out of the top soil layer.

6.9 EFFECTIVE STRESS IN THE ZONE OF CAPILLARY RISE

The general relationship among total stress, effective stress, and pore water pressure was given in Eq. (6.4) as

$$\sigma = \sigma' + u$$

The pore water pressure u at a point in a layer of soil fully saturated by capillary rise is equal to $-\gamma_w h$ (h = height of the point under consideration measured from the ground water table) with the atmospheric pressure taken as datum. If partial saturation is caused by capillary action, it can be approximated as

$$u = -\left(\frac{S}{100}\right)\gamma_w h$$

(6.19)

where S = degree of saturation, in percent.

▼ **EXAMPLE 6.6**

A granular soil deposit is shown in Figure 6.18a. Plot the variation of total stress, pore water pressure, and effective stress with depth. For the granular soil, we are given $e = 0.5$ and G_s (specific gravity of soil solids) = 2.65.

Solution   Unit weight calculation:
Between levels a and b,

$$\gamma_{dry} = \frac{G_s\gamma_w}{1+e} = \frac{(2.65)(9.81)}{1+0.5} = 17.3 \text{ kN/m}^3$$

Between levels b and c,

$$\gamma_{moist} = \frac{(G_s + Se)\gamma_w}{1+e} = \frac{[2.65 + (0.5)(0.5)](9.81)}{1+0.5}$$

$$= 18.97 \text{ kN/m}^3$$

Between levels c and d,

$$\gamma_{sat} = \frac{(G_s + e)\gamma_w}{1+e} = \frac{(2.65 + 0.5)(9.81)}{1+0.5} = 20.6 \text{ kN/m}^3$$

Stress calculation:

	Total stress (kN/m²)	Pore water pressure (kN/m²)	Effective stress (kN/m²)
Level a	**0**	**0**	**0**
Level b	2 × 17.3 = **34.6**	Immediately above level b = 0 Immediately below level b = −0.5 × 9.8 × 1 = **−4.91**	**34.6** **39.51**
Level c	34.6 + (1 × 18.97) = **53.57**	**0**	**53.57**
Level d	53.57 + (2 × 20.6) = **94.77**	2 × 9.81 = **19.62**	**75.15**

The plots of total stress, pore water pressure, and effective stress with depth are given in Figures 6.18b, c, and d. ▼

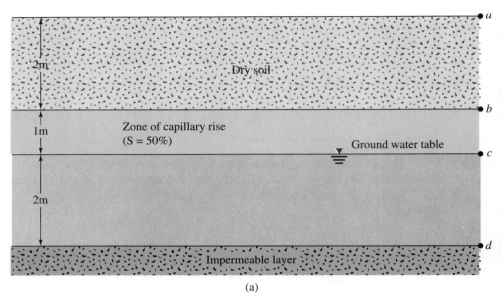

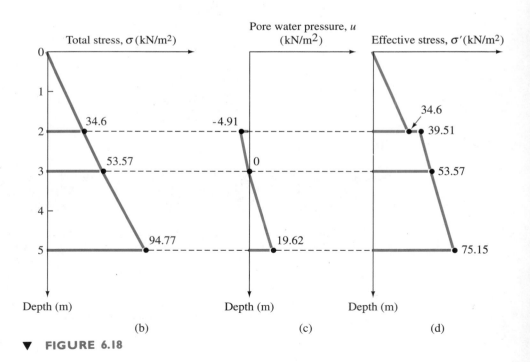

(a)

(b)   (c)   (d)

▼ **FIGURE 6.18**

6.10 GENERAL COMMENTS

The effective stress principle is probably the most important concept in geotechnical engineering. The compressibility and shearing resistance of a soil are dependent to a great extent on the effective stress. Thus, the concept of effective stress is significant in solving geotechnical engineering problems such as the lateral earth pressure on retaining structures (Chapter 10), the load-bearing capacity and settlement of foundations (Chapter 11), and the stability of earth slopes (Chapter 12).

In Eq. (6.2), the effective stress σ' is defined as the sum of the vertical components of all integranular *contact* forces over a unit gross cross-sectional area. This is mostly true in granular soils; however, in fine-grained soils, intergranular contact may not physically be there, since the clay particles are surrounded by tightly held water film. In a more general sense, Eq. (6.3) can be rewritten as

$$\sigma = \sigma_{ig} + u(1 - a'_s) - A' + R' \tag{6.20}$$

where σ_{ig} = intergranular stress
A' = electrical attractive force per unit area of cross-section of soil
R' = electrical repulsive force per unit area of cross-section of soil

For granular soils, silts, and clays of low plasticity, the magnitudes of A' and R' are small. Hence, for all practical purposes,

$$\sigma_{ig} = \sigma' \approx \sigma - u$$

However, if $A' - R'$ is large, then $\sigma_{ig} \neq \sigma'$. Such situations can be encountered in highly plastic, dispersed clay. Many interpretations have been made in the past to distinguish between the intergranular stress and effective stress. In any case, the effective stress principle is an excellent approximation used in solving engineering problems.

PROBLEMS

6.1 A soil profile is shown in Figure 6.19. Calculate the values of σ, u, and σ' at points A, B, C, and D. Plot the variation of σ, u, and σ' with depth. We are given these values:

Layer no.	Thickness (ft)	Unit weight (lb/ft³)
I	$H_1 = 4$	$\gamma_d = 110$
II	$H_2 = 5$	$\gamma_{sat} = 120$
III	$H_3 = 6$	$\gamma_{sat} = 125$

6.2 Repeat Problem 6.1 with the following data:

Layer no.	Thickness (ft)	Unit weight (lb/ft³)
I	$H_1 = 4.5$	$\gamma_d = 95$
II	$H_2 = 10$	$\gamma_{sat} = 116$
III	$H_3 = 8.5$	$\gamma_{sat} = 122$

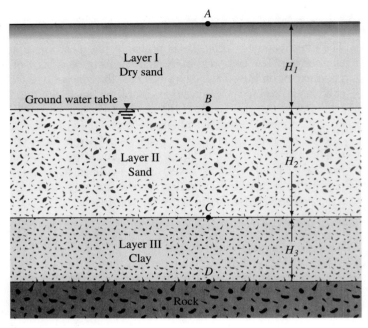

▼   **FIGURE 6.19**

6.3   Repeat Problem 6.1 with the following values:

Layer no.	Thickness (m)	Unit weight (kN/m³)
I	$H_1 = 2$	$\gamma_d = 15.6$
II	$H_2 = 3$	$\gamma_{sat} = 16.2$
III	$H_3 = 4$	$\gamma_{sat} = 17.8$

6.4   Repeat Problem 6.1 with the following data:

Layer no.	Thickness (m)	Soil parameters
I	$H_1 = 3$	$e = 0.4$, $G_s = 2.62$
II	$H_2 = 4$	$e = 0.60$, $G_s = 2.68$
III	$H_3 = 2$	$e = 0.81$, $G_s = 2.73$

6.5   Repeat Problem 6.1 with the following values:

Layer no.	Thickness (ft)	Soil parameters
I	$H_1 = 8$	$e = 0.6$, $G_s = 2.65$
II	$H_2 = 6$	$e = 0.52$, $G_s = 2.68$
III	$H_3 = 3$	$w = 40\%$, $e = 1.1$

6.6 Plot the variation of total stress, pore water pressure, and effective stress with depth for the sand and clay layers shown in Figure 6.20 with $H_1 = 4$ m and $H_2 = 3$ m. Give numerical values.

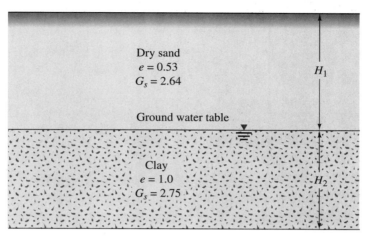

▼ **FIGURE 6.20**

6.7 Repeat Problem 6.6 with $H_1 = 8$ ft and $H_2 = 22$ ft.

6.8 A soil profile is shown in Figure 6.21.
  a. Calculate the total stress, pore water pressure, and effective stress at A, B, and C.
  b. How high should the ground water table rise so that the effective stress at C is $104 \, \text{kN/m}^2$?

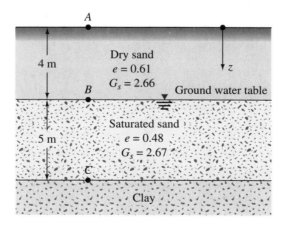

▼ **FIGURE 6.21**

6.9 A sand has $G_s = 2.66$. Calculate the hydraulic gradient that will cause boiling for $e = 0.35$, 0.45, 0.55, 0.7, and 0.8. Plot a graph for i_{cr} versus e.

6.10   A 30-ft thick layer of stiff saturated clay is underlain by a layer of sand (Figure 6.22). The sand is under artesian pressure. Calculate the maximum depth of cut, H, that can be made in the clay.

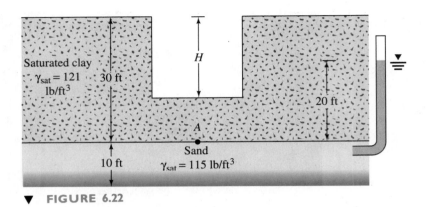

▼ **FIGURE 6.22**

6.11   A cut is made in a stiff saturated clay that is underlain by a layer of sand (Figure 6.23). What should be the height of the water, h, in the cut so that the stability of the saturated clay is not lost?

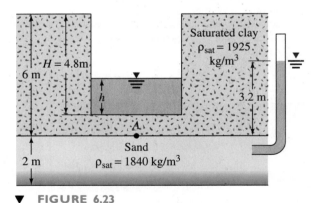

▼ **FIGURE 6.23**

6.12   Refer to Figure 6.3a. Given $H_1 = 2$ ft; $H_2 = 3$ ft; $h = 1.25$ ft; $\gamma_{sat} = 118.5$ lb/ft^3; coefficient of permeability of sand, $k = 0.12$ cm/sec; and area of tank $= 4.8$ ft^2, what is the rate of upward seepage of water (ft^3/min)?

6.13   Refer to Figure 6.3a. Given $H_1 = 0.7$ m, $H_2 = 1.2$ m, $\gamma_{sat} = 18.5$ kN/m^3, area of tank $= 0.5$ m^2, and coefficient of permeability $= 0.1$ cm/sec, what value of h will cause boiling?

6.14   From the sieve analysis of a sand, the effective size was determined to be 0.16. Using Allen Hazen's formula, estimate the range of capillary rise in this sand for a void ratio of 0.55.

6.15 A soil profile is shown in Figure 6.24. Given $H_1 = 6$ ft, $H_2 = 4$ ft, and $H_3 = 9$ ft, plot the variation of σ, u, and σ' with depth.

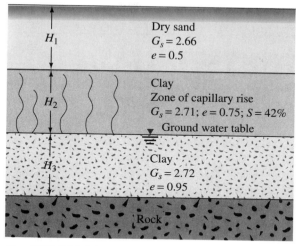

Dry sand
$G_s = 2.66$
$e = 0.5$

Clay
Zone of capillary rise
$G_s = 2.71$; $e = 0.75$; $S = 42\%$
Ground water table

Clay
$G_s = 2.72$
$e = 0.95$

Rock

▼ **FIGURE 6.24**

6.16 Repeat Problem 6.15 with $H_1 = 1.5$ m, $H_2 = 1.75$ m, and $H_3 = 3$ m.

6.17 Find the factor of safety against heave on the downstream side of the single-row sheet pile structure shown in Figure 5.27. (*Note:* The depth of penetration of sheet piles into the permeable layer is 15 ft.) Assume $\gamma_{sat} = 122.4$ lb/ft^3.

6.18 Repeat Problem 6.17, assuming that a filter 4 ft thick is placed on the downstream side. We are given $\gamma_{sat(filter)} = 128$ lb/ft^3.

REFERENCES

Bishop, A. W., Alpen, I., Blight, G. C., and Donald, I. B. (1960). "Factors Controlling the Strength of Partially Saturated Cohesive Soils," *Proceedings*, Research Conference on Shear Strength of Cohesive Soils, ASCE, 500–532.

Hazen, A. (1930). In *American Civil Engineering Handbook*, Wiley, New York.

Skempton, A. W. (1960). "Correspondence," *Geotechnique*, Vol. 10, No. 4, 186.

Terzaghi, K. (1922). "Der Grundbruch an Stauwerken und seine Verhütung," *Die Wasserkraft*, Vol. 17, 445–449.

Terzaghi, K. (1925). *Erdbanmechanik auf Bodenphysikalisher Grundlage*, Dueticke, Vienna.

Terzaghi, K. (1936). "Relation between Soil Mechanics and Foundation Engineering: Presidential Address," *Proceedings*, First International Conference on Soil Mechanics and Foundation Engineering, Boston, Vol. 3, 13–18.

Terzaghi, K., and Peck, R. B. (1948). *Soil Mechanics in Engineering Practice*, Wiley, New York.

CHAPTER SEVEN

STRESSES IN A SOIL MASS

Construction of a foundation causes changes in the stress, usually a net increase. The net stress increase in the soil depends on the load per unit area to which the foundation is subjected, the depth below the foundation at which the stress estimation is desired, and other factors. It is necessary to estimate the net increase of vertical stress in soil that occurs as a result of the construction of a foundation so that settlement can be calculated. The settlement calculation procedure is discussed in more detail in Chapter 8. This chapter discusses the principles of estimation of vertical stress increase in soil caused by various types of loading, based on the theory of elasticity. Although natural soil deposits, in most cases, are not fully elastic, isotropic, or homogeneous materials, calculations for estimating increases in vertical stress yield fairly good results for practical work.

7.1 NORMAL AND SHEAR STRESSES ON A PLANE

Students in a soil mechanics course are familiar with the fundamental principles of the mechanics of deformable solids. This section is a brief review of the basic concepts of normal and shear stresses on a plane that can be found in any course on the mechanics of materials.

Figure 7.1a shows a two-dimensional soil element that is being subjected to normal and shear stresses ($\sigma_y > \sigma_x$). To determine the normal stress and the shear stress on a plane EF that makes an angle θ with the plane AB, we need to consider the free body diagram EFB as shown in Figure 7.1b. Let σ_n and τ_n be the normal stress and the shear stress, respectively, on the plane EF. From geometry,

$$\overline{EB} = \overline{EF} \cos \theta \tag{7.1}$$

and

$$\overline{FB} = \overline{EF} \sin \theta \tag{7.2}$$

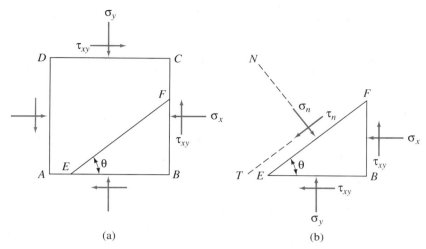

▼ **FIGURE 7.1** (a) A soil element with normal and shear stresses acting on it; (b) free body diagram of *EFB* as shown in (a)

Summing the components of forces that act on the element in the direction of N and T, we have

$$\sigma_n(\overline{EF}) = \sigma_x(\overline{EF}) \sin^2 \theta + \sigma_y(\overline{EF}) \cos^2 \theta + 2\tau_{xy}(\overline{EF}) \sin \theta \cos \theta$$

or

$$\sigma_n = \sigma_x \sin^2 \theta + \sigma_y \cos^2 \theta + 2\tau_{xy} \sin \theta \cos \theta$$

or

$$\sigma_n = \frac{\sigma_y + \sigma_x}{2} + \frac{\sigma_y - \sigma_x}{2} \cos 2\theta + \tau_{xy} \sin 2\theta \qquad (7.3)$$

Again,

$$\tau_n(\overline{EF}) = -\sigma_x(\overline{EF}) \sin \theta \cos \theta + \sigma_y(\overline{EF}) \sin \theta$$
$$\times \cos \theta - \tau_{xy}(\overline{EF}) \cos^2 \theta + \tau_{xy}(\overline{EF}) \sin^2 \theta$$

or

$$\tau_n = \sigma_y \sin \theta \cos \theta - \sigma_x \sin \theta \cos \theta - \tau_{xy}(\cos^2 \theta - \sin^2 \theta)$$

or

$$\tau_n = \frac{\sigma_y - \sigma_x}{2} \sin 2\theta - \tau_{xy} \cos 2\theta \qquad (7.4)$$

From Eq. (7.4), it can be seen that we can choose the value of θ in such a way that τ_n will be equal to zero. Substituting $\tau_n = 0$, we get

$$\tan 2\theta = \frac{2\tau_{xy}}{\sigma_y - \sigma_x} \tag{7.5}$$

For given values of τ_{xy}, σ_x, and σ_y, Eq. (7.5) will give two values of θ that are 90° apart. This means that there are two planes that are at right angles to each other on which the shear stress is zero. Such planes are called *principal planes*. The normal stresses that act on the principal planes are referred to as *principal stresses*. The values of principal stresses can be found by substituting Eq. (7.5) into Eq. (7.3), which yields,

Major principal stress:

$$\sigma_n = \sigma_1 = \frac{\sigma_y + \sigma_x}{2} + \sqrt{\left[\frac{(\sigma_y - \sigma_x)}{2}\right]^2 + \tau_{xy}^2} \tag{7.6}$$

Minor principal stress:

$$\sigma_n = \sigma_3 = \frac{\sigma_y + \sigma_x}{2} - \sqrt{\left[\frac{(\sigma_y - \sigma_x)}{2}\right]^2 + \tau_{xy}^2} \tag{7.7}$$

The normal stress and shear stress that act on any plane can also be determined by plotting a Mohr's circle, as shown in Figure 7.2. The following sign conventions are used in Mohr's circles: compressive normal stresses are taken as positive, and shear stresses are considered positive if they act on opposite faces of the element in such a way that they tend to produce a counterclockwise rotation.

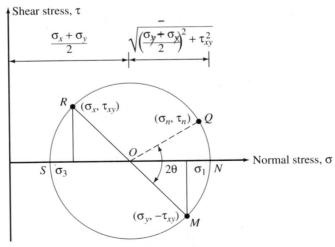

▼ **FIGURE 7.2**   Principles of the Mohr's circle

For plane AD of the soil element shown in Figure 7.1a, normal stress equals $+\sigma_x$ and shear stress equals $+\tau_{xy}$. For plane AB, normal stress equals $+\sigma_y$ and shear stress equals $-\tau_{xy}$.

The points R and M in Figure 7.2 represent the stress conditions on planes AD and AB, respectively. O is the point of intersection of the normal stress axis with the line RM. The circle $MNQRS$ drawn with O as the center and OR as the radius is the Mohr's circle for the stress conditions considered. The radius of the Mohr's circle is equal to

$$\sqrt{\left[\frac{(\sigma_y - \sigma_x)}{2}\right]^2 + \tau_{xy}^2}$$

The stress on plane EF can be determined by moving an angle 2θ (which is twice the angle the plane EF makes in a counterclockwise direction with plane AB in Figure 7.1a) in a counterclockwise direction from the point M along the circumference of the Mohr's circle to reach point Q. The abscissa and ordinate of point Q, respectively, give the normal stress, σ_n, and the shear stress, τ_n, on plane EF.

Because the ordinates (that is, the shear stresses) of points N and S are 0, they represent the stresses on the principal planes. The abscissa of point N is equal to σ_1 [Eq. (7.6)], and the abscissa for point S is σ_3 [Eq. (7.7)].

As a special case, if the planes AB and AD were major and minor principal planes, the normal stress and the shear stress on plane EF could be found by substituting $\tau_{xy} = 0$. Equations (7.3) and (7.4) show that $\sigma_y = \sigma_1$ and $\sigma_x = \sigma_3$ (Figure 7.3a).

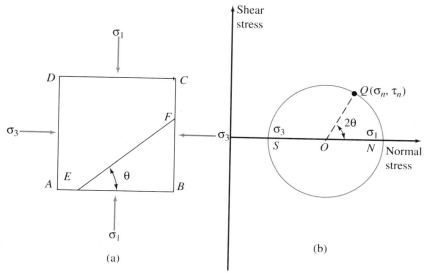

(a)

(b)

▼ **FIGURE 7.3**   (a) Soil element with AB and AD as major and minor principal planes; (b) Mohr's circle for soil element shown in (a).

Thus,

$$\sigma_n = \frac{\sigma_1 + \sigma_3}{2} + \frac{\sigma_1 - \sigma_3}{2} \cos 2\theta \qquad (7.8)$$

$$\tau_n = \frac{\sigma_1 - \sigma_3}{2} \sin 2\theta \qquad (7.9)$$

The Mohr's circle for such stress conditions is shown in Figure 7.3b. The abscissa and the ordinate of point Q give the normal stress and the shear stress, respectively, on the plane EF.

7.2 THE POLE METHOD OF FINDING STRESSES ALONG A PLANE

Another important technique for finding stresses along a plane from a Mohr's circle is the *pole method*, or the method of *origin of planes*. This is illustrated in Figure 7.4. Figure 7.4a is the same stress element shown in Figure 7.1a; Figure 7.4b is the Mohr's circle for the stress conditions indicated. According to the pole method, we draw a line from a known point on the Mohr's circle parallel to the plane on which the state of stress acts. The point of intersection of this line with the Mohr's circle is called the *pole*. This is a unique point for the state of stress under consideration. For example, the point M on the Mohr's circle in Figure 7.4b represents the stresses on the plane AB. The line MP is drawn parallel to AB. So point P is the pole (origin of planes) in this case. If we

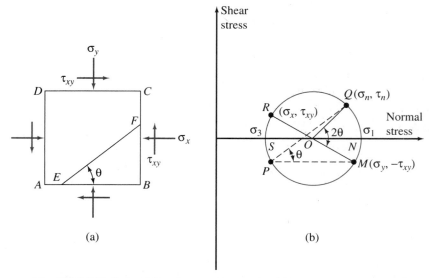

▼ **FIGURE 7.4** (a) Soil element with normal and shear stresses acting on it; (b) use of pole method to find the stresses along a plane

need to find the stresses on a plane *EF*, we draw a line from the pole parallel to *EF*. The point of intersection of this line with the Mohr's circle is *Q*. The coordinates of *Q* give the stresses on the plane *EF*. (*Note:* From geometry, angle *QOM* is twice angle *QPM*.)

▼ EXAMPLE 7.1

For the stressed soil element shown in Figure 7.5a, determine:

- **a.** The major principal stress
- **b.** The minor principal stress
- **c.** The normal and shear stresses on the plane *AE*

Use the pole method.

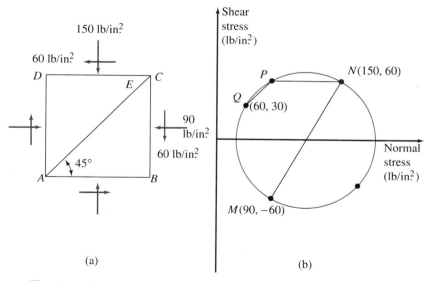

(a)

(b)

▼ FIGURE 7.5

Solution

On plane *AD*: normal stress = 90 lb/in.²
  shear stress = −60 lb/in.²
On plane *AB*: normal stress = 150 lb/in.²
  shear stress = 60 lb/in.²

The Mohr's circle is plotted in Figure 7.5b. From the plot,

- **a.** Major principal stress = **187.1 lb/in.²**
- **b.** Minor principal stress = **52.9 lb/in.²**

c. *NP* is the line drawn parallel to the plane *CD*. *P* is the pole. *PQ* is drawn parallel to *AE* (Figure 7.5a). The coordinates of point *Q* give the stresses on the plane *AE*. Thus,

normal stress = **60 lb/in.²**
shear stress = **30 lb/in.²**   ▼

▼ EXAMPLE 7.2

A soil element is shown in Figure 7.6. The magnitudes of stresses are $\sigma_x = 120$ kN/m², $\tau = 40$ kN/m², $\sigma_y = 300$ kN/m², and $\theta = 20°$. Determine:

a. The magnitudes of the principal stresses.
b. The normal and shear stresses on plane *AB*. Use Eqs. (7.3), (7.4), (7.6), and (7.7).

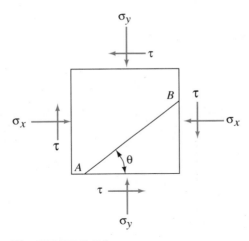

▼ **FIGURE 7.6**

Solution
a. From Eqs. (7.6) and (7.7),

$$\left.\begin{array}{c}\sigma_3\\\sigma_1\end{array}\right\} = \frac{\sigma_y + \sigma_x}{2} \pm \sqrt{\left[\frac{\sigma_y - \sigma_x}{2}\right]^2 + \tau_{xy}^2}$$

$$= \frac{300 + 120}{2} \pm \sqrt{\left[\frac{300 - 120}{2}\right]^2 + (-40)^2}$$

$$\sigma_1 = \textbf{308.5 kN/m}^2$$

$$\sigma_3 = \textbf{111.5 kN/m}^2$$

b. From Eq. (7.3),

$$\sigma_n = \frac{\sigma_y + \sigma_x}{2} + \frac{\sigma_y - \sigma_x}{2} \cos 2\theta + \tau \sin 2\theta$$

$$= \frac{300 + 120}{2} + \frac{300 - 120}{2} \cos (2 \times 20) + (-40) \sin (2 \times 20)$$

$$= \textbf{252.23 kN/m}^2$$

From Eq. (7.4),

$$\tau_n = \frac{\sigma_y - \sigma_x}{2} \sin 2\theta - \tau \cos 2\theta$$

$$= \frac{300 - 120}{2} \sin (2 \times 20) - (-40) \cos (2 \times 20)$$

$$= \textbf{88.49 kN/m}^2 \quad \blacktriangledown$$

7.3 STRESS CAUSED BY A POINT LOAD

Boussinesq (1883) solved the problem of stresses produced at any point in a homogeneous, elastic, and isotropic medium as the result of a point load applied on the surface of an infinitely large half-space. According to Figure 7.7, Boussinesq's solution for

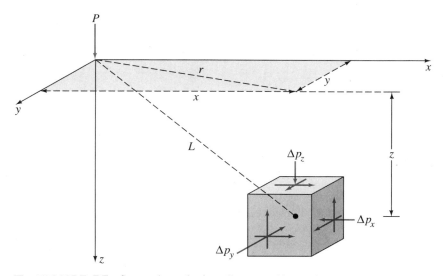

▼ **FIGURE 7.7** Stresses in an elastic medium caused by a point load

normal stresses at a point A caused by the point load P is

$$\Delta p_x = \frac{P}{2\pi} \left\{ \frac{3x^2 z}{L^5} - (1 - 2\mu) \left[\frac{x^2 - y^2}{Lr^2(L + z)} + \frac{y^2 z}{L^3 r^2} \right] \right\} \tag{7.10a}$$

$$\Delta p_y = \frac{P}{2\pi} \left\{ \frac{3y^2 z}{L^5} - (1 - 2\mu) \left[\frac{y^2 - x^2}{Lr^2(L + z)} + \frac{x^2 z}{L^3 r^2} \right] \right\} \tag{7.10b}$$

and

$$\Delta p_z = \frac{3P}{2\pi} \frac{z^3}{L^5} = \frac{3P}{2\pi} \frac{z^3}{(r^2 + z^2)^{5/2}} \tag{7.11}$$

where $r = \sqrt{x^2 + y^2}$
$L = \sqrt{x^2 + y^2 + z^2} = \sqrt{r^2 + z^2}$
$\mu = $ Poisson's ratio

Note that Eqs. (7.10a) and (7.10b), which are the expressions for horizontal normal stresses, are dependent on the Poisson's ratio of the medium. However, the relationship for the vertical normal stress, Δp_z, as given by Eq. (7.11), is independent of Poisson's ratio. The relationship for Δp_z can be rewritten in the following form:

$$\Delta p_z = \frac{P}{z^2} \left\{ \frac{3}{2\pi} \frac{1}{[(r/z)^2 + 1]^{5/2}} \right\} = \frac{P}{z^2} I_1 \tag{7.12}$$

where $I_1 = \frac{3}{2\pi} \frac{1}{[(r/z)^2 + 1]^{5/2}}.$ \hfill (7.13)

The variation of I_1 for various values of r/z is given in Table 7.1.

▼ **TABLE 7.1**   **Variation of I_1**
                   [Eq. (7.13)]

r/z	I_1	r/z	I_1
0	0.4775	0.9	0.1083
0.1	0.4657	1.0	0.0844
0.2	0.4329	1.5	0.0251
0.3	0.3849	1.75	0.0144
0.4	0.3295	2.0	0.0085
0.5	0.2733	2.5	0.0034
0.6	0.2214	3.0	0.0015
0.7	0.1762	4.0	0.0004
0.8	0.1386	5.0	0.00014

▼ **EXAMPLE 7.3**

Consider a point load $P = 1000$ lb (Figure 7.7). Plot the variation of the vertical stress increase Δp_z with depth caused by the point load below the ground surface with $x = 3$ ft and $y = 4$ ft.

Solution

$$r = \sqrt{x^2 + y^2} = \sqrt{3^2 + 4^2} = 5 \text{ ft}$$

The following table can now be prepared:

r (ft)	z (ft)	$\dfrac{r}{z}$	I_1^a	$\Delta p_z = \dfrac{p}{z^2} I_1^b$ (lb/ft²)
5.0	0	∞	0	0
	2	2.5	0.0034	0.85
	4	1.25	0.0424	2.65
	6	0.83	0.1295	3.60
	10	0.5	0.2733	2.73
	15	0.33	0.3713	1.65
	20	0.25	0.4103	1.03

[a] Eq. (7.13)
[b] Eq. (7.12). *Note:* $P = 1000$ lb.

▼

7.4 WESTERGAARD'S SOLUTION FOR VERTICAL STRESS CAUSED BY A POINT LOAD

Westergaard (1938) proposed a solution for determining the vertical stress caused by a point load P in an elastic solid medium in which layers alternate with thin rigid reinforcements (Figure 7.8a). This type of assumption may be an idealization of a clay layer with thin seams of sand. For such an assumption, the vertical stress increase at a point A (Figure 7.8b) can be given by

$$\Delta p_z = \frac{P\eta}{2\pi z^2} \left[\frac{1}{\eta^2 + (r/z)^2} \right]^{3/2} \tag{7.14}$$

where $\eta = \sqrt{\dfrac{1 - 2\mu}{2 - 2\mu}}$ (7.15)

$\mu =$ Poisson's ratio of the solid between the rigid reinforcements

$$r = \sqrt{x^2 + y^2}$$

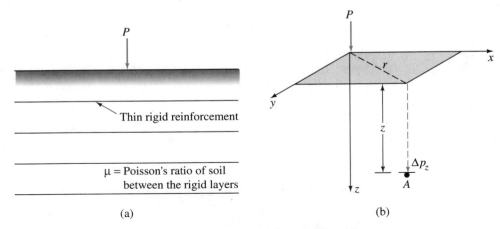

(a)   (b)

▼ **FIGURE 7.8**   Westergaard's solution for vertical stress caused by a point load

Equation (7.14) can be rewritten as

$$\Delta p_z = \left(\frac{P}{z^2}\right) I_2 \qquad\qquad (7.16)$$

where $I_2 = \dfrac{1}{2\pi\eta^2}\left[\left(\dfrac{r}{\eta z}\right)^2 + 1\right]^{-3/2}$. $\qquad\qquad$ (7.17)

   In most practical problems of geotechnical engineering, Boussinesq's solution (Section 7.3) is preferred over Westergaard's solution. For that reason, *further develop-ment of stress calculation under various types of loading will use Boussinesq's solution in this chapter.*

▼ **EXAMPLE 7.4**

Consider Example 7.3. Calculate the stress increase Δp_z using Westergaard's solution and given $\mu = 0.35$.

Solution   From Eq. (7.15),

$$\eta = \sqrt{\frac{1 - 2\mu}{2 - 2\mu}} = \sqrt{\frac{1 - (2)(0.35)}{2 - (2)(0.35)}} = \sqrt{\frac{0.3}{1.3}} = 0.48$$

$$r = 5 \text{ ft}$$

The following table can now be prepared:

r (ft)	z (ft)	$\dfrac{r}{z}$	$\dfrac{r}{\eta z}$	I_2^a	$\Delta p_z = \dfrac{P}{z^2} I_2^b$ (lb/ft^2)
5.0	0	∞	∞	0	0
	2	2.5	5.21	0.00467	1.168
	4	1.25	2.60	0.0323	2.019
	6	0.83	1.63	0.0866	2.406
	10	0.5	1.04	0.230	2.3
	15	0.33	0.69	0.3852	1.712
	20	0.25	0.52	0.4824	1.206

a Eq. (7.17)
b Eq. (7.16)

For ease of comparison, these results are plotted in Figure 7.9 along with those obtained from Example 7.3.

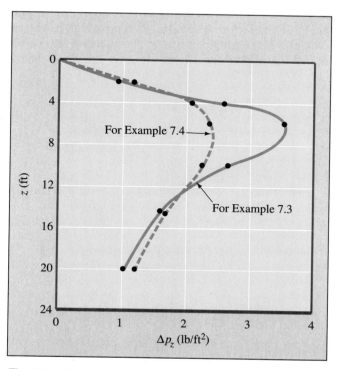

▼ **FIGURE 7.9**   Plot of vertical stress increase with depth   ▼

7.5 VERTICAL STRESS CAUSED BY A LINE LOAD

Figure 7.10 shows a flexible line load of infinite length that has an intensity q/unit length on the surface of a semi-infinite soil mass. The vertical stress increase, Δp, inside the soil mass can be determined by using the principles of the theory of elasticity, or

$$\Delta p = \frac{2qz^3}{\pi(x^2 + z^2)^2} \tag{7.18}$$

The preceding equation can be rewritten in the following form:

$$\Delta p = \frac{2q}{\pi z[(x/z)^2 + 1]^2}$$

or

$$\frac{\Delta p}{(q/z)} = \frac{2}{\pi[(x/z)^2 + 1]^2} \tag{7.19}$$

Note that Eq. (7.19) is in a nondimensional form. Using this equation, we can calculate the variation of $\Delta p/(q/z)$ with x/z. This is given in Table 7.2. The value of Δp calculated by using Eq. (7.19) is the additional stress on soil caused by the line load. The value of Δp does not include the overburden pressure of the soil above the point A.

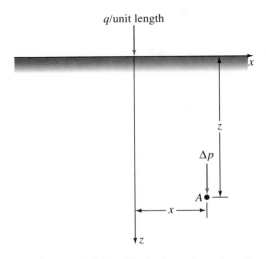

▼ **FIGURE 7.10**   Line load over the surface of a semi-infinite soil mass

▼ **TABLE 7.2** **Variation of**
$\Delta p / (q/z)$
with x/z
[Eq. (7.19)]

x/z	$\dfrac{\Delta p}{q/z}$
0	0.637
0.1	0.624
0.2	0.589
0.3	0.536
0.4	0.473
0.5	0.407
0.6	0.344
0.7	0.287
0.8	0.237
0.9	0.194
1.0	0.159
1.5	0.060
2.0	0.025
3.0	0.006

▼ **EXAMPLE 7.5**

Figure 7.11a shows two line loads on the ground surface. Determine the increase in the stress at point A.

Solution   Refer to Figure 7.11b. The total stress at point A is

$$\Delta p = \Delta p_1 + \Delta p_2$$

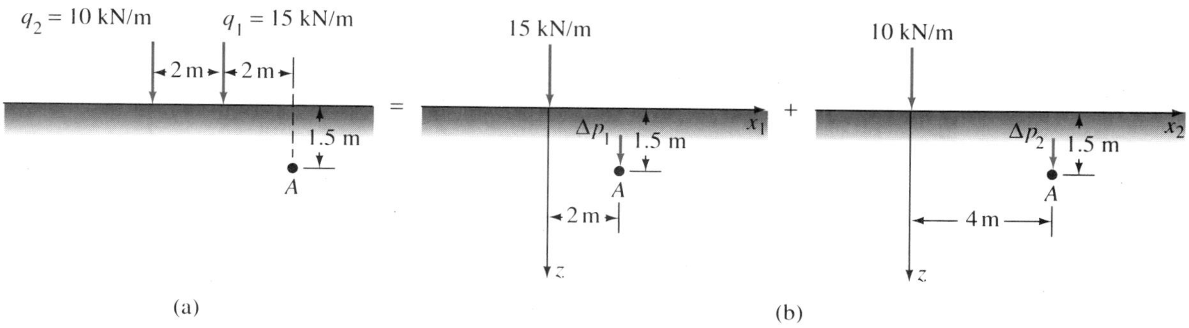

(a)

(b)

▼ **FIGURE 7.11**   (a) Two line loads on the ground surface; (b) use of superposition principle to obtain stress at point A

or

$$\Delta p = \frac{2q_1 z^3}{\pi(x_1^2 + z^2)^2} + \frac{2q_2 z^3}{\pi(x_2^2 + z^2)^2}$$

$$= \frac{(2)(15)(1.5)^3}{\pi[(2)^2 + (1.5)^2]^2} + \frac{(2)(10)(1.5)^3}{\pi[(4)^2 + (1.5)^2]^2}$$

$$= 0.825 + 0.065 = \mathbf{0.89 \ kN/m} \qquad \blacktriangledown$$

7.6 VERTICAL STRESS CAUSED BY A STRIP LOAD (FINITE WIDTH AND INFINITE LENGTH)

The fundamental equation for the vertical stress increase at a point in a soil mass as the result of a line load (Section 7.5) can be used to determine the vertical stress at a point caused by a flexible strip load of width B (Figure 7.12). Let the load per unit area of the strip shown in Figure 7.12 be equal to q. If we consider an elemental strip of width dr, the load per unit length of this strip will be equal to $q \ dr$. This elemental strip can be treated as a line load. Equation (7.18) gives the vertical stress increase, dp, at point A inside the soil mass caused by this elemental strip load. To calculate the vertical stress increase, we need to substitute $q \ dr$ for q and $(x - r)$ for x. So

$$dp = \frac{2(q \ dr)z^3}{\pi[(x - r)^2 + z^2]^2} \qquad (7.20)$$

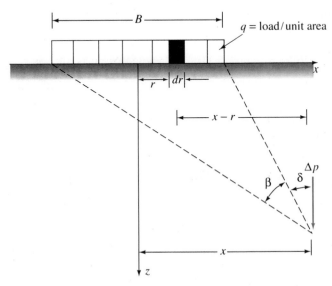

▼ **FIGURE 7.12**   Vertical stress caused by a flexible strip load (*Note:* Angles measured in counter-clockwise direction are taken as positive.)

The total increase in the vertical stress (Δp) at point A caused by the entire strip load of width B can be determined by integration of Eq. (7.20) with limits of r from $-B/2$ to $+B/2$, or

$$\Delta p = \int dp = \int_{-B/2}^{+B/2} \left(\frac{2q}{\pi}\right)\left\{\frac{z^3}{[(x-r)^2 + z^2]^2}\right\} dr$$

$$= \frac{q}{\pi}\left\{\tan^{-1}\left[\frac{z}{x-(B/2)}\right] - \tan^{-1}\left[\frac{z}{x+(B/2)}\right]\right.$$

$$\left. - \frac{Bz[x^2 - z^2 - (B^2/4)]}{[x^2 + z^2 - (B^2/4)]^2 + B^2z^2}\right\} \tag{7.21}$$

▼ **TABLE 7.3** Variation of $\Delta p/q$ with $2z/B$ and $2x/B$*

$2x/B$	$2z/B$	$\Delta p/q$	$2x/B$	$2z/B$	$\Delta p/q$
0	0	1.0000	1.5	1.0	0.2488
	0.5	0.9594		1.5	0.2704
	1.0	0.8183		2.0	0.2876
	1.5	0.6678		2.5	0.2851
	2.0	0.5508			
	2.5	0.4617	2.0	0.25	0.0027
	3.0	0.3954		0.5	0.0194
	3.5	0.3457		1.0	0.0776
	4.0	0.3050		1.5	0.1458
				2.0	0.1847
0.5	0	1.0000		2.5	0.2045
	0.25	0.9787			
	0.5	0.9028	2.5	0.5	0.0068
	1.0	0.7352		1.0	0.0357
	1.5	0.6078		1.5	0.0771
	2.0	0.5107		2.0	0.1139
	2.5	0.4372		2.5	0.1409
1.0	0.25	0.4996	3.0	0.5	0.0026
	0.5	0.4969		1.0	0.0171
	1.0	0.4797		1.5	0.0427
	1.5	0.4480		2.0	0.0705
	2.0	0.4095		2.5	0.0952
	2.5	0.3701		3.0	0.1139
1.5	0.25	0.0177			
	0.5	0.0892			

* After Jurgenson (1934)

Equation (7.21) can be simplified:

$$\Delta p = \frac{q}{\pi} \left[\beta + \sin \beta \cos (\beta + 2\delta) \right]$$

(7.22)

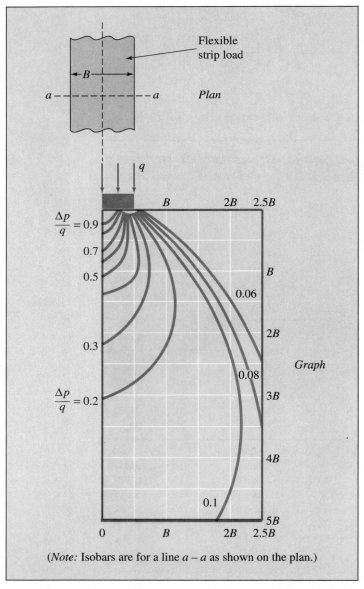

(*Note:* Isobars are for a line $a - a$ as shown on the plan.)

▼ **FIGURE 7.13**   Vertical pressure isobars under a flexible strip load

The angles β and δ are defined in Figure 7.12.

Table 7.3 shows the variation of $\Delta p/q$ with $2z/B$ for $2x/B$ equal to 0, 0.5, 1.0, 1.5, 2.0, 2.5, and 3.0. This table can be used conveniently for the calculation of vertical stress at a point caused by a flexible strip load. The net increase equation as given by Eq. (7.22) can also be used to calculate stresses at various grid points under the load. Then stress *isobars* can be drawn. These are contours of equal stress increase. Such a plot is given in Figure 7.13.

▼ **EXAMPLE 7.6**

With reference to Figure 7.12, we are given $q = 200$ kN/m², $B = 6$ m, and $z = 3$ m. Determine the vertical stress increase at $x = \pm 9$ m, ± 6 m, ± 3 m, and 0 m. Plot a graph of Δp against x.

Solution   The following table can be made:

x (m)	$2x/B$	$2z/B$	$\Delta p/q^a$	Δp^b (kN/m²)
± 9	± 3	1	0.0171	3.42
± 6	± 2	1	0.0776	15.52
± 3	± 1	1	0.4797	95.94
0	0	1	0.8183	163.66

a From Table 7.3
b $q = 200$ kN/m²

The plot of Δp against x is given in Figure 7.14.

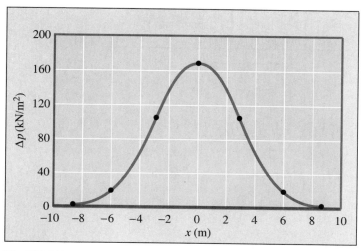

▼ **FIGURE 7.14**   Plot of Δp against distance x

7.7 VERTICAL STRESS CAUSED BY A LINEARLY INCREASING LOAD (FINITE WIDTH AND INFINITE LENGTH)

Figure 7.15 shows a linearly increasing vertical loading of an infinite strip of width B. The load per unit area increases from 0 at $x = 0$ to q at $x = B$. In order to determine the increase in the vertical stress at point A caused by the loading, consider an elemental strip with width dr. For the elemental strip, the load per unit length is $(q/B)r \, dr$. Approximating this as a line load and using Eq. (7.18), we can give the vertical stress increase dp at A caused by the elemental strip as

$$dp = \frac{\left(\frac{2}{\pi}\right)\left[\frac{q}{B} r \, dr\right] z^3}{[(x - r)^2 + z^2]^2}$$

The total increase in the vertical stress at point A can thus be given as

$$\Delta p = \int_{r=0}^{r=B} dp = \frac{2q}{\pi B} \int_0^B \frac{z^3 r \, dr}{[(x - r)^2 + z^2]^2}$$

$$= \frac{q}{\pi}\left(\frac{x}{B} \alpha - \frac{\sin 2\delta}{2}\right) \tag{7.23}$$

The angles α and δ are shown in Figure 7.15. Table 7.4 gives the values of $\Delta p/q$ for various values of $2x/B$ and $2z/B$.

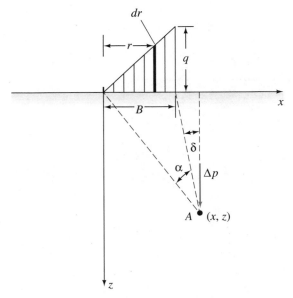

▼ **FIGURE 7.15**   Vertical stress caused by a linearly increasing load

▼ **TABLE 7.4** **Values of $\Delta p/q$ [Eq. (7.23)]**

2x/B					2z/B				
	0	0.5	1.0	1.5	2.0	2.5	3.0	4.0	5.0
−3	0	0.0003	0.0018	0.00054	0.0107	0.0170	0.0235	0.0347	0.0422
−2	0	0.0008	0.0053	0.0140	0.0249	0.0356	0.0448	0.0567	0.0616
−1	0	0.0041	0.0217	0.0447	0.0643	0.0777	0.0854	0.0894	0.0858
0	0	0.0748	0.1273	0.1528	0.1592	0.1553	0.1469	0.1273	0.1098
1	0.5	0.4797	0.4092	0.3341	0.2749	0.2309	0.1979	0.1735	0.1241
2	0.5	0.4220	0.3524	0.2952	0.2500	0.2148	0.1872	0.1476	0.1211
3	0	0.0152	0.0622	0.1010	0.1206	0.1268	0.1258	0.1154	0.1026
4	0	0.0019	0.0119	0.0285	0.0457	0.0596	0.0691	0.0775	0.0776
5	0	0.0005	0.0035	0.0097	0.0182	0.0274	0.0358	0.0482	0.0546

7.8 VERTICAL STRESS BELOW THE CENTER OF A UNIFORMLY LOADED CIRCULAR AREA

Using Boussinesq's solution for vertical stress Δp caused by a point load [Eq. (7.11)], one can also develop an expression for the vertical stress below the center of a uniformly loaded flexible circular area.

From Figure 7.16, let the intensity of pressure on the circular area of radius R be equal to q. The total load on the elemental area (shaded in the figure) $= qr\,dr\,d\alpha$. The

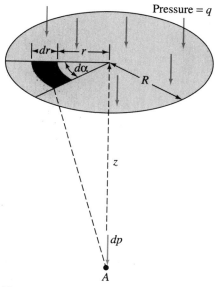

▼ **FIGURE 7.16** Vertical stress below the center of a uniformly loaded flexible circular area

vertical stress, dp, at point A caused by the load on the elemental area (which may be assumed to be a concentrated load) can be obtained from Eq. (7.11):

$$dp = \frac{3(qr \, dr \, d\alpha)}{2\pi} \frac{z^3}{(r^2 + z^2)^{5/2}} \tag{7.24}$$

The increase in the stress at A caused by the entire loaded area can be found by integrating Eq. (7.24), or

$$\Delta p = \int dp = \int_{\alpha=0}^{\alpha=2\pi} \int_{r=0}^{r=R} \frac{3q}{2\pi} \frac{z^3 r}{(r^2 + z^2)^{5/2}} \, dr \, d\alpha$$

So

$$\Delta p = q \left\{ 1 - \frac{1}{[(R/z)^2 + 1]^{3/2}} \right\} \tag{7.25}$$

The variation of $\Delta p/q$ with z/R as obtained from Eq. (7.25) is given in Table 7.5. A plot of this is also shown in Figure 7.17. The value of Δp decreases rapidly with depth, and, at $z = 5R$, it is about 6% of q, which is the intensity of pressure at the ground surface.

▼ **TABLE 7.5**   **Variation of $\Delta p/q$ with z/R [Eq. (7.25)]**

z/R	$\Delta p/q$
0	1
0.02	0.9999
0.05	0.9998
0.10	0.9990
0.2	0.9925
0.4	0.9488
0.5	0.9106
0.8	0.7562
1.0	0.6465
1.5	0.4240
2.0	0.2845
2.5	0.1996
3.0	0.1436
4.0	0.0869
5.0	0.0571

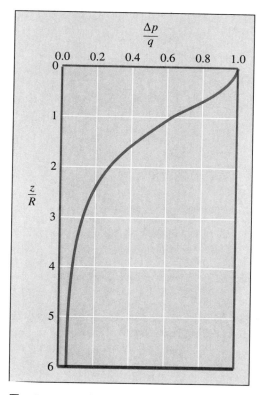

▼ **FIGURE 7.17**   Intensity of stress under the center of a uniformly loaded flexible area

7.9 VERTICAL STRESS CAUSED BY A RECTANGULARLY LOADED AREA

Boussinesq's solution can also be used to calculate the vertical stress increase below a flexible rectangular loaded area, as shown in Figure 7.18. The loaded area is located at the ground surface and has length L and width B. The uniformly distributed load per unit area is equal to q. To determine the increase in the vertical stress Δp at point A located at depth z below the corner of the rectangular area, we need to consider a small elemental area $dx\,dy$ of the rectangle. This is shown in Figure 7.18. The load on this elemental area can be given by

$$dq = q\,dx\,dy \qquad\qquad (7.26)$$

The increase in the stress (dp) at point A caused by the load dq can be determined by using Eq. (7.11). However, we need to replace P with $dq = q\,dx\,dy$ and r^2 with $x^2 + y^2$. Thus,

$$dp = \frac{3q\,dx\,dy\,z^3}{2\pi(x^2 + y^2 + z^2)^{5/2}} \qquad\qquad (7.27)$$

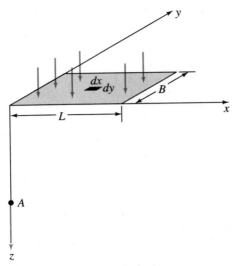

▼ **FIGURE 7.18**   Vertical stress below the corner of a uniformly loaded flexible rectangular area

The increase in the stress Δp at A caused by the entire loaded area can now be determined by integrating the preceding equation:

$$\Delta p = \int dp = \int_{y=0}^{B} \int_{x=0}^{L} \frac{3qz^3(dx\ dy)}{2\pi(x^2 + y^2 + z^2)^{5/2}} = qI_3 \tag{7.28}$$

where
$$I_3 = \frac{1}{4\pi} \left[\frac{2mn\sqrt{m^2 + n^2 + 1}}{m^2 + n^2 + m^2n^2 + 1} \left(\frac{m^2 + n^2 + 2}{m^2 + n^2 + 1} \right) \right.$$

$$\left. + \tan^{-1} \left(\frac{2mn\sqrt{m^2 + n^2 + 1}}{m^2 + n^2 - m^2n^2 + 1} \right) \right] \tag{7.29}$$

$$m = \frac{B}{z} \tag{7.30}$$

$$n = \frac{L}{z} \tag{7.31}$$

The variation of I_3 with m and n is shown in Figure 7.19.

The increase in the stress at any point below a rectangularly loaded area can be found by using Eq. (7.28) and Figure 7.19. This can be explained by reference to Figure 7.20. Let us determine the stress at a point below point A' at depth z. The loaded area can be divided into four rectangles as shown. The point A' is the corner common to all four rectangles. The increase in the stress at depth z below point A' due to each rectangular area can now be calculated by using Eq. (7.28). The total stress increase

THIS FIGURE IS NOT COMPLETELY ACCURATE

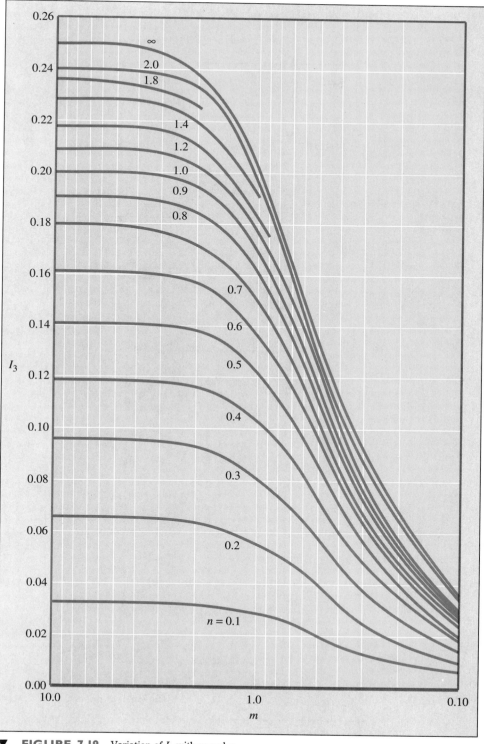

▼ **FIGURE 7.19** Variation of I_3 with m and n

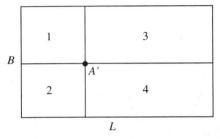

FIGURE 7.20 Increase of stress at any point below a rectangularly loaded flexible area

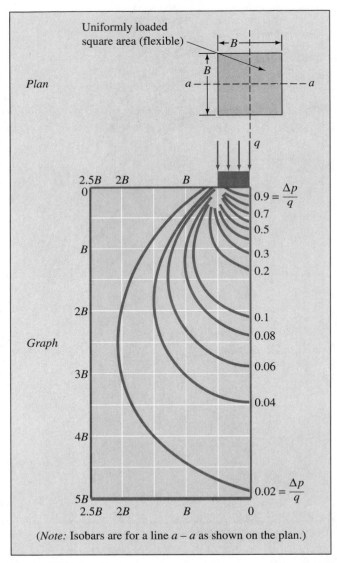

(*Note:* Isobars are for a line $a - a$ as shown on the plan.)

▼ **FIGURE 7.21** Vertical pressure isobars under a uniformly loaded square area

caused by the entire loaded area can now be given by

$$p = q[I_{3(1)} + I_{3(2)} + I_{3(3)} + I_{3(4)}]$$

(7.32)

where $I_{3(1)}, I_{3(2)}, I_{3(3)}$, and $I_{3(4)}$ = values of I_3 for rectangles 1, 2, 3, and 4, respectively.

As shown in Figure 7.13 (which is for a strip-loading case), Eq. (7.28) can be used to calculate the stress increase at various grid points. From those grid points, stress isobars can be plotted. Figure 7.21 shows such a plot for a uniformly loaded square area. Note that the stress isobars are valid for a vertical plane drawn through line *aa* as shown at the top of Figure 7.21. Figure 7.22 is a nondimensional plot of $\Delta p/q$ below the center of a rectangularly loaded area with $L/B = 1$, 1.5, 2, and ∞, which has been calculated by using Eq. (7.28).

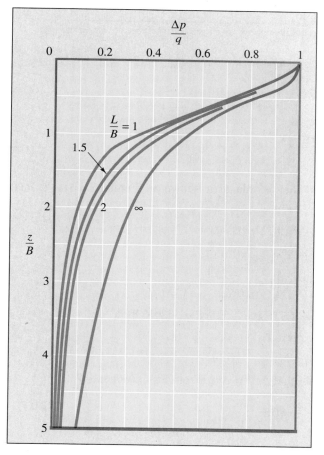

▼ **FIGURE 7.22**   Stress increase under the center of a uniformly loaded rectangular flexible area

▼   **EXAMPLE 7.7**

The flexible area shown in Figure 7.23 is uniformly loaded. Given $q = 3000$ lb/ft^2, determine the veritcal stress increase at point A.

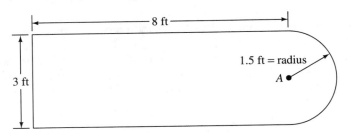

Plan

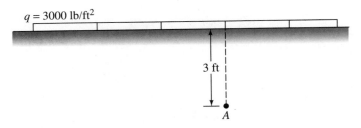

▼   **FIGURE 7.23**

Solution   The flexible area shown in Figure 7.23 is divided into three parts in Figure 7.24. At A,

$$\Delta p = \Delta p_1 + \Delta p_2 + \Delta p_3$$

From Eq. (7.25),

$$\Delta p_1 = (\tfrac{1}{2})q\left\{1 - \frac{1}{[(R/z)^2 + 1]^{3/2}}\right\}$$

We have $R = 1.5$ ft, $z = 3$ ft, and $q = 3000$ lb/ft^2, so

$$\Delta p_1 = \frac{3000}{2}\left\{1 - \frac{1}{[(1.5/3)^2 + 1]^{3/2}}\right\} = 426.7 \text{ lb/ft}^2$$

It can be seen that $\Delta p_2 = \Delta p_3$. From Eqs. (7.30) and (7.31),

$$m = \frac{1.5}{3} = 0.5$$

$$n = \frac{8}{3} = 2.67$$

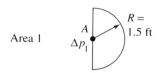

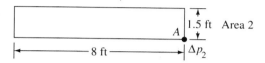

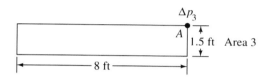

▼ **FIGURE 7.24**

From Figure 7.19, for $m = 0.5$ and $n = 2.67$, the magnitude of $I_3 = 0.138$. Thus, from Eq. (7.28),

$$\Delta p_2 = \Delta p_3 = qI_3 = (3000)(0.138) = 414 \text{ lb/ft}^2$$

so

$$\Delta p_1 = 426.7 + 414 + 414 = \textbf{1254.7 lb/ft}^2 \qquad ▼$$

7.10 INFLUENCE CHART FOR VERTICAL PRESSURE

Equation (7.25) can be rearranged and written in the form

$$\frac{R}{z} = \sqrt{\left(1 - \frac{\Delta p}{q}\right)^{-2/3} - 1} \tag{7.33}$$

Note that R/z and $\Delta p/q$ in the preceding equation are nondimensional quantities. The values of R/z that correspond to various pressure ratios are given in Table 7.6.

Using the values of R/z obtained from Eq. (7.33) for various pressure ratios, Newmark (1942) presented an influence chart that can be used to determine the vertical pressure at any point below a uniformly loaded flexible area of any shape.

▼ **TABLE 7.6** **Values of R/z for Various Pressure Ratios**

$\Delta p/q$	R/z	$\Delta p/q$	R/z
0	0	0.55	0.8384
0.05	0.1865	0.60	0.9176
0.10	0.2698	0.65	1.0067
0.15	0.3383	0.70	1.1097
0.20	0.4005	0.75	1.2328
0.25	0.4598	0.80	1.3871
0.30	0.5181	0.85	1.5943
0.35	0.5768	0.90	1.9084
0.40	0.6370	0.95	2.5232
0.45	0.6997	1.00	∞
0.50	0.7664		

Figure 7.25 shows an influence chart that has been constructed by drawing concentric circles. The radii of the circles are equal to the R/z values corresponding to $\Delta p/q = 0, 0.1, 0.2, \ldots, 1$. (*Note:* For $\Delta p/q = 0$, $R/z = 0$, and for $\Delta p/q = 1$, $R/z = \infty$, so there are nine circles shown.) The unit length for plotting the circles is $\overline{AB}$. The circles are divided by several equally spaced radial lines. The influence value of the chart is given by $1/N$, where N is equal to the number of elements in the chart. In Figure 7.25, there are 200 elements; hence, the influence value is 0.005.

The procedure for obtaining vertical pressure at any point below a loaded area is as follows:

1. Determine the depth z below the uniformly loaded area at which the stress increase is required.
2. Plot the plan of the loaded area with a scale of z equal to the unit length of the chart ($\overline{AB}$).
3. Place the plan (plotted in Step 2) on the influence chart in such a way that the point below which the stress is to be determined is located at the center of the chart.
4. Count the number of elements (M) of the chart enclosed by the plan of the loaded area.

The increase in the pressure at the point under consideration is given by

$$\boxed{\Delta p = (IV)qM} \tag{7.34}$$

where IV = influence value
q = pressure on the loaded area

▼ **FIGURE 7.25**   Influence chart for vertical pressure based on Boussinesq's theory (after Newmark, 1942)

▼ **EXAMPLE 7.8**

The cross-section and plan of a column footing are shown in Figure 7.26. Find the increase in stress produced by the column footing at point A.

Solution   Point A is located at a depth 3 m below the bottom of the footing. The plan of the square footing has been replotted to a scale of $\overline{AB} = 3$ m and placed on the influence chart (Figure 7.27) in such a way that point A on the plan falls directly over the center of the chart. The number of elements inside the outline of the plan is about 48.5. Hence,

$$\Delta p = (IV)qM = 0.005\left(\frac{660}{3 \times 3}\right)48.5 = \textbf{17.78 kN/m}^2$$

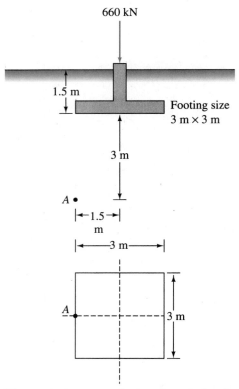

660 kN

1.5 m

Footing size
3 m × 3 m

3 m

A •

|←1.5→|
m

|←——3 m——→|

3 m

▼ **FIGURE 7.26** Cross-section and plan of a column footing

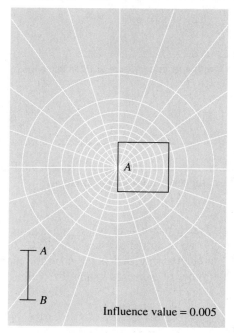

A

A
B

Influence value = 0.005

▼ **FIGURE 7.27** Determination of stress at a point by
use of Newmark's influence chart ▼

7.11 AVERAGE VERTICAL STRESS INCREASE CAUSED BY A RECTANGULARLY LOADED AREA

In Section 7.9, the vertical stress increase below the corner of a uniformly loaded rectangular area was given as (Figure 7.28)

$$\Delta p = qI_3 \tag{7.28}$$

In many cases it is required to determine the average stress increase, Δp_{av}, below the corner of a uniformly loaded rectangular area with limits of $z = 0$ to $z = H$ as shown in Figure 7.28c. This can be evaluated as

$$\Delta p_{av} = \frac{1}{H} \int_0^H (qI_3)\, dz = qI_4 \tag{7.35}$$

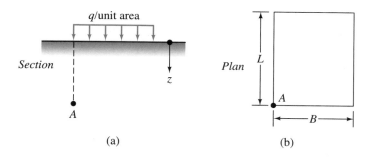

(a)   (b)

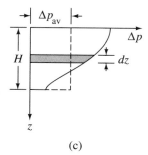

(c)

▼ **FIGURE 7.28**   Average vertical stress increase caused by a rectangularly loaded flexible area: (a) section of a loaded area, (b) plan of the load area, (c) average stress increase calculations

where $I_4 = f(m', n')$ (7.36)

$$m' = \frac{B}{H}$$ (7.37)

$$n' = \frac{L}{H}$$ (7.38)

The variation of I_4 is shown in Figure 7.29 in a slightly different form, as proposed by Griffiths (1984).

In estimating the consolidation settlement under a foundation (Chapter 8), it may be required to determine the average vertical stress increase in only a given layer—that

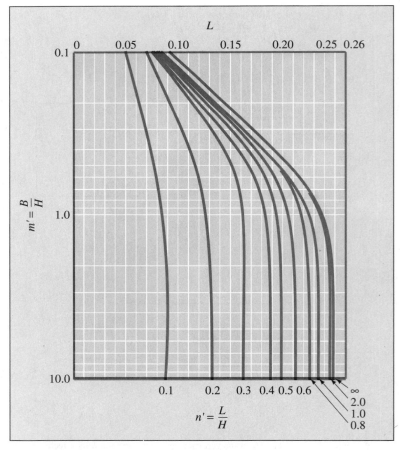

▼ **FIGURE 7.29** Variation of I_4 with m' and n'

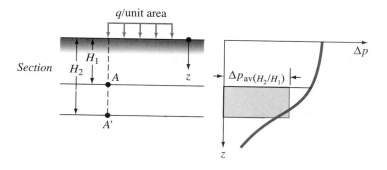

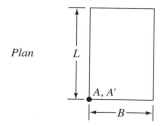

▼   **FIGURE 7.30**   Average pressure increase between $z = H_1$ and $z = H_2$ below the corner of a uniformly loaded rectangular area

is, between $z = H_1$ and $z = H_2$, as shown in Figure 7.30. This can be done as (Griffiths, 1984)

$$\Delta p_{av(H_2/H_1)} = q \left[\frac{H_2 I_{4(H_2)} - H_1 I_{4(H_1)}}{H_2 - H_1} \right]$$

(7.39)

where $\Delta p_{av(H_2/H_1)}$ = average stress increase immediately below the corner of a uniformly loaded rectangular area between depths $z = H_1$ and $z = H_2$

$I_{4(H_2)} = I_4$ for $z = 0$ to $z = H_2$

$$= f\left(m' = \frac{B}{H_2}, n' = \frac{L}{H_2} \right)$$

$I_{4(H_1)} = I_4$ for $z = 0$ to $z = H_1$

$$= f\left(m' = \frac{B}{H_1}, n' = \frac{L}{H_1} \right)$$

▼ **EXAMPLE 7.9**

Refer to Figure 7.31. Determine the *average* stress increase below the center of the loaded area between $z = 5$ ft and $z = 7.5$ ft (that is, between points A and A').

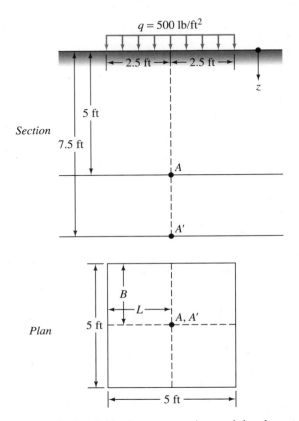

▼ **FIGURE 7.31** Average stress increase below the center of a foundation

Solution Refer to the bottom of Figure 7.31. The loaded area can be divided into four rectangular areas, each measuring 2.5 ft × 2.5 ft ($L \times B$). Using Eq. (7.39), we can give the average stress increase (between the required depths) below the corner of each rectangular area as

$$\Delta p_{av(H_2/H_1)} = q\left[\frac{H_2 I_{4(H_2)} - H_1 I_{4(H_1)}}{H_2 - H_1}\right] = 500\left[\frac{(7.5)I_{4(H_2)} - (5)I_{4(H_1)}}{7.5 - 5}\right]$$

$$\text{For } I_{4(H_2)}: m' = \frac{B}{H_2} = \frac{2.5}{7.5} = 0.33$$

$$n' = \frac{L}{H_2} = \frac{2.5}{7.5} = 0.33$$

Referring to Figure 7.29, for $m' = 0.33 = n'$ and $I_{4(H_2)} = 0.136$, we have:

For $I_{4(H_1)}$: $m' = \dfrac{B}{H_1} = \dfrac{2.5}{5} = 0.5$

$$n' = \dfrac{L}{H_1} = \dfrac{2.5}{5} = 0.5$$

From Figure 7.29, $I_{4(H_1)} = 0.175$, so

$$\Delta p_{av(H_2/H_1)} = 500\left[\frac{(7.5)(0.136) - (5)(0.175)}{7.5 - 5}\right] = \textbf{29.0 lb/ft}^2$$

The average stress increase between $z = 5$ ft and $z = 7.5$ ft below the center of the loaded area is equal to

$$4\,\Delta p_{av(H_2/H_1)} = (4)(29) = \textbf{116 lb/ft}^2 \quad \blacktriangledown$$

7.12 GENERAL COMMENTS

The equations and graphs presented in Sections 7.4–7.11 are for the calculation of vertical stress increases only. They are based on the integration of Boussinesq's equation for the vertical stress increase caused by a point load [Eq. (7.12)]. Similar equations and graphs can also be developed for horizontal stress increases caused by different types of loading by using Eqs. (7.10a) and (7.10b). Many of these solutions can be found in Poulos and Davis (1974).

The equations and graphs presented in this chapter are based entirely on the principles of the theory of elasticity; however, one must realize the limitations of these theories when they are applied to a soil medium. This is because soil deposits, in general, are not homogeneous, perfectly elastic, and isotropic. Hence, some deviations from the theoretical stress calculations can be expected in the field. Only a limited number of field observations are available in the literature at the present time. Based on these results, it appears that one could expect a difference of $\pm 25\%$ to 30% between theoretical estimates and actual field values.

PROBLEMS **7.1–7.7** For the soil elements shown in Figures 7.32–7.38, determine the maximum and minimum principal stresses. Also determine the normal and shear stresses on plane AB. [*Note:* For Problems 7.1 and 7.2, use Eqs. (7.3), (7.4), (7.6), and (7.7); for Problems 7.3, 7.4, and 7.5, use the Mohr's circle; and for Problems 7.6 and 7.7, use the pole method.]

7.8 Refer to Figure 7.7. Given $P = 7000$ lb, determine the vertical stress increase at a point with $x = 5$ ft, $y = 4$ ft, and $z = 6$ ft. Use Boussinesq's solution.

7.9 Repeat Problem 7.8 using Westergaard's solution and given $\mu = 0.4$.

7.10 Refer to Figure 7.10. The magnitude of the line load q is 50 kN/m. Calculate and plot the variation of the vertical stress increase, Δp, between the limits of $x = -8$ m and $+8$ m, given $z = 3$ m.

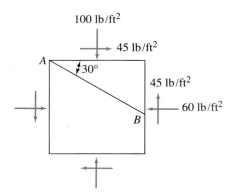

▼   **FIGURE 7.32**   Soil element for Problem 7.1

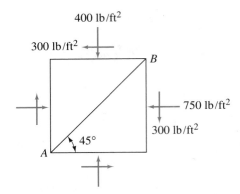

▼   **FIGURE 7.33**   Soil element for Problem 7.2

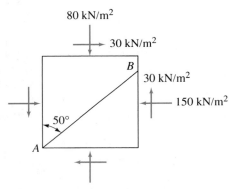

▼   **FIGURE 7.34**   Soil element for Problem 7.3

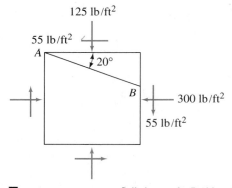

▼   **FIGURE 7.35**   Soil element for Problem 7.4

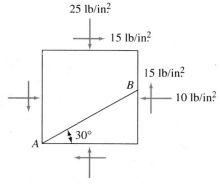

▼   **FIGURE 7.36**   Soil element for Problem 7.5

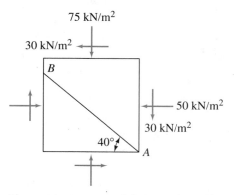

▼   **FIGURE 7.37**   Soil element for Problem 7.6

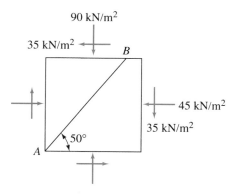

▼ **FIGURE 7.38**   Soil element for Problem 7.7

7.11 Refer to Figure 7.10. Assume $q = 4500$ lb/ft. Point A is located at a depth of 5 ft below the ground surface. Because of the application of the point load, the vertical stress at A increases by 500 lb/ft^2. What is the horizontal distance between the line load and point A?

7.12 Refer to Figure 7.39. Determine the vertical stress increase, Δp, at point A with the following values:

$q_1 = 60$ kN/m $x_1 = 1.5$ m $z = 1.5$ m

$q_2 = 0$ $x_2 = 0.5$ m

7.13 Repeat Problem 7.12 with the following values:

$q_1 = 500$ lb/ft $x_1 = 5$ ft $z = 4$ ft

$q_2 = 300$ lb/ft $x_2 = 3$ ft

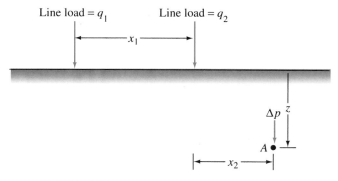

▼ **FIGURE 7.39**   Stress at a point due to two line loads

7.14   Repeat Problem 7.12 with the following data:

$q_1 = 100$ kN/m $x_1 = 3$ m $z = 2.5$ m

$q_2 = 260$ kN/m $x_2 = 2.5$ m

7.15   Refer to Figure 7.12. Given $B = 12$ ft, $q = 400$ lb/ft^2, $x = 5$ ft, and $z = 6$ ft, determine the vertical stress increase, Δp, at point A.

7.16   Repeat Problem 7.15 for $q = 6000$ kN/m^2, $B = 3$ m, $x = 1.5$ m, and $z = 3$ m.

7.17   Refer to Figure 7.15. Given $q = 3500$ lb/ft^2, $B = 5$ ft, and $z = 4$ ft, calculate the increase in the vertical stress, Δp, at $x = 7.5$ ft, 5 ft, -5 ft, and -7.5 ft.

7.18   Consider a circularly loaded flexible area on the ground surface. Given radius of circular area, $R = 6$ ft, and uniformly distributed load, $q = 3500$ lb/ft^2, calculate the vertical stress increase, Δp, at a point located 5 ft below the ground surface (immediately below the center of the circular area).

7.19   Repeat Problem 7.18 with $R = 3$ m, $q = 250$ kN/m^2, and $z = 2.5$ m.

7.20   The plan of a flexible rectangular loaded area is shown in Figure 7.40. The uniformly distributed load on the flexible area, q, is 1800 lb/ft^2. Determine the increase in the vertical stress, Δp, at a depth of $z = 5$ ft below

   a.   point A
   b.   point B
   c.   point C

7.21   Repeat Problem 7.20. Use Newmark's influence chart for vertical pressure distribution.

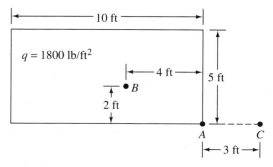

▼   **FIGURE 7.40**

7.22 Refer to Figure 7.41. The circular flexible area is uniformly loaded. Given $q = 250$ kN/m² and using Newmark's chart, determine the vertical stress increase, Δp, at point A.

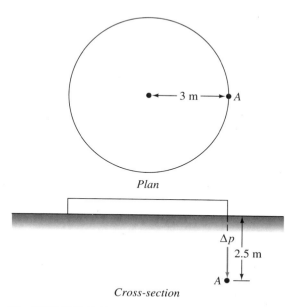

Plan

Cross-section

▼ **FIGURE 7.41**

7.23 Refer to Problem 7.20. Determine the average stress increase, Δp_{av}, below points A and B between depths $z = 0$ and $z = 5$ ft.

REFERENCES

Boussinesq, J. (1883). *Application des Potentials à L'Etude de L'Equilibre et du Mouvement des Solides Elastiques*, Gauthier–Villars, Paris.

Griffiths, D. V. (1984). "A Chart for Estimating the Average Vertical Stress Increase in an Elastic Foundation Below a Uniformly Loaded Rectangular Area," *Canadian Geotechnical Journal*, Vol. 21, No. 4, 710–713.

Jurgenson, L. (1934). "The Application of Theories of Elasticity and Plasticity to Foundation Problems," in *Contribution to Soil Mechanics, 1925–1940*, Boston Society of Civil Engineers, Boston.

Newmark, N. M. (1942). "Influence Charts for Computation of Stresses in Elastic Soil," University of Illinois Engineering Experiment Station, *Bulletin No. 338*.

Poulos, G. H., and Davis, E. H. (1974). *Elastic Solutions for Soil and Rock Mechanics*, Wiley, New York.

Westergaard, H. M. (1938). "A Problem of Elasticity Suggested by a Problem in Soil Mechanics: Soft Material Reinforced by Numerous Strong Horizontal Sheets," in *Contribution to the Mechanics of Solids*, Stephen Timoshenko 60th Anniversary Vol., Macmillan, New York.

Supplementary References for Further Study

Ahlvin, R. B., and Ulery, H. H. (1962). "Tabulated Values for Determining the Complete Pattern of Stresses, Strains, and Deflections Beneath a Uniform Load on a Homogeneous Half Space," Highway Research Record, *Bulletin No. 324*, 1–13.

Das, B. M. (1983). *Advanced Soil Mechanics*, McGraw-Hill, New York.

Giroud, J. P. (1970). "Stresses under Linearly Loaded Rectangular Area," *Journal of the Soil Mechanics and Foundations Division*, ASCE, Vol. 98, No. SM1, 263–268.

CHAPTER EIGHT

COMPRESSIBILITY
OF SOIL

A stress increase caused by the construction of foundations or other loads compresses soil layers. The compression is caused by (a) deformation of soil particles, (b) relocations of soil particles, and (c) expulsion of water or air from the void spaces. In general, the soil settlement caused by load may be divided into three broad categories:

1. *Immediate settlement*, which is caused by the elastic deformation of dry soil and of moist and saturated soils without any change in the moisture content. Immediate settlement calculations are generally based on equations derived from the theory of elasticity.
2. *Primary consolidation settlement*, which is the result of a volume change in saturated cohesive soils because of expulsion of the water that occupies the void spaces.
3. *Secondary consolidation settlement*, which is observed in saturated cohesive soils and is the result of the plastic adjustment of soil fabrics. It follows the primary consolidation settlement with a constant effective stress.

This chapter presents the fundamental principles for estimating the immediate and consolidation settlements of soil layers under superimposed loadings.

8.1 FUNDAMENTALS OF CONSOLIDATION

When a saturated soil layer is subjected to a stress increase, the pore water pressure is suddenly increased. In sandy soils that are highly permeable, the drainage caused by the increase in the pore water pressure is completed immediately. Pore water drainage is accompanied by a reduction in the volume of the soil mass, resulting in settlement. Because of rapid drainage of the pore water in sandy soils, immediate settlement and consolidation take place simultaneously.

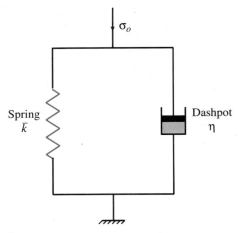

When a saturated compressible clay layer is subjected to a stress increase, elastic settlement occurs immediately. Since the coefficient of permeability of clay is significantly smaller than that of sand, the excess pore water pressure generated by loading gradually dissipates over a long period of time. Thus, the associated volume change (that is, the consolidation) in the clay may continue long after the immediate settlement. The settlement caused by consolidation in clay may be several times greater than the immediate settlement.

The time-dependent deformation of saturated clayey soils can best be understood by first considering a simple rheological model consisting of a linear elastic spring and a dashpot connected in parallel (Kelvin model, Figure 8.1). The stress-strain-time relationship for the spring and the dashpot can be given by

$$\text{Spring: } \sigma = \bar{k}\varepsilon \tag{8.1}$$

$$\text{Dashpot: } \sigma = \eta\,\frac{d\varepsilon}{dt} \tag{8.2}$$

where σ = stress
  ε = strain
  $\bar{k}$ = spring constant
  η = dashpot constant
  t = time

The viscoelastic response for the stress σ_o in Figure 8.1 can be written as

$$\sigma_o = \bar{k}\varepsilon + \eta\,\frac{d\varepsilon}{dt} \tag{8.3}$$

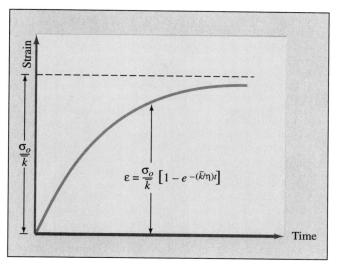

▼ **FIGURE 8.2**   Strain-time diagram for the Kelvin model

If stress, σ_O, is applied at time $t = 0$ and remains constant thereafter, then the equation for strain at any time t can be found by solving the preceding differential equation. Thus,

$$\varepsilon = \frac{\sigma_O}{\bar{k}}(1 - e^{-(\bar{k}/\eta)t}) + \varepsilon_O e^{-(\bar{k}/\eta)t}$$

where ε_O = strain at time $t = 0$. If ε_O is taken to be 0, then

$$\varepsilon = \frac{\sigma_O}{\bar{k}}(1 - e^{-(\bar{k}/\eta)t}) \tag{8.4}$$

The nature of the variation of strain with time represented by Eq. (8.4) is shown in Figure 8.2. At time $t = \infty$, the strain will approach the maximum value $\sigma_O/\bar{k}$. This is the strain that the spring alone would have immediately undergone with the application of the same stress, σ_O, without the dashpot having been attached to it. The distribution of stress at any time t between the spring and dashpot can be evaluated from Eqs. (8.3) and (8.4).

Portion of stress carried by the spring:

$$\sigma_s = \bar{k}\varepsilon = \sigma_O(1 - e^{-(\bar{k}/\eta)t}) \tag{8.5}$$

Portion of stress carried by the dashpot:

$$\sigma_d = \eta\,\frac{d\varepsilon}{dt} = \sigma_O e^{-(\bar{k}/\eta)t} \tag{8.6}$$

(Note: $\sigma_O = \sigma_s + \sigma_d$.)

Figure 8.3 shows the variation of σ_s and σ_d with time. At time $t = 0$, the stress σ_O is totally carried by the dashpot. The share of stress carried by the spring gradually increases with time, and the stress carried by the dashpot decreases at an equal rate. At time $t = \infty$, the stress σ_O is carried entirely by the spring.

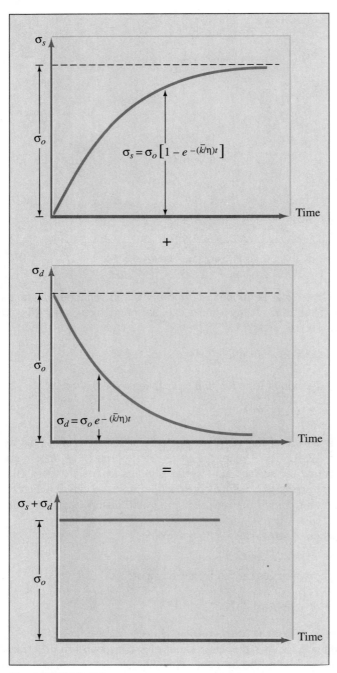

$$\sigma_s = \sigma_o \left[1 - e^{-(\bar{k}/\eta)t} \right]$$

$$\sigma_d = \sigma_o\, e^{-(\bar{k}/\eta)t}$$

▼   **FIGURE 8.3**   Stress-time diagram for spring and dashpot in Kelvin model

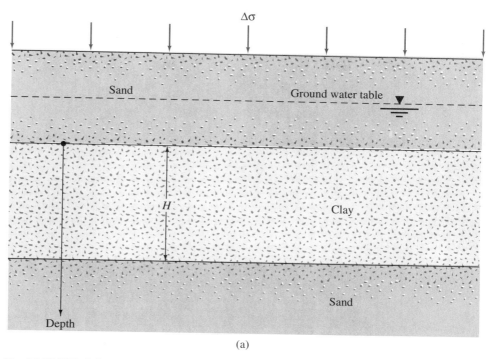

(a)

▼ **FIGURE 8.4** Variation of total stress, pore water pressure, and effective stress in a clay layer drained at top and bottom as the result of an added stress, $\Delta\sigma$

With this in mind, we can analyze the strain of a saturated clay layer subjected to a stress increase (Figure 8.4a). Consider the case where a layer of saturated clay of thickness H that is confined between two layers of sand is being subjected to an instantaneous increase of *total stress* of $\Delta\sigma$. This incremental total stress will be transmitted to the pore water and the soil solids. This means that the total stress $\Delta\sigma$ will be divided in some proportion between effective stress and pore water pressure. The behavior of the effective stress change will be similar to that of the spring in the Kelvin model, and the behavior of the pore water pressure change will be similar to that of the dashpot. From the principle of effective stress (Chapter 6),

$$\Delta\sigma = \Delta\sigma' + \Delta u \tag{8.7}$$

where $\Delta\sigma'$ = increase in the effective stress
$\quad\Delta u$ = increase in the pore water pressure

Since clay has a very low coefficient of permeability and water is incompressible as compared to the soil skeleton, at time $t = 0$, the entire incremental stress, $\Delta\sigma$, will be carried by water ($\Delta\sigma = \Delta u$) at all depths (Figure 8.4b). None will be carried by the soil skeleton (that is, incremental effective stress, $\Delta\sigma' = 0$). This is similar to the behavior of the Kelvin model, where at time $t = 0$, $\sigma_O = \sigma_d$ and $\sigma_s = 0$.

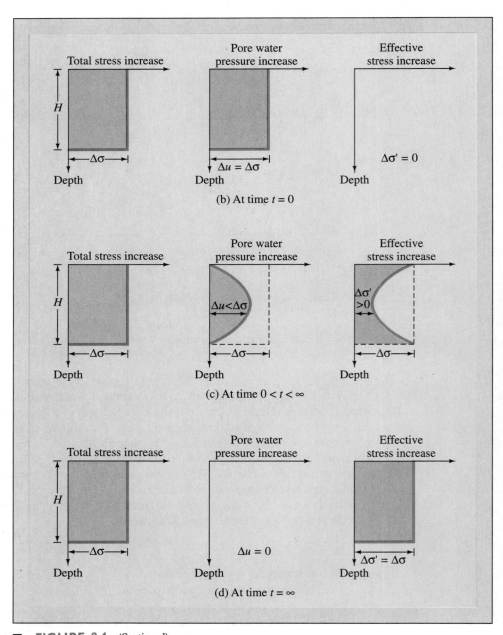

▼ **FIGURE 8.4**   (Continued)

After the application of incremental stress, $\Delta\sigma$, to the clay layer, the water in the void spaces will start to be squeezed out and will drain in both directions into the sand layers. By this process, the excess pore water pressure at any depth in the clay layer will gradually decrease, and the stress carried by the soil solids (effective stress) will increase. Thus, at time $0 < t < \infty$,

$$\Delta\sigma = \Delta\sigma' + \Delta u \qquad (\Delta\sigma' > 0 \text{ and } \Delta u < \Delta\sigma)$$

However, the magnitudes of $\Delta\sigma'$ and Δu at various depths will change (Figure 8.4c), depending on the minimum distance of the drainage path to either the top or bottom sand layer. This is similar to the Kelvin model behavior for $0 < t < \infty$, where the stress carried by the spring increases with a similar reduction in the stress carried by the dashpot.

Theoretically, at time $t = \infty$, the entire excess pore water pressure would be dissipated by drainage from all points of the clay later, thus giving $\Delta u = 0$. Now the total stress increase, $\Delta\sigma$, will be carried by the soil structure (Figure 8.4d), so

$$\Delta\sigma = \Delta\sigma'$$

Again, this is similar to the spring-dashpot behavior, for which, at time $t = \infty$, $\sigma_O = \sigma_s$ and $\sigma_d = 0$.

This gradual process of drainage under an additional load application and the associated transfer of excess pore water pressure to effective stress cause the time-dependent settlement in the clay soil layer.

Several types of rheological models have been used by various investigators for a better representation of the stress-strain-time behavior of soils. Only the Kelvin model has been treated in this section to explain the fundamental concept of consolidation.

8.2 ONE-DIMENSIONAL LABORATORY CONSOLIDATION TEST

The one-dimensional consolidation testing procedure was first suggested by Terzaghi. This test is performed in a consolidometer (sometimes referred to as an oedometer). The schematic diagram of a consolidometer is shown in Figure 8.5. The soil specimen is placed inside a metal ring with two porous stones, one at the top of the specimen and another at the bottom. The specimens are usually $2\frac{1}{2}$ in. (63.5 mm) in diameter and 1 in. (25.4 mm) thick. The load on the specimen is applied through a lever arm, and compression is measured by a micrometer dial gauge. The specimen is kept under water during the test. Each load is usually kept for 24 hours. After that, the load is usually doubled, thus doubling the pressure on the specimen, and the compression measurement is continued. At the end of the test, the dry weight of the test specimen is determined.

The general shape of the plot of deformation of the specimen against time for a given load increment is shown in Figure 8.6. From the plot, it can be observed that there are three distinct stages, which may be described as follows:

Stage I: Initial compression, which is mostly caused by preloading.

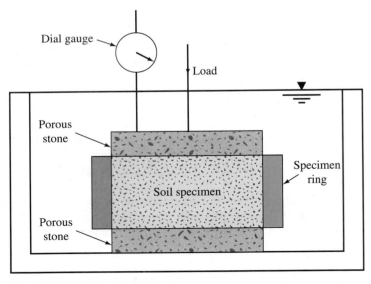

▼   **FIGURE 8.5**   Consolidometer

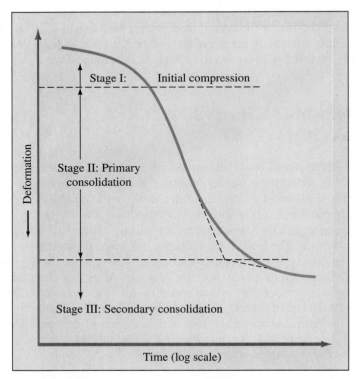

▼   **FIGURE 8.6**   Time-deformation plot during consolidation for a given load increment

Stage II: Primary consolidation, during which excess pore water pressure is gradually transferred into effective stress because of the expulsion of pore water.

Stage III: Secondary consolidation, which occurs after complete dissipation of the excess pore water pressure, when some deformation of the specimen takes place because of the plastic readjustment of soil fabric.

8.3   VOID RATIO-PRESSURE PLOTS

After the time-deformation plots for various loadings are obtained in the laboratory, it is necessary to study the change in the void ratio of the specimen with pressure. Following is a step-by-step procedure for doing that:

1. Calculate the height of solids, H_s, in the soil specimen (Figure 8.7):

$$H_s = \frac{W}{AG_s \gamma_w} \tag{8.8}$$

where W_s = dry weight of the specimen
A = area of the specimen
G_s = specific gravity of soil solids
γ_w = unit weight of water

2. Calculate the initial height of voids, H_v:

$$H_v = H - H_s \tag{8.9}$$

where H = initial height of the specimen.

3. Calculate the initial void ratio, e_O, of the specimen:

$$e_O = \frac{V_v}{V_s} = \frac{H_v A}{H_s A} = \frac{H_v}{H_s} \tag{8.10}$$

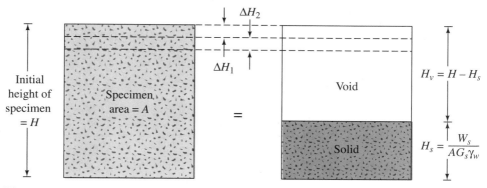

▼   **FIGURE 8.7**   Change of height of specimen in one-dimensional consolidation test

4. For the first incremental loading p_1 (total load/unit area of sample), which causes a deformation ΔH_1, calculate the change in the void ratio Δe_1:

$$\Delta e_1 = \frac{\Delta H_1}{H_s} \tag{8.11}$$

ΔH_1 is obtained from the initial and the final dial readings for the loading.

5. Calculate the new void ratio, e_1, after consolidation caused by the pressure increment p_1:

$$e_1 = e_o - \Delta e_1 \tag{8.12}$$

For the next loading, p_2 (*note:* p_2 equals the cumulative load per unit area of specimen), which causes additional deformation ΔH_2, the void ratio e_2 at the end of consolidation can be calculated as

$$e_2 = e_1 - \frac{\Delta H_2}{H_s} \tag{8.13}$$

Proceeding in a similar manner, one can obtain the void ratios at the end of the consolidation for all load increments.

The total pressures (p) and the corresponding void ratios (e) at the end of consolidation are plotted on semilogarithmic graph paper. The typical shape of such a plot is shown in Figure 8.8.

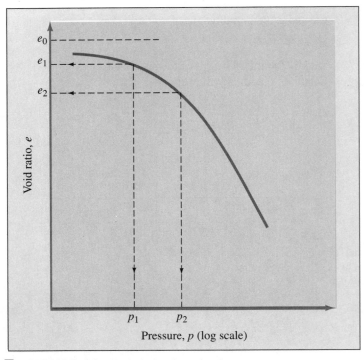

▼ **FIGURE 8.8**   Typical plot of e against log p

▼   **EXAMPLE 8.1**

Following are the results of a laboratory consolidation test on a soil specimen obtained from the field: dry mass of specimen = 116.74 g, height of specimen at beginning of test = 1 in., G_s = 2.72, and diameter of specimen = 2.5 in.

Pressure, p (ton/ft²)	Final height of specimen at end of consolidation (in.)
0	1.000
0.5	0.9917
1.0	0.9844
2.0	0.9562
4.0	0.9141
8.0	0.8686

Perform the necessary calculations and draw an e-log p curve.

Solution   From Eq. (8.8),

$$H_s = \frac{W_s}{AG_s \gamma_w} = \frac{116.74 \text{ g}}{\left[\frac{\pi}{4}(2.5 \times 2.54)^2\right](2.72)(1 \text{ g/cm}^3)}$$

$$= 1.356 \text{ cm} = 0.539 \text{ in.}$$

Now the following table can be prepared:

Pressure, p (ton/ft²)	Height at end of consolidation, H (in.)	$H_v = H - H_s$ (in.)	$e = H_v/H_s$
0	1.000	0.461	0.855
0.5	0.9917	0.4527	0.840
1.0	0.9844	0.4454	0.826
2.0	0.9562	0.4172	0.774
4.0	0.9141	0.3751	0.696
8.0	0.8686	0.3296	0.612

The e-log p plot is shown in Figure 8.9.

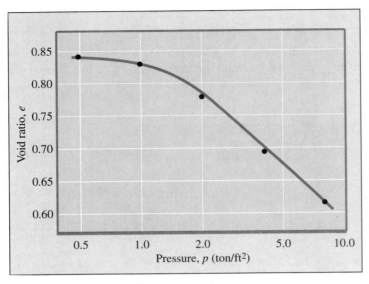

▼   **FIGURE 8.9**   ▼

8.4   NORMALLY CONSOLIDATED AND OVERCONSOLIDATED CLAYS

Figure 8.8 showed that the upper part of the e-log p plot is somewhat curved with a flat slope, followed by a linear relationship for the void ratio, with log p having a steeper slope. This can be explained in the following manner.

A soil in the field at some depth has been subjected to a certain maximum effective past pressure in its geologic history. This maximum effective past pressure may be equal to or less than the existing overburden pressure at the time of sampling. The reduction of pressure in the field may be caused by natural geologic processes or human processes. During the soil sampling, the existing overburden pressure is also released, resulting in some expansion. When this specimen is subjected to a consolidation test, a small amount of compression (that is, a small change in void ratio) will occur when the total pressure applied is less than the maximum effective overburden pressure in the field to which the soil has been subjected in the past. When the total applied pressure on the specimen is greater than the maximum effective past pressure, the change in the void ratio is much larger, and the e-log p relationship is practically linear with a steeper slope.

This relationship can be verified in the laboratory by loading the specimen to exceed the maximum overburden pressure, and then unloading and reloading again. The e-log p plot for such cases is shown in Figure 8.10, in which cd represents unloading and dfg represents the reloading process.

This leads us to the two basic definitions of clay based on stress history:

1. *Normally consolidated*, whose present effective overburden pressure is the maximum pressure that the soil has been subjected to in the past.

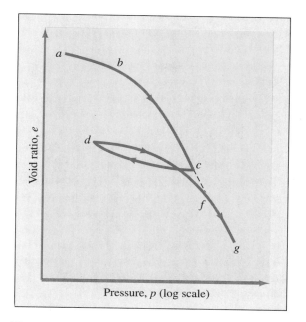

▼ **FIGURE 8.10** Plot of e against log p showing loading, unloading, and reloading branches

2. *Overconsolidated*, whose present effective overburden pressure is less than that which the soil has experienced in the past. The maximum effective past pressure is called the *preconsolidation pressure*.

Casagrande (1936) suggested a simple graphic construction to determine the preconsolidation pressure, p_c, from the laboratory e-log p plot. The procedure is as follows (see Figure 8.11):

1. By visual observation, establish point a at which the e-log p plot has a minimum radius of curvature.
2. Draw a horizontal line ab.
3. Draw the line ac tangent at a.
4. Draw the line ad, which is the bisector of the angle bac.
5. Project the straight-line portion gh of the e-log p plot back to intersect ad at f. The abscissa of point f is the preconsolidation pressure, p_c.

The overconsolidation ratio (*OCR*) for a soil can now be defined as

$$OCR = \frac{p_c}{p}$$

where p_c = preconsolidation pressure of a specimen
p = present effective vertical pressure

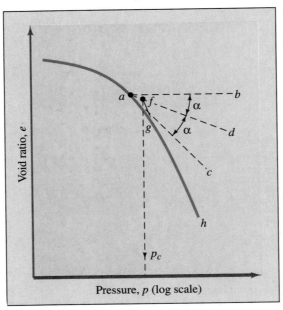

Void ratio, e

Pressure, p (log scale)

▼   **FIGURE 8.11**   Graphic procedure for determining preconsolidation pressure

8.5 EFFECT OF DISTURBANCE ON VOID RATIO-PRESSURE RELATIONSHIP

A soil specimen will be remolded when it is subjected to some degree of disturbance. This will affect the void ratio-pressure relationship for the soil. For a normally consolidated clayey soil of low to medium sensitivity (Figure 8.12) under an effective overburden pressure of p_o and with a void ratio of e_o, the change in the void ratio with an increase of pressure in the field will be roughly as shown by curve 1. This is the *virgin compression curve*, which is approximately a straight line on a semilogarithmic plot. However, the laboratory consolidation curve for a fairly undisturbed specimen of the same soil (curve 2) will be located to the left of curve 1. If the soil is completely remolded and a consolidation test is conducted on it, the general position of the e-log p plot will be represented by curve 3. Curves 1, 2, and 3 will intersect approximately at a void ratio of $e = 0.4e_o$ (Terzaghi and Peck, 1967).

For an overconsolidated clayey soil of low to medium sensitivity that has been subjected to a preconsolidation pressure of p_c (Figure 8.13) and for which the present effective overburden pressure and the void ratio are p_o and e_o, respectively, the field consolidation curve will take a path represented approximately by *cbd*. Note that *bd* is a part of the virgin compression curve. The laboratory consolidation test results on a specimen subjected to moderate disturbance will be represented by curve 2. Schmertmann (1953) concluded that the slope of line *cb*, which is the field recompression path, has approximately the same slope as the laboratory rebound curve *fg*. The empirical procedure to estimate void ratio-pressure relationships is given in Appendix E.

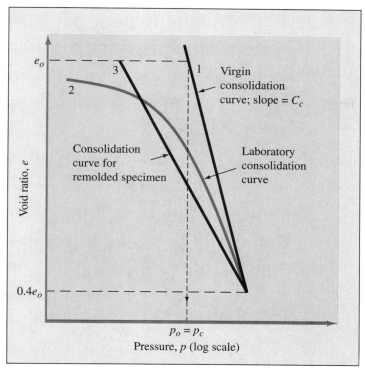

▼ **FIGURE 8.12**
Consolidation characteristics
of normally consolidated
clay of low to medium
sensitivity

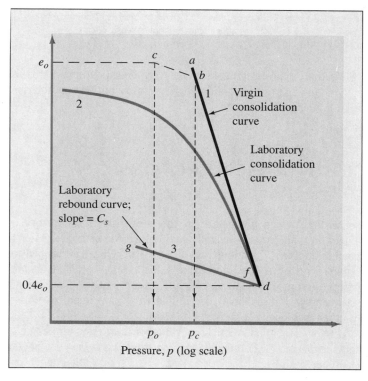

▼ **FIGURE 8.13**
Consolidation characteristics
of overconsolidated clay of
low to medium sensitivity

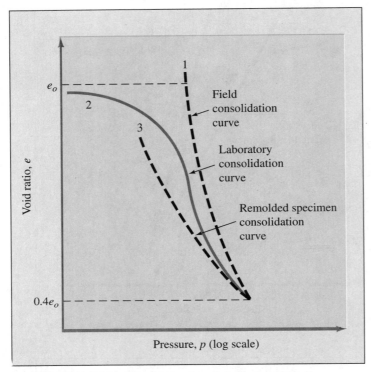

▼ FIGURE 8.14   Consolidation characteristics of sensitive clays

Soils that exhibit high sensitivity have flocculent structures. In the field, they are generally somewhat overconsolidated. The consolidation characteristics of such soils are shown in Figure 8.14.

8.6 INFLUENCE OF OTHER FACTORS ON e-log p RELATIONSHIP

In Section 8.2, it was noted that in the conventional laboratory consolidation test, a given load on a specimen is usually maintained for 24 hours. After that, the load on the specimen is doubled. Questions may arise about what will happen to the e-log p curve if (a) a given load on a specimen is kept for a time $t \neq 24$ hours, and (b) other factors remaining the same, the load increment ratio $\Delta p/p$ (Δp = load increment per unit area of specimen and p = initial load per unit area of specimen) on the specimen is kept at values other than one.

Crawford (1964) conducted several laboratory tests on Leda clay in which the load on the specimens was doubled each time (that is, $\Delta p/p = 1$); however, the duration of each load maintained on the specimens was varied. The e-log p curves obtained from

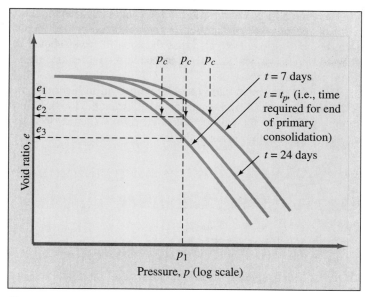

▼   **FIGURE 8.15**   Effect of load duration on *e*-log *p* curve

such tests are shown in Figure 8.15. From this plot, it may be seen that when the duration of the load maintained on a specimen is increased, the *e*-log *p* curve gradually moves to the left. This means that, for a given load per unit area on the specimen (p), the void ratio at the end of consolidation will decrease as the time t is increased. For example, in Figure 8.15, at $p = p_1$, $e = e_2$ for $t = 24$ hours and $e = e_3$ for $t = 7$ days; however, $e_3 < e_2$.

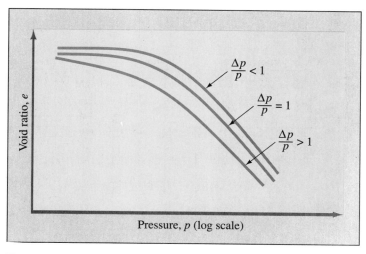

▼   **FIGURE 8.16**   Effect of load increment ratio on *e*-log *p* curve

The reason for such variation in the e-log p curve is that as time t is increased, the amount of secondary consolidation of the specimen is also increased. This will tend to reduce the void ratio e. Note also that the e-log p curves shown in Figure 8.15 will give slightly different values for the preconsolidation pressure (p_c). The value of p_c will increase with the decrease of t.

The load increment ratio $(\Delta p/p)$ also has an influence on the e-log p curves. This was discussed in detail by Leonards and Altschaeffl (1964). Figure 8.16 shows the variation of e with log p for various values of $\Delta p/p$. When $\Delta p/p$ is gradually increased, the e-log p curve gradually moves to the left.

8.7 CALCULATION OF SETTLEMENT FROM ONE-DIMENSIONAL PRIMARY CONSOLIDATION

With the knowledge gained from the analysis of consolidation test results, we can now proceed to calculate the probable settlement caused by primary consolidation in the field, assuming one-dimensional consolidation.

Let us consider a saturated clay layer of thickness H and cross-sectional area A under an existing average effective overburden pressure p_o. Because of an increase of pressure, Δp, let the primary settlement be S. Thus, the change in volume (Figure 8.17) can be given by

$$\Delta V = V_0 - V_1 = HA - (H - S)A = SA \tag{8.14}$$

where V_0 and V_1 are the initial and final volumes, respectively. However, the change in the total volume is equal to the change in the volume of voids, ΔV_v. Thus,

$$\Delta V = SA = V_{v0} - V_{v1} = \Delta V_v \tag{8.15}$$

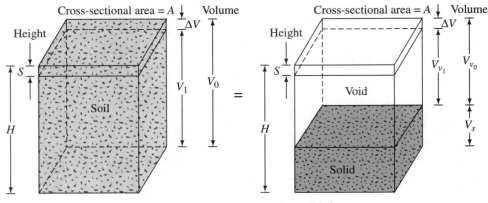

▼ **FIGURE 8.17**   Settlement caused by one-dimensional consolidation

where V_{v0} and V_{v1} are the initial and final void volumes, respectively. From the definition of void ratio,

$$\Delta V_v = \Delta e V_s \tag{8.16}$$

where Δe = change of void ratio. But

$$V_s = \frac{V_0}{1 + e_o} = \frac{AH}{1 + e_o} \tag{8.17}$$

where e_o = initial void ratio at volume V_0. Thus, from Eqs. (8.14), (8.15), (8.16), and (8.17),

$$\Delta V = SA = \Delta e V_s = \frac{AH}{1 + e_o} \Delta e$$

or

$$S = H \frac{\Delta e}{1 + e_o} \tag{8.18}$$

For normally consolidated clays that exhibit a linear e-log p (Figure 8.12) relationship,

$$\Delta e = C_c[\log (p_o + \Delta p) - \log p_o] \tag{8.19}$$

where C_c = slope of the e-log p plot and is defined as the compression index. Substitution of Eq. (8.19) in Eq. (8.18) gives

$$S = \frac{C_c H}{1 + e_o} \log \left(\frac{p_o + \Delta p}{p_o}\right) \tag{8.20}$$

For a thicker clay layer, it is more accurate if the layer is divided into a number of sublayers and calculations for settlement are made separately for each sublayer. Thus, the total settlement for the entire layer can be given as

$$S = \sum \left[\frac{C_c H_i}{1 + e_o} \log \left(\frac{p_{o(i)} + \Delta p_{(i)}}{p_{o(i)}}\right)\right]$$

where H_i = thickness of sublayer i
   $p_{o(i)}$ = initial average effective overburden pressure for sublayer i
   $\Delta p_{(i)}$ = increase of vertical pressure for sublayer i

In overconsolidated clays (Figure 8.13), for $p_o + \Delta p \le p_c$ field e-log p variation will be along the line cb, the slope of which will be approximately equal to that for the laboratory rebound curve. The slope of the rebound curve, C_s, is referred to as the *swell index*, so

$$\Delta e = C_s[\log (p_o + \Delta p) - \log p_o] \tag{8.21}$$

From Eqs. (8.18) and (8.21),

$$S = \frac{C_s H}{1 + e_o} \log\left(\frac{p_o + \Delta p}{p_o}\right)$$

(8.22)

If $p_o + \Delta p > p_c$, then

$$S = \frac{C_s H}{1 + e_o} \log\frac{p_c}{p_o} + \frac{C_c H}{1 + e_o} \log\left(\frac{p_o + \Delta p}{p_c}\right)$$

(8.23)

However, if the e-log p curve is given, it is possible ~~simply~~ to pick Δe off the plot for the appropriate range of pressures. This figure may be substituted into Eq. (8.18) for the calculation of settlement, S.

8.8 COMPRESSION INDEX (C_c)

The compression index for the calculation of field settlement caused by consolidation can be determined by graphic construction (as shown in Figure 8.12) after obtaining laboratory test results for void ratio and pressure.

Terzaghi and Peck (1967) suggested the following empirical expressions for compression index:

For undisturbed clays:

$$C_c = 0.009(LL - 10)$$

(8.24)

For remolded clays:

$$C_c = 0.007(LL - 10)$$

(8.25)

where LL = liquid limit, in percent.

In the absence of laboratory consolidation data, Eq. (8.24) is often used for an approximate calculation of primary consolidation in the field.

Several other correlations for the compression index are also available now. They have been developed by tests on various clays. Some of these correlations are given in Section E.2 (Appendix E).

Based on observations on several natural clays, Rendon–Herrero (1983) gave the relationship for the compression index in the form

$$C_c = 0.141 G_s^{1.2} \left(\frac{1 + e_0}{G_s} \right)^{2.38}$$

(8.26)

Nagaraj and Murty (1985) expressed the compression index as

$$C_c = 0.2343 \left[\frac{LL(\%)}{100} \right] G_s$$

(8.27)

8.9 SWELL INDEX (C_s)

The swell index is appreciably smaller in magnitude than the compression index and can generally be determined from laboratory tests. In most cases,

$$C_s \simeq \frac{1}{5} \quad \text{to} \quad \frac{1}{10} C_c$$

(8.28)

The swell index was expressed by Nagaraj and Murty (1985) as

$$C_s = 0.0463 \left[\frac{LL(\%)}{100} \right] G_s$$

(8.29)

Typical values of the liquid limit, plastic limit, virgin compression index, and swell index for some natural soils are given in Table 8.1.

▼ **TABLE 8.1 Compression and Swell of Natural Soils**

Soil	Liquid limit	Plastic limit	Compression index, C_c	Swell index, C_s
Boston Blue clay	41	20	0.35	0.07
Chicago clay	60	20	0.4	0.07
Ft. Gordon clay, Georgia	51	26	0.12	—
New Orleans clay	80	25	0.3	0.05
Montana clay	60	28	0.21	0.05

▼ **EXAMPLE 8.2**

Refer to the e-log p curve obtained in Example 8.1.

    a. Determine the preconsolidation pressure, p_c.
    b. Find the compression index, C_c.

Solution

a. The e-log p plot shown in Figure 8.9 has been replotted in Figure 8.18. Using the procedure shown in Figure 8.11, we determine the preconsolidation pressure. From the plot, $p_c = \mathbf{1.6\ ton/ft^2}$.

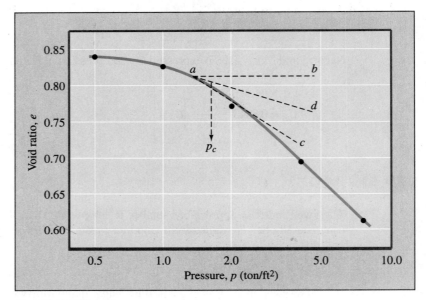

▼   **FIGURE 8.18**

b. From the e-log p plot,

$$p_2 = 8\ \text{ton/ft}^2 \qquad e_2 = 0.612$$
$$p_1 = 4\ \text{ton/ft}^2 \qquad e_1 = 0.696$$

So

$$C_c = \frac{e_1 - e_2}{\log (p_2/p_1)} = \frac{0.696 - 0.612}{\log (8/4)} = \mathbf{0.279} \qquad ▼$$

▼   **EXAMPLE 8.3**

Refer to Examples 8.1 and 8.2. For the clay, what will be the void ratio for a pressure of 10 ton/ft²? (*Note:* $p_c = 1.6$ ton/ft².)

Solution   From Example 8.1, we find the following values:

$$e_1 = 0.696 \qquad p_1 = 4\ \text{ton/ft}^2$$
$$e_2 = 0.612 \qquad p_2 = 8\ \text{ton/ft}^2$$

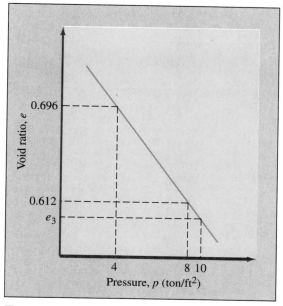

▼ FIGURE 8.19

Also, from Example 8.2, $C_c = 0.279$. Referring to Figure 8.19, we have

$$C_c = \frac{e_1 - e_3}{\log p_3 - \log p_1}$$

$$e_3 = e_1 - C_c \log\left(\frac{p_3}{p_1}\right) = 0.696 - 0.279 \log\left(\frac{10}{4}\right) = \mathbf{0.585} \qquad \blacktriangledown$$

▼ EXAMPLE 8.4

A soil profile is shown in Figure 8.20. If a uniformly distributed load Δp is applied at the ground surface, what will be the settlement of the clay layer caused by primary consolidation? We are given that p_c for the clay is 2600 lb/ft^2 and $C_s = \frac{1}{6}C_c$.

Solution   The average effective stress at the middle of the clay layer is

$$p_O = 8\gamma_{dry(sand)} + (23 - 8)[\gamma_{sat(sand)} - \gamma_w] + \left(\frac{17}{2}\right)[\gamma_{sat(clay)} - \gamma_w]$$

or

$$p_O = (8)(105) + (15)(120 - 62.4) + (8.5)(122.4 - 62.4)$$

$$= 2214 \text{ lb/ft}^2$$

$$p_c = 2600 \text{ lb/ft}^2 > 2214 \text{ lb/ft}^2$$

$$p_O + \Delta p = 2214 + 1000 = 3214 \text{ lb/ft}^2 > p_c$$

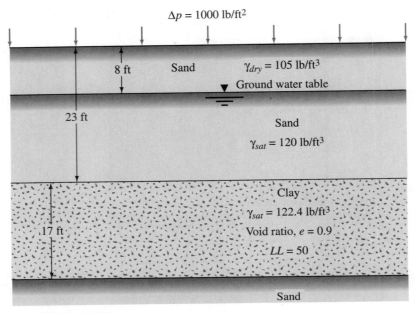

$\Delta p = 1000$ lb/ft^2

8 ft   Sand $\gamma_{dry} = 105$ lb/ft^3

Ground water table

23 ft

Sand

$\gamma_{sat} = 120$ lb/ft^3

Clay

$\gamma_{sat} = 122.4$ lb/ft^3

17 ft

Void ratio, $e = 0.9$

$LL = 50$

Sand

▼   **FIGURE 8.20**

So Eq. (8.23) needs to be used:

$$S = \frac{C_s H}{1 + e_o} \log \left(\frac{p_c}{p_o}\right) + \frac{C_c H}{1 + e_o} \log \left(\frac{p_o + \Delta p}{p_c}\right)$$

We have $H = 17$ ft and $e_o = 0.9$. From Eq. (8.24),

$$C_c = 0.009(LL - 10) = 0.009(50 - 10) = 0.36$$

$$C_s = \tfrac{1}{6}C_c = \frac{0.36}{6} = 0.06$$

Thus,

$$S = \frac{17}{1 + 0.9} \left[0.06 \log \left(\frac{2600}{2214}\right) + 0.36 \log \left(\frac{2214 + 1000}{2600}\right) \right]$$

$$= 0.334 \text{ ft} \approx \textbf{4 in.} \quad \blacktriangledown$$

▼   **EXAMPLE 8.5**

A soil profile is shown in Figure 8.21a. Laboratory consolidation tests were conducted on a sample collected from the middle of the clay layer. The field consolidation curve interpolated from the laboratory test results (as shown in Figure 8.13) is shown in Figure 8.21b. Calculate the settlement in the field caused by primary consolidation for a surcharge of 48 kN/m^2 applied at the ground surface.

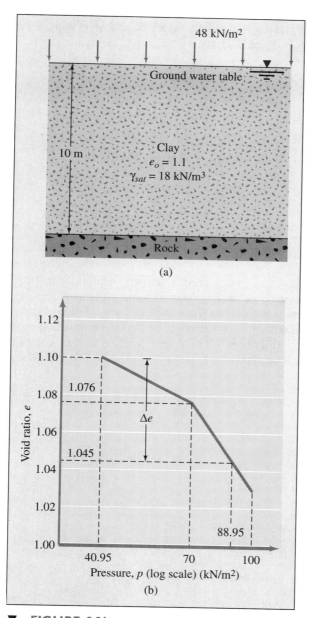

▼ **FIGURE 8.21** (a) Soil profile; (b) field consolidation curve

Solution

$$p_o = (5)(\gamma_{sat} - \gamma_w) = 5(18.0 - 9.81)$$
$$= 40.95 \text{ kN/m}^2$$
$$e_o = 1.1$$
$$\Delta p = 48 \text{ kN/m}^2$$
$$p_o + \Delta p = 40.95 + 48 = 88.95 \text{ kN/m}^2$$

The void ratio corresponding to 88.95 kN/m² (Figure 8.21b) is 1.045. Hence, $\Delta e = 1.1 - 1.045 = 0.055$. We have

settlement, $S = H \dfrac{\Delta e}{1 + e_o}$ [Eq. (8.18)]

so

$$S = 10 \, \frac{(0.055)}{1 + 1.1} = 0.262 \text{ m} = \mathbf{262 \text{ mm}} \qquad \blacktriangledown$$

8.10 SETTLEMENT FROM SECONDARY CONSOLIDATION

Section 8.2 showed that at the end of primary consolidation (that is, after complete dissipation of excess pore water pressure) some settlement is observed because of the plastic adjustment of soil fabrics. This stage of consolidation is called *secondary consolidation*. During secondary consolidation, the plot of deformation against the log of time is practically linear (Figure 8.6). The variation of the void ratio e with time t for a given load increment will be similar to that shown in Figure 8.6. This is shown in Figure 8.22.

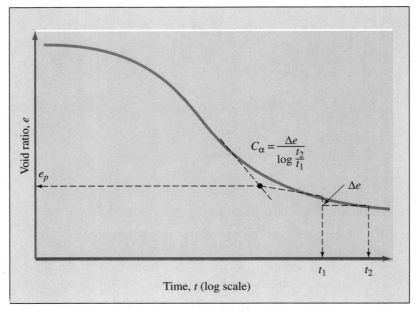

▼ **FIGURE 8.22**   Variation of e with log t under a given load increment, and definition of secondary consolidation index

The secondary compression index can be defined from Figure 8.22 as

$$C_\alpha = \frac{\Delta e}{\log t_2 - \log t_1} = \frac{\Delta e}{\log (t_2/t_1)} \tag{8.30}$$

where C_α = secondary compression index
Δe = change of void ratio
t_1, t_2 = time

The magnitude of the secondary consolidation can be calculated as

$$S_s = C'_\alpha H \log \left(\frac{t_1}{t_2}\right) \tag{8.31}$$

where

$$C'_\alpha = \frac{C_\alpha}{1 + e_p} \tag{8.32}$$

e_p = void ratio at the end of primary consolidation (Figure 8.22)
H = thickness of clay layer

The general magnitudes of C'_α as observed in various natural deposits are given in Figure 8.23.

Secondary consolidation settlement is more important than primary consolidation in organic and highly compressible inorganic soils. In overconsolidated inorganic clays, the secondary compression index is very small and of less practical significance.

There are several factors that might affect the magnitude of secondary consolidation, some of which are not very clearly understood (Mesri, 1973). The ratio of secondary to primary compression for a given thickness of soil layer is dependent on the ratio of the stress increment (Δp) to the initial effective stress (p). For small $\Delta p/p$ ratios, the secondary-to-primary compression ratio is larger.

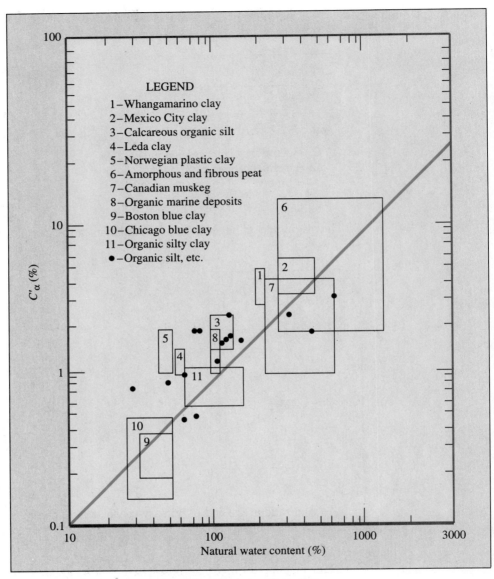

▼   FIGURE 8.23 C'_α for natural soil deposits (after Mesri, 1973)

Refer to Example 8.4. Assume that the primary consolidation will be complete in 3.5 years. Estimate the secondary consolidation that would occur from 3.5 years to 10 years after the load application. Given $C_\alpha = 0.022$, what is the total consolidation settlement after 10 years?

Solution   From Eq. (8.32),

$$C'_\alpha = \frac{C_\alpha}{1 + e_p}$$

The value of e_p can be calculated as

$$e_p = e_0 - \Delta e_{primary}$$

Combining Eqs. (8.18) and (8.23),

$$\Delta e = C_s \log \left(\frac{p_c}{p_0}\right) + C_s \log \left(\frac{p_0 + \Delta p}{p_c}\right)$$

$$= 0.06 \log \left(\frac{2600}{2214}\right) + 0.36 \log \left(\frac{3214}{2600}\right)$$

$$= 0.0042 + 0.033 = 0.0372$$

We are given $e_0 = 0.9$, so

$$e_p = 0.9 - 0.0372 = 0.8628$$

Hence,

$$C'_\alpha = \frac{0.022}{1 + 0.8628} = 0.012$$

$$S_\alpha = C'_\alpha H \log \left(\frac{t_2}{t_1}\right) = (0.012)(17 \times 12) \log \left(\frac{10}{3.5}\right) \approx 1.12 \text{ in.}$$

Total consolidation settlement = primary consolidation settlement (S) + secondary consolidation settlement (S_s). From Example 8.4, $S = 4.0$ in., so

total consolidation settlement = $4.0 + 1.12 =$ **5.12 in.**   ▼

8.11   TIME RATE OF CONSOLIDATION

The total settlement caused by primary consolidation resulting from an increase in the stress on a soil layer can be calculated by the use of one of the three equations: [(8.20), (8.22), or (8.23)] given in Section 8.7. However, they do not provide any information regarding the rate of primary consolidation. Terzaghi (1925) proposed the first theory to consider the rate of one-dimensional consolidation for saturated clay soils. The mathematical derivations are based on the following assumptions (also see Taylor, 1948):

1. The clay-water system is homogeneous.
2. Saturation is complete.
3. Compressibility of water is negligible.
4. Compressibility of soil grains is negligible (but soil grains rearrange).
5. The flow of water is in one direction only (that is, in the direction of compression).
6. Darcy's law is valid.

Figure 8.24a shows a layer of clay of thickness $2H_{dr}$ that is located between two highly permeable sand layers. If the clay layer is subjected to an increased pressure of Δp, the pore water pressure at any point A in the clay layer will increase. For one-dimensional consolidation, water will be squeezed out in the vertical direction toward the sand layer.

Figure 8.24b shows the flow of water through a prismatic element at A. For the soil element shown,

$$\text{rate of outflow of water} - \text{rate of inflow of water} = \text{rate of volume change}$$

Thus,

$$\left(v_z + \frac{\partial v_z}{\partial z}\,dz\right)dx\,dy - v_z\,dx\,dy = \frac{\partial V}{\partial t}$$

where V = volume of the soil element
       v_z = velocity of flow in z direction

or

$$\frac{\partial v_z}{\partial z}\,dx\,dy\,dz = \frac{\partial V}{\partial t} \tag{8.33}$$

Using Darcy's law, we have

$$v_z = ki = -k\,\frac{\partial h}{\partial z} = -\frac{k}{\gamma_w}\frac{\partial u}{\partial z} \tag{8.34}$$

where u = excess pore water pressure caused by the increase of stress.
        From Eqs. (8.33) and (8.34),

$$-\frac{k}{\gamma_w}\frac{\partial^2 u}{\partial z^2} = \frac{1}{dx\,dy\,dz}\frac{\partial V}{\partial t} \tag{8.35}$$

During consolidation, the rate of change in the volume of the soil element is equal to the rate of change in the volume of voids. So

$$\frac{\partial V}{\partial t} = \frac{\partial V_v}{\partial t} = \frac{\partial(V_s + eV_s)}{\partial t} = \frac{\partial V_s}{\partial t} + V_s\frac{\partial e}{\partial t} + e\frac{\partial V_s}{\partial t} \tag{8.36}$$

where V_s = volume of soil solids
       V_v = volume of voids

But (assuming that soil solids are incompressible),

$$\frac{\partial V_s}{\partial t} = 0$$

and

$$V_s = \frac{V}{1 + e_o} = \frac{dx\,dy\,dz}{1 + e_o}$$

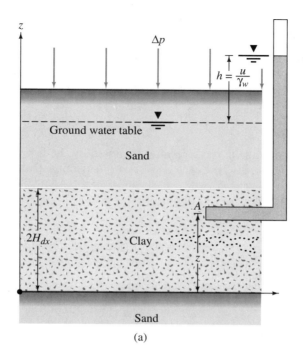

(a)

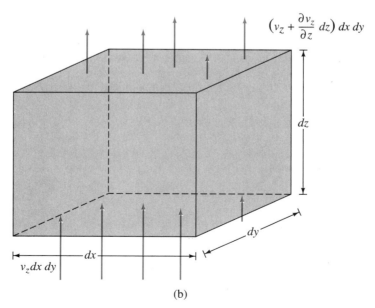

(b)

▼   **FIGURE 8.24**   (a) Clay layer undergoing consolidation; (b) flow of water at A during consolidation

Substitution for $\partial V_s/\partial t$ and V_s in Eq. (8.36) yields

$$\frac{\partial V}{\partial t} = \frac{dx \, dy \, dz}{1 + e_o} \frac{\partial e}{\partial t} \tag{8.37}$$

where e_o = initial void ratio.

Combining Eqs. (8.35) and (8.37) gives

$$-\frac{k}{\gamma_w} \frac{\partial^2 u}{\partial z^2} = \frac{1}{1 + e_o} \frac{\partial e}{\partial t} \tag{8.38}$$

The change in the void ratio is caused by the increase of effective stress (that is, decrease of excess pore water pressure). Assuming that they are linearly related, we have

$$\partial e = a_v \, \partial(\Delta p') = -a_v \, \partial u \tag{8.39}$$

where $\partial(\Delta p')$ = change in effective pressure

a_v = coefficient of compressibility (a_v can be considered to be constant for a narrow range of pressure increase)

Combining Eqs. (8.38) and (8.39) gives

$$-\frac{k}{\gamma_w} \frac{\partial^2 u}{\partial z^2} = -\frac{a_v}{1 + e_o} \frac{\partial u}{\partial t} = -m_v \frac{\partial u}{\partial t}$$

where m_v = coefficient of volume compressibility = $a_v/(1 + e_o)$, or

$$\frac{\partial u}{\partial t} = c_v \frac{\partial^2 u}{\partial z^2} \tag{8.40}$$

where c_v = coefficient of consolidation = $k/(\gamma_w m_v)$.

Eq. (8.40) is the basic differential equation of Terzaghi's consolidation theory and can be solved with the following boundary conditions:

$$z = 0, \quad u = 0$$

$$z = 2H_{dr}, \quad u = 0$$

$$t = 0, \quad u = u_o$$

The solution yields

$$u = \sum_{m=0}^{m=\infty} \left[\frac{2u_o}{M} \sin\left(\frac{Mz}{H_{dr}}\right) \right] e^{-M^2 T_v} \tag{8.41}$$

where m is an integer

$$M = \frac{\pi}{2}(2m + 1)$$

u_o = initial excess pore water pressure

$$T_v = \frac{c_v t}{H_{dr}^2} = \text{time factor}$$

The time factor is a nondimensional number.

Because consolidation progresses by the dissipation of excess pore water pressure, the degree of consolidation at a distance z at any time t is

$$U_z = \frac{u_O - u_z}{u_O} = 1 - \frac{u_z}{u_O} \tag{8.42}$$

where u_z = excess pore water pressure at time t.

Equations (8.41) and (8.42) can be combined to obtain the degree of consolidation at any depth z. This is shown in Figure 8.25.

The average degree of consolidation for the entire depth of the clay layer at any time t can be written from Eq. (8.42) as

$$U = \frac{S_t}{S} = 1 - \frac{\left(\dfrac{1}{2H_{dr}}\right) \displaystyle\int_0^{2H_{dr}} u_z \, dz}{u_O} \tag{8.43}$$

where U = average degree of consolidation
  S_t = settlement of the layer at time t
  S = ultimate settlement of the layer from primary consolidation

Substitution of the expression for excess pore water pressure, u_z, given in Eq. (8.42) into Eq. (8.43) gives

$$U = I - \sum_{m=0}^{m=\infty} \frac{2}{M^2} e^{-M^2 T_v} \tag{8.44}$$

The variation in the average degree of consolidation with the nondimensional time factor, T_v, is given in Table 8.2, which represents the case where u_O is the same for the entire depth of the consolidating layer (see also Figure 8.26).

Table 8.3 gives the values of T_v for linear variation of the initial excess pore water pressure in the clay layer with drainage in one direction.

The values of the time factor and their corresponding average degrees of consolidation for the case presented in Table 8.2 may also be approximated by the following simple relationship:

$$\text{For } U = 0\% \text{ to } 60\%, \quad T_v = \frac{\pi}{4}\left(\frac{U\%}{100}\right)^2 \tag{8.45}$$

$$\text{For } U > 60\%, \quad T_v = 1.781 - 0.933 \log (100 - U\%) \tag{8.46}$$

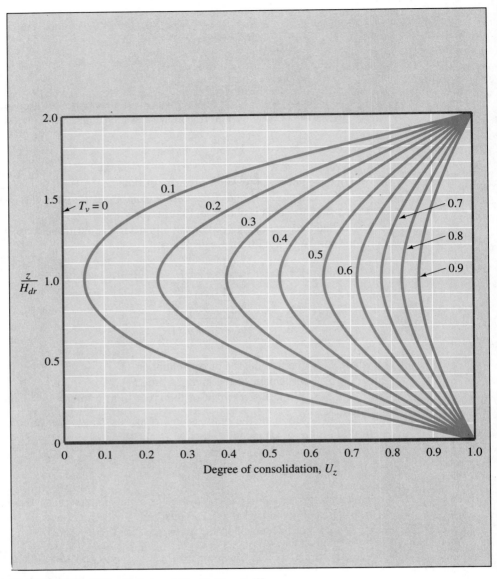

▼ **FIGURE 8.25**   Variation of U_z with T_v and z/H_{dr}

▼ **TABLE 8.2** **Variation of Time Factor with Degree of Consolidation**[a]

Degree of consolidation, $U\%$	Time factor, T_v
0	0
10	0.008
20	0.031
30	0.071
40	0.126
50	0.197
60	0.287
70	0.403
80	0.567
90	0.848
100	∞

[a] u_o is constant with depth.

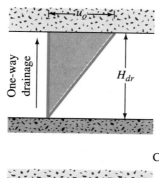

Different types of drainage with u_o constant

▼ **TABLE 8.3** **Variation of Time Factor with Degree of Consolidation**

Degree of consolidation, $U\%$	Time factor, T_v	
	Case I	Case II
0	0	0
10	0.003	0.047
20	0.009	0.100
30	0.024	0.158
40	0.048	0.221
50	0.092	0.294
60	0.160	0.383
70	0.271	0.500
80	0.440	0.665
90	0.720	0.940
100	∞	∞

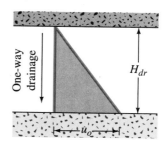

Case I

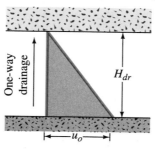

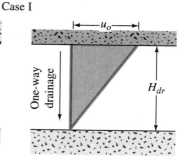

Case II

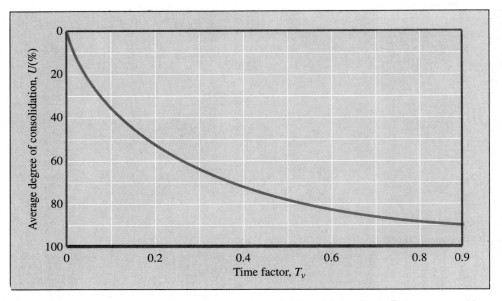

▼ **FIGURE 8.26**   Variation of average degree of consolidation with time factor, T_v (u_O constant with depth)

Sivaram and Swamee (1977) gave the following empirical relationships for U and T_v (for U varying from 0% to 100%):

$$\frac{U\%}{100} = \frac{\left(\dfrac{4T_v}{\pi}\right)^{0.5}}{\left[1 + \left(\dfrac{4T_v}{\pi}\right)^{2.8}\right]^{0.179}}$$

(8.47)

and

$$T_v = \frac{\left(\dfrac{\pi}{4}\right)\left(\dfrac{U\%}{100}\right)^2}{\left[1 - \left(\dfrac{U\%}{100}\right)^{5.6}\right]^{0.357}}$$

(8.48)

8.12 COEFFICIENT OF CONSOLIDATION

The coefficient of consolidation, c_v, generally decreases as the liquid limit of soil increases. The range of variation of c_v for a given liquid limit of soil is rather wide.

For a given load increment on a specimen, there are two commonly used graphic methods for determining c_v from laboratory one-dimensional consolidation tests. One of them is the *logarithm-of-time method* proposed by Casagrande and Fadum (1940), and the other is the *square-root-of-time method* suggested by Taylor (1942). The general procedures for obtaining c_v by the two methods are given below.

Sridharan and Prakash (1985) proposed a semi-empirical approach to determine the coefficient of consolidation for a given load increment. This is called the *hyperbola method*. The procedure for obtaining c_v according to the hyperbola method is also described below.

Logarithm-of-Time Method

For a given incremental loading of the laboratory test, the specimen deformation against log-of-time plot is shown in Figure 8.27. The following constructions are needed to determine c_v:

1. Extend the straight-line portions of primary and secondary consolidations to intersect at A. The ordinate of A is represented by d_{100}—that is, the deformation at the end of 100% primary consolidation.

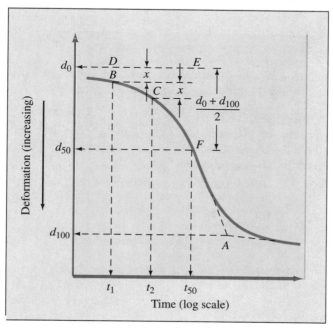

▼ **FIGURE 8.27** Logarithm-of-time method for determining coefficient of consolidation

2. The initial curved portion of the plot of deformation versus log t is approximated to be a parabola on the natural scale. Select times t_1 and t_2 on the curved portion such that $t_2 = 4t_1$. Let the difference of sample deformation during time $(t_2 - t_1)$ be equal to x.

3. Draw a horizontal line DE such that the vertical distance BD is equal to x. The deformation corresponding to the line DE is d_0 (that is, deformation at 0% consolidation).

4. The ordinate of point F on the consolidation curve represents the deformation at 50% primary consolidation, and its abscissa represents the corresponding time (t_{50}).

5. For 50% average degree of consolidation, $T_v = 0.197$ (Table 8.2);

$$T_{50} = \frac{c_v t_{50}}{H_{dr}^2}$$

or

$$c_v = \frac{0.197 H_{dr}^2}{t_{50}} \tag{8.49}$$

where H_{dr} = average longest drainage path during consolidation.

For specimens drained at both top and bottom, H_{dr} equals one-half of the average height of the specimen during consolidation. For specimens drained on only one side, H_{dr} equals the average height of the specimen during consolidation.

Square-Root-of-Time Method

In this method, a plot of deformation against the square root of time is made for the incremental loading (Figure 8.28). Other graphic constructions required are as follows:

1. Draw a line AB through the early portion of the curve.

2. Draw a line AC such that $\overline{OC} = 1.15\,\overline{OB}$. The abscissa of point D, which is the intersection of AC and the consolidation curve, gives the square root of time for 90% consolidation $(\sqrt{t_{90}})$.

3. For 90% consolidation, $T_{90} = 0.848$ (Table 8.2), so

$$T_{90} = 0.848 = \frac{c_v t_{90}}{H_{dr}^2}$$

or

$$c_v = \frac{0.848 H_{dr}^2}{t_{90}} \tag{8.50}$$

H_{dr} in Eq. (8.50) is determined in a manner similar to the logarithm-of-time method.

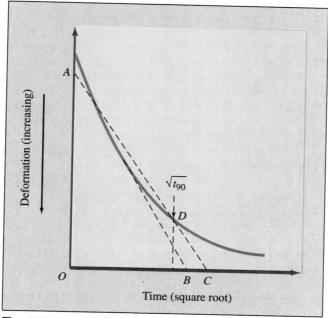

FIGURE 8.28 Square-root-of-time fitting method

Hyperbola Method

In the hyperbola method, the following procedure is recommended for the determination of c_v:

1. Obtain the time (t) and the specimen deformation (ΔH) from the laboratory consolidation test.
2. Plot the graph of $t/\Delta H$ against t as shown in Figure 8.29.
3. Identify the straight-line portion bc and project it back to point d. Determine the intercept D.
4. Determine the slope m of the line bc.
5. Calculate c_v as

$$c_v = 0.3\left(\frac{mH_{dr}^2}{D}\right)$$

(8.51)

Note that the unit of D is time/length and the unit of m is (time/length)/time = 1/length, so the unit of c_v is

$$\frac{\left(\dfrac{1}{\text{length}}\right)(\text{length})^2}{\left(\dfrac{\text{time}}{\text{length}}\right)} = \frac{(\text{length})^2}{\text{time}}$$

The hyperbola method is fairly simple to use, and it gives good results for $U = 60\%$ to 90%.

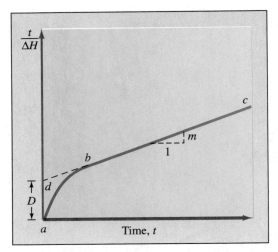

▼ **FIGURE 8.29**   Hyperbola method for determination of c_v

---

▼ **EXAMPLE 8.7**

A soil profile is shown in Figure 8.30. A surcharge load of 2500 lb/ft² is applied on the ground surface.

    a.   How high will the water rise in the piezometer immediately after the application of load?

    b.   What is the degree of consolidation at point A when $h = 20$ ft?

    c.   Find h when the degree of consolidation at A is 60%.

Solution

    a.   Assuming a uniform increase in the initial excess pore water pressure through the 12-ft depth of the clay layer, we have

$$u_O = \Delta p = 2500 \text{ lb/ft}^2$$

$$h = \frac{2500}{62.4} = \textbf{40.06 ft}$$

    b. $U_A\% = \left(1 - \frac{u_A}{u_O}\right)100 = \left(1 - \frac{20 \times 62.4}{40.06 \times 62.4}\right)100 = \textbf{50.1\%}$

    c. $U_A = 0.6 = \left(1 - \frac{u_A}{u_O}\right)$

    or

$$0.6 = \left(1 - \frac{u_A}{2500}\right)$$

$$u_A = (1 - 0.6)2500 = 1000 \text{ lb/ft}^2$$

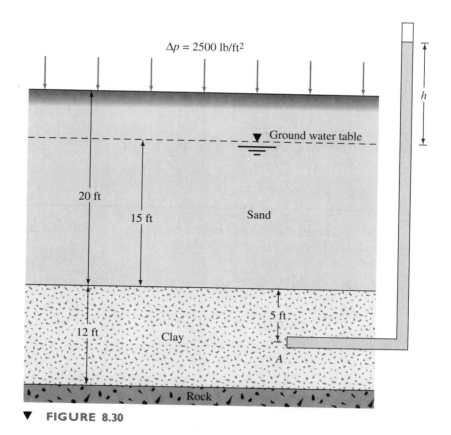

$\Delta p = 2500 \text{ lb/ft}^2$

Ground water table

20 ft

15 ft

Sand

h

12 ft   Clay

5 ft

A

Rock

▼ **FIGURE 8.30**

Hence

$$h = \frac{1000}{62.4} = \textbf{16.03 ft} \qquad \blacktriangledown$$

▼ **EXAMPLE 8.8**

The time required for 50% consolidation of a 25-mm-thick clay layer (drained at both top and bottom) in the laboratory is 2 min 20 sec. How long (in days) will it take for a 3-m-thick clay layer of the same clay in the field under the same pressure increment to reach 50% consolidation? In the field, there is a rock layer at the bottom of the clay.

Solution

$$T_{50} = \frac{c_v t_{\text{lab}}}{H^2_{dr(\text{lab})}} = \frac{c_v t_{\text{field}}}{H^2_{dr(\text{field})}}$$

or

$$\frac{t_{lab}}{H^2_{dr(lab)}} = \frac{t_{field}}{H^2_{dr(field)}}$$

$$\frac{140 \text{ sec}}{\left(\dfrac{0.025 \text{ m}}{2}\right)^2} = \frac{t_{field}}{(3 \text{ m})^2}$$

$$t_{field} = 8{,}064{,}000 \text{ sec} = \textbf{93.33 days} \qquad \blacktriangledown$$

▼ **EXAMPLE 8.9**

Refer to Example 8.8. How long (in days) will it take in the field for 30% primary consolidation to occur? Use Eq. (8.45).

Solution   From Eq. (8.45),

$$\frac{c_v t_{field}}{H^2_{dr(lab)}} = T_v \propto U^2$$

So

$$t \propto U^2$$

$$\frac{t_1}{t_2} = \frac{U^2_1}{U^2_2}$$

or

$$\frac{93.33 \text{ days}}{t_2} = \frac{50^2}{30^2}$$

$$t_2 = \textbf{33.6 days} \qquad \blacktriangledown$$

▼ **EXAMPLE 8.10**

For a normally consolidated clay,

$$p_o = 2 \text{ ton/ft}^2 \qquad e = e_o = 1.22$$

$$p_o + \Delta p = 4 \text{ ton/ft}^2 \qquad e = 0.98$$

The coefficient of permeability, k, of the clay for the loading range is 2.0×10^{-4} ft/day.

    a. How long (in days) will it take for a 12-ft-thick clay layer (drained on one side) in the field to reach 60% consolidation?

    b. What is the settlement at that time (that is, at 60% consolidation)?

Solution

a. The coefficient of volume compressibility is

$$m_v = \frac{a_v}{1 + e_{av}} = \frac{\left(\dfrac{\Delta e}{\Delta p}\right)}{1 + e_{av}}$$

$$\Delta e = 1.22 - 0.98 = 0.24$$

$$\Delta p = 4 - 2 = 2 \text{ ton/ft}^2$$

$$e_{av} = \frac{1.22 + 0.98}{2} = 1.1$$

So

$$m_v = \frac{\dfrac{0.24}{2}}{1 + 1.1} = 0.057 \text{ ft}^2/\text{ton}$$

$$c_v = \frac{k}{m_v \gamma_w} = \frac{2.0 \times 10^{-4} \text{ ft/day}}{(0.057 \text{ ft}^2/\text{ton})\left(\dfrac{62.4}{2000} \text{ ton/ft}^3\right)} = 0.112 \text{ ft}^2/\text{day}$$

$$T_{60} = \frac{c_v t_{60}}{H_{dr}^2}$$

$$t_{60} = \frac{T_{60} H_{dr}^2}{c_v}$$

From Table 8.2, for $U = 60\%$, the value of T_{60} is 0.287, so

$$t_{60} = \frac{(0.287)(12)^2}{0.112} = \textbf{369 days}$$

b. $C_c = \dfrac{e_1 - e_2}{\log (p_2/p_1)} = \dfrac{1.22 - 0.98}{\log (4/2)} = 0.797$

From Eq. (8.20),

$$S = \frac{C_c H}{1 + e_o} \log \left(\frac{p_o + \Delta p}{p_o}\right)$$

$$= \frac{(0.797)(12)}{1 + 1.22} \log \left(\frac{4}{2}\right) = 1.08 \text{ ft}$$

S at $60\% = (0.6)(1.08) \approx \textbf{0.65 ft}$ ▼

▼   **EXAMPLE 8.11**

For a laboratory consolidation test on a soil specimen (drained on both sides), the following results were obtained:

Thickness of the clay specimen = 25 mm

$$p_1 = 50 \text{ kN/m}^2 \qquad e_1 = 0.92$$

$$p_2 = 120 \text{ kN/m}^2 \qquad e_2 = 0.78$$

Time for 50% consolidation = 2.5 min

Determine the coefficient of permeability of the clay for the loading range.

Solution

$$m_v = \frac{a_v}{1 + e_{av}} = \frac{\left(\dfrac{\Delta e}{\Delta p}\right)}{1 + e_{av}}$$

$$= \frac{\dfrac{0.92 - 0.78}{120 - 50}}{1 + \dfrac{0.92 + 0.78}{2}} = 0.00108 \text{ m}^2/\text{kN}$$

$$c_v = \frac{T_{50} H_{dr}^2}{t_{50}}$$

From Table 8.2, for $U = 50\%$, the value of $T_v = 0.197$, so

$$c_v = \frac{(0.197)\left(\dfrac{0.025 \text{ m}}{2}\right)^2}{2.5 \text{ min}} = 1.23 \times 10^{-5} \text{ m}^2/\text{min}$$

$$k = c_v m_v \gamma_w = (1.23 \times 10^{-5})(0.00108)(9.81)$$

$$= \mathbf{1.303 \times 10^{-7} \text{ m/min}} \qquad ▼$$

8.13   CALCULATION OF CONSOLIDATION SETTLEMENT UNDER A FOUNDATION

Chapter 7 showed that the increase in the vertical stress in soil caused by a load applied over a limited area decreases with depth z measured from the ground surface downward. Hence, to estimate the one-dimensional settlement of a foundation, we can use Eq. (8.20), (8.22), or (8.23). However, the increase of stress Δp in these equations should be the average increase in the pressure below the center of the foundation. The values can be determined by using the procedure described in Chapter 7 (Section 7.11; also see Example 7.9).

Assuming the pressure increase varies parabolically, we can estimate the value of Δp_{av} as (Simpson's rule)

$$\Delta p_{av} = \frac{\Delta p_t + 4 \, \Delta p_m + \Delta p_b}{6}$$

(8.52)

where Δp_t, Δp_m, and Δp_b represent the increase in the pressure at the top, middle, and bottom of the layer, respectively. The values can be determined using the procedure described in Chapter 7.

▼ **EXAMPLE 8.12**

Calculate the settlement of the 10-ft-thick clay layer (Figure 8.31) that will result from the load carried by a 5-ft square footing. The clay is normally consolidated.

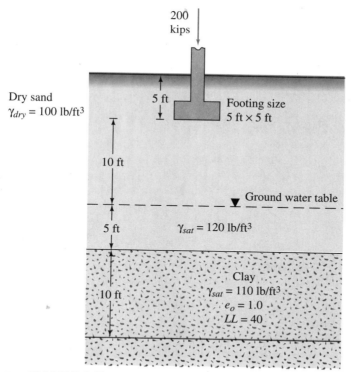

200 kips

Dry sand
$\gamma_{dry} = 100 \text{ lb/ft}^3$

5 ft

Footing size
5 ft × 5 ft

10 ft

Ground water table

5 ft

$\gamma_{sat} = 120 \text{ lb/ft}^3$

Clay
$\gamma_{sat} = 110 \text{ lb/ft}^3$
$e_o = 1.0$
$LL = 40$

10 ft

▼ **FIGURE 8.31**

Solution For normally consolidated clay, from Eq. (8.20),

$$S = \frac{C_c H}{1 + e_o} \log \left(\frac{p_o + \Delta p}{p_o} \right)$$

where $C_c = 0.009(LL - 10) = 0.009(40 - 10) = 0.27$

$H = 10 \times 12 = 120 \text{ in.}$

$e_O = 1.0$

$p_O = 15 \text{ ft} \times \gamma_{dry(sand)} + 5 \text{ ft}[\gamma_{sat(sand)} - 62.4] + \dfrac{10}{2}[\gamma_{sat(clay)} - 62.4]$

$= 15 \times 100 + 5(120 - 62.4) + 5(110 - 62.4) = 2026 \text{ lb/ft}^2$

As shown in Example 7.9, Δp_{av} below the center of the footing between $z = 15$ ft and $z = 25$ ft can be given as

$$\Delta p_{av} = 4p\left[\frac{H_2 I_{4(H_2)} - H_1 I_{4(H_1)}}{H_2 - H_1}\right]$$

where

$$I_{4(H_2)} = f\left(m' = \frac{B}{H_2}, \quad n' = \frac{L}{H_2}\right)$$

$$B = L = \frac{5}{2} = 2.5 \text{ ft}$$

$$H_2 = 25 \text{ ft}$$

$$H_1 = 15 \text{ ft}$$

$$m' = \frac{B}{H_2} = \frac{2.5}{25} = 0.1$$

$$n' = \frac{L}{H_2} = \frac{2.5}{25} = 0.1$$

Also

$$m' = \frac{B}{H_1} = \frac{2.5}{15} = 0.167$$

$$n' = \frac{L}{H_1} = \frac{2.5}{15} = 0.167$$

From Figure 7.29, for $m' = n' = 0.1$, $I_{4(H_2)} = 0.05$ and for $m' = n' = 0.167$, $I_{4(H_1)} = 0.075$. Also

$$p = \frac{200}{5 \times 5} = 8 \text{ kips/ft}^2$$

So

$$\Delta p_{av} = (4)(8)\left[\frac{(25)(0.05) - (15)(0.075)}{25 - 15}\right] = (32)(0.0125)$$

$$= 0.4 \text{ kips/ft}^2 = 400 \text{ lb/ft}^2$$

Substituting these values into the settlement equation gives

$$S = \frac{(0.27)(120)}{1 + 1} \log \left(\frac{2026 + 400}{2026} \right) = \textbf{1.27 in.} \qquad \blacktriangledown$$

8.14   IMMEDIATE SETTLEMENT CALCULATION BASED ON ELASTIC THEORY

Immediate or elastic settlement of foundations occurs immediately after the application of load without a change in the moisture content. The magnitude of the contact settlement will depend on the flexibility of the foundation and the type of material on which it is resting.

A uniformly loaded, perfectly flexible foundation resting on an elastic material such as saturated clay will have a sagging profile, as shown in Figure 8.32a because of elastic settlement. However, if the foundation is rigid and is resting on an elastic material such as clay, it will undergo uniform settlement and the contact pressure will be redistributed (Figure 8.32b).

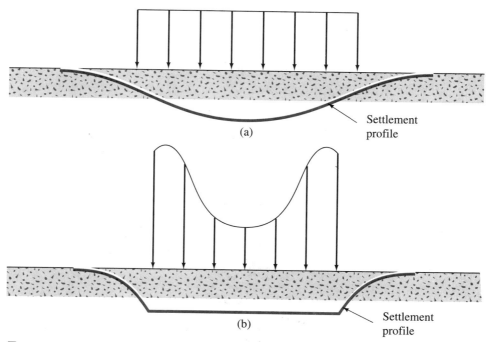

▼   **FIGURE 8.32**   Immediate settlement profile and contact pressure in clay: (a) flexible foundation; (b) rigid foundation

The settlement profile and contact-pressure distribution described above are true for soils in which the modulus of elasticity is fairly constant with depth. In the case of cohesionless sand, the modulus of elasticity increases with depth. Additionally, there is a lack of lateral confinement at the ground surface on the edge of the foundation. The sand at the edge of a flexible foundation is pushed outward, and the deflection curve of the foundation takes a concave downward shape.

Immediate settlement for foundations that rest on elastic material (of infinite thickness) can be calculated from equations derived by using the principles of the theory of elasticity. They are of the form

$$\rho_i = pB \frac{1 - \mu^2}{E} I_\rho \qquad (8.53)$$

where ρ_i = elastic settlement
p = net pressure applied
B = width of the foundation ($=$ diameter of circular foundation)
μ = Poisson's ratio
E = modulus of elasticity of soil
I_ρ = nondimensional influence factor

Schleicher (1926) expressed the influence factor for the corner of a flexible rectangular footing as

$$I_\rho = \frac{1}{\pi} \left[m_1 \ln \left(\frac{1 + \sqrt{m_1^2 + 1}}{m_1} \right) + \ln (m_1 + \sqrt{m_1^2 + 1}) \right] \qquad (8.54)$$

where $m_1 = \dfrac{\text{length of the foundation}}{\text{width of the foundation}}$.

Table 8.4 gives the influence factors for rigid and flexible foundations. Representative values of the modulus of elasticity and Poisson's ratio for different types of soils are given in Tables 8.5 and 8.6.

Note that Eq. (8.53) is based on the assumption that the pressure p is applied at the ground surface. In practice, foundations are placed at a certain depth below the ground surface. Deeper foundation embedment has a tendency to reduce the magnitude of the foundation settlement, ρ_i. However, if Eq. (8.53) is used to calculate settlement, it results in a conservative estimate.

▼ **TABLE 8.4** **Influence Factors for Foundations [Eq. (8.54)]**

Shape	m_1	I_ρ Flexible Center	Corner	Rigid
Circle	—	1.00	0.64	0.79
Rectangle	1	1.12	0.56	0.88
	1.5	1.36	0.68	1.07
	2	1.53	0.77	1.21
	3	1.78	0.89	1.42
	5	2.10	1.05	1.70
	10	2.54	1.27	2.10
	20	2.99	1.49	2.46
	50	3.57	1.8	3.0
	100	4.01	2.0	3.43

▼ **TABLE 8.5** **Representative Values of the Modulus of Elasticity**

Type of soil	Modulus of Elasticity psi	kN/m² [a]
Soft clay	250–500	1380–3450
Hard clay	850–2000	5865–13,800
Loose sand	1500–4000	10,350–27,600
Dense sand	5000–10,000	34.500–69,000

[a] 1 psi = 6.9 kN/m²

▼ **TABLE 8.6** **Representative Values of Poisson's Ratio**

Type of soil	Poisson's ratio, μ
Loose sand	0.2–0.4
Medium sand	0.25–0.4
Dense sand	0.3–0.45
Silty sand	0.2–0.4
Soft clay	0.15–0.25
Medium clay	0.2–0.5

8.15 TOTAL FOUNDATION SETTLEMENT

The total settlement of a foundation can be given as

$$S_T = S + S_s + \rho_i \tag{8.55}$$

where S_T = total settlement
S = primary consolidation settlement
S_s = secondary consolidation settlement
ρ_i = immediate settlement

When foundations are constructed on very compressible clays, the consolidation settlement can be several times greater than the immediate settlement (ρ_i).

▼ **EXAMPLE 8.13**

Estimate the immediate settlement of a column footing 4 ft in diameter that is constructed on an unsaturated clay layer, given total load carried by the column footing = 19 tons, E_{clay} = 1000 lb/in.^{2}, and μ = 0.2. Assume the footing to be rigid.

Solution　Using Eq. (8.53), we have

$$\rho_i = pB\,\frac{1 - \mu^2}{E}\,I_\rho$$

$$p = \frac{(19)(2000)}{\dfrac{\pi}{4}\,(4)^2} = 3023.95 \text{ lb/ft}^2$$

From Table 8.4, for a circular rigid foundation, I_ρ = 0.79, so

$$\rho_i = (3023.95)(4)\left[\frac{1 - 0.2^2}{(1000)(144)}\right](0.79) = 0.0637 \text{ ft} = \mathbf{0.764 \text{ in.}} \qquad ▼$$

8.16 SETTLEMENT CAUSED BY A PRELOAD FILL FOR CONSTRUCTION OF TAMPA VA HOSPITAL

There are several case histories in the literature for which the fundamental principles of soil compressibility have been used to predict the actual settlement of soil profiles under superimposed loading. In some cases, the actual and predicted settlements have remarkable agreement; in many others, the predicted settlements deviate to a large extent from the actual settlement observed. The disagreement in the latter cases may have several causes:

1. Improper evaluation of soil properties
2. Nonhomogeneity and irregularity of soil profiles

3. Error in the evaluation of the net stress increase with depth, which induces settlement

The following case history involves compressible clay layers. This case is presented to familiarize the readers with the variations that often exist between theory and practice.

Wheeless and Sowers (1972) presented the field measurements of settlement caused by a preload fill used in the construction of the Tampa Veterans Administration Hospital. Figure 8.33 shows the simplified general subsoil conditions at the building site. In general, the subsoil consisted of 15 to 20 ft (4.57 to 6.1 m) of subangular quartz sand at the top followed by clayey soil of varying thicknesses. The void ratio of the clayey soil varied from 0.7 to 1.4. The silt and clay content of the clayey soil varied from 5% to 75%. The Tampa limestone underlying the clay layer was a complex assortment of chalky, poorly consolidated calcereous deposits. The ground water table was located at a depth of about 15 ft (4.57 m) below the ground surface (elevation +25 ft). Figure 8.34 shows the consolidation curves obtained in the laboratory for clayey sand and sandy clay samples collected from various depths at the site.

The plan of the hospital building is shown in Figure 8.35 (broken lines). Figure 8.33 also shows the cross-section of the building. For a number of reasons, it was decided that the hospital should be built with a mat foundation. As can be seen from Figure 8.33, some soil had to be excavated to build the mat. As reported by Wheeless and Sowers, preliminary calculations indicated that the average building load in the eight-story area would be equal to the weight of the soil to be excavated for the construction of the mat. In that case, the consolidation settlement of the clay layer under

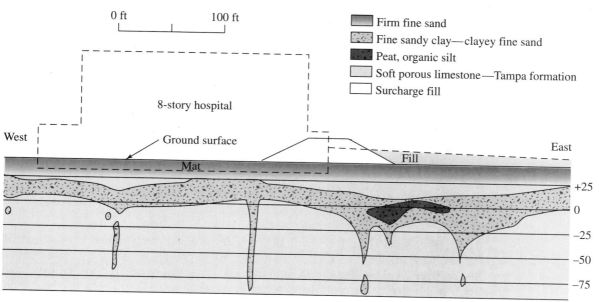

▼ **FIGURE 8.33**   Simplified general subsoil conditions at the site of Tampa VA Hospital (after Wheeless and Sowers, 1972)

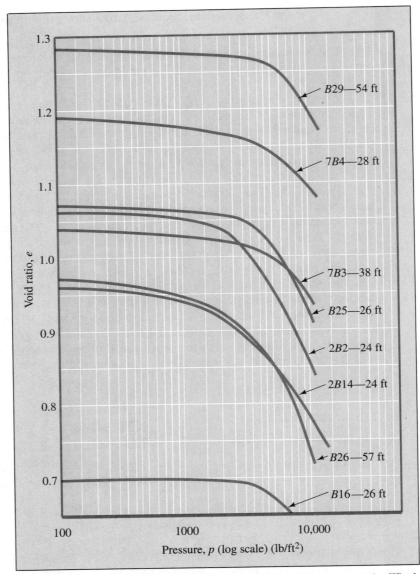

FIGURE 8.34   Consolidation curves of clayey sands and sandy clays (after Wheeless and Sowers, 1972)

the building would be rather small. However, the grading plan required a permanent fill of 16 ft (4.88 m) over the original ground surface to provide access to the main floor on the east side. This is also shown in Figure 8.33. Preliminary calculations indicated that the weight of this fill could be expected to produce a soil settlement of about 4 in. (101.6 mm) near the east side of the building. This settlement would produce undue bending and overstressing of the mat foundation. For that reason, it was decided to

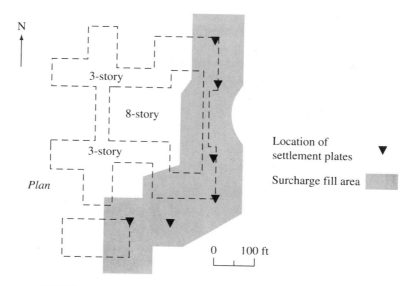

▼ **FIGURE 8.35**   Plan of the Tampa VA Hospital (after Wheeless and Sowers, 1972)

build a temporary fill that was 26 ft (7.93 m) high and limited to the front area of the proposed building. The fill area is shown in Figures 8.33 and 8.35. This temporary fill was built because the vertical stress that it induced in the clay layer would be greater than the stress induced by the permanent fill of 16 ft (4.88 m) as required by the grading plan. This would produce faster consolidation settlement. In about four months, the settlement would be approximately 4 in. (101.6 mm), which is the magnitude of maximum settlement expected from the required permanent fill of 16 ft (4.88 m). At that time, if the excess fill material is removed (Figure 8.36) and the building constructed, the settlement of the mat on the east side will be negligible. This technique of achieving the probable settlement of soil before construction is called *preloading*.

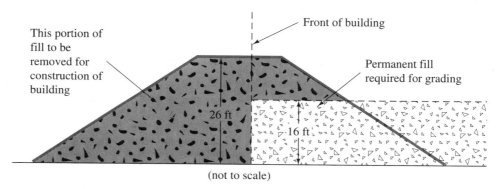

▼ **FIGURE 8.36**

Figure 8.35 shows the locations of eight settlement plates placed on the original ground surface before the construction of the temporary fill. Figure 8.37 shows the time-settlement records beneath the surcharge fill area as observed from the settlement plates. The following table is a comparison of the total estimated and observed consolidation settlements caused by preloading.

Settlement plate location	Observed settlement (in.)	Estimated consolidation settlement (in.)
3	2.6	2.9
4	2.5	2.9
6	2.9	3.0
7	3.4	3.8

Wheeless and Sowers did not give the detailed calculations by which the estimated consolidation settlements were obtained. However, we can make some preliminary calculations to check the order of magnitude. By observing the section shown in Figure 8.33, an approximate soil profile can be drawn as given in Figure 8.38. In this

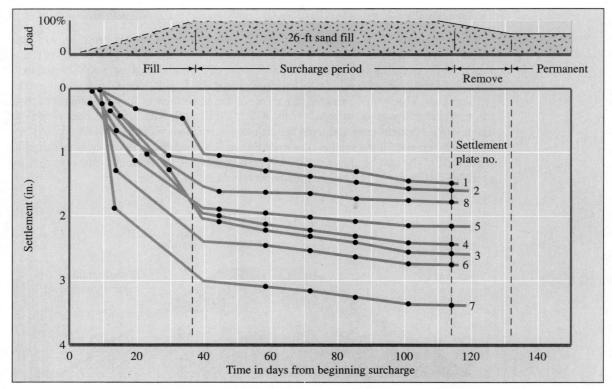

▼ FIGURE 8.37   Settlement-time curves beneath the surcharge fill area for the construction of Tampa VA Hospital (after Wheeless and Sowers, 1972)

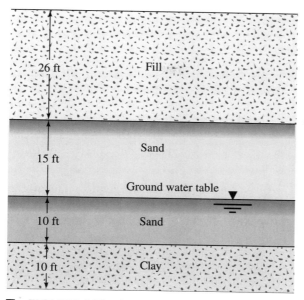

26 ft — Fill

15 ft — Sand

Ground water table

10 ft — Sand

10 ft — Clay

section, the approximate thickness of the clay layer is taken to be about 10 ft (3.05 m). Figure 8.34 shows the e-log p curves for various soils at the site. If we assume the e-log p curve marked as 2B14-24 ft to be the average one, it will have a compression index (C_c) of about 0.22. Assuming the clay to be normally consolidated, we have

$$S = \frac{C_c H}{1 + e_o} \log \left(\frac{p_o + \Delta p}{p_o} \right)$$

In Figure 8.38, assuming $\gamma_{dry(sand)} \simeq 115$ lb/ft^3 and $\gamma'_{(clay)} = \gamma'_{(sand)} \simeq 60$ lb/ft^3 gives

$$p_o = (15)(115) + (10)(60) + \left(\frac{10}{2} \right)(60)$$

$$= 1725 + 600 + 300 = 2625 \text{ lb/ft}^2$$

$$p_o + \Delta p = (26 + 15)(115) + (10)(60) + \left(\frac{10}{2} \right)(60)$$

$$= 4715 + 600 + 300 = 5615 \text{ lb/ft}^2$$

For the preceding value of $p_o = 2625$ lb/ft^2, the consolidation curve gives a value of $e_o \simeq 0.9$, so

$$S = \frac{(0.22)(10)}{1 + 0.9} \log \left(\frac{5615}{2625} \right) = 0.38 \text{ ft} \simeq 4.6 \text{ in.}$$

Although several assumptions have been made, the preceding value of $S \simeq 4.6$ in. compares rather favorably with the estimated settlement of 3 to 4 in. as given by

Wheeless and Sowers (1972). From the preceding comparisons of observed and estimated settlements given by Wheeless and Sowers and Figure 8.37, the following conclusions can be drawn:

1. In all cases, the estimated settlement exceeded the observed settlement.
2. Most of the settlement was complete in about 90 days.
3. The difference between the estimated and observed settlements varied from 3% to 16%, with an average of 13%.
4. Two-thirds to four-fifths of the total observed settlement was completed during the period of fill construction. The rate of consolidation was much faster than anticipated.

Wheeless and Sowers suggested that the accelerated rate of consolidation may be caused primarily by irregular sandy seams within the clay stratum. In Section 8.11, it was shown that the average degree of consolidation is related to the time factor, T_v. Also

$$t = \frac{T_v H_{dr}^2}{c_v}$$

For similar values of T_v (or average degree of consolidation) and c_v, the time t will be less if the maximum length of the drainage path (H_{dr}) is less. The presence of irregular sandy seams in the clay layer tends to reduce the magnitude of H_{dr}. This is the reason a faster rate of consolidation was attained in this area.

The structural load of the VA hospital was completed in the early part of 1970. No noticeable foundation movement has occurred.

8.17   GENERAL COMMENTS

In this chapter, we discussed the following topics:

1. Laboratory consolidation test procedures and interpretation of the results
2. Primary and secondary consolidation settlement calculation procedures
3. Time rate of consolidation (primary)
4. Immediate settlement calculation based on elastic theory

In calculating the immediate settlement, ρ_i, using Eqs. (8.53) and (8.54), one needs to keep in mind that these equations were developed using the theory of elasticity with the assumption that the soil is homogeneous, elastic, and isotropic. Most soil profiles in real life do not satisfy these assumptions, however. As outlined in Section 8.14, Eq. (8.53) is based on the assumption that the base of the foundation is located on the ground surface. In most practical cases, this is not true. The depth of embedment of the foundation has a tendency to reduce the magnitude of ρ_i. Also, Eq. (8.53) has been developed by integration of the vertical strain as (see Figure 8.39).

$$\rho_i = \int_0^\infty \varepsilon_z \, dz \tag{8.56}$$

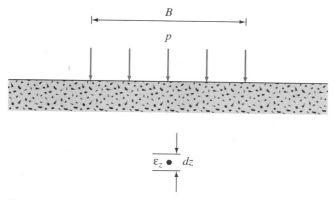

▼ **FIGURE 8.39**   Derivation of Eq. (8.56)

where ε_z = vertical strain at a depth z measured from the ground surface. If an *incompressible layer* is located at a depth $z \leq 2B$ to $3B$, it could reduce the magnitude of the immediate settlement.

It is very important that the estimated value of the modulus of elasticity of soil, E, be realistic. In granular soils, the modulus of elasticity increases with depth because of the increase in the average confining pressure. In that case, a representative value of E will be that obtained at a depth of about $1.5B$ to $2B$ below the center of the foundation.

PROBLEMS   **8.1**   Figure 8.40 shows a soil profile. The uniformly distributed load on the ground surface is Δp.

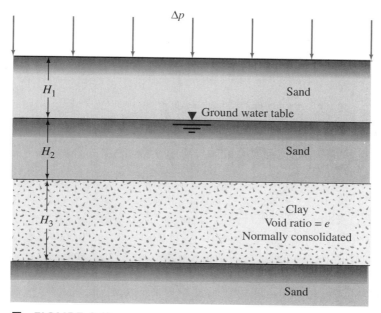

▼ **FIGURE 8.40**

Estimate the primary settlement of the clay layer given:

$H_1 = 4$ ft, $H_2 = 5$ ft, $H_3 = 4.5$ ft

Sand: $e = 0.62$, $G_s = 2.62$

Clay: $e = 0.98$, $G_s = 2.75$, $LL = 50$

$\Delta p = 2200$ lb/ft^2

8.2 Repeat Problem 8.1 with the following values:

$H_1 = 1.5$ m, $H_2 = 2$ m, $H_3 = 2$ m

Sand: $e = 0.55$, $G_s = 2.67$

Clay: $e = 1.1$, $G_s = 2.73$, $LL = 45$

$\Delta p = 120$ kN/m^2

8.3 Repeat Problem 8.1 for these data:

$\Delta p = 87$ kN/m^2

$H_1 = 1$ m, $H_2 = 3$ m, $H_3 = 3.2$ m

Sand: $\gamma_{dry} = 14.6$ kN/m^3, $\gamma_{sat} = 17.3$ kN/m^3

Clay: $\gamma_{sat} = 19.3$ kN/m^3, $LL = 38$, $e = 0.75$

8.4 If the clay layer in Problem 8.3 is preconsolidated and the average preconsolidation pressure is 80 kN/m^2, what will be the expected primary consolidation settlement if $C_s = \frac{1}{5}C_c$.

8.5 A soil profile is shown in Figure 8.41. The average preconsolidation pressure of the clay is 3400 lb/ft^2. Estimate the primary consolidation settlement that will take place as a result of a surcharge of 2200 lb/ft^2 if $C_s = \frac{1}{6}C_c$.

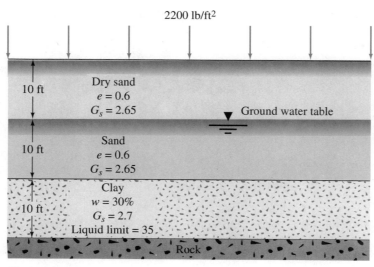

▼ **FIGURE 8.41**

8.6 The results of a laboratory consolidation test on a clay specimen are given here.

Pressure, p (lb/ft^2)	Total height of specimen at end of consolidation (in.)
500	0.6947
1,000	0.6850
2,000	0.6705
4,000	0.6520
8,000	0.6358
16,000	0.6252

Initial height of specimen = 0.748 in., G_s = 2.68, mass of dry specimen = 95.2 g, and area of specimen = 4.91 in.^{2}.
  a. Draw the e-log p graph.
  b. Determine the preconsolidation pressure.
  c. Determine the compression index, C_c.

8.7 Following are the results of a consolidation test:

e	Pressure, p (ton/ft^2)
1.1	0.25
1.085	0.5
1.055	1.0
1.01	2.0
0.94	4.0
0.79	8.0
0.63	16.0

  a. Plot the e-log p curve.
  b. Using Casagrande's method, determine the preconsolidation pressure.
  c. Calculate the compression index, C_c..

8.8 The coordinates of two points on a virgin compression curve are as follows:
$$e_1 = 1.75 \qquad p_1 = 190 \text{ kN/m}^2$$
$$e_2 = 1.49 \qquad p_2 = 385 \text{ kN/m}^2$$
Determine the void ratio that will correspond to a pressure of 600 kN/m^2.

8.9 The laboratory consolidation data for an undisturbed clay specimen are as follows:
$$e_1 = 1.1 \qquad p_1 = 95 \text{ kN/m}^2$$
$$e_2 = 0.9 \qquad p_2 = 475 \text{ kN/m}^2$$
What will be the void ratio for a pressure of 600 kN/m^2? (*Note: $p_c < 95$ kN/m^2.*)

8.10   Given here are the relationships of e and p for a clay soil:

e	p (ton/ft^2)
1.0	0.2
0.97	0.5
0.85	1.8
0.75	3.2

For this clay soil in the field, the following values are given: $H = 7.5$ ft, $p_o = 0.6$ ton/ft^2, and $p_o + \Delta p = 2.1$ ton/ft^2. Calculate the expected settlement caused by primary consolidation.

8.11   Consider the virgin compression curve described in Problem 8.8.
   a.   Find the coefficient of volume compressibility for the pressure range stated.
   b.   If the coefficient of consolidation for the pressure range is 0.0023 cm^2/sec, find the coefficient of permeability (cm/sec) of the clay corresponding to the average void ratio.

8.12   Refer to Problem 8.1. Given $c_v = 0.003$ cm^2/sec, how long will it take for 50% primary consolidation to take place?

8.13   Refer to Problem 8.2. Given $c_v = 2.62 \times 10^{-6}$ m^2/min, how long will it take for 65% primary consolidation to take place?

8.14   Laboratory tests on a 25-mm-thick clay specimen drained at both the top and bottom show that 50% consolidation takes place in 8.5 min.
   a.   How long will it take for a similar clay layer in the field, 3.2 m thick and drained at the top only, to undergo 50% consolidation?
   b.   Find the time required for the clay layer in the field as described in part (a) to reach 65% consolidation.

8.15   A 10-ft-thick layer (two-way drainage) of saturated clay under a surcharge loading underwent 90% primary consolidation in 75 days. Find the coefficient of consolidation of clay for the pressure range.

8.16   For a 1.2-in.-thick undisturbed clay specimen described in Problem 8.15, how long will it take to undergo 90% consolidation in the laboratory for a similar consolidation pressure range? The laboratory test's specimen will have two-way drainage.

8.17   A normally consolidated clay layer is 15 ft thick (one-way drainage). From the application of a given pressure, the total anticipated primary consolidation settlement will be 6.4 in.
   a.   What is the average degree of consolidation for the clay layer when the settlement is 2 in.?
   b.   If the average value of c_v for the pressure range is 0.003 cm^2/sec, how long will it take for 50% settlement to occur?
   c.   How long will it take for 50% consolidation to occur if the clay layer is drained at both top and bottom?

8.18 In laboratory consolidation tests on a clay specimen (drained on both sides), the following results were obtained:

Thickness of clay layer $= 1$ in.

$p_1 = 1000 \ \text{lb/ft}^2 \qquad e_1 = 0.75$

$p_2 = 2000 \ \text{lb/ft}^2 \qquad e_2 = 0.61$

Time for 50% consolidation $(t_{50}) = 3.1$ min

Determine the coefficient of permeability of the clay for the loading range.

8.19 During a laboratory consolidation test, the time and dial gauge readings obtained from an increase of pressure on the specimen from 0.5 ton/ft^2 to 1 ton/ft^2 are given here:

Time (min)	Dial gauge reading (in. × 10⁴)	Time (min)	Dial gauge reading (in. × 10⁴)
0	1565	16.0	1800
0.1	1607	30.0	1865
0.25	1615	60.0	1938
0.5	1625	120.0	2000
1.0	1640	240.0	2050
2.0	1663	480.0	2080
4.0	1692	960.0	2100
8.0	1740	1440.0	2112

a. Find the time for 50% primary consolidatation (t_{50}) using the logarithm-of-time method.

b. Find the time for 90% primary consolidation (t_{90}) using the square-root-of-time method.

c. If the average height of the specimen during consolidation caused by this incremental loading was 0.88 in. and it was drained at both the top and bottom, calculate the coefficient of consolidation using t_{50} and t_{90} obtained from parts (a) and (b).

d. Discuss the possible reasons for the difference in the values of c_v obtained in part (c).

8.20 Refer to the laboratory test results given in Problem 8.19. Using the hyperbola method, determine c_v. The average height of the specimen during consolidation was 0.88 in., and it was drained at both the top and bottom.

8.21 A continuous footing is shown in Figure 8.42. Using Newmark's influence chart (see Chapter 7), find the vertical stresses at A, B, and C caused by the load carried by the footing.

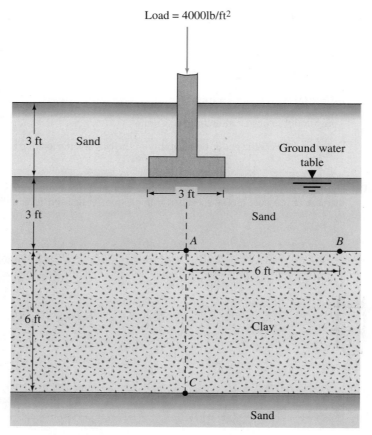

Load = 4000lb/ft²

3 ft   Sand

Ground water table

|← 3 ft →|

3 ft   Sand

A   B

6 ft

6 ft   Clay

C

Sand

▼   **FIGURE 8.42**

8.22   Use Eq. (8.52) to calculate the settlement of the footing described in Problem 8.21 from consolidation of the clay layer given:

Sand: $e = 0.6$, $G_s = 2.65$; degree of saturation of sand above ground water table is 30%

Clay: $e = 0.85$, $G_s = 2.75$, $LL = 45$; the clay is normally consolidated

8.23   Consider a rectangular column foundation 6 ft × 3 ft in a plan located on a sand layer that extends to a great depth. Assume $E = 140$ ton/ft² and $\mu = 0.4$. If the net pressure increase, p, on the foundation is 3000 lb/ft², estimate the elastic settlement assuming that the foundation is rigid.

8.24   Estimate the elastic settlement of a rigid 3-m-square footing constructed over a loose sand layer given: load carried by the footing = 711 kN, $\mu = 0.32$, and $E_{sand} = 16,200$ kN/m².

REFERENCES

Casagrande, A. (1936). "Determination of the Preconsolidation Load and Its Practical Signifi-cance," *Proceedings*, 1st International Conference on Soil Mechanics and Foundation Engin-eering, Cambridge, Mass., Vol. 3, 60–64.

Casagrande, A., and Fadum, R. E. (1940). "Notes on Soil Testing for Engineering Purposes," *Harvard University Graduate School Engineering Publication No. 8.*

Crawford, C. B. (1964). "Interpretation of the Consolidation Tests," *Journal of the Soil Mechanics and Foundations Division*, ASCE, Vol. 90, No. SM5, 93–108.

Leonards, G. A., and Altschaeffl, A. G. (1964). "Compressibility of Clay," *Journal of the Soil Mechanics and Foundations Division*, ASCE, Vol. 90, No. SM5, 133–156.

Mesri, G. (1973). "Coefficient of Secondary Compression," *Journal of the Soil Mechanics and Foundations Division*, ASCE, Vol. 99, No. SM1, 122–137.

Nagaraj, T., and Murty, B. R. S. (1985). "Prediction of the Preconsolidation Pressure and Recompression Index of Soils," *Geotechnical Testing Journal*, Vol. 8, No. 4, 199–202.

Rendon-Herrero, O. (1983). "Universal Compression Index Equation," *Discussion, Journal of Geotechnical Engineering*, ASCE, Vol. 109, No. 10, 1349.

Schleicher, F. (1926). "Zur Theorie des Baugrundes," *Bauingenieur*, Vol. 7, 931–935, 949–952.

Schmertmann, J. H. (1953). "Undisturbed Consolidation Behavior of Clay," *Transactions*, ASCE, Vol. 120, 1201.

Sivaram, B., and Swamee, A. (1977). "A Computational Method for Consolidation Coefficient," *Soils and Foundations*, Vol. 17, No. 2, 48–52.

Sridharan, A., and Prakash, K. (1985). "Improved Rectangular Hyperbola Method for the Determination of Coefficient of Consolidation," *Geotechnical Testing Journal*, ASTM, Vol. 8, No. 1, 37–40.

Taylor, D. W. (1942). "Research on Consolidation of Clays," *Serial No. 82*, Department of Civil and Sanitary Engineering, Massachusetts Institute of Technology, Cambridge, Mass.

Taylor, D. W. (1948). *Fundamentals of Soil Mechanics*, Wiley, New York.

Terzaghi, K. (1925). *Erdbaumechanik auf Bodenphysikalischer*, Deutichke, Vienna.

Terzaghi, K., and Peck, R. B. (1967). *Soil Mechanics in Engineering Practice*, 2nd ed., Wiley, New York.

Wheeless, L. D., and Sowers, G. F. (1972). "Mat Foundation and Preload Fill, VA Hospital, Tampa," *Proceedings*, Speciality Conference on Performance of Earth and Earth-Supported Structures, ASCE, Vol. 1, Part 2, 939–951.

Supplementary References for Further Study

Christian, J. T., and Carrier, W. D., III (1978). "Janbu, Bjerrum, and Kjaernsli's Chart Reinterpreted," *Canadian Geotechnical Journal*, Vol. 15, No. 1, 124–128.

Das, B. M. (1983). *Advanced Soil Mechanics*, Hemisphere, Washington, D.C.

Lambe, T. W. (1964). "Methods of Estimating Settlement," *Journal of the Soil Mechanics and Foundations Division*, ASCE, Vol. 90, No. SM5, 43–69.

Lowe, J., III, Jonas, E., and Obrician, V. (1969). "Controlled Gradient Consolidation Test," *Journal of the Soil Mechanics and Foundations Division*, ASCE, Vol. 95, No. SM1, 77–98.

Smith, R. E., and Wahls, H. E. (1969). "Consolidation under Constant Rate of Strain," *Journal of the Soil Mechanics and Foundations Division*, ASCE, Vol. 95, No. SM2, 519–538.

SHEAR STRENGTH OF SOIL

The *shear strength* of a soil mass is the internal resistance per unit area that the soil mass can offer to resist failure and sliding along any plane inside it. One must understand the nature of shearing resistance in order to analyze soil stability problems such as bearing capacity, slope stability, and lateral pressure on earth-retaining structure.

9.1 MOHR–COULOMB FAILURE CRITERIA

Mohr (1900) presented a theory for rupture in materials that a material fails because of a critical combination of normal stress and shearing stress, and not from either maximum normal or shear stress alone. Thus, the functional relationship between normal stress and shear stress on a failure plane can be expressed in the form (Figure 9.1a)

$$\tau_f = f(\sigma) \tag{9.1}$$

The failure envelope defined by Eq. (9.1) is a curved line, as shown in Figure 9.1b. For most soil mechanics problems, it is sufficient to approximate the shear stress on the failure plane as a linear function of the normal stress (Couiomb, 1776). This can be written as

$$\tau_f = c + \sigma \tan \phi \tag{9.2}$$

where c = cohesion
ϕ = angle of internal friction

The preceding relationship is called the *Mohr–Coulomb failure criteria*.

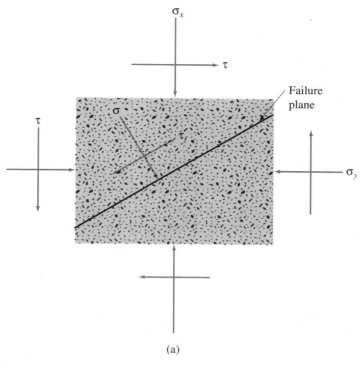

(a)

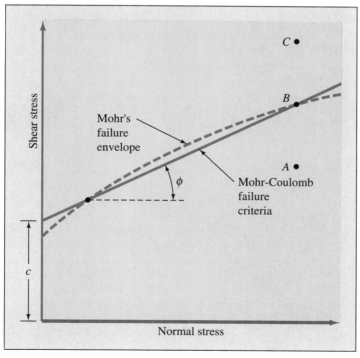

(b)

▼ **FIGURE 9.1** Mohr's failure envelope and the Mohr–Coulomb failure criteria

The significance of the failure envelope can be explained as follows. If the normal stress and the shear stress on a plane in a soil mass are such that they plot as point A in Figure 9.1b, then shear failure will not occur along that plane. If the normal stress and the shear stress on a plane plot as point B (which falls on the failure envelope), then shear failure will occur along that plane. A state of stress on a plane represented by point C cannot exist, since it plots above the failure envelope, and shear failure in a soil would have occurred already.

Inclination of the Plane of Failure Caused by Shear

As stated by the Mohr–Coulomb failure criteria, failure from shear will occur when the shear stress on a plane reaches a value given by Eq. (9.2). To determine the inclination of the failure plane with the major principal plane, refer to Figure 9.2, where σ_1 and σ_3 are, respectively, the major and minor principal stresses. The failure plane EF makes an angle θ with the major principal plane. In order to determine the angle θ and the relationship between σ_1 and σ_3, refer to Figure 9.3, which is a plot of the Mohr's circle for the state of stress shown in Figure 9.2 (see Chapter 7). In Figure 9.3, fgh is the failure envelope defined by the relationship $s = c + \sigma \tan \phi$. The radial line ab defines the major principal plane (CD in Figure 9.2), and the radial line ad defines the failure plane (EF in Figure 9.2). It can be shown that $\angle bad = 2\theta = 90 + \phi$, or

$$\theta = 45 + \frac{\phi}{2} \tag{9.3}$$

Again, from Figure 9.3,

$$\frac{\overline{ad}}{\overline{fa}} = \sin \phi \tag{9.4}$$

$$\overline{fa} = fO + Oa = c \cot \phi + \frac{\sigma_1 + \sigma_3}{2} \tag{9.5a}$$

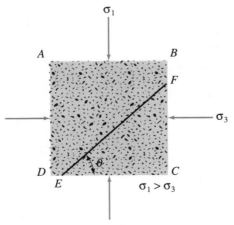

▼ **FIGURE 9.2**   Inclination of failure plane in soil with major principal plane

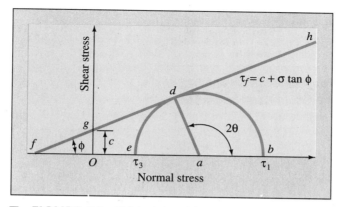

▼ **FIGURE 9.3**   Mohr's circle and failure envelope

Also

$$\overline{ad} = \frac{\sigma_1 - \sigma_3}{2}$$

(9.5b)

Substituting Eqs. (9.5a) and (9.5b) into Eq. (9.4), we obtain

$$\sin \phi = \frac{\dfrac{\sigma_1 - \sigma_3}{2}}{c \cot \phi + \dfrac{\sigma_1 + \sigma_3}{2}}$$

or

$$\sigma_1 = \sigma_3 \left(\frac{1 + \sin \phi}{1 - \sin \phi}\right) + 2c\left(\frac{\cos \phi}{1 - \sin \phi}\right)$$

(9.6)

However,

$$\frac{1 + \sin \phi}{1 - \sin \phi} = \tan^2 \left(45 + \frac{\phi}{2}\right)$$

and

$$\frac{\cos \phi}{1 - \sin \phi} = \tan \left(45 + \frac{\phi}{2}\right)$$

Thus,

$$\sigma_1 \doteq \sigma_3 \tan^2 \left(45 + \frac{\phi}{2}\right) + 2c \tan \left(45 + \frac{\phi}{2}\right)$$

(9.7)

▼ **TABLE 9.1   Typical Values of Drained Angle of Friction for Sands and Silts**

Soil type	ϕ (deg)
Sand: Rounded grains:	
Loose	27–30
Medium	30–35
Dense	35–38
Sand: Angular grains	
Loose	30–35
Medium	35–40
Dense	40–45
Gravel with some sand	34–48
Silts	26–35

Shear Failure Law in Saturated Soil

In saturated soil, the total normal stress at a point is the sum of the effective stress and the pore water pressure, or

$$\sigma = \sigma' + u$$

The effective stress, σ', is carried by the soil solids. So, to apply to soil mechanics, Eq. (9.2) needs to be rewritten as

$$\tau_f = c + (\sigma - u) \tan \phi = c + \sigma' \tan \phi \qquad (9.8)$$

The value of c for sand and inorganic silt is 0. For normally consolidated clays, c can be approximated at 0. Overconsolidated clays have values of c that are greater than 0. The angle of friction, ϕ, is sometimes referred to as the *drained angle of friction*. Typical values of ϕ for some granular soils are given in Table 9.1.

9.2 DETERMINATION OF SHEAR STRENGTH PARAMETERS FOR SOILS IN THE LABORATORY

The shear strength parameters of a soil can be determined in the laboratory by primarily two types of tests: direct shear test and triaxial test. The procedures for conducting each of these tests are explained in some detail in the following sections.

Direct Shear Test

This is the oldest and simplest form of shear test arrangement. A diagram of the direct shear test apparatus is shown in Figure 9.4. The test equipment consists of a metal shear box in which the soil specimen is placed. The soil specimens may be square or circular in plan. The size of the specimens generally used is about 3 or 4 in.² (1935.48 or 2580.64 mm²) across and about 1 in. (25.4 mm) high. The box is split horizontally into halves. Normal force on the specimen is applied from the top of the shear box. The normal stress on the specimens can be as great as 150 lb/in.² (1034.2 kN/m²). Shear force is applied by moving one half of the box relative to the other to cause failure in the soil specimen.

Depending on the equipment, the shear test can be either stress-controlled or strain-controlled.

In stress-controlled tests, the shear force is applied in equal increments until the specimen fails. The failure takes place along the plane of split of the shear box. After the application of each incremental load, the shear displacement of the top half of the box is measured by a horizontal dial gauge. The change in the height of the specimen (and thus the volume change of the specimen) during the test can be obtained from the readings of a dial gauge that measures the vertical movement of the upper loading plate.

In strain-controlled tests, a constant rate of shear displacement is applied to one half of the box by a motor that acts through gears. The constant rate of shear displacement is observed by a horizontal dial gauge. The resisting shear force of the soil corresponding to any shear displacement can be measured by a horizontal proving ring or load cell. The volume change of the specimen during the test is obtained in a manner

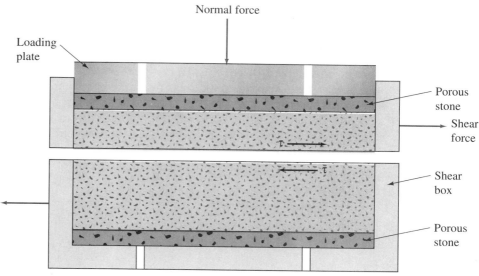

▼ **FIGURE 9.4**   Diagram of direct shear test arrangement

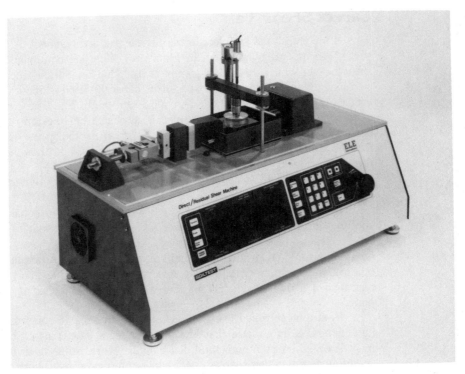

similar to the stress-controlled tests. Figure 9.5 shows a photograph of strain-controlled direct shear test equipment.

The advantage of the strain-controlled tests is that, in the case of dense sand, peak shear resistance (that is, at failure) as well as lesser shear resistance (that is, at a point after failure called *ultimate strength*) can be observed and plotted. In stress-controlled tests, only the peak shear resistance can be observed and plotted. Note that the peak shear resistance in stress-controlled tests can be only approximated. This is because failure occurs at a stress level somewhere between the prefailure load increment and the failure load increment. Nevertheless, stress-controlled tests probably model real field situations better than strain-controlled tests.

For a given test, the normal stress can be calculated as

$$\sigma = \text{normal stress} = \frac{\text{normal force}}{\text{area of cross-section of the specimen}} \tag{9.9}$$

The resisting shear stress for any shear displacement can be calculated as

$$\tau = \text{shear stress} = \frac{\text{resisting shear force}}{\text{area of cross-section of the specimen}} \tag{9.10}$$

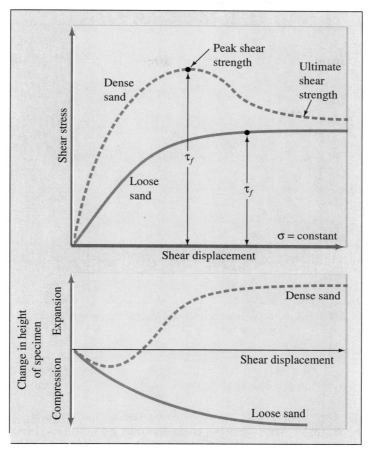

▼ **FIGURE 9.6**   Plot of shear stress and change in height of specimen against shear displacement for loose and dense dry sand (direct shear test)

Figure 9.6 shows a typical plot of shear stress and change in the height of the specimen against shear displacement for loose and dense sands. These observations were obtained from a strain-controlled test. The following generalizations can be developed from Figure 9.6 regarding the variation of resisting shear stress with shear displacement:

1. In loose sand, the resisting shear stress increases with shear displacement until a failure shear stress of τ_f is reached. After that, the shear resistance remains approximately constant for any further increase in the shear displacement.

2. In dense sand, the resisting shear stress increases with shear displacement until it reaches a failure stress of τ_f. This τ_f is called the *peak shear strength*. After failure stress is attained, the resisting shear stress gradually decreases

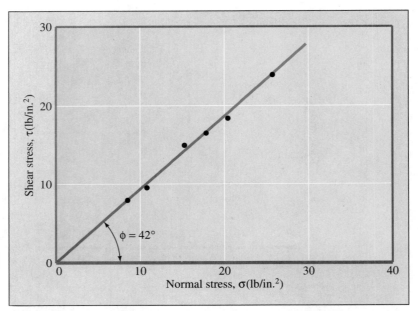

▼ **FIGURE 9.7**   Determination of shear strength parameters for a dry sand using the results of direct shear tests

as shear displacement increases until it finally reaches a constant value called the *ultimate shear strength.*

Direct shear tests are repeated on similar specimens at various normal stresses. The normal stresses and the corresponding values of τ_f obtained from a number of tests are plotted on a graph from which the shear strength parameters are determined. Figure 9.7 shows such a plot for tests on a dry sand. The equation for the average line obtained from experimental results is

$$\tau_f = \sigma \tan \phi \qquad\qquad (9.11)$$

(*Note:* $c = 0$ for sand and $\sigma = \sigma'$ for dry conditions.) So the friction angle

$$\phi = \tan^{-1}\left(\frac{\tau_f}{\sigma}\right)$$

It is important to note that *in situ* cemented sands may show a c intercept.

Drained Direct Shear Test on Saturated Sand and Clay

The shear box that contains the soil specimen is generally kept inside a container that can be filled with water to saturate the specimen. A *drained test* is made on a saturated soil specimen by keeping the rate of loading slow enough so that the excess pore water pressure generated in the soil is completely dissipated by drainage. Pore water from the specimen is drained through two porous stones (see Figure 9.4).

Since the coefficient of permeability of sand is high, the excess pore water pressure generated because of loading (normal and shear) is dissipated quickly. Hence, for an ordinary loading rate, essentially full drainage conditions exist. The friction angle ϕ obtained from a drained direct shear test of saturated sand will be the same as that for a similar specimen of dry sand.

The coefficient of permeability of clay is very small compared with that of sand. When a normal load is applied to a clay soil specimen, a sufficient length of time must elapse for full consolidation—that is, for dissipation of excess pore water pressure. For that reason, the shearing load has to be applied at a very slow rate. The test may last from 2 to 5 days. Figure 9.8 shows the results of a drained direct shear test on an overconsolidated clay. Figure 9.9 shows the plot of τ_f against σ' obtained from a number of drained direct shear tests on a normally consolidated and an overconsolidated clay. Note that $\sigma = \sigma'$ and the value of $c \simeq 0$ for a normally consolidated clay.

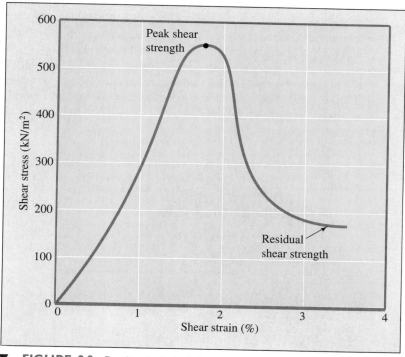

▼ **FIGURE 9.8**   Results of a drained direct shear test on an overconsolidated clay

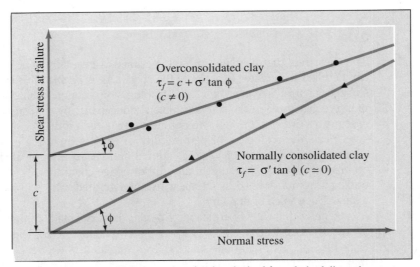

▼ **FIGURE 9.9**   Failure envelope for clay obtained from drained direct shear tests

General Comments on Direct Shear Test

The direct shear test is rather simple to perform, but it has some inherent shortcomings. The reliability of the results may be questioned. This is because in this test the soil is not allowed to fail along the weakest plane but is forced to fail along the plane of split of the shear box. Also, the shear stress distribution over the shear surface of the specimen is not uniform. In spite of these shortcomings, the direct shear test is the simplest and most economical for a dry or saturated sandy soil.

In many foundation design problems, it will be necessary to determine the angle of friction between the soil and the material in which the foundation is constructed (Figure 9.10). The foundation material may be concrete, steel, or wood. The shear strength along the surface of contact of the soil and the foundation can be given as

$$\tau_f = c_a + \sigma' \tan \delta \tag{9.12}$$

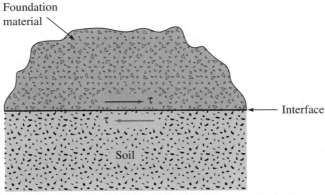

▼ **FIGURE 9.10**   Interface of a foundation material and soil

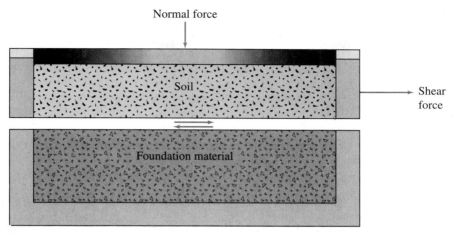

▼ **FIGURE 9.11**   Direct shear test to determine interface friction angle

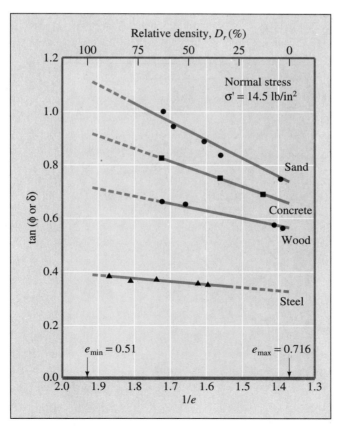

▼ **FIGURE 9.12**   Variation of tan ϕ and tan δ with $1/e$. *Note:* e = void ratio, σ' = 14.5 lb/in.²; quartz sand (after Acar, Durgunoglu, and Tumay, 1982)

where c_a = adhesion

δ = angle of friction between the soil and the foundation material

Note that the preceding equation is similar in form to Eq. (9.8). The shear strength parameters between a soil and a foundation material can be conveniently determined by a direct shear test. This is a great advantage of the direct shear test. The foundation material can be placed in the bottom part of the direct shear test box and then the soil can be placed above it (that is, in the top part of the box), as shown in Figure 9.11, and the test can be conducted in the usual manner.

Figure 9.12 shows the results of direct shear tests conducted in this manner with a quartz sand and concrete, wood, and steel as foundation materials with $\sigma' = 14.5$ lb/in.² (100 kN/m²). Figure 9.13 shows the variation of δ and ϕ between the quartz sand (relative density = 45%) and the same foundation materials as a function of σ' ($c_a = 0$). It is important to realize that the magnitude of δ and ϕ decreases with the increase of normal stress, σ'. That can be explained by referring to Figure 9.14. It was mentioned in Section 9.1 and Figure 9.1 that Mohr's failure envelope is actually curved, and Eqs. (9.2), (9.8), and (9.12) are only approximations. If a direct shear test is conducted with $\sigma' = \sigma'_{(1)}$, the shear strength will be $\tau_{f(1)}$, so

$$\delta_1 = \tan^{-1}\left[\frac{\tau_{f(1)}}{\sigma'_{(1)}}\right]$$

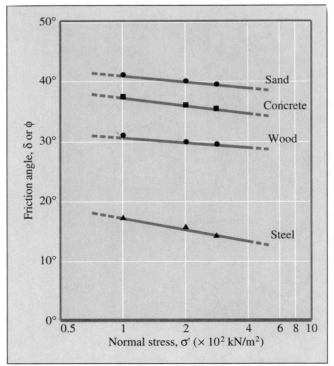

▼ **FIGURE 9.13**   Variation of ϕ and δ with σ'. *Note:* Relative density = 45%; quartz sand (after Acar, Durgunoglu, and Tumay, 1982)

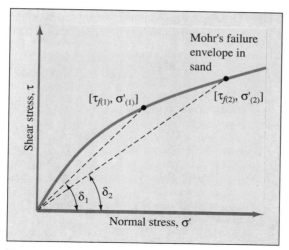

▼   **FIGURE 9.14**   Curvilinear nature of Mohr's failure envelope in sand

This is shown in Figure 9.14. In a similar manner, if the test is conducted with $\sigma' = \sigma'_{(2)}$, then

$$\delta = \delta_2 = \tan^{-1}\left[\frac{\tau_{f(2)}}{\sigma'_{(2)}}\right]$$

As can be seen from Figure 9.14, $\delta_2 < \delta_1$ since $\sigma'_{(2)} > \sigma'_{(1)}$. Keeping this in mind, one must realize that the values of ϕ given in Table 9.1 are only average values.

▼   **EXAMPLE 9.1**

Direct shear tests were performed on a dry, sandy soil. The size of the specimen was 2 in. × 2 in. × 0.75 in. Test results were as follows:

Test no.	Normal force (lb)	Normal stress[a] $\sigma = \sigma'$ (lb/ft²)	Shear force at failure (lb)	Shear stress at failure[b] τ_f (lb/ft²)
1	20	720	12.0	432.0
2	30	1080	18.3	658.8
3	70	2520	42.1	1515.6
4	100	3600	60.1	2163.6

[a] $\sigma = \dfrac{\text{normal force}}{\text{area of specimen}} = \dfrac{(\text{normal force})(144)}{(2 \text{ in.})(2 \text{ in.})}$

[b] $\tau_f = \dfrac{\text{shear force}}{\text{area of specimen}} = \dfrac{(\text{shear force})(144)}{(2 \text{ in.})(2 \text{ in.})}$

Find the shear stress parameters.

Solution   The shear stresses, τ_f, obtained from the tests are plotted against the normal stresses in Figure 9.15, from which we find $c = 0$, $\phi = 32°$.

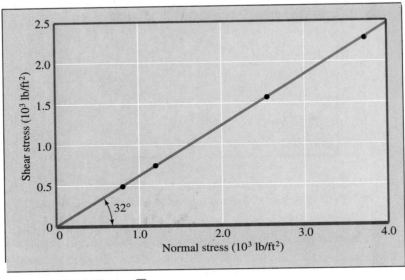

▼   **FIGURE 9.15**   ▼

9.3 TRIAXIAL SHEAR TEST

The triaxial shear test is one of the most reliable methods now available for determining shear strength parameters. It is widely used for research and conventional testing. A diagram of the triaxial test layout is shown in Figure 9.16.

In this test, a soil specimen about 1.4 in. (35.6 mm) in diameter and 3 in. (76.2 mm) in length is generally used. The specimen is encased by a thin rubber membrane and placed inside a plastic cylindrical chamber that is usually filled with water or glycerine. The specimen is subjected to a confining pressure by compression of the fluid in the chamber. (Note that air is sometimes used as a compression medium.) To cause shear failure in the specimen, axial stress is applied through a vertical loading ram (sometimes called *deviator stress*). This can be done in one of two ways:

1. Application of dead weights or hydraulic pressure in equal increments until the specimen fails. (Axial deformation of the specimen resulting from the load applied through the ram is measured by a dial gauge.)

2. Application of axial deformation at a constant rate by means of a geared or hydraulic loading press. This is a strain-controlled test.

The axial load applied by the loading ram corresponding to a given axial deformation is measured by a proving ring or load cell attached to the ram.

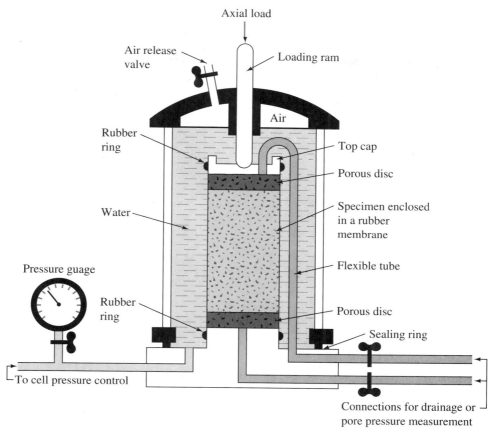

▼ FIGURE 9.16 Diagram of triaxial test equipment (after Bishop and Bjerrum, 1960)

Connections to measure drainage into or out of the specimen, or to measure pressure in the pore water (as per the test conditions), are also provided. The following three standard types of triaxial tests are generally conducted:

1. consolidated-drained test or drained test (CD test)
2. consolidated-undrained test (CU test)
3. unconsolidated-undrained test or undrained test (UU test)

The general procedures and implications for each of the tests in *saturated soils* are described in the following sections.

Consolidated-Drained Test

In this test, the specimen is first subjected to an all-around confining pressure, σ_3, by compression of the chamber fluid (Figure 9.17a). As confining pressure is applied, the pore water pressure of the specimen increases by u_c. This increase in the pore water

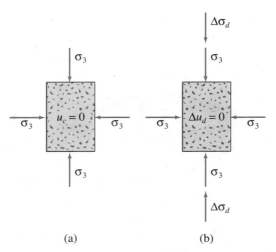

(a)   (b)

▼ **FIGURE 9.17**   Consolidated-drained triaxial test: (a) specimen under chamber confining pressure; (b) deviator stress application

pressure can be expressed in the form of a nondimensional parameter:

$$B = \frac{u_c}{\sigma_3} \tag{9.13}$$

where B = Skempton's pore pressure parameter (Skempton, 1954).

For saturated soft soils, B is approximately equal to 1; however, for saturated stiff soils, the magnitude of B can be less than 1. Black and Lee (1973) gave the theoretical values of B for various soils at complete saturation. These values are listed in Table 9.2.

If the connection to drainage is kept open, dissipation of the excess pore water pressure, and thus consolidation, will occur. With time, u_c will become equal to 0. In saturated soil, the change in the volume of the specimen (ΔV_c) that takes place during consolidation can be obtained from the volume of pore water drained (Figure 9.18a).

▼ **TABLE 9.2**   **Theoretical Values of B at Complete Saturation**

Type of soil	Theoretical value
Normally consolidated soft clay	0.9998
Lightly overconsolidated soft clays and silts	0.9988
Overconsolidated stiff clays and sands	0.9877
Very dense sands and very stiff clays at high confining pressures	0.9130

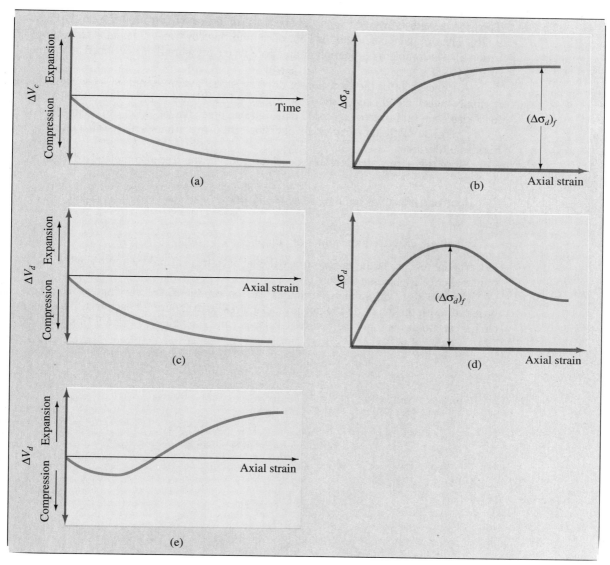

FIGURE 9.18 Consolidated-drained triaxial test: (a) volume change of specimen caused by chamber confining pressure; (b) plot of deviator stress against strain in the vertical direction for loose sand and normally consolidated clay; (c) volume change in loose sand and normally consolidated clay during deviator stress application; (d) plot of deviator stress against strain in the vertical direction for dense sand and over-consolidated clay; (e) volume change in dense sand and overconsolidated clay during deviator stress application

Then the deviator stress, $\Delta\sigma_d$, on the specimen is increased at a very slow rate (Figure 9.17b). The drainage connection is kept open, and the slow rate of deviator stress application allows complete dissipation of any pore water pressure that developed as a result ($\Delta u_d = 0$).

A typical plot of the variation of deviator stress against strain in loose sand and normally consolidated clay is shown in Figure 9.18b. Figure 9.18d shows a similar plot for dense sand and overconsolidated clay. The volume change, ΔV_d, of specimens that occurs because of the application of deviator stress in various soils is also shown in Figures 9.18c and e.

Since the pore water pressure developed during the test is completely dissipated, we have

total and effective confining stress $= \sigma_3 = \sigma'_3$

and

total and effective axial stress at failure $= \sigma_3 + (\Delta\sigma_d)_f = \sigma_1 = \sigma'_1$

In a triaxial test, σ'_1 is the major principal effective stress at failure and σ'_3 is the minor principal effective stress at failure.

Several tests on similar specimens can be conducted by varying the confining pressure. With the major and minor principal stresses at failure for each test, the Mohr's circles can be drawn and the failure envelopes can be obtained. Figure 9.19 shows the

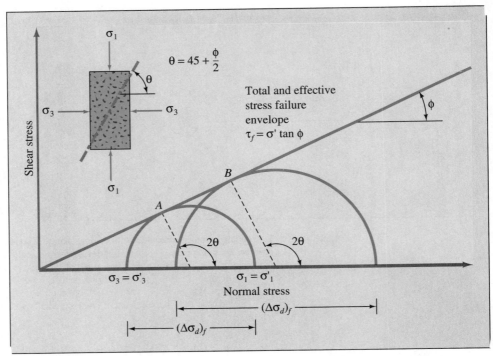

▼ **FIGURE 9.19**   Effective stress failure envelope from drained tests in sand and normally consolidated clay

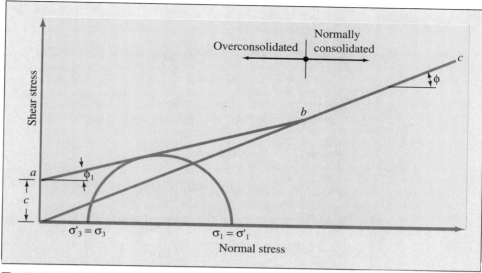

▼ **FIGURE 9.20** Effective stress failure envelope for overconsolidated clay

type of effective stress failure envelope obtained for tests in sand and normally consolidated clay. The coordinates of the point of tangency of the failure envelope with a Mohr's circle (that is, point A) give the stresses (normal and shear) on the failure plane of that test specimen.

Overconsolidation results when a clay is initially consolidated under an all-around chamber pressure of $\sigma_c \, (= \sigma'_c)$ and is allowed to swell by reducing the chamber pressure to $\sigma_3 \, (= \sigma'_3)$. The failure envelope obtained from drained triaxial tests of such overconsolidated clay specimens shows two distinct branches (ab and bc in Figure 9.20). The portion ab has a flatter slope with a cohesion intercept, and the shear strength equation for this branch can be written as

$$\tau_f = c + \sigma' \tan \phi_1 \tag{9.14}$$

The portion bc of the failure envelope represents a normally consolidated stage of soil and follows the equation $\tau_f = \sigma' \tan \phi$.

A consolidated-drained triaxial test on a clayey soil may take several days to complete. This is because it is necessary to apply deviator stress at a very slow rate to ensure full drainage from the soil specimen. For that reason, the CD type of triaxial test is not common.

▼ **EXAMPLE 9.2**

For a normally consolidated clay, the results of a drained triaxial test are as follows:

Chamber confining pressure = 16 lb/in.²

Deviator stress at failure = 25 lb/in.²

a. Find the angle of friction, ϕ.

b. Determine the angle θ that the failure plane makes with the major principal plane.

Solution   For a normally consolidated soil, the failure envelope equation is

$$\tau_f = \sigma' \tan \phi \qquad (\text{since } c = 0)$$

For the triaxial test, the effective major and minor principal stresses at failure are:

$$\sigma'_1 = \sigma_1 = \sigma_3 + (\Delta\sigma_d)_f = 16 + 25 = 41 \text{ lb/in.}^2$$

and

$$\sigma'_3 = \sigma_3 = 16 \text{ lb/in.}^2$$

a. The Mohr's circle and the failure envelope are shown in Figure 9.21, from which

$$\sin \phi = \frac{AB}{OA} = \frac{\left(\dfrac{\sigma'_1 - \sigma'_3}{2}\right)}{\left(\dfrac{\sigma'_1 + \sigma'_3}{2}\right)}$$

or

$$\sin \phi = \frac{\sigma'_1 - \sigma'_3}{\sigma'_1 + \sigma'_3} = \frac{41 - 16}{41 + 16} = 0.438$$

$$\phi = \mathbf{26°}$$

b. $\theta = 45 + \dfrac{\phi}{2} = 45° + \dfrac{26}{2} = \mathbf{58°}$

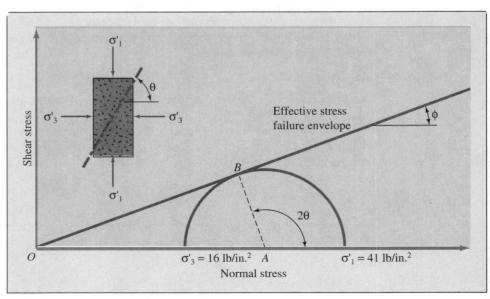

▼   **FIGURE 9.21**   ▼

▼   **EXAMPLE 9.3**

Refer to Example 9.2.

   a. Find the normal stress σ' and the shear stress τ_f on the failure plane.

   b. Determine the effective normal stress on the plane of maximum shear stress.

Solution

   a. From Eqs. (7.8) and (7.9),

$$\sigma' \text{ (on the failure plane)} = \frac{\sigma_1' + \sigma_3'}{2} + \frac{\sigma_1' - \sigma_3'}{2} \cos 2\theta$$

and

$$\tau_f = \frac{\sigma_1' - \sigma_3'}{2} \sin 2\theta$$

Substituting the values of $\sigma_1' = 41$ lb/in.^{2}, $\sigma_3' = 16$ lb/in.^{2}, and $\theta = 58°$ into the preceding equations, we get

$$\sigma' = \frac{41 + 16}{2} + \frac{41 - 16}{2} \cos (2 \times 58) = \textbf{23.0 lb/in.}^2$$

and

$$\tau_f = \frac{41 - 16}{2} \sin (2 \times 58) = \textbf{11.2 lb/in.}^2$$

   b. From Eq. (7.9), it can be seen that the maximum shear stress will occur on the plane with $\theta = 45°$. From Eq. (7.8),

$$\sigma' = \frac{\sigma_1' + \sigma_3'}{2} + \frac{\sigma_1' - \sigma_3'}{2} \cos 2\theta$$

Substituting $\theta = 45°$ into the preceding equation gives

$$\sigma' = \frac{41 + 16}{2} + \frac{41 - 16}{2} \cos 90 = \textbf{28.5 lb/in.}^2 \qquad ▼$$

▼   **EXAMPLE 9.4**

The equation of the effective stress failure envelope for normally consolidated clayey soil is $\tau_f = \sigma' \tan 30°$. A drained triaxial test was conducted with the same soil at a chamber confining pressure of 10 lb/in.^{2}. Calculate the deviator stress at failure.

Solution   For normally consolidated clay, $c = 0$. Thus, from Eq. (9.7),

$$\sigma_1' = \sigma_3' \tan^2 \left(45 + \frac{\phi}{2} \right)$$

$$\phi = 30°$$

$$\sigma_1' = 10 \tan^2 \left(45 + \frac{30}{2} \right) = 30 \text{ lb/in.}^2$$

so

$$(\Delta\sigma_d)_f = \sigma_1' - \sigma_3' = 30 - 10 = \textbf{20 lb/in.}^2 \qquad \blacktriangledown$$

▼ **EXAMPLE** 9.5

The results of two drained triaxial tests on a saturated clay are given here:

Specimen I: $\sigma_3 = 10$ lb/in.^{2}

$(\Delta\sigma_d)_f = 24.7$ lb/in.^{2}

Specimen II: $\sigma_3 = 15$ lb/in.^{2}

$(\Delta\sigma_d)_f = 33.5$ lb/in.^{2}

Determine the shear strength parameters.

Solution   Refer to Figure 9.22. For specimen I, the principal stresses at failure are

$$\sigma_3' = \sigma_3 = 10 \text{ lb/in.}^2$$

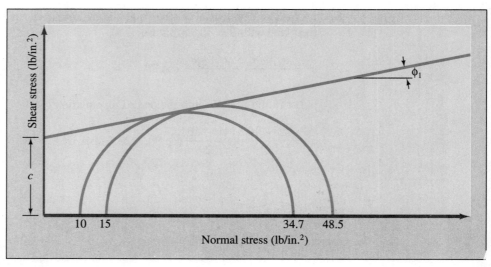

▼ **FIGURE 9.22**

and

$$\sigma_1' = \sigma_1 = \sigma_3 + (\Delta\sigma_d)_f = 10 + 24.7 = 34.7 \text{ lb/in.}^2$$

Similarly, the principal stresses at failure for specimen II are

$$\sigma_3' = \sigma_3 = 15 \text{ lb/in.}^2$$

and

$$\sigma_1' = \sigma_1 = \sigma_3 + (\Delta\sigma_d)_f = 15 + 33.5 = 48.5 \text{ lb/in.}^2$$

Using the relationship given by Eq. (9.7), we have

$$\sigma_1' = \sigma_3' \tan^2 \left(45 + \frac{\phi_1}{2} \right) + 2c \tan \left(45 + \frac{\phi_1}{2} \right)$$

Thus, for specimen I,

$$34.7 = 10 \tan^2 \left(45 + \frac{\phi_1}{2} \right) + 2c \tan \left(45 + \frac{\phi_1}{2} \right)$$

and for specimen II,

$$48.5 = 15 \tan^2 \left(45 + \frac{\phi_1}{2} \right) + 2c \tan \left(45 + \frac{\phi_1}{2} \right)$$

Solving the two preceding equations, we obtain

$$\phi = 28° \qquad c = 2.1 \text{ lb/in.}^2 \qquad \blacktriangledown$$

Drained Angle of Friction for Normally Consolidated Clay

The drained angle of friction, ϕ, generally decreases with the plasticity index of soil. This is shown in Figure 9.23 for a number of clays as reported by Kenney (1959). Although there is a considerable scattering, the general pattern seems to hold.

Consolidated-Undrained Test

The consolidated-undrained test is the most common type of triaxial test. In this test, the saturated soil specimen is first consolidated by an all-round chamber fluid pressure, σ_3, that results in drainage. After the pore water pressure generated by the application of confining pressure is completely dissipated (that is, $u_c = B\sigma_3 = 0$), the deviator stress, $\Delta\sigma_d$, on the specimen is increased to cause shear failure. During this phase of the test, the drainage line from the specimen is kept closed. Since drainage is not permitted, the pore water pressure, Δu_d, will increase. During the test, simultaneous measurements of $\Delta\sigma_d$ and Δu_d are made. The increase in the pore water pressure, Δu_d, can be

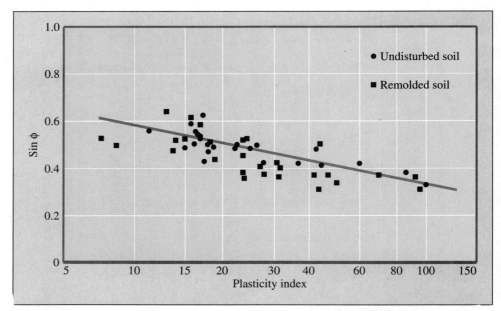

▼   **FIGURE 9.23**   Variation of sin ϕ with plasticity index for a number of soils (after Kenney, 1959)

expressed in a nondimensional form as

$$\bar{A} = \frac{\Delta u_d}{\Delta \sigma_d}$$

(9.15)

where $\bar{A}$ = Skempton's pore pressure parameter (Skempton, 1954).

The general patterns of variation of $\Delta \sigma_d$ and Δu_d with axial strain for sand and clay soils are shown in Figures 9.24d, e, f, and g. In loose sand and normally consolidated clay, the pore water pressure increases with strain. In dense sand and overconsolidated clay, the pore water pressure increases with strain up to a certain limit, beyond which it decreases and becomes negative (with respect to the atmospheric pressure). This is because of a tendency of the soil to dilate.

Unlike the consolidated-drained test, the total and effective principal stresses are not the same in the consolidated-undrained test. Since the pore water pressure at failure is measured in this test, the principal stresses may be analyzed as follows:

▶   Major principal stress at failure (total):

$\sigma_3 + (\Delta \sigma_d)_f = \sigma_1$

▶   Major principal stress at failure (effective):

$\sigma_1 - (\Delta u_d)_f = \sigma_1'$

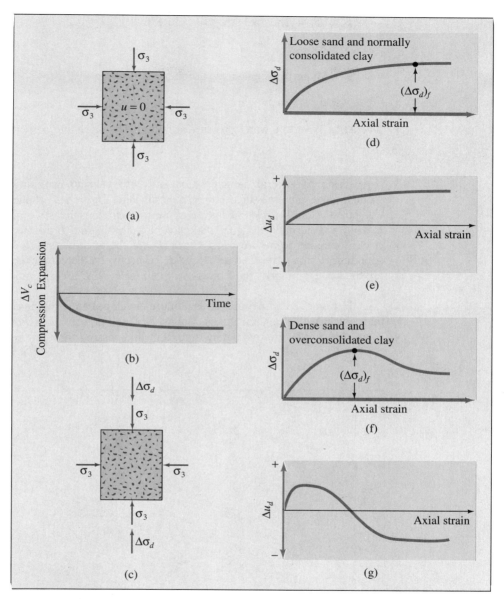

▼ FIGURE 9.24   Consolidated-undrained test: (a) specimen under chamber confining pressure; (b) volume change in specimen caused by confining pressure; (c) deviator stress application; (d) deviator stress against axial strain; (e) associated variation of pore water pressure for loose sand and normally consolidated clay; (f) deviator stress against axial strain; (g) associated variation of pore water pressure for dense sand and overconsolidated clay

▶ Minor principal stress at failure (total):

σ_3

▶ Minor principal stress at failure (effective):

$\sigma_3 - (\Delta u_d)_f = \sigma'_3$

where $(\Delta u_d)_f$ = pore water pressure at failure. The preceding derivations show that

$$\sigma_1 - \sigma_3 = \sigma'_1 - \sigma'_3$$

Tests on several similar specimens with varying confining pressures may be made to determine the shear strength parameters. Figure 9.25 shows the total and effective stress Mohr's circles at failure obtained from consolidated-undrained triaxial tests in sand and normally consolidated clay. Note that A and B are two total stress Mohr's circles obtained from two tests. C and D are the effective stress Mohr's circles corresponding to total stress circles A and B, respectively. The diameters of circles A and C are the same; similarly, the diameters of circles B and D are the same.

In Figure 9.25, the total stress failure envelope can be obtained by drawing a line that touches all the total stress Mohr's circles. For sand and normally consolidated clays, this will be approximately a straight line passing through the origin and may be

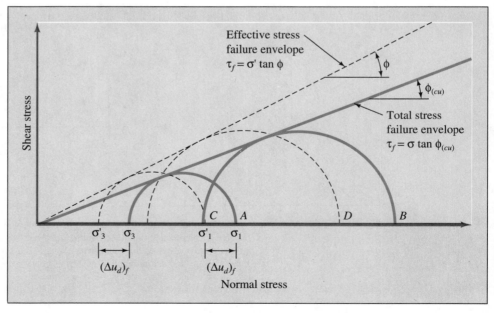

▼ **FIGURE 9.25**   Total and effective stress failure envelopes for consolidated-undrained triaxial tests. (*Note:* The figure assumes that no back pressure is applied.)

expressed by the equation

$$\tau_f = \sigma \tan \phi_{(cu)} \tag{9.16}$$

where σ = total stress

$\phi_{(cu)}$ = the angle that the total stress failure envelope makes with the normal stress axis, also known as the consolidated-undrained angle of shearing resistance

Equation (9.16) is seldom used for practical considerations.

Again referring to Figure 9.25, we see that the failure envelope that is tangent to all the effective stress Mohr's circles can be represented by the equation $\tau_f = \sigma' \tan \phi$, which is the same as that obtained from consolidated-drained tests (see Figure 9.19).

In overconsolidated clays, the total stress failure envelope obtained from consolidated-undrained tests will take the shape shown in Figure 9.26. The straight line $a'b'$ is represented by the equation

$$\tau_f = c_{(cu)} + \sigma \tan \phi_{1(cu)} \tag{9.17}$$

and the straight line $b'c'$ follows the relationship given by Eq. (9.16). The effective stress failure envelope drawn from the effective stress Mohr's circles will be similar to that shown in Figure 9.26.

Consolidated-drained tests on clay soils take considerable time. For that reason, consolidated-undrained tests can be conducted on such soils with pore pressure measurements to obtain the drained shear strength parameters. Since drainage is not allowed in these tests during the application of deviator stress, they can be performed rather quickly.

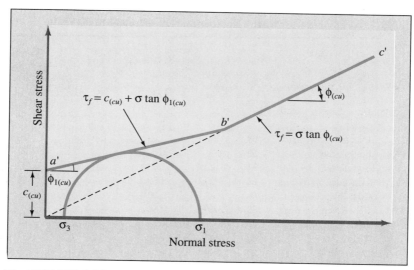

▼ **FIGURE 9.26**   Total stress failure envelope obtained from consolidated-undrained tests in overconsolidated clay

Skempton's pore water pressure parameter $\bar{A}$ has been defined in Eq. (9.15). At failure, the parameter $\bar{A}$ can be written as

$$\bar{A} = \bar{A}_f = \frac{(\Delta u_d)_f}{(\Delta \sigma_d)_f} \qquad\qquad (9.18)$$

The general range of $\bar{A}_f$ values in most clay soils is as follows:

▶ Normally consolidated clays: 0.5 to 1
▶ Overconsolidated clays: -0.5 to 0

Table 9.3 gives the values of $\bar{A}_f$ for some normally consolidated clays as obtained by the Norwegian Geotechnical Institute. Figure 9.27 shows the variation of $\bar{A}_f$ with the overconsolidation ratio of a silty clay as obtained by the author from triaxial tests conducted in the laboratory. For these tests, each saturated soil specimen was initially consolidated by a chamber confining pressure $\sigma_3 = \sigma'_3 = \sigma'_c$. After initial consolidation, the confining pressure was reduced while the drainage line was kept open. Thus, at this time, $\sigma_3 = \sigma'_3 = \sigma'_{3(\text{test})}$ and the overconsolidation ratio can be given as $\sigma'_c / \sigma'_{3(\text{test})}$. Once equilibrium was reached, the drainage line was closed and deviator stress was applied, accompanied by pore water pressure measurement.

▼ **TABLE 9.3** **Triaxial Test Results for Some Normally Consolidated Clays Obtained by the Norwegian Geotechnical Institute***

Location	Liquid limit	Plastic limit	Liquidity index	Sensitivity[a]	Drained friction angle, ϕ (deg)	$\bar{A}_f$
Seven Sisters, Canada	127	35	0.28		19	0.72
Sarpborg	69	28	0.68	5	25.5	1.03
Lilla Edet, Sweden	68	30	1.32	50	26	1.10
Fredrikstad	59	22	0.58	5	28.5	0.87
Fredrikstad	57	22	0.63	6	27	1.00
Lilla Edet, Sweden	63	30	1.58	50	23	1.02
Göta River, Sweden	60	27	1.30	12	28.5	1.05
Göta River, Sweden	60	30	1.50	40	24	1.05
Oslo	48	25	0.87	4	31.5	1.00
Trondheim	36	20	0.50	2	34	0.75
Drammen	33	18	1.08	8	28	1.18

* After Bjerrum and Simons (1960)
[a] See Section 9.7 for the definition of sensitivity.

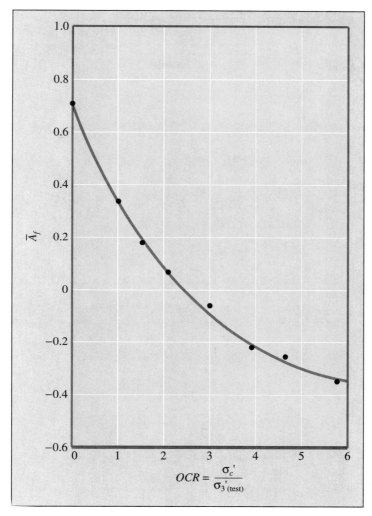

▼ **FIGURE 9.27** Variation of $\bar{A}_f$ with the overconsolidation ratio for a silty clay

▼ **EXAMPLE 9.6**

A consolidated-undrained test on a normally consolidated clay yielded the following results:

$\sigma_3 = 12 \ \text{lb/in.}^2$

Deviator stress, $(\Delta\sigma_d)_f = 9.1 \ \text{lb/in.}^2$

Pore pressure, $(\Delta u_d)_f = 6.8 \ \text{lb/in.}^2$

Calculate the consolidated-undrained friction angle and the drained friction angle.

Solution   Refer to Figure 9.28:

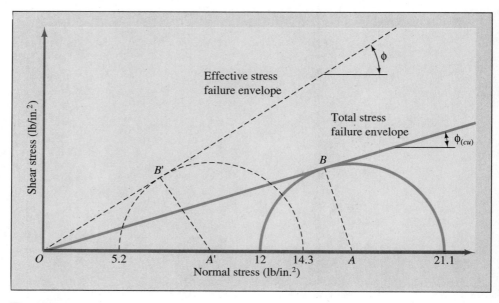

▼   **FIGURE 9.28**

$$\sigma_3 = 12 \text{ lb/in.}^2$$

$$\sigma_1 = \sigma_3 + (\Delta\sigma_d)_f = 12 + 9.1 = 21.1 \text{ lb/in.}^2$$

$$\sigma_1 = \sigma_3 \tan^2\left(45 + \frac{\phi_{cu}}{2}\right)$$

$$21.1 = 12 \tan^2\left(45 + \frac{\phi_{cu}}{2}\right)$$

$$\phi_{cu} = 2\left[\tan^{-1}\left(\frac{21.1}{12}\right)^{0.5} - 45\right] = \mathbf{16°}$$

Again,

$$\sigma'_3 = \sigma_3 - (\Delta\sigma_d)_f = 12 - 6.8 = 5.2 \text{ lb/in.}^2$$

$$\sigma'_1 = \sigma_1 - (\Delta\sigma_d)_f = 21.1 - 6.8 = 14.3 \text{ lb/in.}^2$$

$$\sigma'_1 = \sigma'_3 \tan^2\left(45 + \frac{\phi}{2}\right)$$

$$14.3 = 5.2 \tan^2\left(45 + \frac{\phi}{2}\right)$$

$$\phi = 2\left[\tan^{-1}\left(\frac{14.3}{5.2}\right)^{0.5} - 45\right] = \mathbf{27.8°}   ▼$$

Unconsolidated-Undrained Test

In unconsolidated-undrained tests, drainage from the soil specimen is not permitted during the application of chamber pressure, σ_3. The test specimen is sheared to failure by the application of deviator stress, $\Delta\sigma_d$, without allowing drainage. Since drainage is not allowed at any stage, the test can be performed very quickly. Because of the application of chamber confining pressure, σ_3, the pore water pressure in the soil specimen will increase by u_c. There will be a further increase in the pore water pressure (Δu_d) because of the deviator stress application. Hence, the total pore water pressure, u, in the specimen at any stage of deviator stress application can be given as

$$u = u_c + \Delta u_d \tag{9.19}$$

From Eqs. (9.13) and (9.15), $u_c = B\sigma_3$ and $\Delta u_d = \bar{A}\,\Delta\sigma_d$, so

$$u = B\sigma_3 + \bar{A}\,\Delta\sigma_d = B\sigma_3 + \bar{A}(\sigma_1 - \sigma_3) \tag{9.20}$$

This test is usually conducted on clay specimens and depends on a very important strength concept for saturated cohesive soils. The added axial stress at failure $(\Delta\sigma_d)_f$ is practically the same regardless of the chamber confining pressure. This is shown in Figure 9.29. The failure envelope for the total stress Mohr's circles becomes a horizontal line and hence is called a $\phi = 0$ condition, and

$$\tau_f = c_u \tag{9.21}$$

where c_u is the undrained shear strength and is equal to the radius of the Mohr's circles.

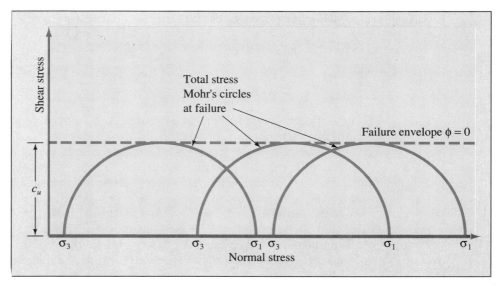

▼ FIGURE 9.29   Total stress Mohr's circles and failure envelope ($\phi = 0$) obtained from unconsolidated-undrained triaxial tests

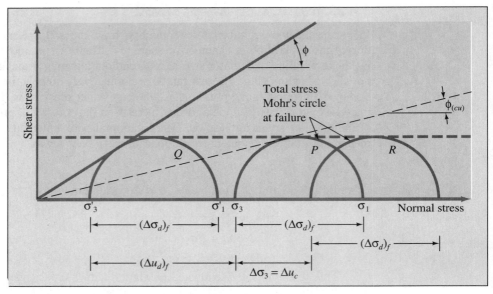

▼   **FIGURE 9.30**   The $\phi = 0$ concept

The reason for obtaining the same added axial stress $(\Delta\sigma_d)_f$ regardless of the confining pressure can be explained as follows. If a clay specimen (no. 1) is consolidated at a chamber pressure σ_3 and then sheared to failure without allowing drainage, the total stress conditions at failure can be represented by the Mohr's circle P in Figure 9.30. The pore pressure developed in the specimen at failure is equal to $(\Delta u_d)_f$. Thus, the major and minor principal effective stresses at failure are

$$\sigma_1' = [\sigma_3 + (\Delta\sigma_d)_f] - (\Delta u_d)_f = \sigma_1 - (\Delta u_d)_f$$

and

$$\sigma_3' = \sigma_3 - (\Delta u_d)_f$$

Q is the effective stress Mohr's circle drawn with the preceding principal stresses. Note that the diameters of circles P and Q are the same.

Now let us consider another similar clay specimen (no. 2) that has been consolidated under a chamber pressure σ_3. If the chamber pressure is increased by $\Delta\sigma_3$ without allowing drainage, the pore water pressure will increase by an amount Δu_c. For saturated soils, under isotropic stresses, the pore water pressure increase is equal to the total stress increase, so $\Delta u_c = \Delta\sigma_3$. At this time the effective confining pressure is equal to $\sigma_3 + \Delta\sigma_3 - \Delta u_c = \sigma_3 + \Delta\sigma_3 - \Delta\sigma_3 = \sigma_3$. This is the same as the effective confining pressure of specimen no. 1 before the application of deviator stress. Hence, if specimen no. 2 is sheared to failure by increasing the axial stress, it should fail at the same deviator stress $(\Delta\sigma_d)_f$ that was obtained for specimen no. 1. The total stress Mohr's circle at failure will be R (Figure 9.30). The added pore pressure increase caused by the application of $(\Delta\sigma_d)_f$ will be $(\Delta u_d)_f$.

At failure the minor principal effective stress:

$$[\sigma_3 + \Delta\sigma_3] - [\Delta u_c + (\Delta u_c)_f] = \sigma_3 - (\Delta u_d)_f = \sigma'_3$$

and the major principal effective stress:

$$[\sigma_3 + \Delta\sigma_3 + (\Delta\sigma_d)_f] - [\Delta u_c + (\Delta u_d)_f] = [\sigma_3 + (\Delta\sigma_d)_f] - (\Delta u_d)_f$$

$$= \sigma_1 - (\Delta u_d)_f = \sigma'_1$$

Thus, the effective stress Mohr's circle will still be Q since strength is a function of effective stress. Note that the diameters of circles P, Q, and R are all the same.

Any value of $\Delta\sigma_3$ could have been chosen for testing specimen no. 2. In any case, the deviator stress $(\Delta\sigma_d)_f$ to cause failure would have been the same.

9.4   UNCONFINED COMPRESSION TEST OF SATURATED CLAY

This is a special type of unconsolidated-undrained test that is commonly used for clay specimens. In this test, the confining pressure σ_3 is 0. An axial load is rapidly applied to the specimen to cause failure. At failure the total minor principal stress is 0 and the total major principal stress is σ_1 (Figure 9.31). Since the undrained shear strength is independent of the confining pressure,

$$\tau_f = \frac{\sigma_1}{2} = \frac{q_u}{2} = c_u \tag{9.22}$$

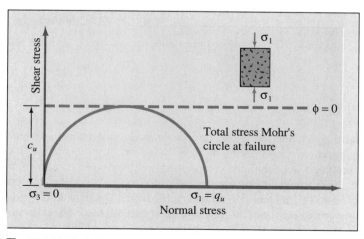

▼ **FIGURE 9.31**   Unconfined compression test

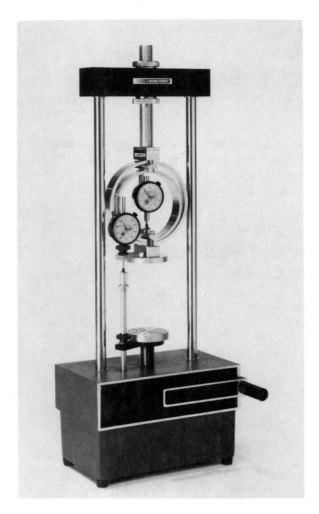

▼ **FIGURE 9.32**   Unconfined compression test equipment (courtesy of Soiltest, Inc., Lake Bluff, Illinois)

where q_u is referred to as the *unconfined compression strength*. Table 9.4 gives the approximate consistencies of clays based on their unconfined compression strength. A photograph of unconfined compression test equipment is shown in Figure 9.32.

Theoretically, for similar saturated clay specimens, the unconfined compression tests and the unconsolidated-undrained triaxial tests should yield the same values of c_u. In practice, however, unconfined compression tests on saturated clays yield slightly lower values of c_u than those obtained from unconsolidated-undrained tests. This fact is demonstrated in Figure 9.33.

▼ **TABLE 9.4** **General Relationship of Consistency and Unconfined Compression Strength of Clays**

| Consistency | q_u | |
	ton/ft^2	kN/m^2 [a]
Very soft	0–0.25	0–23.94 $\simeq$ 24
Soft	0.25–0.5	24–48
Medium	0.5–1	48–96
Stiff	1–2	96–192
Very stiff	2–4	192–383
Hard	>4	>383

[a] Conversion factor: 1 lb/ft^2 = 47.88 N/m^2
Note: Values are rounded off to the nearest whole number.

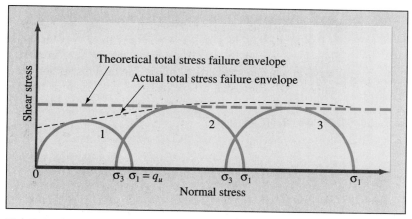

▼ **FIGURE 9.33** Comparison of results of unconfined compression tests and unconsolidated-undrained tests for a saturated clay soil. (*Note:* Mohr's circle no. 1 is for unconfined compression tests; Mohr's circles no. 2 and 3 are for unconsolidated-undrained triaxial tests.)

9.5 GENERAL COMMENTS ON TRIAXIAL TESTS

The following general observations can be made regarding triaxial tests:

1. In contrast to the direct shear test, the shear failure planes of specimens are not predetermined in triaxial tests.

2. From the discussion of various types of triaxial tests, it is clear that the shear strength of any soil is dependent on the pore water pressure generated during loading. The pore water pressure dissipates with drainage. In the field, the shear strength of soil will thus depend on the rate of application of load and drainage.

   For field conditions, in a granular soil, full drainage is likely to occur if the rate of load application is moderate. Under those circumstances, the drained shear strength parameters will govern. In contrast, for normally consolidated clays ($k \simeq 10^{-6}$ cm/sec), the time required for dissipation of excess pore water pressure developed by the construction (for example, a footing) may be too long, and undrained conditions may exist during and immediately after construction. Hence, the $\phi = 0$ condition may be more appropriate for use. For excavation in overconsolidated clay, however, the drained case is most critical.

3. The triaxial test gives a greater ability to control and know stress conditions on the specimen as compared with the direct shear test.

9.6 STRESS PATH

Results of triaxial tests can be represented by diagrams called *stress paths*. A stress path is a line that connects a series of points with each point representing a successive stress state experienced by a soil specimen during the progress of a test. There are several ways in which a stress path can be drawn. This section covers one of them.

Lambe (1964) suggested a type of stress path representation that plots q' against p' (where p' and q' are the coordinates of the top of the Mohr's circle). Thus, relationships for p' and q' are as follows:

$$p' = \frac{\sigma'_1 + \sigma'_3}{2} \tag{9.23}$$

$$q' = \frac{\sigma'_1 - \sigma'_3}{2} \tag{9.24}$$

This type of stress path plot can be explained with the aid of Figure 9.34. Let us consider a normally consolidated clay specimen subjected to an isotropically consolidated-drained triaxial test. At the beginning of the application of deviator stress, $\sigma'_1 = \sigma'_3 = \sigma_3$, so

$$p' = \frac{\sigma'_3 + \sigma'_3}{2} = \sigma'_3 = \sigma_3 \tag{9.25}$$

and

$$q' = \frac{\sigma'_3 - \sigma'_3}{2} = 0 \tag{9.26}$$

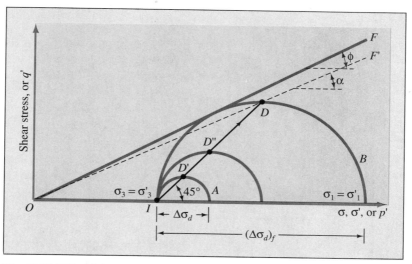

▼ FIGURE 9.34 Stress path—plot of q' against p' for a consolidated-drained triaxial test on a normally consolidated clay

For this condition, p' and q' will plot as a point (that is, I in Figure 9.34). At some other time during deviator stress application, $\sigma'_1 = \sigma'_3 + \Delta\sigma_d = \sigma_3 + \Delta\sigma_d$; $\sigma'_3 = \sigma_3$. The Mohr's circle marked A in Figure 9.34 corresponds to this state of stress on the soil specimen. The values of p' and q' for this stress condition are

$$p' = \frac{\sigma'_1 + \sigma'_3}{2} = \frac{(\sigma'_3 + \Delta\sigma_d) + \sigma'_3}{2} = \sigma'_3 + \frac{\Delta\sigma_d}{2} = \sigma_3 + \frac{\Delta\sigma_d}{2} \tag{9.27}$$

$$q' = \frac{(\sigma'_3 + \Delta\sigma_d) - \sigma'_3}{2} = \frac{\Delta\sigma_d}{2} \tag{9.28}$$

If these values of p' and q' are plotted on Figure 9.34, it will be represented by point D' at the top of the Mohr's circle. So, if the p' and q' values at various stages of the deviator stress application are plotted and these points are joined, a straight line like ID will result. The straight line ID is referred to as the stress path in a q'–p' plot for a consolidated-drained triaxial test. Note that the line ID makes an angle of $45°$ with the horizontal. The point D represents the failure condition of the soil specimen in the test. Also, it may be seen that Mohr's circle B represents the failure stress condition.

For normally consolidated clays, the failure envelope can be given by $\tau_f = \sigma' \tan \phi$. This is the line OF in Figure 9.34 (also see Figure 9.19). A modified failure envelope can now be defined by line OF'. This is commonly called the K_f line. The equation of the K_f line can be expressed as

$$q' = p' \tan \alpha \tag{9.29}$$

where α = the angle that the modified failure envelope makes with the horizontal.

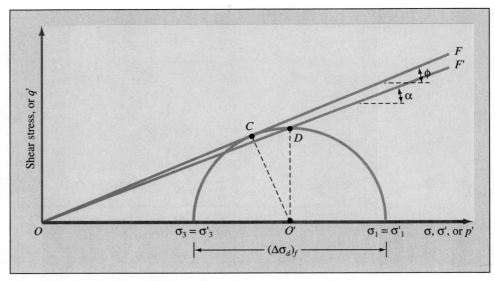

▼   **FIGURE 9.35**   Relationship between ϕ and α

The relationship between the angles ϕ and α can be determined by referring to Figure 9.35 in which, for clarity, the Mohr's circle at failure (that is, circle B) and lines OF and OF' as shown in Figure 9.34 have been redrawn. Note that O' is the center of the Mohr's circle at failure. Now

$$\frac{DO'}{OO'} = \tan \alpha$$

so

$$\tan \alpha = \frac{\dfrac{\sigma'_1 - \sigma'_3}{2}}{\dfrac{\sigma'_1 + \sigma'_3}{2}} = \frac{\sigma'_1 - \sigma'_3}{\sigma'_1 + \sigma'_3} \tag{9.30}$$

Again

$$\frac{CO'}{OO'} = \sin \phi$$

or

$$\sin \phi = \frac{\dfrac{\sigma'_1 - \sigma'_3}{2}}{\dfrac{\sigma'_1 + \sigma'_3}{2}} = \frac{\sigma'_1 - \sigma'_3}{\sigma'_1 + \sigma'_3} \tag{9.31}$$

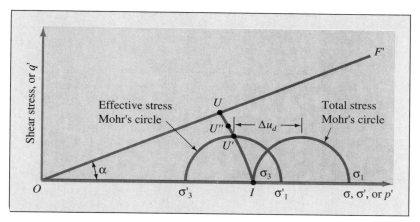

Stress path—plot of q' against p' for a consolidated-undrained triaxial test on a normally consolidated clay

Comparing Eqs. (9.30) and (9.31), we see that

$$\boxed{\sin \phi = \tan \alpha} \tag{9.32}$$

or

$$\phi = \sin^{-1} (\tan \alpha) \tag{9.33}$$

Figure 9.36 shows a q'-p' plot for a normally consolidated clay specimen subjected to an isotropically consolidated-undrained triaxial test. At the beginning of the application of deviator stress, $\sigma_1' = \sigma_3' = \sigma_3$. Hence, $p' = \sigma_3'$ and $q' = 0$. This is represented by point I. At some other stage of the deviator stress application,

$$\sigma_1' = \sigma_3 + \Delta\sigma_d - \Delta u_d$$

and

$$\sigma_3' = \sigma_3 - \Delta u_d$$

So

$$p' = \frac{\sigma_1' + \sigma_3'}{2} = \sigma_3 + \frac{\Delta\sigma_d}{2} - \Delta u_d \tag{9.34}$$

and

$$q' = \frac{\sigma_1' - \sigma_3'}{2} = \frac{\Delta\sigma_d}{2} \tag{9.35}$$

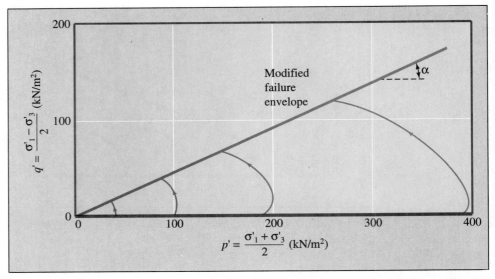

▼ FIGURE 9.37   Stress path for Lagunillas clay—plot of q' against p' as obtained from a number of consolidated-undrained triaxial tests (redrawn after Lambe, 1964)

The preceding values of p' and q' will plot as point U' in Figure 9.36. Points such as U'' represent values of p' and q' as the test progresses. At failure of the soil specimen,

$$p' = \sigma_3 + \frac{(\Delta\sigma_d)_f}{2} - (\Delta u_d)_f \qquad (9.36)$$

and

$$q' = \frac{(\Delta\sigma_d)_f}{2} \qquad (9.37)$$

The values of p' and q' given by Eqs. (9.36) and (9.37) will plot as point U. Hence, the effective stress path for a consolidated-undrained test can be given by the curve $IU'U$. Note that point U will fall on the modified failure envelope OF' (see Figure 9.35) that is inclined at an angle α to the horizontal. Figure 9.37 shows a number of stress paths obtained from consolidated-undrained triaxial tests of Lagunillas clay. Lambe (1964) proposed a technique to evaluate the elastic and consolidation settlements of foundations on clay soils by using the stress paths determined in this manner.

▼ EXAMPLE 9.7

For a normally consolidated clay, the failure envelope is given by the equation $\tau_f = \sigma' \tan \phi$. The corresponding modified failure envelope (q'-p' plot) is given by Eq.

(9.29) as $q' = p' \tan \alpha$. In a similar manner, if the failure envelope is $\tau_f = c + \sigma' \tan \phi$, then the corresponding modified failure envelope is a q'-p' plot that can be expressed as $q' = m + p' \tan \alpha$. Express α as a function of ϕ, and give m as a function of c and ϕ.

Solution   Refer to Figure 9.38:

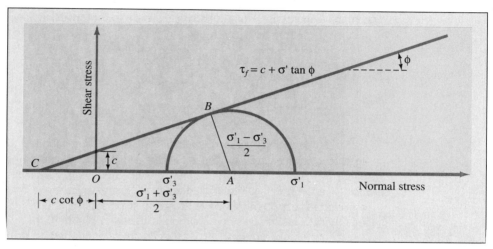

▼ **FIGURE 9.38**

$$\sin \phi = \frac{AB}{AC} = \frac{AB}{CO + OA} = \frac{\left(\dfrac{\sigma'_1 - \sigma'_3}{2}\right)}{c \cot \phi + \left(\dfrac{\sigma'_1 + \sigma'_3}{2}\right)}$$

$$\frac{\sigma'_1 - \sigma'_3}{2} = c \cos \phi + \left(\frac{\sigma'_1 + \sigma'_3}{2}\right) \sin \phi \qquad (a)$$

or

$$q' = m + p' \tan \alpha \qquad (b)$$

Comparing Eqs. (a) and (b), we find

$$m = c \, \cos \phi$$

and

$$\tan \alpha = \sin \phi$$

or

$$\alpha = \tan^{-1} (\sin \phi) \qquad ▼$$

9.7 SENSITIVITY AND THIXOTROPY OF CLAY

For many naturally deposited clay soils, it is observed that the unconfined compression strength is greatly reduced when the soils are tested after remolding without any change in the moisture content, as shown in Figure 9.39. This property of clay soils is called *sensitivity*. The degree of sensitivity may be defined as the ratio of the unconfined compression strength in an undisturbed state to that in a remolded state, or

$$S_t = \frac{q_{u(\text{undisturbed})}}{q_{u(\text{remolded})}} \tag{9.38}$$

The sensitivity ratio of most clays ranges from about 1 to 8; however, highly flocculent marine clay deposits may have sensitivity ratios ranging from about 10 to 80. There are some clays that turn to viscous fluids upon remolding. These are found mostly in the previously glaciated areas of North America and Scandinavia. These clays are referred to as "quick" clays. Rosenqvist (1953) classified clays on the basis of their sensitivity. This general classification is shown in Figure 9.40.

The loss of strength of clay soils from remolding is primarily caused by the destruction of the clay particle structure that was developed during the original process of sedimentation.

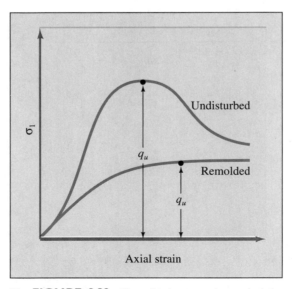

▼ **FIGURE 9.39**   Unconfined compression strength for undisturbed and remolded clay

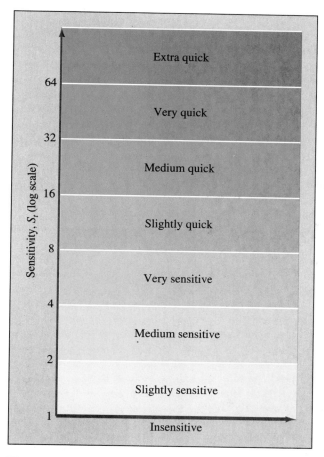

▼ FIGURE 9.40   Classification of clays based on sensitivity

   If, however, after remolding, a soil specimen is kept in an undisturbed state (that is, without any change in the moisture content), it will continue to gain strength with time. This phenomenon is referred to as *thixotropy*. Thixotropy is a time-dependent reversible process in which materials under constant composition and volume soften when remolded. This loss of strength is gradually regained with time when the materials are allowed to rest. This is shown in Figure 9.41a.

   Most soils are partially thixotropic, however; that is, part of the strength loss caused by remolding is never regained with time. The nature of the strength-time variation for partially thixotropic materials is shown in Figure 9.41b. For soils, the difference between the undisturbed strength and the strength after thixotropic hardening can be attributed to the destruction of the clay-particle structure that was developed during the original process of sedimentation.

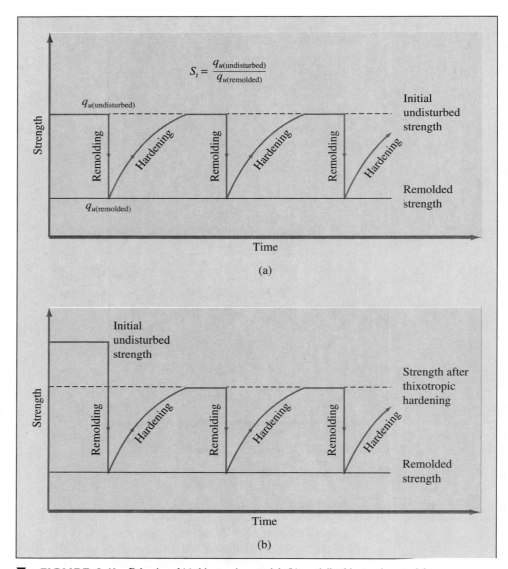

9.8 UNDRAINED COHESION OF NORMALLY CONSOLIDATED AND OVERCONSOLIDATED DEPOSITS

In normally consolidated clay deposits, the undrained shear strength, c_u, increases with the effective overburden pressure. Skempton (1957) gave a statistical relationship for the undrained shear strength, the effective overburden pressure (σ'), and the plasticity index (*PI*) of the soil in the form

$$\frac{c_u}{\sigma'} = 0.11 + 0.0037(PI) \tag{9.39}$$

where *PI* is expressed as a percentage.

Equation (9.39) is an extremely useful relationship. If the plasticity index of a normally consolidated clay deposit is known, then the variation of undrained shear strength with depth can be estimated.

Ladd and colleagues (1977) demonstrated that, for overconsolidated clays, the following relationship holds approximately true:

$$\frac{(c_u/\sigma')_{\text{overconsolidated}}}{(c_u/\sigma')_{\text{normally consolidated}}} = (OCR)^{0.8} \tag{9.40}$$

where *OCR* = overconsolidation ratio.

The overconsolidation ratio was defined in Chapter 8 as

$$OCR = \frac{\sigma'_c}{\sigma'} \tag{9.41}$$

where σ'_c = preconsolidation pressure.

▼ EXAMPLE 9.8

A soil profile is shown in Figure 9.42. The clay is normally consolidated. Its liquid limit is 68 and its plastic limit is 27. Estimate the unconfined compression strength of the clay of a depth of 30 ft measured from the ground surface.

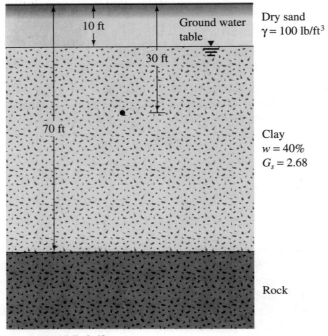

Solution   For the saturated clay layer, the void ratio is

$$e = wG_s = (2.68)(0.4) = 1.07$$

The effective unit weight is

$$\gamma'_{clay} = \left(\frac{G_s - 1}{1 + e}\right)\gamma_w = \frac{(2.68 - 1)(62.4)}{1 + 1.07} = 50.6 \text{ lb/ft}^3$$

The effective stress at a depth of 30 ft from the gound surface is

$$\sigma' = 10\gamma_{sand} + 20\gamma'_{clay} = (10)(100) + (20)(50.6)$$

$$= 2012 \text{ lb/ft}^2$$

From Eq. (9.39),

$$\frac{c_u}{\sigma'} = 0.11 + 0.0037(PI)$$

$$\frac{c_u}{2012} = 0.11 + 0.0037(68 - 27)$$

$$c_u = 526.5 \text{ lb/ft}^2$$

The unconfined compression strength is therefore

$$q_u = 2c_u = (2)(526.5) = \textbf{1053.0 lb/ft}^2 \quad \blacktriangledown$$

▼ **EXAMPLE 9.9**

A clay has a plasticity index of 23. Estimate the undrained cohesion of this clay if the effective overburden pressure, σ', is 1600 lb/ft² and the overconsolidation ratio is 3.2.

Solution From Eq. (9.40),

$$\frac{(c_u/\sigma')_{\text{overconsolidated}}}{(c_u/\sigma')_{\text{normally consolidated}}} = (OCR)^{0.8}$$

However,

$$(c_u/\sigma')_{\text{normally consolidated}} = 0.11 + 0.0037(PI)$$
$$= 0.11 + 0.0037(23) = 0.1951$$
$$OCR = 3.2$$

So

$$(c_u/\sigma')_{\text{overconsolidated}} = (3.2)^{0.8}(0.1951) = 0.495$$

and

$$c_u = (0.495)(\sigma') = (0.495)(1600) = \textbf{792 lb/ft}^2 \qquad ▼$$

9.9 VANE SHEAR TEST

Fairly reliable results for the undrained shear strength c_u ($\phi = 0$ concept), of very plastic cohesive soils may be obtained directly from vane shear tests. The shear vane usually consists of four thin, equal-sized steel plates welded to a steel torque rod (Figure 9.43). First, the vane is pushed into the soil. Then torque is applied at the top of the torque rod to rotate the vane at a uniform speed. A cylinder of soil of height h and diameter d will resist the torque until the soil fails. The undrained shear strength of the soil can be calculated as follows.

If T is the maximum torque applied at the head of the torque rod to cause failure, it should be equal to the sum of the resisting moment of the shear force along the side surface of the soil cylinder (M_s) and the resisting moment of the shear force at each end (M_e) (Figure 9.44a):

$$T = M_s + \underbrace{M_e + M_e}_{\text{Two ends}} \tag{9.42}$$

The resisting moment M_s can be given as

$$M_s = \underbrace{(\pi dh)c_u}_{\substack{\text{Surface} \\ \text{area}}} \underbrace{(d/2)}_{\substack{\text{Moment} \\ \text{arm}}} \tag{9.43}$$

where d = diameter of the shear vane
h = height of the shear vane

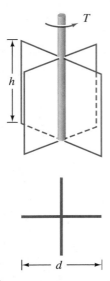

▼   **FIGURE 9.43**   Diagram of vane shear equipment

For the calculation of M_e, investigators have assumed several types of distribution of shear strength mobilization at the ends of the soil cylinder:

1. *Triangular:* Shear strength mobilization of c_u at the periphery of the soil cylinder and decreases linearly to 0 at the center.
2. *Uniform:* Shear strength mobilization is constant (that is, c_u) from the periphery to the center of the soil cylinder.
3. *Parabolic:* Shear strength mobilization is c_u at the periphery of the soil cylinder and decreases parabolically to 0 at the center.

These variations in shear strength mobilization are shown in Figure 9.44b. In general, the torque T at failure can be expressed as

$$T = \pi c_u \left[\frac{d^2 h}{2} + \beta \frac{d^3}{4} \right] \tag{9.44}$$

or

$$c_u = \frac{T}{\pi \left[\dfrac{d^2 h}{2} + \beta \dfrac{d^3}{4} \right]} \tag{9.45}$$

where $\beta = 1/2$ for triangular mobilization of undrained shear strength
$\beta = 2/3$ for uniform mobilization of undrained shear strength
$\beta = 3/5$ for parabolic mobilization of undrained shear strength
[Note that Eq. (9.45) is usually referred to as Calding's equation.]

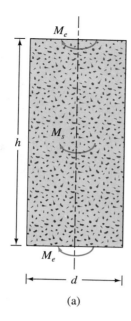

(a)

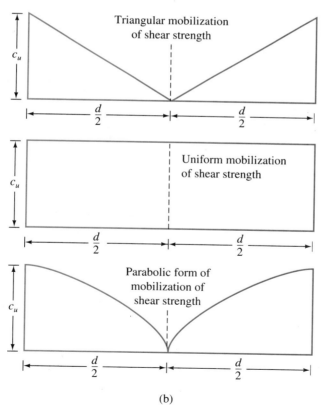

(b)

▼ **FIGURE 9.44** Derivation of Eq. (9.43): (a) resisting moment of shear force; (b) variations in shear strength mobilization

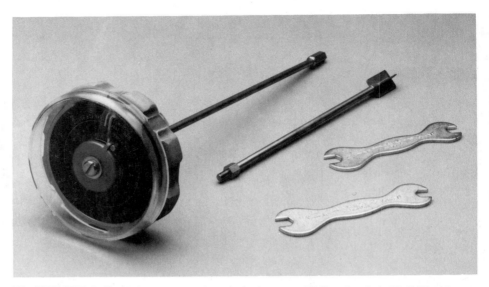

▼ **FIGURE 9.45**   Laboratory vane shear device (courtesy of Soiltest, Inc., Lake Bluff, Illinois)

Vane shear tests can be conducted in the laboratory and also in the field during soil exploration. The laboratory shear vane has dimensions of about $\frac{1}{2}$ in. (12.7 mm) in diameter and 1 in. (25.4 mm) in height. Figure 9.45 shows a photograph of laboratory vane shear equipment. Field shear vanes with the following dimensions are used by the Bureau of Reclamation:

$d = 2$ in. (50.8 mm); $h = 4$ in. (101.6 mm)

$d = 3$ in. (76.2 mm); $h = 6$ in. (152.4 mm)

$d = 4$ in. (101.6 mm); $h = 8$ in. (203.2 mm)

In the field, where there is a considerable variation in the undrained shear strength with depth, vane shear tests are extremely useful. In a short period of time, it is possible to establish a reasonable pattern of the change of c_u with depth. However, if the clay deposit at a given site is more or less uniform, a few unconsolidated-undrained triaxial tests on undisturbed samples will allow a reasonable estimation of soil parameters for design work. Vane shear tests are also limited by the strength of soils in which they can be used. The undrained shear strength obtained from a vane shear test is also dependent on the rate of application of torque T.

Bjerrum (1974) showed that as the plasticity of soils increases, c_u obtained from vane shear tests may give results that are unsafe for foundation design. For that reason, he suggested the following correction:

$$c_{u(\text{design})} = \lambda c_{u(\text{vane shear})}$$

(9.46)

▼ **TABLE 9.5**   **Variation of**
λ with PI
[Eq. (9.46)]

PI	λ
5	1.32
10	1.16
15	1.07
20	0.997
25	0.945
30	0.902
35	0.866
40	0.835

where λ = correction factor = $1.7 - 0.54 \log(PI)$ (9.47)
         PI = plasticity index

Table 9.5 shows the variation of λ with PI.

Effect of the Rate of Rotation of Vane on Undrained Shear Strength

In conducting a field vane shear test, the vane is rotated at an approximate rate of 6° per min. However, it has been shown that the undrained cohesion, c_u, as determined from the vane shear test may be a function of the clay type and the rate of angular rotation (ω) of the vane. Figures 9.46 and 9.47 show the laboratory vane shear test

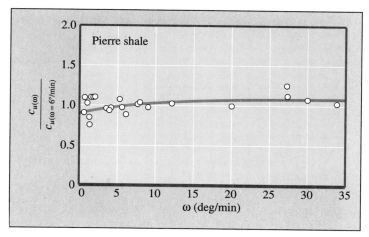

▼ **FIGURE 9.46**   Variation of $[c_{u(\omega)}]/[c_{u(\omega = 6°/\text{min})}]$ with ω for Pierre shale (after Sharifounnasab and Ullrich, 1985)

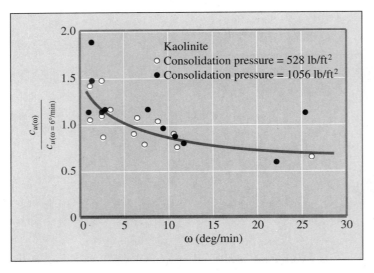

▼ **FIGURE 9.47**   Variation of $[c_{u(\omega)}]/[c_{u(\omega=6°/min)}]$ with ω for kaolinite (after Sharifounnasab and Ullrich, 1985)

results on Pierre shale and kaolinite, respectively. In these figures, the plot has been made between $c_{u(\omega)}/c_{u(\omega=6°/min)}$ and ω. From these figures, the following observations can be made:

1. The rate of rotation of the vane has some influence on the magnitude of c_u. Based on the results of Sharifounnasab and Ullrich (1985),

$$\frac{c_{u(\omega)}}{c_{u(\omega=6°/min)}} = \alpha'\left(\frac{\omega°/min}{6°/min}\right)^{\beta'} \tag{9.48}$$

For Pierre shale, $\alpha' = 1.03$ and $\beta' = 0.05$. Similarly, for kaolinite, $\alpha' = 0.99$ and $\beta' = -0.16$.

2. The reason β' is positive for Pierre shale and negative for kaolinite is that Pierre shale is highly plastic. In contrast, in clays with low plasticity, like kaolinite, partial drainage may occur during the vane test, which results in a lower value of c_u.

9.10 OTHER METHODS FOR DETERMINATION OF UNDRAINED SHEAR STRENGTH OF COHESIVE SOILS

A modified form of the vane shear apparatus is a *torvane* (Figure 9.48), which is a hand-held device with a calibrated spring. This can be used for determining c_u for tube specimens collected from the field during soil exploration, and it can also be used in the field. The torvane is pushed into the soil and then rotated until the soil fails. The undrained shear strength can be read off the top of the calibrated dial.

▼ **FIGURE 9.48** Torvane (courtesy of Soiltest, Inc., Lake Bluff, Illinois)

Figure 9.49 shows a *pocket penetrometer*, which is pushed directly into the soil. The unconfined compression strength (q_u) is measured by a calibrated spring. This can be used both in the laboratory and in the field.

9.11   SHEAR STRENGTH OF UNSATURATED COHESIVE SOILS

The relationship among total stress, effective stress, and pore water pressure for unsaturated soils was briefly introduced in Chapter 6 as

$$\sigma' = \sigma - u_a + \chi(u_a - u_w) \tag{9.49}$$

where σ' = effective stress
σ = total stress
u_a = pore air pressure
u_w = pore water pressure

When the expression for σ' is substituted into the shear strength equation [Eq. (9.8)], which is based on effective stress parameters, we get

$$\tau_f = c + [\sigma - u_a + \chi(u_a - u_w)] \tan \phi \tag{9.50}$$

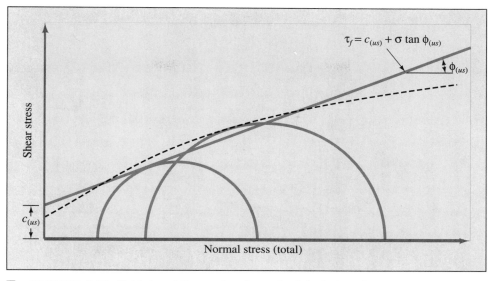

▼ **FIGURE 9.50** Total stress failure envelope for unsaturated cohesive soils

As pointed out before, the values of χ depend primarily on the degree of saturation. With ordinary triaxial equipment used for laboratory testing, it is not possible to determine accurately the effective stresses in unsaturated soil specimens, so the common practice is to conduct undrained triaxial tests on unsaturated specimens and measure only the total stress. Figure 9.50 shows a total stress failure envelope obtained from a number of undrained triaxial tests conducted with a given initial degree of saturation. The failure envelope is generally curved. Higher confining pressure causes higher compression of the air in void spaces; thus, the solubility of void air in void water is increased. For design purposes, the curved envelope is sometimes approximated as a straight line as shown in Figure 9.50 with an equation

$$\tau_f = c_{(us)} + \sigma \tan \phi_{(us)} \tag{9.51}$$

Note that $c_{(us)}$ and $\phi_{(us)}$ in the preceding equation are empirical constants.

Figure 9.51 shows the variation of the total stress envelopes with change of the initial degree of saturation obtained from undrained tests on an inorganic clay. Note that for these tests the specimens were prepared with approximately the same initial dry unit weight of about 106 lb/ft³ (16.7 kN/m³). For a given total normal stress, the shear stress to cause failure decreases as the degree of saturation increases. When the degree of saturation reaches 100%, the total stress failure envelope becomes a horizontal line that is the same as with the $\phi = 0$ concept.

In practical cases where there is a likelihood that a cohesive soil deposit might become saturated because of rainfall or a rise in the ground water table, the strength of partially saturated clay should not be used for design considerations. Instead, the unsaturated soil specimens collected from the field must be saturated in the laboratory and the undrained strength determined.

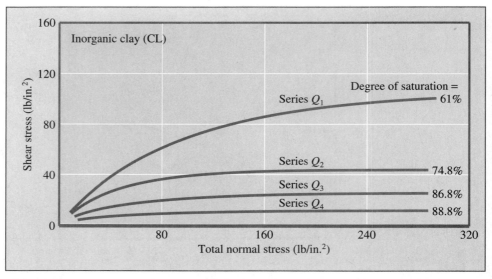

▼ **FIGURE 9.51** Variation of the total stress envelope with change of initial degree of saturation obtained from undrained tests of an inorganic clay (after Casagrande and Hirschfeld, 1960)

9.12 GENERAL COMMENTS

In this chapter, the shear strengths of grandular and cohesive soils were examined. Laboratory procedures for determining the shear strength parameters were described.

In textbooks, determination of the shear strength parameters of cohesive soils appears fairly simple. However, in practice, the proper choice of these parameters for design and stability checks of various earth, earth-retaining, and earth-supported structures is very difficult and requires experience and an appropriate theoretical background in geotechnical engineering. In this chapter, three types of strength parameters (that is, consolidated-drained, consolidated-undrained, and unconsolidated-undrained) were introduced. Their use depends on drainage conditions.

Consolidated-drained strength parameters can be used to determine the long-term stability of structures such as earth embankments and cut slopes. Consolidated-undrained shear strength parameters can be used to study stability problems relating to cases where the soil initially is fully consolidated and then there is rapid loading. An excellent example of this is the stability of slopes of earth dams after rapid drawdown. The unconsolidated-undrained shear strength of clays can be used to evaluate the end-of-construction stability of saturated cohesive soils with the assumption that the load caused by construction has been applied rapidly and there has been little time for drainage to take place. The bearing capacity of foundations on soft saturated clays and the stability of the base of embankments on soft clays are examples of this condition.

The unconsolidated-undrained shear strength of some saturated clays can vary depending on the direction of load application; this is referred to as *anisotropy with*

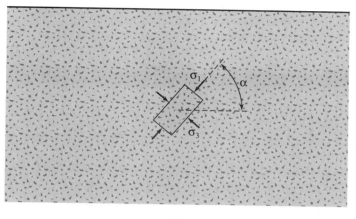

(a)

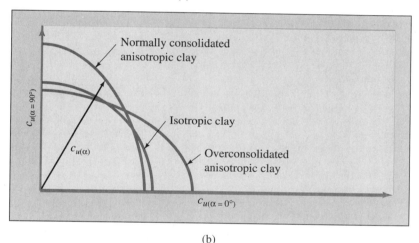

(b)

▼ **FIGURE 9.52** Strength anisotropy in clay

respect to strength. Anisotropy is primarily caused by the nature of the deposition of the cohesive soils, and subsequent consolidation makes the clay particles orient perpendicular to the direction of the major principal stress. Parallel orientation of the clay particles can cause the strength of clay to vary with direction. Figure 9.52a shows an element of saturated clay in a deposit with the major principal stress making an angle α with respect to the horizontal. For anisotropic clays, the magnitude of c_u will be a function of α. Casagrande and Carrillo (1944) proposed the following relationship for the directional variation of undrained shear strength:

$$c_{u(\alpha)} = c_{u(\alpha = 0°)} + [c_{u(\alpha = 90°)} - c_{u(\alpha = 0°)}] \sin^2 \alpha \tag{9.52}$$

For normally consolidated clays, $c_{u(\alpha = 90°)} > c_{u(\alpha = 0°)}$; for overconsolidated clays, $c_{u(\alpha = 90°)} < c_{u(\alpha = 0°)}$. Figure 9.52b shows the directional variation for $c_{u(\alpha)}$ based on Eq. (9.52). The anisotropy with respect to strength for clays can have an important

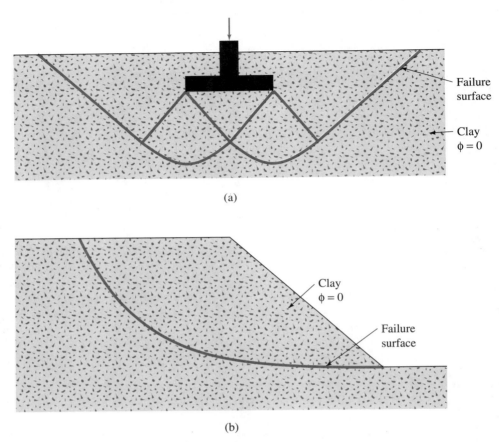

▼ FIGURE 9.53   Potential failure surface in soil (a) under a shallow foundation; (b) in an earth embankment

effect on the load-bearing capacity of foundations and the stability of earth embankment (Figure 9.53) because the direction of the major principal stress along the potential failure surfaces changes.

The sensitivity of clays was discussed in Section 9.7. It is imperative that sensitive clay deposits are properly identified. For instance, when machine foundations (which are subjected to vibratory loading) are constructed over sensitive clays, the clay may substantially lose its load-bearing capacity, and failure may occur.

PROBLEMS

9.1   A direct shear test was conducted on a specimen of dry sand with a normal stress of 20 lb/in.^{2}. Failure occurred at a shear stress of 13.5 lb/in.^{2}. The size of the specimen tested was 2 in. × 2 in. × 1 in. (height). Determine the angle of friction, ϕ. For a normal stress of 12 lb/in.^{2}, what shear force would be required to cause failure in the specimen?

9.2   The size of a sand specimen in a direct shear test was 2 in. × 2 in. × 1.2 in. (height). It is known that, for the sand, tan $\phi = 0.65/e$ (where e = void ratio) and the specific gravity of solids $G_s = 2.65$. During the test, a normal stress of 20 lb/in.^{2} was applied. Failure occurred at a shear stress of 15 lb/in.^{2}. What was the weight of the sand specimen?

9.3 The angle of friction of a compacted dry sand is 38°. In a direct shear test on the sand, a normal stress of 12 lb/in.^{2} was applied. The size of the specimen was 2 in. × 2 in. × 1.2 in. (height). What shear force (in lb) will cause failure?

9.4 Repeat Problem 9.3 with the following changes:

Friction angle = 37°

Normal stress = 150 kN/m^2

9.5 Following are the results of four drained direct shear tests on a normally consolidated clay:

Diameter of specimen = 50 mm

Height of specimen = 25 mm

Test no.	Normal force (N)	Shear force at failure (N)
1	271	120.6
2	406.25	170.64
3	474	204.1
4	541.65	244.3

Draw a graph for the shear stress at failure against the normal stress. Determine the drained angle of friction from the graph.

9.6 Repeat Problem 9.5 with the following data:

Test no.	Normal force (N)	Shear force at failure (N)
1	150	110
2	250	183
3	350	255
4	550	400

9.7 The relationship between the relative density, D_r, and the angle of friction, ϕ, of a sand can be given as $\phi° = 25 + 0.18D_r$ (D_r in %). A drained triaxial test on the same sand was conducted with a chamber confining pressure of 15 lb/in.^{2}. The relative density of compaction was 45%. Calculate the major principal stress at failure.

9.8 Consider the triaxial test described in Problem 9.7.
a. Estimate the angle that the failure plane makes with the major principal plane.
b. Determine the normal and shear stresses (when the specimen failed) on a plane that makes an angle of 30° with the major principal plane.

9.9 The effective stress failure envelope of a sand can be given as $\tau_f = \sigma' \tan 41°$. A drained triaxial test was conducted on the same sand. The specimen failed when the deviator stress was 57.22 lb/in.^{2}. What was the chamber confining pressure during the test?

9.10   Refer to Problem 9.9.

    a.   Estimate the angle that the failure plane makes with the minor principal plane.

    b.   Determine the normal stress and the shear stress on a plane that makes an angle of 35° with the minor principal plane.

9.11   For a normally consolidated clay, the results of a drained triaxial test are as follows:

Chamber confining pressure = 150 kN/m²

Deviator stress at failure = 275 kN/m²

Determine the soil friction angle, ϕ.

9.12   For a normally consolidated clay, we are given $\phi = 25°$. In a drained triaxial test, the specimen failed at a deviator stress of 22 lb/in.². What was the chamber confining pressure, σ_3?

9.13   A consolidated-drained triaxial test was conducted on a normally consolidated clay. The results were as follows:

$\sigma_3 = 276$ kN/m²

$(\Delta\sigma_d)_f = 276$ kN/m²

Determine:

    a.   The angle of friction, ϕ

    b.   The angle θ that the failure plane makes with the major principal stress

    c.   The normal stress σ' and the shear stress τ_f on the failure plane

9.14   Refer to Problem 9.13.

    a.   Determine the effective normal stress on the plane of maximum shear stress.

    b.   Explain why the shear failure took place along the plane as determined in part (b) and not along the plane of maximum shear stress.

9.15   The results of two drained triaxial tests on a saturated clay are given here:

Specimen I: chamber confining pressure = 69 kN/m²
                    deviator stress at failure = 213 kN/m²

Specimen II: chamber confining pressure = 120 kN/m²
                    deviator stress at failure = 258.7 kN/m²

Calculate the shear strength parameters of the soil.

9.16   If the specimen of clay described in Problem 9.15 is tested in a triaxial apparatus with a chamber confining pressure of 200 kN/m², what will be the major principal stress at failure? Assume a full drained condition during the test.

9.17   A sandy soil has a drained angle of friction of 36°. In a drained triaxial test on the same soil, the deviator stress at failure is 2.8 ton/ft². What is the chamber confining pressure?

9.18   A consolidated-undrained test was conducted on a normally consolidated specimen with a chamber confining pressure of 20 lb/in.². The specimen failed while the deviator stress was 18 lb/in.². The pore water pressure in the specimen at that time was 10.9 lb/in.². Determine the consolidated-undrained and the drained friction angles.

9.19   Repeat Problem 9.18 with the following values:

$\sigma_3 = 12$ lb/in.²

$(\Delta\sigma_d)_f = 8.38$ lb/in.²

$(\Delta u_d)_f = 5.6$ lb/in.²

9.20   The shear strength of a normally consolidated clay can be given by the equation $\tau_f = \sigma'$

tan 28°. A consolidated-undrained triaxial test was conducted on the clay. Following are the results of the test:

Chamber confining pressure = 105 kN/m²

Deviator stress at failure = 97 kN/m²

Determine:
a. The consolidated-undrained friction angle (ϕ_{cu})
b. The pore water pressure developed in the clay specimen at failure

9.21 For the clay specimen described in Problem 9.20, what would have been the deviator stress at failure if a drained test had been conducted with the same chamber confining pressure (that is, $\sigma_3 = 105$ kN/m²)?

9.22 For a clay soil, we are given $\phi = 28°$ and $\phi_{cu} = 18°$. A consolidated-undrained triaxial test was conducted on this clay soil with a chamber confining pressure of 15 lb/in.². Determine the deviator stress and the pore water pressure at failure.

9.23 Repeat Problem 9.22 with $\phi_{cu} = 19°$, $\phi = 28°$, and $\sigma_3 = 120$ kN/m².

9.24 During a consolidated-undrained triaxial test on a clayey soil specimen, the minor and major principal stresses at failure were 2000 lb/ft² and 3900 lb/ft², respectively. What will be the axial stress at failure if a similar specimen is subjected to an unconfined compression test?

9.25 The friction angle, ϕ, of a normally consolidated clay specimen collected during field exploration was determined from drained triaxial tests to be 22°. The unconfined compression strength, q_u, of a similar specimen was found to be 120 kN/m². Determine the pore water pressure at failure for the unconfined compression test.

9.26 Repeat Problem 9.25 with $\phi = 25°$ and $q_u = 1.25$ ton/ft².

9.27 A 30-ft-thick normally consolidated clay deposit is shown in Figure 9.54. The plasticity index of the clay is 28%. Estimate the undrained cohesion at the middle of the clay layer.

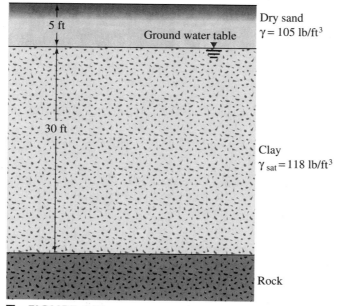

5 ft

Ground water table

Dry sand
$\gamma = 105$ lb/ft³

30 ft

Clay
$\gamma_{sat} = 118$ lb/ft³

Rock

▼ **FIGURE 9.54**

9.28   Assume that the clay described in Problem 9.27 is overconsolidated and the overconsolidation ratio is 3. Estimate the unconfined compression strength of a clay specimen taken from the middle of the clay layer.

9.29   The results of two consolidated-drained triaxial tests on a clayey soil are given here:

Test no.	σ_3' (lb/in.^{2})	$\sigma_{1(\text{failure})}'$ (lb/in.^{2})
1	26.6	73.4
2	11.96	48.04

Use the failure envelope equation given in Example 9.7—that is, $q' = m + p' \tan \alpha$. (Do not plot the graph.)

a.   Find m and α.
b.   Find c and ϕ.

REFERENCES

Acar, Y. B., Durgunoglu, H. T., and Tumay, M. T. (1982). "Interface Properties of Sand," *Journal of the Geotechnical Engineering Division*, ASCE, Vol. 108, No. GT4, 648–654.

Bishop, A. W., and Bjerrum, L. (1960). "The Relevance of the Triaxial Test to the Solution of Stability Problems," *Proceedings*, Research Conference on Shear Strength of Cohesive Soils, ASCE, 437–501.

Bjerrum, L. (1974). "Problems of Soil Mechanics and Construction on Soft Clays," Norwegian Geotechnical Institute, *Publications No 110*, Oslo.

Bjerrum, L., and Simons, N. E. (1960). "Compression of Shear Strength Characteristics of Normally Consolidated Clay," *Proceedings*, Research Conference on Shear Strength of Cohesive Soils, ASCE, 711–726.

Black, D. K., and Lee, K. L. (1973). "Saturating Laboratory Samples by Back Pressure," *Journal of the Soil Mechanics and Foundations Division*, ASCE, Vol. 99, No. SM1, 75–93.

Casagrande, A., and Carrillo, N. (1944). "Shear Failure of Anisotropic Materials," in *Contribution to Soil Mechanics 1941–1953*, Boston Society of Civil Engineers, Boston.

Casagrande, A., and Hirschfeld, R. C. (1960). "Stress Deformation and Strength Characteristics of a Clay Compacted to a Constant Dry Unit Weight," *Proceedings*, Research Conference on Shear Strength of Cohesive Soils, ASCE, 359–417.

Coulomb, C. A. (1776). "Essai sur une application des regles de Maximums et Minimis á quelques Problèmes de Statique, relatifs á l'Architecture," *Memoires de Mathematique et de Physique*, Présentés, á l'Academie Royale des Sciences, Paris, Vol. 3, 38.

Kenney, T. C. (1959). Discussion, *Proceedings*, ASCE, Vol. 85, No. SM3, 67–79.

Ladd, C. C., Foote, R., Ishihara, K., Schlosser, F., and Poulos, H. G. (1977). "Stress Deformation and Strength Characteristics," *Proceedings*, 9th International Conference on Soil Mechanics and Foundation Engineering, Tokyo, Vol. 2, 421–494.

Lambe, T. W. (1964). "Methods of Estimating Settlement," *Journal of the Soil Mechanics and Foundations Division*, ASCE, Vol. 90, No. SM5, 47–74.

Mohr, O. (1900). "Welche Umstände Bedingen die Elastizitätsgrenze und den Bruch eines Materiales?," *Zeitschrift des Vereines Deutscher Ingenieure*, Vol. 44, 1524–1530, 1572–1577.

Rosenqvist, I. Th. (1953). "Considerations on the Sensitivity of Norwegian Quick Clays," *Geotechnique*, Vol. 3, No. 5, 195–200.

Sharifounnasab, M., and Ullrich, C. R. (1985). "Rate of Shear Vane Strength," *Journal of Geotechnical Engineering*, ASCE, No. 111, No. GT1, 135–140.

Skempton, A. W. (1954). "The Pore Water Coefficients *A* and *B*," *Geotechnique*, Vol. 4, 143–147.

Skempton, A. W. (1957). "Discussion: The Planning and Design of New Hong Kong Airport," *Proceedings*, Institute of Civil Engineers, London, Vol. 7, 305–307.

Supplementary Reference for Further Study

Bishop, A. W., and Henkel, D. J. (1957). *The Measurement of Soil Properties in the Triaxial Test*, Edward Arnold (Publishers) Ltd., London.

LATERAL EARTH PRESSURE

Retaining structures such as retaining walls, basement walls, and bulkheads are commonly encountered in foundation engineering, and they may support slopes of earth masses. Proper design and construction of these structures require a thorough knowledge of the lateral forces that act between the retaining structures and the soil masses being retained. These lateral forces are caused by lateral earth pressure. This chapter is devoted to the study of various earth pressure theories.

10.1 EARTH PRESSURE AT REST

Let us consider the mass of soil shown in Figure 10.1. The mass is bounded by a frictionless wall AB that extends to an infinite depth. A soil element located at a depth z will be subjected to a vertical pressure σ_v and a horizontal pressure σ_h. For the case considered here, σ_v and σ_h are both effective and total pressures, and there are no shear stresses on the vertical and horizontal planes.

If the wall AB is static—that is, if it does not move either to the right or to the left of its initial position—the soil mass will be in a state of *elastic equilibrium*; that is, the horizontal strain is zero. The ratio of the horizontal stress to the vertical stress is called the *coefficient of earth pressure at rest, K_o*, or

$$K_o = \frac{\sigma_h}{\sigma_v} \tag{10.1}$$

Since $\sigma_v = \gamma z$, we have

$$\sigma_h = K_o(\gamma z) \tag{10.2}$$

For coarse-grained soils, the coefficient of earth pressure at rest can be estimated by the empirical relationship (Jaky, 1944)

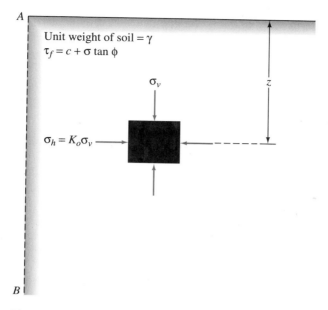

▼ **FIGURE 10.1** Earth pressure at rest

$$K_o = 1 - \sin \phi \qquad (10.3)$$

where ϕ = drained friction angle.

For fine-grained, normally consolidated soils, Massarsch (1979) suggested the following equation for K_o:

$$K_o = 0.44 + 0.42 \left[\frac{PI(\%)}{100} \right] \qquad (10.4)$$

For overconsolidated clays, the coefficient of earth pressure at rest can be approximated as

$$K_{o(\text{overconsolidated})} = K_{o(\text{normally consolidated})} \sqrt{OCR} \qquad (10.5)$$

where OCR = overconsolidation ratio. The overconsolidation ratio was defined in Chapter 8 as

$$OCR = \frac{\text{preconsolidation pressure}}{\text{present effective overburden pressure}} \qquad (10.6)$$

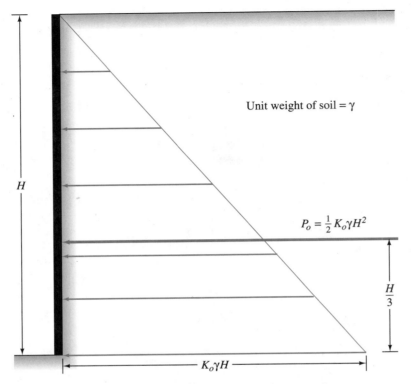

Unit weight of soil $= \gamma$

$P_o = \frac{1}{2} K_o \gamma H^2$

$\frac{H}{3}$

$K_o \gamma H$

▼   **FIGURE 10.2**   Distribution of earth pressure at rest on a wall

Figure 10.2 shows the distribution of earth pressure at rest on a wall of height H. The total force per unit length of the wall, P_o, is equal to the area of the pressure diagram, so

$$P_o = \frac{1}{2} K_o \gamma H^2 \tag{10.7}$$

Earth Pressure at Rest for Partially Submerged Soil

Figure 10.3a shows a wall of height H. The ground water table is located at a depth H_1 below the ground surface, and there is no compensating water on the other side of the wall. For $z \le H_1$, the lateral earth pressure at rest can be given as $\sigma_h = K_o \gamma z$. The variation of σ_h with depth is shown by triangle ACE in Figure 10.3a. However, for $z \ge H_1$ (that is, below the ground water table), the pressure on the wall is found from the effective stress and pore water pressure components in the following manner:

$$\text{effective vertical pressure} = \sigma'_v = \gamma H_1 + \gamma'(z - H_1) \tag{10.8}$$

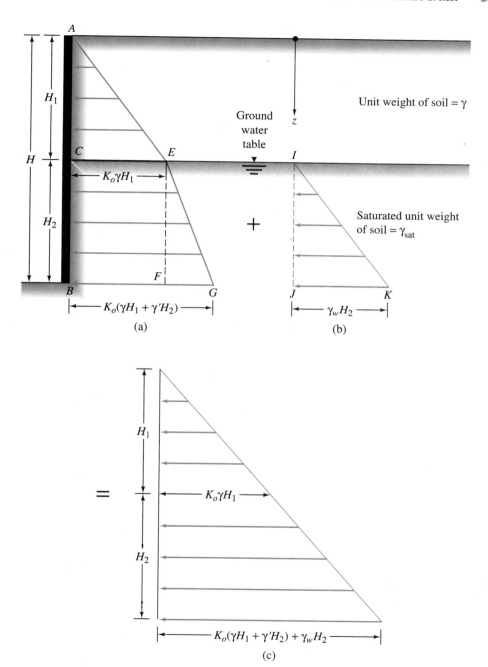

FIGURE 10.3   Distribution of earth pressure at rest for partially submerged soil

where $\gamma' = \gamma_{sat} - \gamma_w =$ the effective unit weight of soil. So, the effective lateral pressure at rest is

$$\sigma_h' = K_o \sigma_v' = K_o[\gamma H_1 + \gamma'(z - H_1)] \tag{10.9}$$

The variation of σ_h' with depth is shown by *CEGB* in Figure 10.3a. Again, the lateral pressure from pore water is

$$u = \gamma_w(z - H_1) \tag{10.10}$$

The variation of u with depth is shown in Figure 10.3b.

Hence, the total lateral pressure from earth and water at any depth $z \geq H_1$ is equal to

$$\sigma_h = \sigma_h' + u$$
$$= K_o[\gamma H_1 + \gamma'(z - H_1)] + \gamma_w(z - H_1) \tag{10.11}$$

The force per unit width of the wall can be found from the sum of the areas of the pressure diagrams in Figures 10.3a and b and is equal to

$$P_o = \underbrace{\frac{1}{2} K_o \gamma H_1^2}_{\substack{\text{Area} \\ ACE}} + \underbrace{K_o \gamma H_1 H_2}_{\substack{\text{Area} \\ CEFB}} + \underbrace{\frac{1}{2}(K_o \gamma' + \gamma_w)H_2^2}_{\substack{\text{Areas} \\ EFG \text{ and } IJK}} \tag{10.12}$$

10.2 COMMENTS ON EARTH PRESSURE INCREASE CAUSED BY COMPACTION

While designing a wall that may be subjected to lateral earth pressure at rest, care must be taken in evaluating the value of K_o. Sherif, Fang, and Sherif (1984), based on their laboratory tests, showed that Jaky's equation for K_o [Eq. (10.3)] gives good results when the backfill is loose sand. However, for a dense sand backfill, Eq. (10.3) grossly underestimates the lateral earth pressure at rest. This is because of the process of compaction of backfill. For that reason, they recommended the following design relationship:

$$K_o = (1 - \sin \phi) + \left[\frac{\gamma_d}{\gamma_{d(min)}} - 1\right]5.5 \tag{10.13}$$

where $\gamma_d =$ actual compacted dry unit weight of the sand behind the wall
$\gamma_{d(min)} =$ dry unit weight of the sand in the loosest state (Chapter 2)

10.3 RANKINE'S THEORY OF ACTIVE AND PASSIVE EARTH PRESSURES

The term *plastic equilibrium* in soil refers to the condition where every point in a soil mass is on the verge of failure. Rankine (1857) investigated the stress conditions in soil at a state of plastic equilibrium. The following sections deal with Rankine's theory of earth pressure.

Rankine's Active State

Figure 10.4a shows the same soil mass that was considered in Figure 10.1. It is bounded by a frictionless wall, AB, that extends to an infinite depth. The vertical and horizontal principal stresses (total and effective) on a soil element at a depth z are σ_v and σ_h, respectively. As we have seen in Section 10.1, if the wall AB is not allowed to move at all, then $\sigma_h = K_o \sigma_v$. The stress condition in the soil element can be represented by the Mohr's circle a in Figure 10.4b. However, if the wall AB is allowed to move away from the soil mass gradually, then the horizontal principal stress will decrease. Ultimately a state will be reached when the stress condition in the soil element can be represented by the Mohr's circle b, the state of plastic equilibrium, and failure in soil occurs. This represents *Rankine's active state*, and the pressure σ_a on the vertical plane (which is a principal plane) is Rankine's *active earth pressure*. Following is the derivation for expressing σ_a in terms of γ, z, c, and ϕ. From Figure 10.4b,

$$\sin \phi = \frac{CD}{AC} = \frac{CD}{AO + OC}$$

but

$$CD = \text{the radius of the failure circle} = \frac{\sigma_v - \sigma_a}{2}$$

$$AO = c \cot \phi$$

and

$$OC = \frac{\sigma_v + \sigma_a}{2}$$

so

$$\sin \phi = \frac{\dfrac{\sigma_v - \sigma_a}{2}}{c \cot \phi + \dfrac{\sigma_v + \sigma_a}{2}}$$

or

$$c \cos \phi + \frac{\sigma_v + \sigma_a}{2} \sin \phi = \frac{\sigma_v - \sigma_a}{2}$$

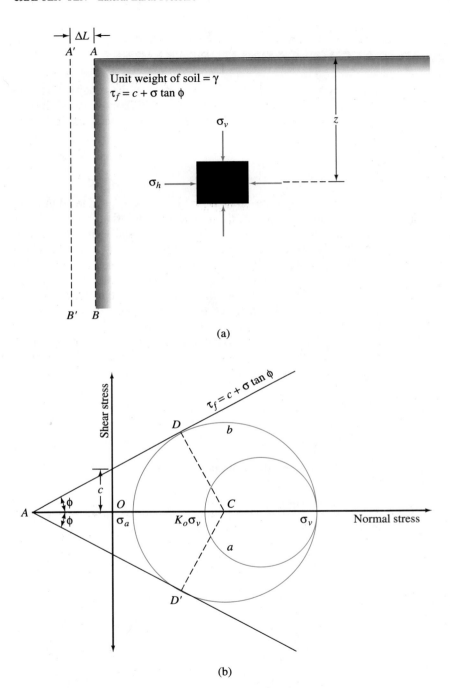

(a)

(b)

▼ **FIGURE 10.4**   Rankine's active earth pressure

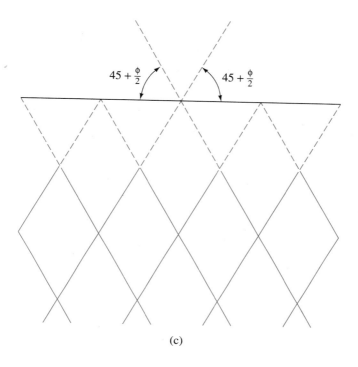

(c)

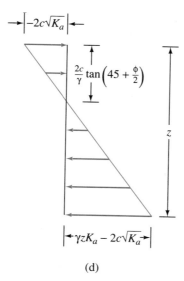

(d)

▼   **FIGURE 10.4**   (*Continued*)

or

$$\sigma_a = \sigma_v \frac{1 - \sin \phi}{1 + \sin \phi} - 2c \frac{\cos \phi}{1 + \sin \phi} \tag{10.14}$$

But

σ_v = the vertical effective overburden pressure = γz

$$\frac{1 - \sin \phi}{1 + \sin \phi} = \tan^2 \left(45 - \frac{\phi}{2} \right)$$

and

$$\frac{\cos \phi}{1 + \sin \phi} = \tan \left(45 - \frac{\phi}{2} \right)$$

Substituting the above into Eq. (10.14), we get

$$\sigma_a = \gamma z \tan^2 \left(45 - \frac{\phi}{2} \right) - 2c \tan \left(45 - \frac{\phi}{2} \right) \tag{10.15}$$

The variation of σ_a with depth is shown in Figure 10.4d. For cohesionless soils, $c = 0$ and

$$\sigma_a = \sigma_v \tan^2 \left(45 - \frac{\phi}{2} \right) \tag{10.16}$$

The ratio of σ_a to σ_v is called the *coefficient of active earth pressure, K_a*, or

$$K_a = \frac{\sigma_a}{\sigma_v} = \tan^2 \left(45 - \frac{\phi}{2} \right) \tag{10.17}$$

Again, from Figure 10.4b we can see that the failure planes in the soil make $\pm(45 + \phi/2)$-degree angles with the direction of the major principal plane—that is, the horizontal. These are called *slip planes*. The slip planes are shown in Figure 10.4c.

Rankine's Passive State

Rankine's passive state is explained in Figure 10.5. AB is a frictionless wall that extends to an infinite depth. The initial stress condition on a soil element is represented by the Mohr's circle a in Figure 10.5b. If the wall is gradually pushed into the soil mass, the principal stress σ_h will increase. Ultimately the wall will reach a situation where the stress condition for the soil element can be expressed by the Mohr's circle b. At this time, failure of the soil will occur. This is referred to as *Rankine's passive state*. The

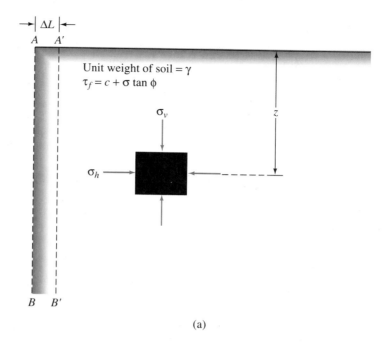

(a)

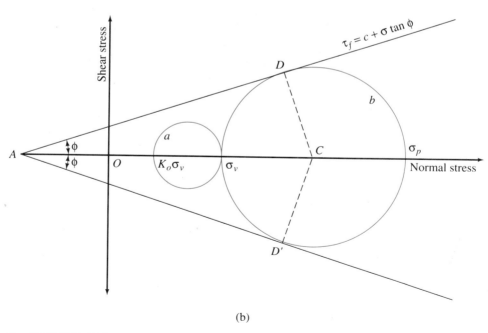

(b)

▼ **FIGURE 10.5** Rankine's passive earth pressure

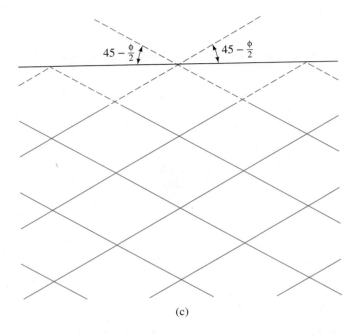

(c)

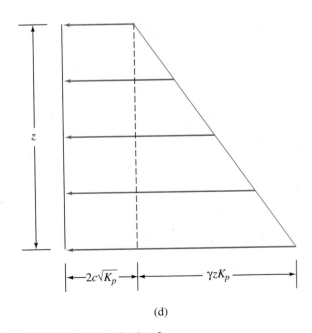

(d)

▼ **FIGURE 10.5** (*Continued*)

lateral earth pressure σ_p, which is the major principal stress, is called *Rankine's passive earth pressure*. From Figure 10.5b, it can be shown that

$$\sigma_p = \sigma_v \tan^2\left(45 + \frac{\phi}{2}\right) + 2c \tan\left(45 + \frac{\phi}{2}\right)$$

$$= \gamma z \tan^2\left(45 + \frac{\phi}{2}\right) + 2c \tan\left(45 + \frac{\phi}{2}\right) \qquad (10.18)$$

The derivation is similar to that for Rankine's active state.

Figure 10.5d shows the variation of passive pressure with depth. For cohensionless soils ($c = 0$),

$$\sigma_p = \sigma_v \tan^2\left(45 + \frac{\phi}{2}\right)$$

or

$$\frac{\sigma_p}{\sigma_v} = K_p = \tan^2\left(45 + \frac{\phi}{2}\right) \qquad (10.19)$$

K_p in the preceding equation is referred to as the *coefficient of Rankine's passive earth pressure*.

The points D and D' on the failure circle (Figure 10.5b) correspond to the slip planes in the soil. For Rankine's passive state, the slip planes make $\pm(45 - \phi/2)$-degree angles with the direction of the minor principal plane—that is, in the horizontal direction. Figure 10.5c shows the distribution of slip planes in the soil mass.

Effect of Wall Yielding

We have seen in the preceding discussion that sufficient movement of the wall is necessary to achieve a state of plastic equilibrium. However, the distribution of lateral earth pressure against a wall is very much influenced by the manner in which the wall actually yields. In most simple retaining walls (see Figure 10.6), movement may occur by simple translation or, more frequently, by rotation about the bottom.

For preliminary theoretical analysis, let us consider a frictionless retaining wall represented by a plane AB as shown in Figure 10.7a. If the wall AB rotates sufficiently about its bottom to a position $A'B$, then a triangular soil mass ABC' adjacent to the wall will reach Rankine's active state. Since the slip planes in Rankine's active state make angles of $\pm(45 + \phi/2)$ degrees with the major principal plane, the soil mass in the state of plastic equilibrium is bounded by the plane BC', which makes an angle of $(45 + \phi/2)$ degrees with the horizontal. The soil inside the zone ABC' undergoes the same unit deformation in the horizontal direction everywhere and is equal to $\Delta L_a/L_a$.

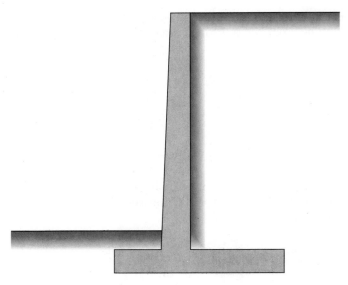

▼ **FIGURE 10.6**   Cantilever retaining wall

The lateral earth pressure on the wall at any depth z from the ground surface can be calculated by Eq. (10.15).

In a similar manner, if the frictionless wall AB (Figure 10.7b) rotates sufficiently into the soil mass to a position $A''B$, then the triangular mass of soil ABC'' will reach Rankine's passive state. The slip plane BC'' bounding the soil wedge that is at a state of plastic equilibrium will make an angle of $(45 - \phi/2)$ degrees with the horizontal. Every point of the soil in the triangular zone ABC'' will undergo the same unit deformation in the horizontal direction and is equal to $\Delta L_p/L_p$. The passive pressure on the wall at any depth z can be evaluated by using Eq. (10.18).

Typical values of maximum wall tilt (ΔL_a and ΔL_p) required for achieving Rankine's state are given in Table 10.1. Figure 10.8 shows the variation of lateral earth pressure with wall tilt.

▼ **TABLE 10.1**   **Typical Values of $\Delta L_a/H$ and $\Delta L_p/H$ for Rankine's State**

Soil type	$\Delta L_a/H$	$\Delta L_p/H$
Loose sand	0.001–0.002	0.01
Dense sand	0.0005–0.001	0.005
Soft clay	0.02	0.04
Stiff clay	0.01	0.02

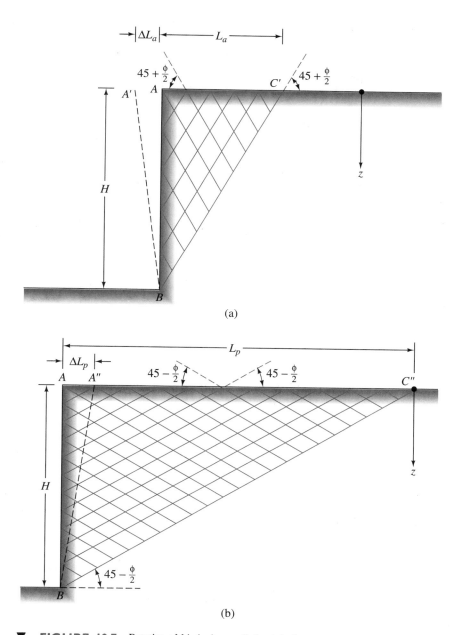

▼ **FIGURE 10.7** Rotation of frictionless wall about the bottom

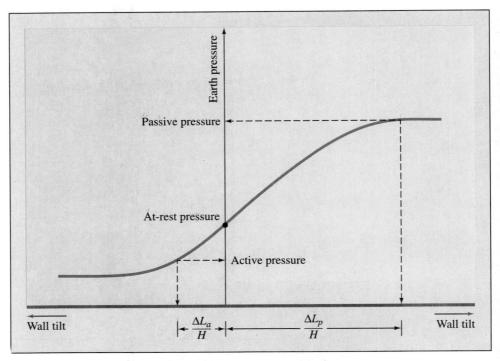

▼ **FIGURE 10.8**   Variation of the magnitude of lateral earth pressure with wall tilt

10.4 DIAGRAMS FOR LATERAL EARTH PRESSURE DISTRIBUTION AGAINST RETAINING WALLS

Backfill—Cohesionless Soil with Horizontal Ground Surface

Active Case: Figure 10.9a shows a retaining wall with cohensionless soil backfill that has a horizontal ground surface. The unit weight and the angle of friction of the soil are γ and ϕ, respectively.

For Rankine's active state, the earth pressure at any depth against the retaining wall can be given by Eq. (10.15):

$$\sigma_a = K_a \gamma z \qquad (\textit{Note: } c = 0)$$

σ_a increases linearly with depth, and at the bottom of the wall, it will be

$$\sigma_a = K_a \gamma H \qquad (10.20)$$

The total force, P_a, per unit length of the wall is equal to the area of the pressure diagram, so

$$P_a = \frac{1}{2} K_a \gamma H^2 \tag{10.21}$$

Passive Case: The lateral pressure distribution against a retaining wall of height H for Rankine's passive state is shown in Figure 10.9b. The lateral earth pressure at any depth z [Eq. (10.19), $c = 0$] is

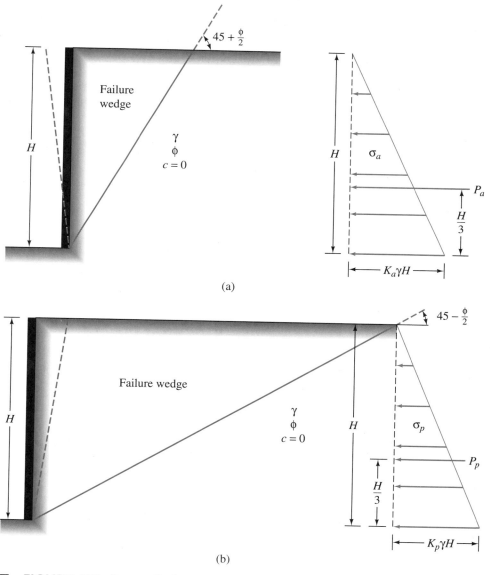

(a)

(b)

▼ **FIGURE 10.9**   Pressure distribution against retaining wall for cohensionless soil backfill with horizontal ground surface: (a) Rankine's active state; (b) Rankine's passive state

$$\sigma_p = K_p \gamma H \tag{10.22}$$

The total force, P_p, per unit length of the wall is

$$P_p = \frac{1}{2} K_p \gamma H^2 \tag{10.23}$$

Backfill—Partially Submerged Cohesionless Soil Supporting Surcharge

Active Case: Figure 10.10a shows a frictionless retaining wall of height H and a backfill of cohesionless soil. The ground water table is located at a depth of H_1 below the ground surface, and the backfill is supporting a surcharge pressure of q per unit area. From Eq. (10.17), the effective active earth pressure at any depth can be given by

$$\sigma'_a = K_a \sigma'_v \tag{10.24}$$

where σ'_v and σ'_a = the effective vertical pressure and lateral pressure, respectively. At $z = 0$,

$$\sigma_v = \sigma'_v = q \tag{10.25}$$

and

$$\sigma_a = \sigma'_a = K_a q \tag{10.26}$$

At depth $z = H_1$,

$$\sigma_v = \sigma'_v = (q + \gamma H_1) \tag{10.27}$$

and

$$\sigma_a = \sigma'_a = K_a(q + \gamma H_1) \tag{10.28}$$

At depth $z = H$,

$$\sigma'_v = (q + \gamma H_1 + \gamma' H_2) \tag{10.29}$$

and

$$\sigma'_a = K_a(q + \gamma H_1 + \gamma' H_2) \tag{10.30}$$

where $\gamma' = \gamma_{sat} - \gamma_w$. The variation of σ'_a with depth is shown in Figure 10.10b.

The lateral pressure on the wall from the pore water between $z = 0$ and H_1 is 0, and for $z > H_1$, it increases linearly with depth (Figure 10.10c). At $z = H$,

$$u = \gamma_w H_2 \tag{10.31}$$

The total lateral pressure diagram (Figure 10.10d) is the sum of the pressure diagrams shown in Figures 10.10b and c. The total active force per unit length of the wall is the area of the total pressure diagram. Thus,

$$P_a = K_a q H + \frac{1}{2} K_a \gamma H_1^2 + K_a \gamma H_1 H_2 + \frac{1}{2} (K_a \gamma' + \gamma_w) H_2^2 \tag{10.32}$$

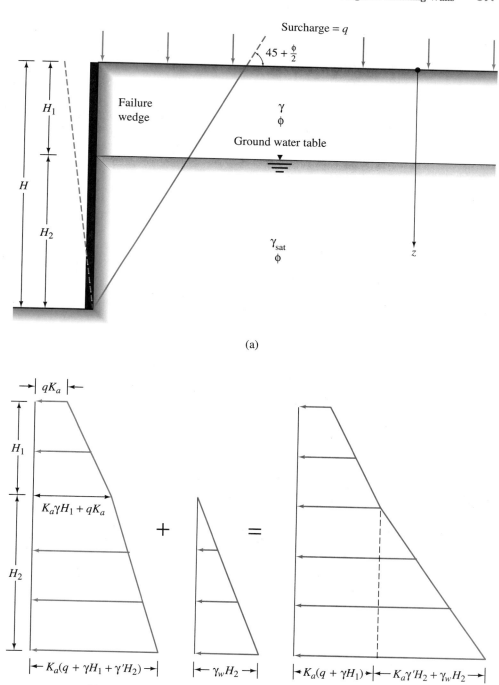

(a)

(b)   (c)   (d)

▼ **FIGURE 10.10** Rankine's active earth pressure distribution against retaining wall with partially submerged cohesionless soil backfill supporting surcharge

Passive Case: Figure 10.11a shows the same retaining wall as was shown in Figure 10.10a. Rankine's passive pressure (effective) at any depth against the wall can be given by Eq. (10.19):

$$\sigma'_p = K_p \sigma'_v$$

Using the preceding equation, we can determine the variation of σ'_p with depth as shown in Figure 10.11b. The variation of the pressure on the wall from water with depth is shown in Figure 10.11c. Figure 10.11d shows the distribution of the total pressure σ_p with depth. The total lateral passive force per unit length of the wall is the area of the diagram given in Figure 10.11d, or

$$P_p = K_p qH + \frac{1}{2} K_p \gamma H_1^2 + K_p \gamma H_1 H_2 + \frac{1}{2}(K_p \gamma' + \gamma_w)H_2^2 \tag{10.33}$$

Backfill—Cohesive Soil with Horizontal Backfill

Active Case: Figure 10.12a shows a frictionless retaining wall with a cohesive soil backfill. The active pressure against the wall at any depth below the ground surface can be expressed as [Eq. (10.15)]

$$\sigma_a = K_a \gamma z - 2\sqrt{K_a}\, c$$

The variation of $K_a \gamma z$ with depth is shown in Figure 10.12b, and the variation of $2\sqrt{K_a}\, c$ with depth is shown in Figure 10.12c. Note that $2\sqrt{K_a}\, c$ is not a function of z, and hence Figure 10.12c is a rectangle. The variation of the net value of σ_a with depth is plotted in Figure 10.12d. Also note that, because of the effect of cohesion, σ_a is negative in the upper part of the retaining wall. The depth z_o at which the active pressure becomes equal to 0 can be found from Eq. (10.15) as

$$K_a \gamma z_o - 2\sqrt{K_a}\, c = 0$$

or

$$z_o = \frac{2c}{\gamma \sqrt{K_a}} \tag{10.34}$$

For the undrained condition—that is, $\phi = 0$, $K_a = \tan^2 45 = 1$, and $c = c_u$ (undrained cohesion),

$$z_o = \frac{2c_u}{\gamma} \tag{10.35}$$

So, with time, tensile cracks at the soil-wall interface will develop up to a depth z_o.

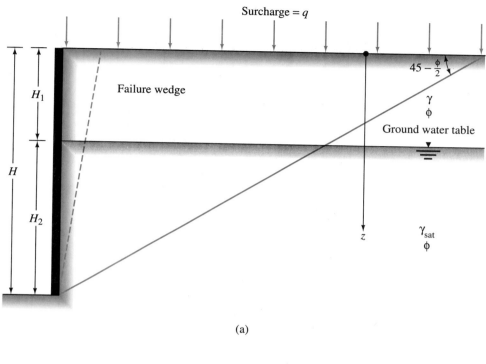

(a)

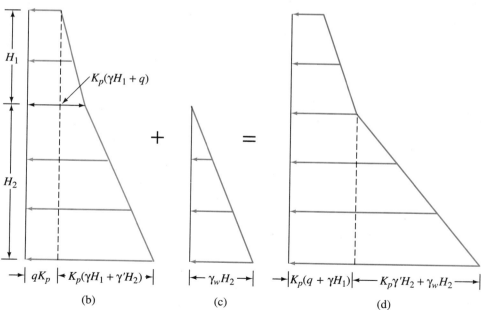

▼ **FIGURE 10.11**   Rankine's passive earth pressure distribution against retaining wall with partially submerged cohesionless soil backfill supporting surcharge

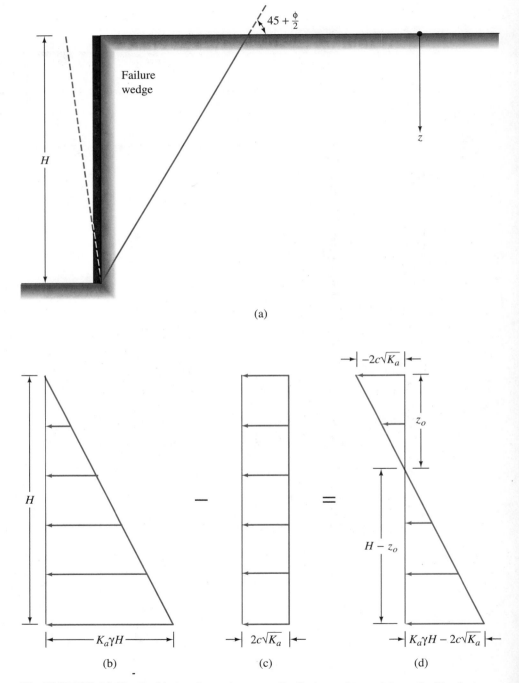

(a)

(b)   (c)   (d)

▼ **FIGURE 10.12**   Rankine's active earth pressure distribution against retaining wall with cohesive soil backfill

The total active force per unit length of the wall can be found from the area of the total pressure diagram (Figure 10.12d), or

$$P_a = \frac{1}{2} K_a \gamma H^2 - 2\sqrt{K_a} cH \tag{10.36}$$

For $\phi = 0$ condition,

$$P_a = \frac{1}{2} \gamma H^2 - 2c_u H \tag{10.37}$$

For calculation of the total active force, it is common practice to take the tensile cracks into account. Since there is no contact between the soil and the wall up to a depth of z_o after the development of tensile cracks, the active pressure distribution against the wall between $z = 2c/(\gamma\sqrt{K_a})$ and H (Figure 10.12d) only is considered. In that case,

$$\begin{aligned}
P_a &= \frac{1}{2} (K_a \gamma H - 2\sqrt{K_a} c)\left(H - \frac{2c}{\gamma\sqrt{K_a}} \right) \\
&= \frac{1}{2} K_a \gamma H^2 - 2\sqrt{K_a} cH + 2\frac{c^2}{\gamma}
\end{aligned} \tag{10.38}$$

For the $\phi = 0$ condition,

$$P_a = \frac{1}{2} \gamma H^2 - 2c_u H + 2\frac{c_u^2}{\gamma} \tag{10.39}$$

Passive Case: Figure 10.13a shows the same retaining wall with backfill similar to that considered in Figure 10.12a. Rankine's passive pressure against the wall at depth z can be given by [Eq. (10.18)]

$$\sigma_p = K_p \gamma z + 2\sqrt{K_p} c$$

At $z = 0$,

$$\sigma_p = 2\sqrt{K_p} c \tag{10.40}$$

and at $z = H$,

$$\sigma_p = K_p \gamma H + 2\sqrt{K_p} c \tag{10.41}$$

The variation of σ_p with depth is shown in Figure 10.13b. The passive force per unit length of the wall can be found from the area of the pressure diagrams as

$$P_p = \frac{1}{2} K_p \gamma H^2 + 2\sqrt{K_p} cH \tag{10.42}$$

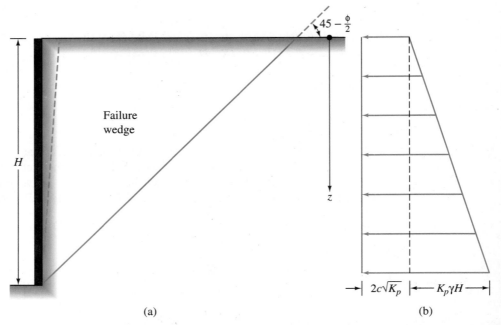

▼ **FIGURE 10.13**   Rankine's passive earth pressure distribution against retaining wall with cohesive soil backfill

For the $\phi = 0$ condition, $K_p = 1$ and

$$P_p = \frac{1}{2} \gamma H^2 + 2c_u H \qquad (10.43)$$

▼ **EXAMPLE 10.1**

Calculate the Rankine active and passive forces per unit length of the wall shown in Figure 10.14a, and also determine the location of the resultant.

Solution   To determine the active force, since $c = 0$,

$$\sigma_a = K_a \sigma_v = K_a \gamma z$$

$$K_a = \frac{1 - \sin \phi}{1 + \sin \phi} = \frac{1 - \sin 30°}{1 + \sin 30°} = \frac{1}{3}$$

At $z = 0$, $\sigma_a = 0$; at $z = 15$ ft, $\sigma_a = (1/3)(100)(15) = 500$ lb/ft².
    The active pressure distribution diagram is shown in Figure 10.14b:

$$\text{active force } P_a = \frac{1}{2}\,(15)(500)$$

$$= \textbf{3750 lb/ft}$$

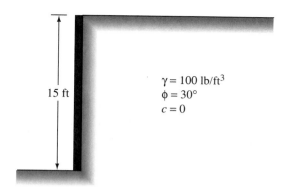

$\gamma = 100 \text{ lb/ft}^3$
$\phi = 30°$
$c = 0$

(a)

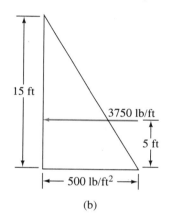

15 ft

3750 lb/ft

5 ft

500 lb/ft²

(b)

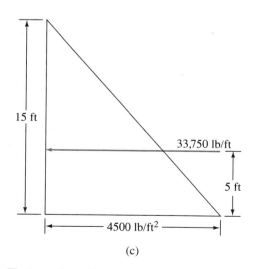

15 ft

33,750 lb/ft

5 ft

4500 lb/ft²

(c)

▼ **FIGURE 10.14**

The pressure distribution is triangular, and so P_a will act at a distance of $15/3 = 5$ ft above the bottom of the wall.

To determine the passive force, we are given $c = 0$, so

$$\sigma_p = K_p \sigma_v = K_p \gamma z$$

$$K_p = \frac{1 + \sin \phi}{1 - \sin \phi} = \frac{1 + 0.5}{1 - 0.5} = 3$$

At $z = 0$, $\sigma_p = 0$; at $z = 15$ ft, $\sigma_p = 3(100)(15) = 4500$ lb/ft^2.

The passive pressure distribution against the wall is shown in Figure 10.14c. Now

$$P_p = \frac{1}{2}(15)(4500) = \textbf{33,750 lb/ft}$$

The resultant will act at a distance of $15/3 = 5$ ft above the bottom of the wall.   ▼

▼   **EXAMPLE   10.2**

If the retaining wall shown in Figure 10.14a is restrained from moving, what will be the lateral force per unit length of the wall?

Solution   If the wall is restrained from moving, the backfill will exert at-rest earth pressure. Thus,

$$\sigma_h = K_o \sigma_v = K_o(\gamma z) \quad \text{[Eq. (10.2)]}$$

$$K_o = 1 - \sin \phi \quad \text{[Eq. (10.3)]}$$

or

$$K_o = 1 - \sin 30° = 0.5$$

and at $z = 0$, $\sigma_h = 0$; at 15 ft, $\sigma_h = (0.5)(15)(100) = 750$ lb/ft^2.

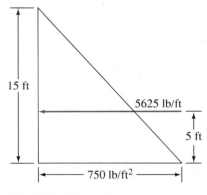

15 ft

5625 lb/ft

5 ft

750 lb/ft^2

▼   **FIGURE 10.15**

The pressure distribution diagram is shown in Figure 10.15.

$$P_o = \frac{1}{2}(15)(750) = \textbf{5625 lb/ft} \qquad \blacktriangledown$$

▼ **EXAMPLE 10.3**

A retaining wall that has a soft, saturated clay backfill is shown in Figure 10.16. For the undrained condition ($\phi = 0$) of the backfill, determine:

a. The maximum depth of the tensile crack
b. P_a before the tensile crack occurs
c. P_a after the tensile crack occurs

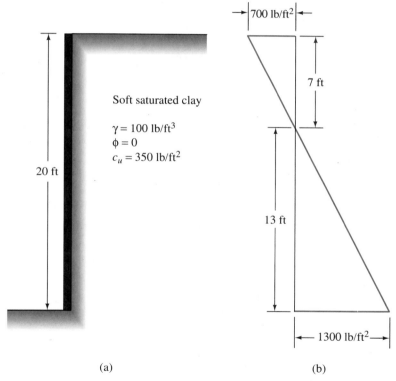

(a)   (b)

▼ **FIGURE 10.16**

Solution For $\phi = 0$, $K_a = \tan^2 45 = 1$ and $c = c_u$. From Eq. (10.15),

$$\sigma_a = \gamma z - 2c_u$$

At $z = 0$,

$$\sigma_a = -2c_u = -(2)(350) = 700 \text{ lb/ft}^2$$

At $z = 20$ ft,

$$\sigma_a = (100)(20) - (2)(350) = 1300 \text{ lb/ft}^2$$

The variation of σ_a with depth is shown in Figure 10.16b.

a. From Eq. (10.35), the depth of the tensile crack equals

$$z_o = \frac{2c_u}{\gamma} = \frac{(2)(350)}{100} = \mathbf{7 \text{ ft}}$$

b. Before the tensile crack occurs [Eq. (10.37)],

$$P_a = \frac{1}{2} \gamma H^2 - 2c_u H$$

or

$$P_a = \frac{1}{2} (100)(20)^2 - 2(350)(20) = \mathbf{6000 \text{ lb/ft}}$$

c. After the tensile crack occurs,

$$P_a = \frac{1}{2} (20 - 7)(1300) = \mathbf{8450 \text{ lb/ft}}$$

Note: The preceding P_a can also be obtained by substituting the proper values into Eq. (10.39).   ▼

▼ **EXAMPLE 10.4**

A frictionless retaining wall is shown in Figure 10.17a. Determine:

a. The active force P_a after the tensile crack occurs
b. The passive force P_p

Solution

a. Given $\phi = 26°$, we have

$$K_a = \frac{1 - \sin \phi}{1 + \sin \phi} = \frac{1 - \sin 26°}{1 + \sin 26°} = 0.39$$

From Eq. (10.15),

$$\sigma_a = K_a \sigma_v - 2c\sqrt{K_a}$$

At $z = 0$,

$$\sigma_a = (0.39)(10) - (2)(8)\sqrt{0.39} = 3.9 - 9.99 = -6.09 \text{ kN/m}^2$$

At $z = 4$ m,

$$\sigma_a = (0.39)[10 + (4)(15)] - (2)(8)\sqrt{0.39} = 27.3 - 9.99$$
$$= 17.31 \text{ kN/m}^2$$

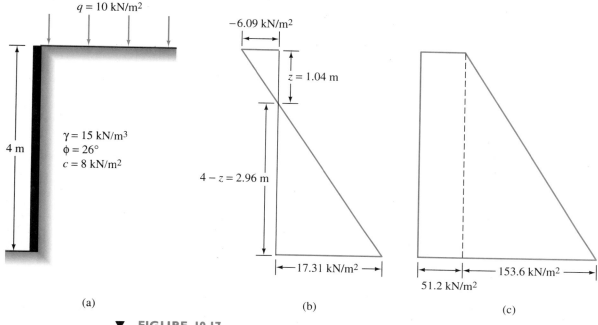

$q = 10$ kN/m^2

$\gamma = 15$ kN/m^3
$\phi = 26°$
$c = 8$ kN/m^2

4 m

-6.09 kN/m^2

$z = 1.04$ m

$4 - z = 2.96$ m

$\leftarrow$ 17.31 kN/m^2 $\rightarrow$

$\leftarrow$ 153.6 kN/m^2 $\rightarrow$
51.2 kN/m^2

(a)   (b)   (c)

▼ **FIGURE 10.17**

The pressure distribution is shown in Figure 10.17b. From this diagram,

$$\frac{6.09}{z} = \frac{17.31}{4 - z}$$

or

$z = 1.04$ m

After the tensile crack occurs,

$$P_a = \frac{1}{2}(4 - z)(17.31) = \left(\frac{1}{2}\right)(2.96)(17.31) = \mathbf{25.62 \ kN/m}$$

b. Given $\phi = 26°$, we have

$$K_p = \frac{1 + \sin \phi}{1 - \sin \phi} = \frac{1 + \sin 26°}{1 - \sin 26°} = \frac{1.4384}{0.5616} = 2.56$$

From Eq. (10.18),

$$\sigma_p = K_p \sigma_v + 2\sqrt{K_p}\,c$$

At $z = 0$, $\sigma_v = 10$ kN/m^2 and

$$\sigma_p = (2.56)(10) + 2\sqrt{2.56}\,(8)$$
$$= 25.6 + 25.6 = 51.2 \ \text{kN/m}^2$$

Again, at $z = 4$ m, $\sigma_v = (10 + 4 \times 15) = 70$ kN/m² and

$$\sigma_p = (2.56)(70) + 2\sqrt{2.56}\,(8)$$

$$= 204.8 \text{ kN/m}^2$$

The pressure distribution is shown in Figure 10.17c. The passive resistance per unit length of the wall is

$$P_p = (51.2)(4) + \frac{1}{2}\,(4)(153.6)$$

$$= 204.8 + 307.2 = \textbf{512 kN/m} \quad \blacktriangledown$$

▼ **EXAMPLE 10.5**

A retaining wall is shown in Figure 10.18a. Determine Rankine's active force P_a per unit length of the wall. Also determine the location of the resultant.

Solution Given $c = 0$, we know that $\sigma'_a = K_a \sigma'_v$. For the upper layer of the soil, Rankine's active earth pressure coefficient is

$$K_a = K_{a(1)} = \frac{1 - \sin 30°}{1 + \sin 30°} = \frac{1}{3}$$

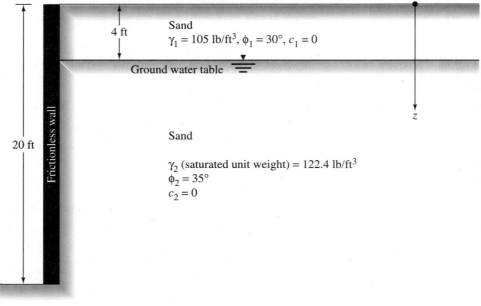

Sand
$\gamma_1 = 105$ lb/ft³, $\phi_1 = 30°$, $c_1 = 0$

Ground water table

4 ft

Sand

γ_2 (saturated unit weight) $= 122.4$ lb/ft³
$\phi_2 = 35°$
$c_2 = 0$

20 ft

Frictionless wall

z

(a)

▼ **FIGURE 10.18**

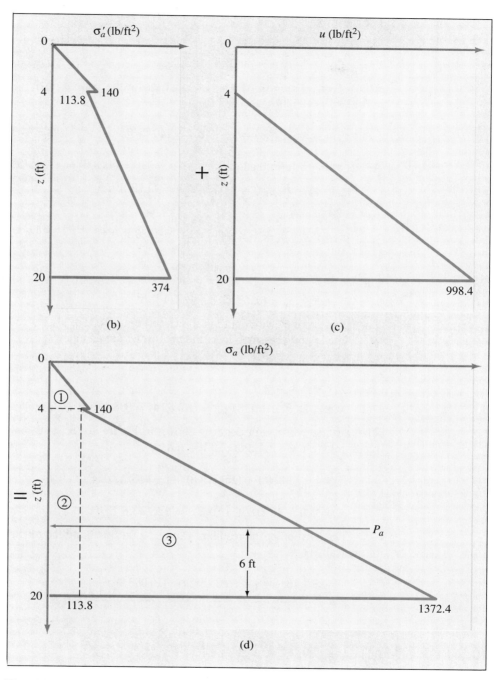

▼ **FIGURE 10.18** (*Continued*)

For the lower layer,

$$K_a = K_{a(2)} = \frac{1 - \sin 35°}{1 + \sin 35°} = \frac{0.4264}{1.5736} = 0.271$$

At $z = 0$, $\sigma_v = \sigma_v' = 0$. At $z = 4$ ft (just inside the bottom of the upper layer), $\sigma_v = \sigma_v' = (4)(105) = 420$ lb/ft². So

$$\sigma_a = \sigma_a' = K_{a(1)}\sigma_v' = \left(\frac{1}{3}\right)(420) = 140 \text{ lb/ft}^2$$

Again, at $z = 4$ ft (in the lower layer), $\sigma_v = \sigma_v' = (4)(105) = 420$ lb/ft², and

$$\sigma_a = \sigma_a' = K_{a(2)}\sigma_v' = (0.271)(420) = 113.8 \text{ lb/ft}^2$$

At $z = 20$ ft,

$$\sigma_v' = (4)(105) + 16(122.4 - 62.4) = 1380 \text{ lb/ft}^2$$
$$\uparrow$$
$$\gamma_w$$

and

$$\sigma_a' = K_{a(2)}\sigma_v' = (0.271)(1380) = 374 \text{ lb/ft}^2$$

The variation of σ_a' with depth is shown in Figure 10.18b.

The lateral pressures from the pore water are as follows:

At $z = 0$, $u = 0$

At $z = 4$ ft, $u = 0$

At $z = 20$ ft, $u = (16)(\gamma_w) = (16)(62.4) = 998.4$ lb/ft²

The variation of u with depth is shown in Figure 10.18c, and that for σ_a (total active pressure) is shown in Figure 10.18d. Thus,

$$P_a = \left(\frac{1}{2}\right)(140)(4) + (16)(113.8) + \left(\frac{1}{2}\right)(16)(1372.4 - 113.8)$$

$$= 280 + 1820.8 + 10{,}068.8 = \mathbf{12{,}169.6 \text{ lb/ft}}$$

The location of the resultant can be found by taking the moment about the bottom of the wall. Thus,

$$\bar{z} = \frac{280\left(16 + \dfrac{4}{3}\right) + (1820.8)(8) + (10{,}068.8)\left(\dfrac{16}{3}\right)}{12{,}169.6} = \mathbf{6 \text{ ft}} \quad \blacktriangledown$$

10.5 RETAINING WALLS WITH FRICTION

So far in our study of active and passive earth pressures, we have considered the case of frictionless walls. In reality, retaining walls are rough, and shear forces develop between the face of the wall and the backfill. To understand the effect of wall friction on the failure surface, let us consider a rough retaining wall *AB* with a horizontal granular backfill as shown in Figure 10.19.

In the active case (Figure 10.19a), when the wall *AB* moves to a position *A'B*, the soil mass in the active zone will be stretched outward. This will cause a downward motion of the soil relative to the wall. This motion causes a downward shear on the wall (Figure 10.19b), and it is called a *positive wall friction in the active case*. If δ is the angle

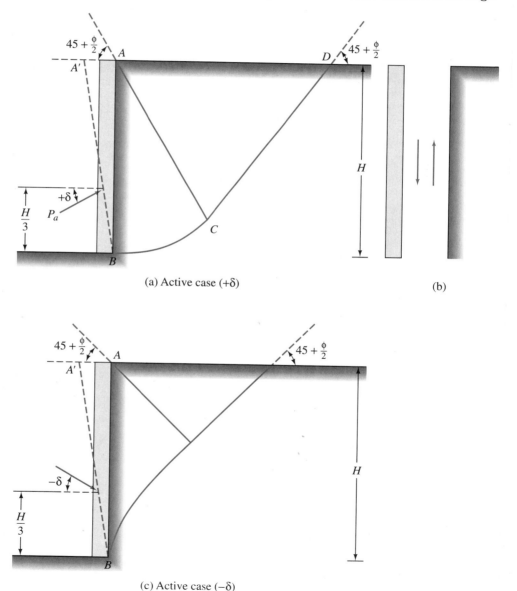

(a) Active case $(+\delta)$

(b)

(c) Active case $(-\delta)$

▼ **FIGURE 10.19** Effect of wall friction on failure surface of friction between the wall and the backfill, then the resultant active force P_a will be inclined at an angle δ to the normal drawn to the back face of the retaining wall. Advanced studies show that the failure surface in the backfill can be represented by *BCD*, as shown in Figure 10.19a. The portion *BC* is curved, and the portion *CD* of the failure surface is a straight line. Rankine's active state exists in the zone *ACD*.

Under certain conditions, if the wall shown in Figure 10.19a is forced downward with reference to the backfill, the direction of the active force P_a will change as shown in Figure 10.19c. This is a situation of negative wall friction $(-\delta)$ in the active case. Figure 10.19c also shows the nature of the failure surface in the backfill.

The effect of wall friction for the passive state is shown in Figures 10.19d and e. When the wall *AB* is pushed to a position $A''B$ (Figure 10.19d), the soil in the passive zone will be compressed. The result is an upward motion relative to the wall. The

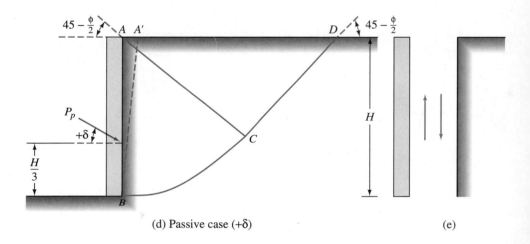

(d) Passive case $(+\delta)$ (e)

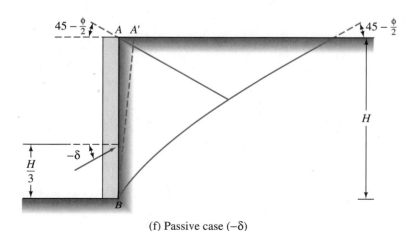

(f) Passive case $(-\delta)$

▼ **FIGURE 10.19** (*Continued*)

upward motion of the soil will cause an upward shear on the retaining wall (Figure 10.19e). This is referred to as *positive wall friction in the passive case*. The resultant passive force, P_p, will be inclined at an angle δ to the normal drawn to the back face of the wall. The failure surface in the soil has a curved lower portion BC and a straight upper portion CD. Rankine's passive state exists in the zone ACD.

If the wall shown in Figure 10.19d is forced upward relative to the backfill by a force, then the direction of the passive force P_p will change as shown in Figure 10.19f. This is *negative wall friction in the passive case* $(-\delta)$. Figure 10.19f also shows the nature of the failure surface in the backfill under such a condition.

For practical considerations, in the case of loose granular backfill, the angle of wall friction δ is taken to be equal to the angle of friction of soil, ϕ. For dense granular backfills, δ is smaller than ϕ and is in the range of $\phi/2 \leq \delta \leq (2/3)\phi$ (also see the discussion in Section 10.2).

10.6 COULOMB'S EARTH PRESSURE THEORY

About 200 years ago, Coulomb (1776) presented a theory for active and passive earth pressures against retaining walls. In this theory, Coulomb assumed that the failure surface is a plane. The wall friction was taken into consideration. The general principles of the derivation of Coulomb's earth pressure theory for a cohesionless backfill (shear strength defined by the equation $\tau_f = \sigma \tan \phi$) are given here.

Active Case

Let AB (Figure 10.20a) be the back face of a retaining wall supporting a granular soil, the surface of which is constantly sloping at an angle α with the horizontal. BC is a trial failure surface. In the stability consideration of the probable failure wedge ABC, the following forces are involved (per unit length of the wall):

1. W, the weight of the soil wedge.

2. F, the resultant of the shear and normal forces on the surface of failure, BC. This is inclined at an angle of ϕ to the normal drawn to the plane BC.

3. P_a, the active force per unit length of the wall. The direction of P_a is inclined at an angle δ to the normal drawn to the face of the wall that supports the soil. δ is the angle of friction between the soil and the wall.

The force triangle for the wedge is shown in Figure 10.20b. From the law of sines, we have

$$\frac{W}{\sin(90 + \theta + \delta - \beta + \phi)} = \frac{P_a}{\sin(\beta - \phi)} \tag{10.44}$$

or

$$P_a = \frac{\sin(\beta - \phi)}{\sin(90 + \theta + \delta - \beta + \phi)} W \tag{10.45}$$

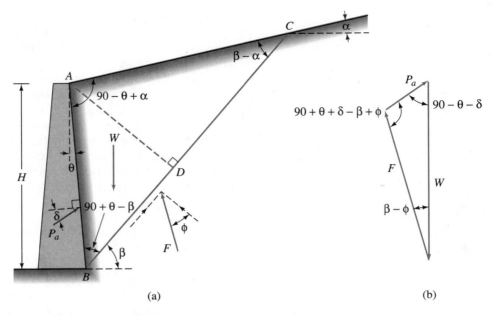

▼ **FIGURE 10.20**   Coulomb's active pressure: (a) trial failure wedge; (b) force polygon

The preceding equation can be written in the form

$$P_a = \frac{1}{2}\gamma H^2 \left[\frac{\cos(\theta - \beta)\cos(\theta - \alpha)\sin(\beta - \phi)}{\cos^2\theta \sin(\beta - \alpha)\sin(90 + \theta + \delta - \beta + \phi)} \right]$$ (10.46)

where γ = unit weight of the backfill. The values of γ, H, θ, α, ϕ, and δ are constants, and β is the only variable. To determine the critical value of β for maximum P_a, we have

$$\frac{dP_a}{d\beta} = 0$$ (10.47)

After solving Eq. (10.47), when the relationship of β is substituted into Eq. (10.46), we obtain Coulomb's active earth pressure as

$$P_a = \frac{1}{2}K_a\gamma H^2$$ (10.48)

where K_a is Coulomb's active earth pressure coefficient and is given by

$$K_a = \frac{\cos^2 (\phi - \theta)}{\cos^2 \theta \cos (\delta + \theta) \left[1 + \sqrt{\dfrac{\sin (\delta + \phi) \sin (\phi - \alpha)}{\cos (\delta + \theta) \cos (\theta - \alpha)}} \right]^2} \tag{10.49}$$

Note that when $\alpha = 0°$, $\theta = 0°$, and $\delta = 0°$, Coulomb's active earth pressure coefficient becomes equal to $(1 - \sin \phi)/(1 + \sin \phi)$, which is the same as Rankine's earth pressure coefficient given earlier in this chapter.

The variation of the values of K_a for retaining walls with a vertical back ($\theta = 0$) and horizontal backfill ($\alpha = 0$) is given in Table 10.2. From this table note that for a given value of ϕ, the effect of wall friction is to reduce somewhat the active earth pressure coefficient.

▼ **TABLE 10.2**   Values of K_a [Eq. (10.49)] for $\theta = 0°$, $\alpha = 0°$

↓ϕ (deg)	δ (deg) → 0	5	10	15	20	25
28	0.3610	0.3448	0.3330	0.3251	0.3203	0.3186
30	0.3333	0.3189	0.3085	0.3014	0.2973	0.2956
32	0.3073	0.2945	0.2853	0.2791	0.2755	0.2745
34	0.2827	0.2714	0.2633	0.2579	0.2549	0.2542
36	0.2596	0.2497	0.2426	0.2379	0.2354	0.2350
38	0.2379	0.2292	0.2230	0.2190	0.2169	0.2167
40	0.2174	0.2089	0.2045	0.2011	0.1994	0.1995
42	0.1982	0.1916	0.1870	0.1841	0.1828	0.1831

Passive Case

Figure 10.21a shows a retaining wall with a sloping cohensionless backfill similar to that considered in Figure 10.20a. The force polygon for equilibrium of the wedge ABC for the passive state is shown in Figure 10.21b. P_p is the notation for the passive force. Other notations used are the same as those for the active case considered in this section. In a procedure similar to the one we followed in the active case, we get

$$P_p = \frac{1}{2} K_p \gamma H^2 \tag{10.50}$$

where K_p = coefficient of passive earth pressure for Coulomb's case, or

$$K_p = \frac{\cos^2(\phi + \theta)}{\cos^2\theta\,\cos(\delta - \theta)\left[1 - \sqrt{\dfrac{\sin(\phi + \delta)\,\sin(\phi + \alpha)}{\cos(\delta - \theta)\,\cos(\alpha - \theta)}}\right]^2} \qquad (10.51)$$

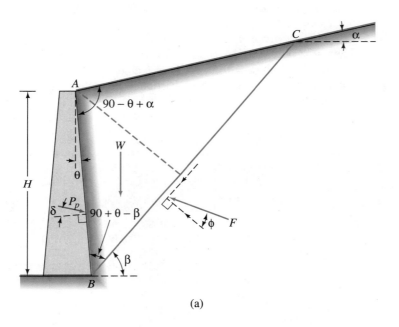

(a)

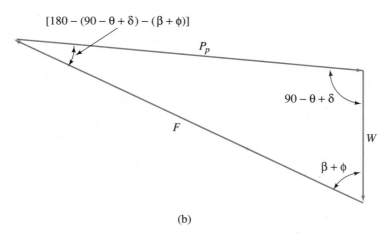

(b)

▼ **FIGURE 10.21**   Coulomb's passive pressure: (a) trial failure wedge; (b) force polygon

▼ **TABLE 10.3** Values of K_p [Eq. (10.51)] for $\theta = 0°$ and $\alpha = 0°$

↓ϕ (deg)	δ (deg) →				
	0	5	10	15	20
15	1.698	1.900	2.130	2.405	2.735
20	2.040	2.313	2.636	3.030	3.525
25	2.464	2.830	3.286	3.855	4.597
30	3.000	3.506	4.143	4.977	6.105
35	3.690	4.390	5.310	6.854	8.324
40	4.600	5.590	6.946	8.870	11.772

For a frictionless wall with the vertical back face supporting granular soil backfill with a horizontal surface (that is, $\theta = 0°$, $\alpha = 0°$, and $\delta = 0°$), Eq. (10.51) yields

$$K_p = \frac{1 + \sin \phi}{1 - \sin \phi} = \tan^2 \left(45 + \frac{\phi}{2} \right)$$

This is the same relationship that was obtained for the passive earth pressure coefficient in Rankine's case given by Eq. (10.19).

The variation of K_p with ϕ and δ (for $\theta = 0$ and $\alpha = 0$) is given in Table 10.3. It can be observed from this table that, for given values of α and ϕ, the value of K_p increases with the wall friction. *Note that making the assumption that the failure surface is a plane in Coulomb's theory grossly overestimates the passive resistance of walls, particularly for $\delta > \phi/2$.* This is somewhat unsafe for all design purposes. The technique of calculating the passive resistance using a curved failure surface is given in Section 10.11.

10.7 GRAPHIC SOLUTION FOR COULOMB'S ACTIVE EARTH PRESSURE

An expedient method for creating a graphic solution of Coulomb's earth pressure theory was given by Culmann (1875). Culmann's solution can be used for any wall friction, regardless of irregularity of backfill and surcharges. Hence, it provides a very powerful technique for estimating lateral earth pressure. The steps in Culmann's solution of active pressure with granular backfill ($c = 0$) are described below with reference to Figure 10.22a.

1. Draw the features of the retaining wall and the backfill to a convenient scale.
2. Determine the value of ψ (degrees) $= 90 - \theta - \delta$, where $\theta =$ the inclination of the back face of the retaining wall with the vertical, and $\delta =$ angle of wall friction.
3. Draw a line BD that makes an angle ϕ with the horizontal.
4. Draw a line BE that makes an angle ψ with line BD.
5. To consider some trial failure wedges, draw lines $BC_1, BC_2, BC_3, \ldots, BC_n$.

6. Find the areas of ABC_1, ABC_2, ABC_3, ..., ABC_n.

7. Determine the weight of soil, W, per unit length of the retaining wall in each of the trial failure wedges as follows:

$$W_1 = (\text{area of } ABC_1) \times (\gamma) \times (1)$$
$$W_2 = (\text{area of } ABC_2) \times (\gamma) \times (1)$$
$$W_3 = (\text{area of } ABC_3) \times (\gamma) \times (1)$$
$$\vdots$$
$$W_n = (\text{area of } ABC_n) \times (\gamma) \times (1)$$

8. Adopt a convenient load scale and plot the weights W_1, W_2, W_3, ..., W_n determined from Step 7 on line BD. (*Note:* $Bc_1 = W_1$, $Bc_2 = W_2$, $Bc_3 = W_3$, ..., $Bc_n = W_n$.)

9. Draw c_1c_1', c_2c_2', c_3c_3', ..., c_nc_n' parallel to the line BE. (*Note:* c_1', c_2', c_3', ..., c_n' are located on lines BC_1, BC_2, BC_3, ..., BC_n, respectively.)

10. Draw a smooth curve through points c_1', c_2', c_3', ..., c_n'. This is called the *Culmann line*.

11. Draw a tangent $B'D'$ to the smooth curve drawn in Step 10. $B'D'$ is parallel to line BD. Let c_a' be the point of tangency.

12. Draw a line $c_a c_a'$ parallel to the line BE.

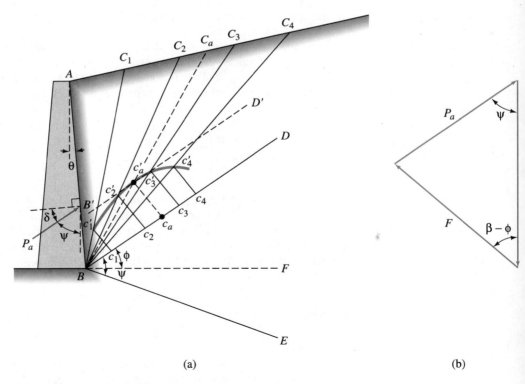

(a)   (b)

▼ **FIGURE 10.22** Culmann's solution for active earth pressure

13. Determine the active force per unit length of wall as

$$P_a = (\text{length of } c_a c'_a) \times (\text{load scale})$$

14. Draw a line $Bc'_a C_a$. ABC_a is the desired failure wedge.

Note that the construction procedure entails, in essence, drawing a number of force polygons for a number of trial wedges and finding the maximum value of the active force that the wall can be subjected to. For example, Figure 10.22b shows the force polygon for the failure wedge ABC_a (similar to that in Figure 10.20b), in which

$W = $ weight of the failure wedge of soil ABC_a

$P_a = $ active force on the wall

$F = $ the resultant of the shear and normal forces acting along BC

$\beta = \angle C_a BF$ (the angle that the failure wedge makes with the horizontal)

The force triangle (Figure 10.22b) is simply rotated in Figure 10.22a and is represented by the triangle $Bc_a c'_a$. Similarly, the force triangles $Bc_1 c'_1$, $Bc_2 c'_2$, $Bc_3 c'_3$, ..., $Bc_n c'_n$ correspond to the trial wedges ABC_1, ABC_2, ABC_3, ..., ABC_n.

The preceding graphic procedure is given in a step-by-step manner to facilitate the basic understanding of the readers. These problems can be easily and effectively solved by the use of computer programs.

The Culmann solution provides us with only the magnitude of the active force per unit length of the retaining wall—not with the point of application of the resultant. The analytical procedure used to find the point of application of the resultant can be tedious. For that reason, without sacrificing much accuracy an approximate method can be used. This is demonstrated in Figure 10.23, in which ABC is the failure wedge determined by

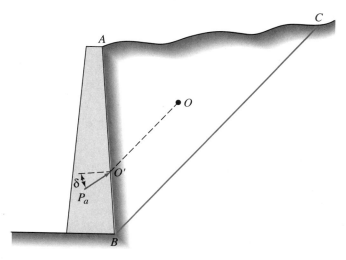

▼ **FIGURE 10.23** Approximate method for finding the point of application of the resultant active force

Culmann's method. O is the center of gravity of the wedge ABC. If a line OO' is drawn parallel to the surface of sliding BC, the point of intersection of this line with the back face of the wall will give the point of application of P_a. Thus, P_a acts at O' inclined at angle δ with the normal drawn to the back face of the wall.

▼ EXAMPLE 10.6

A 15-ft high retaining wall with a granular soil backfill is shown in Figure 10.24. Given $\gamma = 100$ lb/ft³, $\phi = 35°$, and $\delta = 10°$, determine the active thrust per foot of length of the wall.

Solution For this problem, $\psi = 90 - \theta - \delta = 90° - 5° - 10° = 75°$. The graphic construction is shown in Figure 10.24. The weights of the wedges considered are as follows:

Wedge	Weight of wedge
ABC_1	$\frac{1}{2} (\overline{AA'})(\overline{BC_1})\gamma = \frac{1}{2}(4.25)(17.75) \times 100 = 3772$ lb
ABC_2	Weight of ABC_1 + weight of C_1BC_2
	$= 3772 + \frac{1}{2}(17.5)(2.5) \times 100$
	$= 3772 + 2187.5 = 5959.5$ lb
ABC_3	Weight of ABC_2 + weight of $C_2 BC_3$
	$= 5959.5 + 2187.5 = 8147$ lb
ABC_4	Weight of ABC_3 + weight of $C_3 BC_4$
	$= 8147 + 2187.5 = 10{,}334.5$ lb
ABC_5	Weight of ABC_4 + weight of $C_4 BC_5$
	$= 10{,}334.5 + 2187.5 = 12{,}522$ lb

In Figure 10.24,

$$\overline{Bc_1} = 3772 \text{ lb}$$

$$\overline{Bc_2} = 5959.5 \text{ lb}$$

$$\overline{Bc_3} = 8147 \text{ lb}$$

$$\overline{Bc_4} = 10{,}334.5 \text{ lb}$$

$$\overline{Bc_5} = 12{,}522 \text{ lb}$$

The active thrust per unit length of the wall is **4200 lb**.
Note: This problem can be solved easily by a computer program.

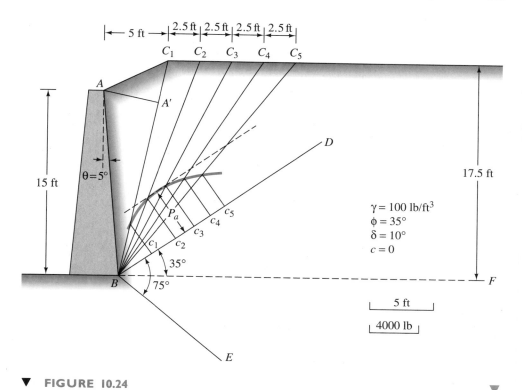

$\gamma = 100 \text{ lb/ft}^3$
$\phi = 35°$
$\delta = 10°$
$c = 0$

▼ FIGURE 10.24

10.8 APPROXIMATE ANALYSIS OF ACTIVE FORCE ON RETAINING WALLS

In practical design considerations, the calculation for active force on a retaining wall can be made by using either Coulomb's or Rankine's method. The procedure for such calculations for a gravity retaining wall with granular backfill is shown in Figure 10.25.

Figure 10.25a shows a gravity retaining wall with its backfill that has a horizontal ground surface. If Coulomb's method is used, the active thrust per unit length of the wall, P_a, can be determined by Eq. (10.48) (or Culmann's analysis). This force will act at an angle of δ to the normal drawn to the back face of the wall. If Rankine's method is used, the active thrust is calculated on a vertical plane drawn through the heel of the wall [Eq. (10.21)]:

$$P_a = \frac{1}{2} K_a \gamma H^2$$

where $K_a = \dfrac{1 - \sin \phi}{1 + \sin \phi} = \tan^2 \left(45 - \dfrac{\phi}{2} \right).$

In such a case, $P_{a(Rankine)}$ is added vectorally to the weight of the wedge of soil, W_s, for the stability analysis.

Figure 10.25b shows a similar retaining wall with a granular backfill that has an inclined ground surface. Equation (10.48) or Culmann's solution may be used to determine the active force on a vertical plane through the heel of the wall, which can then be

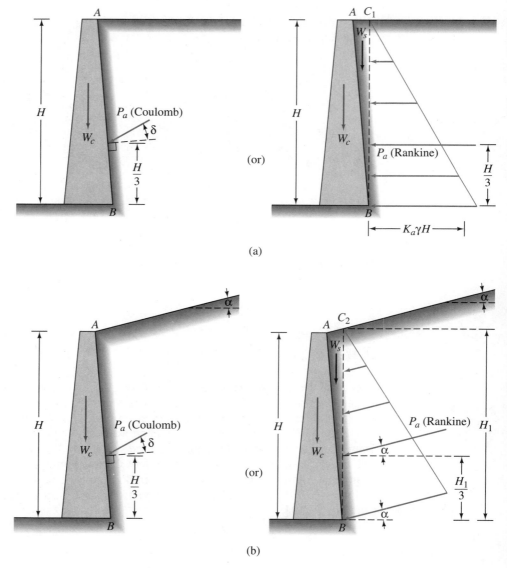

▼ **FIGURE 10.25**   Approximate analyses of active force on gravity retaining walls with granular backfill

vectorally added to the weight of the wedge of soil ABC_2 for stability analysis. Note, however, that the direction of Rankine's active force in this case is not horizontal anymore, and the vertical plane BC_2 is not the minor principal plane. The value of $P_{a(Rankine)}$ can be given by the relationship

$$P_a = \frac{1}{2} K_a \gamma H_1^2$$

(10.52)

where $H_1 = \overline{BC_2}$ and

$$K_a = \text{Rankine's active pressure coefficient}$$
$$= \cos \alpha \, \frac{\cos \alpha - \sqrt{\cos^2 \alpha - \cos^2 \phi}}{\cos \alpha + \sqrt{\cos^2 \alpha - \cos^2 \phi}}$$

(10.53)

where α = slope of the ground surface.

P_a obtained from Eq. (10.52) acts at a distance of $H_1/3$ measured vertically from B and inclined at an angle α with the horizontal. The values of K_a defined by Eq. (10.53) for various slope angles and soil friction angles are given in Table 10.4. For a horizontal ground surface (that is, $\alpha = 0$), Eq. (10.53) translates to

$$K_a = \frac{1 - \sin \phi}{1 + \sin \phi} = \tan^2 \left(45 - \frac{\phi}{2} \right)$$

▼ **TABLE 10.4**   **Values of K_a [Eq. (10.53)]**

↓α (deg)	ϕ (deg) →						
---	28	30	32	34	36	38	40
0	0.361	0.333	0.307	0.283	0.260	0.238	0.217
5	0.366	0.337	0.311	0.286	0.262	0.240	0.219
10	0.380	0.350	0.321	0.294	0.270	0.246	0.225
15	0.409	0.373	0.341	0.311	0.283	0.258	0.235
20	0.461	0.414	0.374	0.338	0.306	0.277	0.250
25	0.573	0.494	0.434	0.385	0.343	0.307	0.275

10.9   ACTIVE FORCE ON RETAINING WALLS WITH EARTHQUAKE FORCES

Coulomb's analysis for an active force on retaining walls can be conveniently extended to include earthquake forces. To do that, let us consider a retaining wall of height H with a sloping granular backfill as shown in Figure 10.26a. ABC is a trial failure wedge. The forces that act on the wedge are as follows:

1. Weight of the soil in the wedge, W
2. Resultant of the shear and the normal forces on the failure surface BC, F
3. Active force per unit length of the wall, P_{ae}
4. Horizontal inertia force, $k_h W$
5. Vertical inertia force, $k_v W$

Note that

$$k_h = \frac{\text{horizontal component of earthquake acceleration}}{g} \qquad (10.54)$$

$$k_v = \frac{\text{vertical component of earthquake acceleration}}{g} \qquad (10.55)$$

where g = acceleration from gravity.

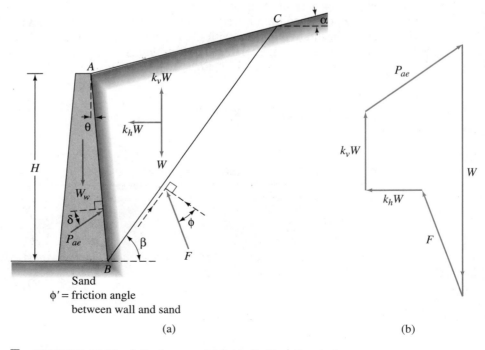

ϕ' = friction angle between wall and sand

(a)   (b)

▼ **FIGURE 10.26** Active force on retaining wall with earthquake forces

The force polygon with the above-stated forces is shown in Figure 10.26b. The relation for active force, P_{ae}, can be determined as

$$P_{ae} = \frac{1}{2} \gamma H^2 (1 - k_v) K_a'$$

(10.56)

where

$$K_a' = \frac{\cos^2 (\phi - \theta - \bar{\beta})}{\cos^2 \theta \cos \bar{\beta} \cos (\delta + \theta + \bar{\beta}) \left\{ 1 + \left[\dfrac{\sin (\delta + \phi) \sin (\phi - \alpha - \bar{\beta})}{\cos (\delta + \theta + \bar{\beta}) \cos (\theta - \alpha)} \right]^{1/2} \right\}^2 }$$

(10.57)

and

$$\bar{\beta} = \tan^{-1} \left(\frac{k_h}{1 - k_v} \right)$$

(10.58)

Note that, with no inertia forces from earthquakes, $\bar{\beta}$ will be equal to 0. Hence, $K_a' = K_a$ as given in Eq. (10.49). Equations (10.56) and (10.57) are generally referred to as the Mononobe–Okabe equations (Mononobe, 1929; Okabe, 1926). The variation of K_a' with $\theta = 0$ and $k_v = 0$ is given in Table 10.5.

Location of the Line of Action of the Resultant Force, P_{ae}

Seed and Whitman (1970) proposed a simple procedure to determine the location of the line of action of the resultant, P_{ae}. Their method is as follows:

1. Let

    $$P_{ae} = P_a + \Delta P_{ae}$$

    (10.59)

    where P_a = Coulomb's active force as determined from Eq. (10.48)
    ΔP_{ae} = additional active force caused by the earthquake effect
2. Calculate P_a [Eq. (10.48)].
3. Calculate P_{ae} [Eq. (10.56)].
4. Calculate $\Delta P_{ae} = P_{ae} - P_a$.
5. According to Figure 10.27, P_a will act at a distance of $H/3$ from the base of the wall. Also, ΔP_{ae} will act at a distance of $0.6H$ from the base of the wall.

k_h	δ (deg)	α (deg)	ϕ (deg) 28	30	35	40	45
0.1	0	0	0.427	0.397	0.328	0.268	0.217
0.2			0.508	0.473	0.396	0.382	0.270
0.3			0.611	0.569	0.478	0.400	0.334
0.4			0.753	0.697	0.581	0.488	0.409
0.5			1.005	0.890	0.716	0.596	0.500
0.1	0	5	0.457	0.423	0.347	0.282	0.227
0.2			0.554	0.514	0.424	0.349	0.285
0.3			0.690	0.635	0.522	0.431	0.356
0.4			0.942	0.825	0.653	0.535	0.442
0.5			—	—	0.855	0.673	0.551
0.1	0	10	0.497	0.457	0.371	0.299	0.238
0.2			0.623	0.570	0.461	0.375	0.303
0.3			0.856	0.748	0.585	0.472	0.383
0.4			—	—	0.780	0.604	0.486
0.5			—	—	—	0.809	0.624
0.1	$\phi/2$	0	0.396	0.368	0.306	0.253	0.207
0.2			0.485	0.452	0.380	0.319	0.267
0.3			0.604	0.563	0.474	0.402	0.340
0.4			0.778	0.718	0.599	0.508	0.433
0.5			1.115	0.972	0.774	0.648	0.552
0.1	$\phi/2$	5	0.428	0.396	0.326	0.268	0.218
0.2			0.537	0.497	0.412	0.342	0.283
0.3			0.699	0.640	0.526	0.438	0.367
0.4			1.025	0.881	0.690	0.568	0.475
0.5			—	—	0.962	0.752	0.620
0.1	$\phi/2$	10	0.472	0.433	0.352	0.285	0.230
0.2			0.616	0.562	0.454	0.371	0.303
0.3			0.908	0.780	0.602	0.487	0.400
0.4			—	—	0.857	0.656	0.531
0.5			—	—	—	0.944	0.722
0.1	$\frac{2}{3}\phi$	0	0.393	0.366	0.306	0.256	0.212
0.2			0.486	0.454	0.384	0.326	0.276
0.3			0.612	0.572	0.486	0.416	0.357
0.4			0.801	0.740	0.622	0.533	0.462
0.5			1.177	1.023	0.819	0.693	0.600
0.1	$\frac{2}{3}\phi$	5	0.427	0.395	0.327	0.271	0.224
0.2			0.541	0.501	0.418	0.350	0.294
0.3			0.714	0.655	0.541	0.455	0.386
0.4			1.073	0.921	0.722	0.600	0.509
0.5			—	—	1.034	0.812	0.679
0.1	$\frac{2}{3}\phi$	10	0.472	0.434	0.354	0.290	0.237
0.2			0.625	0.570	0.463	0.381	0.317
0.3			0.942	0.807	0.624	0.509	0.423
0.4			—	—	0.909	0.699	0.573
0.5			—	—	—	1.037	0.800

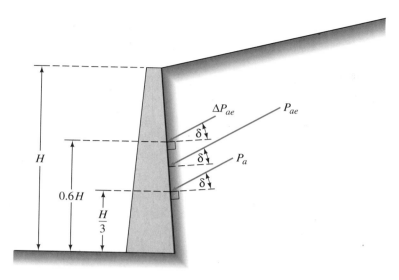

▼ FIGURE 10.27 Location of the line of action of P_{ae}

6. Calculate the location of P_{ae} as

$$\bar{z} = \frac{P_a\left(\dfrac{H}{3}\right) + \Delta P_{ae}(0.6H)}{P_{ae}} \tag{10.60}$$

where $\bar{z}$ = distance of the line of action of P_{ae} from the base of the wall.

Note that the line of action of P_{ae} will be inclined at an angle of δ to the normal drawn to the back face of the retaining wall. It is very important to realize that this method of determining P_{ae} is approximate and does not actually model the soil dynamics.

▼ EXAMPLE 10.7

For a retaining wall with a cohesionless soil backfill, $\gamma = 15.5$ kN/m³, $\phi = 30°$, $\delta = 15°$, $\theta = 0°$, $\alpha = 0°$, $H = 4$ m, $k_v = 0$, and $k_h = 0.2$. Determine P_{ae}. Also determine the location of the resultant line of action of P_{ae}—that is, $\bar{z}$.

Solution   To determine P_{ae}, we use Eq. (10.56):

$$P_{ae} = \frac{1}{2}\gamma H^2 (1 - k_v)K'_a$$

We are given $\phi = 30°$ and $\delta = 15°$, so

$$\delta = \frac{\phi}{2}$$

Also $\theta = 0°$, $\alpha = 0°$, and $k_h = 0.2$. From these values and Table 10.5, we find the magnitude of k_h is equal to 0.452. Hence,

$$P_{ae} = \frac{1}{2}(15.5)(4)^2(1 - 0)(0.452) = \mathbf{56.05\ kN/m}$$

We now locate the resultant line of action. From Eq. (10.48),

$$P_a = \frac{1}{2}K_a\gamma H^2$$

For $\phi = 30°$ and $\delta = 15°$, $K_a = 0.3014$ (Table 10.2), so

$$P_a = \frac{1}{2}(15.5)(4)^2(0.3014) = 37.37\ kN/m$$

Hence, $\Delta P_{ae} = 56.05 - 37.37 = 18.68$ kN/m. From Eq. (10.60),

$$\bar{z} = \frac{P_a\left(\dfrac{H}{3}\right) + \Delta P_{ae}(0.6H)}{P_{ae}} = \frac{(37.37)\left(\dfrac{4}{3}\right) + (18.68)(2.4)}{56.05} = \mathbf{1.69\ m} \qquad \blacktriangledown$$

10.10 DESIGN OF RETAINING WALL BASED ON TOLERABLE LATERAL DISPLACEMENT

It has been shown by several investigators that even under mild earthquake conditions, there is some lateral displacement of a retaining wall. Richards and Elms (1979) proposed a procedure for designing gravity retaining walls for earthquake conditions which allows limited lateral displacement of the walls. This procedure takes into consideration the inertia effect of the wall. Considering the equilibrium of the wall, we can show that

$$W_w = \left[\frac{1}{2}\gamma H^2(1 - k_v)K_a\right]C_{IE} \qquad (10.61)$$

where

$$C_{IE} = \frac{\cos(\theta + \delta) - \sin(\theta + \delta)\tan\phi'}{(1 - k_v)(\tan\phi' - \tan\bar{\beta})} \qquad (10.62)$$

W_w = self-weight of the retaining wall (see Figure 10.26a)

ϕ' = friction angle between the wall and the soil on which it is resting (see Figure 10.26a)

Based on Eqs. (10.61) and (10.62), the following procedure may be adopted to determine the weight of the retaining wall, W_w, for tolerable displacement that may take place during an earthquake:

1. Determine the tolerable displacement, Δ.
2. Obtain a design value of k_h by using the equation

$$k_h = A_a \left(\frac{0.2 A_v^2}{A_a \Delta} \right)^{0.25}$$

where A_a and A_v = effective acceleration coefficients. The magnitudes of A_a and A_v are given by the Applied Technology Council (1978) for various regions of the United States.
3. Assume k_v to be 0, and with the value of k_h calculated in Step 2, obtain K_a' [Eq. (10.57)].
4. Using the above value of K_a', calculate the weight of the retaining wall, W_w [Eqs. (10.61) and (10.62)].
5. Apply a safety factor to the value of W_w obtained in Step 4.

10.11 PASSIVE FORCE ON RETAINING WALLS WITH EARTHQUAKE FORCES

Figure 10.28 shows the failure wedge analysis for a passive force against a retaining wall of height H with a granular backfill which takes into consideration the earthquake forces. As in Figure 10.26, the failure surface is assumed to be a plane. P_{pe} is the

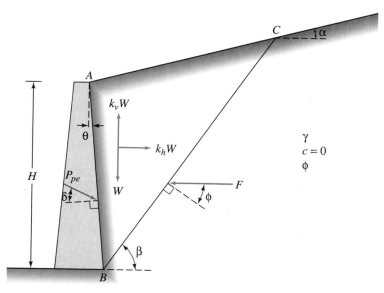

▼ **FIGURE 10.28** Passive force on retaining wall with earthquake force

passive force. All other notations in Figure 10.28 are the same as those in Figure 10.26. Following a procedure similar to that used in Section 10.10, we can obtain (Kapila, 1962)

$$P_{pe} = \frac{1}{2} \gamma H^2 (1 - k_v) K'_p \tag{10.63}$$

where $K'_p = \dfrac{\cos^2(\phi + \theta - \bar{\beta})}{\cos^2\theta \cos^2\bar{\beta} \cos(\delta - \theta + \bar{\beta})\left[1 - \left(\dfrac{\sin(\delta + \phi)\sin(\phi + \alpha - \bar{\beta})}{\cos(\delta - \theta + \bar{\beta})\cos(\alpha - \theta)}\right)^{1/2}\right]^2}$

$$\tag{10.64}$$

Figure 10.29 shows the variation of K'_p with k_h for various values of ϕ with $k_v = \alpha = \theta = \delta = 0$.

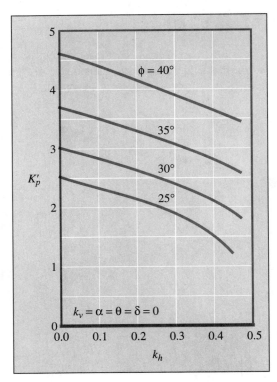

▼ **FIGURE 10.29**   Variation of K'_p with k_h for $k_v = \alpha = \theta = \delta = 0$ (after Davies, Richards, and Chen, 1986)

10.12 PASSIVE EARTH PRESSURE AGAINST RETAINING WALLS WITH CURVED FAILURE SURFACE

Section 10.6 pointed out that assuming the surface of rupture is planar for a failure wedge leads to an overestimation of the passive pressure. In this section, we discuss the trial wedge solution of the passive resistance developed against a wall for cohesionless backfill (that is, $\tau_f = \sigma \tan \phi$) with a horizontal ground surface using a curved failure surface.

In Figure 10.19d, we saw a failure surface developed in a passive state. Generally, the curved lower portion, *BC*, of the failure surface is assumed to be an arc of a logarithmic spiral, the center of which lies on the line *CA* (not necessarily within the limits of points *C* and *A*). The upper straight portion *CD* is a straight line that makes an angle of $(45 - \phi/2)$ degrees with the horizontal. The soil in the zone *ACD* is in Rankine's passive state.

To proceed with the trial wedge solution, we need to know some of the basic properties of the arc of a logarithmic spiral. These are described in the following section.

Properties of a Logarithmic Spiral

The equation of the logarithmic spiral used in solving the problems of passive resistance against a wall is

$$r = r_o e^{\theta \tan \phi}$$

(10.65)

where r = radius of the spiral
$\quad r_o$ = starting radius at $\theta = 0$
$\quad \phi$ = angle of friction of soil
$\quad \theta$ = angle between r and r_o

The basic parameters of a logarithmic spiral are shown in Figure 10.30, in which O is the center of the spiral. The area, A, of the sector *OAB* can be given by

$$A = \int_0^{\theta_1} \frac{1}{2} r(r\, d\theta)$$

(10.66)

Substituting the values of r from Eq. (10.56) into Eq. (10.66), we get

$$A = \int_0^{\theta_1} \frac{1}{2} r_o^2 e^{2\theta \tan \phi} \, d\theta$$
$$= \frac{r_1^2 - r_o^2}{4 \tan \phi}$$

(10.67)

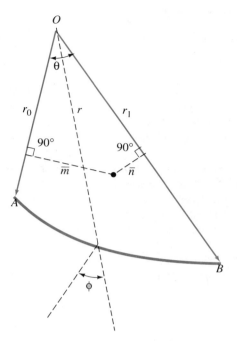

▼ **FIGURE 10.30**   General parameters of a logarithmic spiral

The location of the centroid can be defined by the distances $\bar{m}$ and $\bar{n}$ (Figure 10.30) measured from OA and OB, respectively, and can be given by (Hijab, 1956)

$$\bar{m} = \frac{4}{3}r_o \frac{\tan\phi}{(9\tan^2\phi + 1)} \left[\frac{\left(\dfrac{r_1}{r_o}\right)^3 (3\tan\phi\sin\theta - \cos\theta) + 1}{\left(\dfrac{r_1}{r_o}\right)^2 - 1} \right] \tag{10.68}$$

$$\bar{n} = \frac{4}{3}r_o \frac{\tan\phi}{(9\tan^2\phi + 1)} \left[\frac{\left(\dfrac{r_1}{r_o}\right)^3 - 3\tan\phi\sin\theta - \cos\theta}{\left(\dfrac{r_1}{r_o}\right)^2 - 1} \right] \tag{10.69}$$

Another important property of the logarithmic spiral defined by Eq. (10.65) is that any radial line makes an angle ϕ with the normal to the curve drawn at the point where the radial line and the spiral intersect. This basic property is particularly useful in solving the passive pressure problems that will be shown later.

Trial Wedge Procedure

Figure 10.31 explains the procedure of evaluating the passive resistance by trial wedges (Terzaghi and Peck, 1967). The retaining wall is first drawn to scale as shown in Figure 10.31a. The line C_1A is drawn in such a way that it makes an angle of $(45 - \phi/2)$ degrees with the surface of the backfill. BC_1D_1 is a trial wedge in which BC_1 is the arc of a logarithmic spiral. The arc BC_1 can be traced out by using a trial-and-error procedure by superimposing the worksheet on another sheet of paper on which a logarithmic spiral is drawn. According to the equation $r_1 = r_o e^{\theta \tan \phi}$, O_1 is the center of the spiral. (*Note:* $\overline{O_1B} = r_o$ and $\overline{O_1C_1} = r$ and $\angle BO_1C_1 = \theta_1$; refer to Figure 10.30.)

Now let us consider the stability of the soil mass ABC_1C_1' (Figure 10.31b). For equilibrium, the following forces per unit length of the wall are to be considered:

1. Weight of the soil in zone $ABC_1C_1' = W_1 = (\gamma)(\text{area of } ABC_1C_1')(1)$.
2. The vertical face C_1C_1' is in the zone of Rankine's passive state; hence, the force $P_{d(1)}$ acting on this face is

$$P_{d(1)} = \frac{1}{2} \gamma (d_1)^2 \tan^2 \left(45 + \frac{\phi}{2} \right) \tag{10.70}$$

where $d_1 = \overline{C_1C_1'}$. $P_{d(1)}$ acts horizontally at a distance of $d_1/3$ measured vertically upward from C_1.

3. F_1 is the resultant of the shear and normal forces that act along the surface of sliding BC_1. At any point of the curve, according to the property of the logarithmic spiral, a radial line makes an angle ϕ with the normal. Since the resultant F_1 makes an angle ϕ with the normal to the spiral at its point of application, its line of application will coincide with a radial line and will pass through the point O_1.

4. P_1 is the passive force per unit length of the wall. It acts at a distance of $H/3$ measured vertically from the bottom of the wall. The direction of the force P_1 is inclined at an angle δ with the normal drawn to the back face of the wall.

Now, taking the moments of W_1, $P_{d(1)}$, F_1, and P_1 about the point O_1, for equilibrium we have

$$W_1[l_{W(1)}] + P_{d(1)}[l_1] + F_1[0] = P_1[l_{P(1)}] \tag{10.71}$$

or

$$P_1 = \frac{1}{l_{P(1)}} [W_1 l_{w(1)} + P_{d(1)} l_1] \tag{10.72}$$

where $l_{W(1)}$, l_1, and $l_{P(1)}$ are moment arms for the forces W_1, $P_{d(1)}$, and P_1, respectively.

For determination of P_1 in Eq. (10.72), the values of d_1, l_1, and $l_{P(1)}$ can be found from graphic construction. $P_{d(1)}$ can be obtained from Eq. (10.70) after d_1 is known. To determine the centroid of the area ABC_1C_1' and thus $l_{W(1)}$, the properties of the spiral section O_1BC_1 [Eqs. (10.67), (10.68), and (10.69)] may be combined with the properties of triangles O_1AB and AC_1C_1'. The position of the centroid can also be determined mechanically by trimming a cardboard the size ABC_1C_1' and hanging it by a thread at its corners.

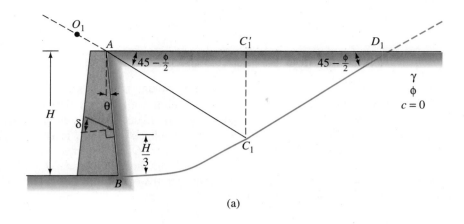

(a)

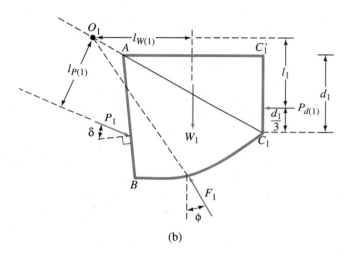

(b)

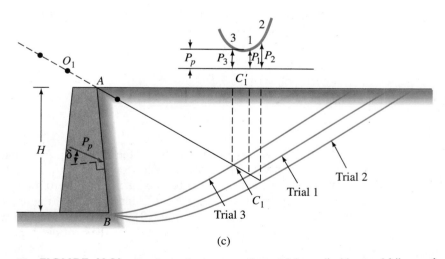

(c)

▼ FIGURE 10.31   Passive earth pressure against retaining wall with curved failure surface

The preceding procedure for finding the trial passive force per unit length of the wall is repeated for several trial wedges such as are shown in Figure 10.31c. Let P_1, P_2, P_3, ..., P_n be the forces that correspond to trial wedges 1, 2, 3, ..., n, respectively. The forces are plotted to some scale as shown in the upper part of the figure. A smooth curve is plotted through the points 1, 2, 3, ..., n. The low point of the smooth curve defines the actual passive force P_p per unit length of the wall.

Comparison of Other Trial Wedge Solutions to Determine Passive Pressure

In the preceding section, the trial wedge solution procedure explained in detail assumes that the curved failure surface BC as shown in Figure 10.19d is an arc of a logarithmic spiral as described by Terzaghi and Peck (1967) and Janbu (1957). Caquot and Kerisel (1948) determined values of the passive earth pressure coefficient for retaining walls with granular backfill by assuming that the curved failure surface is to be an arc of an ellipse. A similar analysis that assumed the curve failure surface to be the arc of a circle was conducted by Packshaw (1969). A comparison of the results obtained by these differing procedures is shown in Figure 10.32. Note that these results are for a retaining wall with vertical back ($\theta = 0$) and horizontal granular backfill ($c = 0$) material. The passive pressure for such cases can be given by

$$P_p = \frac{1}{2} \gamma H^2 K_p \tag{10.73}$$

or

$$K_p = \text{passive pressure coefficient} = \frac{P_p}{0.5\gamma H^2}$$

A comparison of the results obtained by various methods as shown in Figure 10.32 does not show a wide variation in the values of K_p.

Passive Pressure by the Method of Slices

Shields and Tolunay (1973) improved the trial wedge solution by using the *method of slices* for consideration of the stability of the trial soil wedge such as ABC_1C_1' in Figure 10.31a. The details of the analysis are beyond the scope of this text. However, the values of K_p (passive earth pressure coefficient) obtained by this method are given in Table 10.6, and they seem to be as good as any other set of values available currently; the author recommends using these values. Note that the values of K_p shown in Table 10.6 are for retaining walls with a vertical back (that is, $\theta = 0$ in Figure 10.31) supporting a granular backfill (that is, $c = 0$) with a horizontal ground surface. The passive pressure for such a case can be given as

$$P_p = \frac{1}{2} \gamma H^2 K_p$$

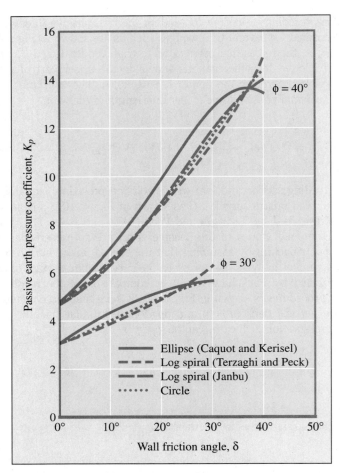

▼ **FIGURE 10.32**   Variation of K_p [Eq. (10.73)] obtained by various methods (after Shields and Tolunay, 1973)

▼ **TABLE 10.6**   **Values of K_p from the Method of Slices***

ϕ (deg)	δ (deg)									
	0	5	10	15	20	25	30	35	40	45
20	2.04	2.26	2.43	2.55	2.70					
25	2.46	2.77	3.03	3.23	3.39	3.63				
30	3.00	3.43	3.80	4.13	4.40	4.64	5.03			
35	3.69	4.29	4.84	5.34	5.80	6.21	6.59	7.25		
40	4.69	5.44	6.26	7.05	7.80	8.51	9.18	9.83	11.03	
45	5.83	7.06	8.30	9.55	10.80	12.04	13.26	14.46	15.60	18.01

ᵏ After Shields and Tolunay (1973)

▼ **EXAMPLE 10.8**

A retaining wall is shown in Figure 10.31a. Assume $\theta = 0$. Given $\gamma = 115$ lb/ft³, $H = 12$ ft, and $\phi = 35°$, estimate the passive force per unit length of the wall for $\delta = \phi/2$ by:

  a. Using Table 10.6 (curved failure surface assumption)
  b. Applying Coulomb's theory (plane failure surface assumption)

Compare the results obtained.

Solution

  a. For the curved failure surface assumption,

$$P_p = \frac{1}{2}\gamma H^2 K_p$$

  From Table 10.6, for $\phi = 35°$ and $\delta = 17.5°$, $K_p \approx 5.6$, so

$$P_p \text{ (for } \delta = 17.5°) = \frac{1}{2}(115)(12)^2(5.6) = \textbf{46,368 lb/ft}$$

  b. For the plane failure surface assumption,

$$P_p = \frac{1}{2}\gamma H^2 K_p$$

  K_p can be obtained from Table 10.3 or Eq. (10.51):

  $K_p(\delta = 17.5°) \approx 7.6$

  Thus,

$$P_p \text{ (for } \delta = 17.5°) = \frac{1}{2}(115)(12)^2(7.6) = \textbf{62,928 lb/ft}$$

The plane failure surface assumption yields a higher value of P_p, which is somewhat unconservative.   ▼

10.13 LATERAL PRESSURE ON RETAINING WALLS FROM SURCHARGES BASED ON THEORY OF ELASTICITY

Point Load Surcharge

The equations for normal stresses inside a homogeneous, elastic, and isotropic medium produced from a point load on the surface were given in Chapter 7 [Eqs. (7.10) and (7.11)].

  We now apply Eq. (7.10a) to determine the lateral pressure on a retaining wall caused by the concentrated point load Q placed at the surface of the backfill as shown in Figure 10.33a. If the load Q is placed on the plane of the section shown, we can substitute $y = 0$ in Eq. (7.10a). Also assuming $\mu = 0.5$, we can write

$$\Delta p_x = \sigma_x = \frac{Q}{2\pi}\left(\frac{3x^2z}{L^5}\right) \tag{10.74}$$

where $L = \sqrt{x^2 + z^2}$. Substituting $x = mH$ and $z = nH$ into Eq. (10.74), we have

$$\sigma_x = \frac{3Q}{2\pi H^2}\frac{m^2 n}{(m^2 + n^2)^{5/2}} \tag{10.75}$$

The horizontal stress expressed by Eq. (10.75) does not include the restraining effect of the wall. This was investigated by Gerber (1929) and Spangler (1938) with large-scale tests. Based on the experimental findings, Eq. (10.75) has been modified as follows to agree with the real conditions:

For $m > 0.4$,

$$\sigma_x = \frac{1.77Q}{H^2}\frac{m^2 n^2}{(m^2 + n^2)^3} \tag{10.76}$$

and for $m \leq 0.4$,

$$\sigma_x = \frac{0.28Q}{H^2}\frac{n^2}{(0.16 + n^2)^3} \tag{10.77}$$

A nondimensional plotting of n against $(\sigma_x H^2)/Q$ for $m \leq 0.4$ and $m = 0.5$ and 0.7 is given in Figure 10.33b.

Line Load Surcharge

Figure 10.34a shows the distribution of lateral pressure against the vertical back face of the wall caused by a line load surcharge placed parallel to the crest. The modified forms of the equations [similar to Eqs. (10.76) and (10.77) for the case of point load surcharge] for line load surcharges are as follows:

$$\sigma_x = \frac{4q}{\pi H}\frac{m^2 n}{(m^2 + n^2)^2} \qquad \text{(for } m > 0.4\text{)} \tag{10.78}$$

$$\sigma_x = \frac{0.203q}{H}\frac{n}{(0.16 + n^2)^2} \qquad \text{(for } m \leq 0.4\text{)} \tag{10.79}$$

where $q =$ load per unit length of the surcharge.

A nondimensional plot of n against $(\sigma_x H)/q$ is given in Figure 10.34b.

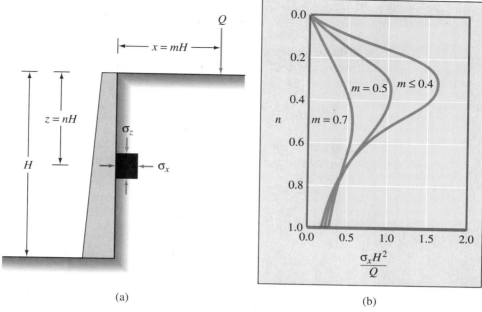

(a)

(b)

▼ **FIGURE 10.33** (a) Lateral pressure against a retaining wall caused by a point load—based on theory of elasticity; (b) plot of n against $(\sigma_x H^2)/Q$ [Eqs. (10.76) and (10.77)]

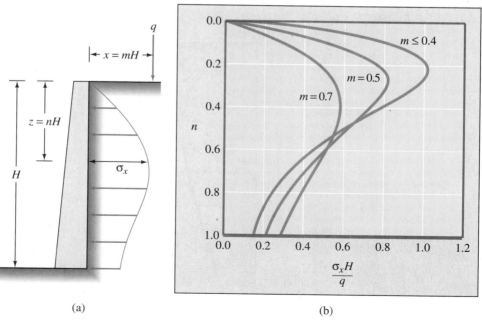

(a)

(b)

▼ **FIGURE 10.34** (a) Lateral pressure on a retaining wall caused by a line load surcharge; (b) plot of n against $(\sigma_x H)/q$ [Eqs. (10.78) and (10.79)]

Strip Load Surcharge

Figure 10.35 shows a strip load surcharge with an intensity of q per unit area located at a distance m_1 from a wall of height H. Based on the theory of elasticity, the horizontal stress σ_x at a depth z on a retaining structure can be given as

$$\sigma_x = \frac{q}{H} (\beta - \sin \beta \cos 2\alpha) \qquad (10.80)$$

The angles α and β are defined in Figure 10.35. For actual soil behavior (from wall restraining effect), the preceding equation can be modified to

$$\sigma_x = \frac{2q}{H} (\beta - \sin \beta \cos 2\alpha) \qquad (10.81)$$

The nature of the distribution of σ_x with depth is shown in Figure 10.35. The force, P, per unit length of the wall caused by the strip load alone can be obtained by integration

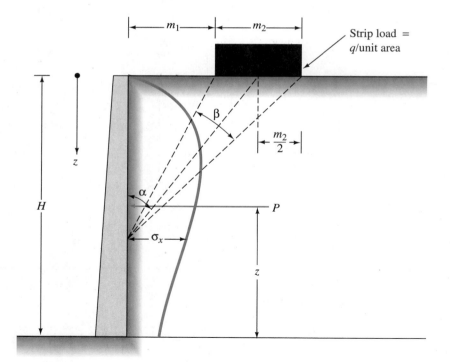

▼   **FIGURE 10.35**   Lateral pressure on a retaining wall caused by a strip load surcharge

of σ_x with limits of z from 0 to H. Jarquio (1981) expressed P in the following form:

$$P = \frac{q}{90} [H(\theta_2 - \theta_1)]$$
(10.82)

$$\text{where } \theta_1 \text{ (deg)} = \tan^{-1} \left(\frac{m_1}{H} \right)$$
(10.83)

$$\theta_2 \text{ (deg)} = \tan^{-1} \left(\frac{m_1 + m_2}{H} \right)$$
(10.84)

▼ **EXAMPLE 10.9**

Consider the retaining wall shown in Figure 10.36a where $H = 10$ ft. A line load of 800 lb/ft is placed on the ground surface parallel to the crest at a distance of 5 ft from the back face of the wall. Determine the increase in the lateral force per unit length of the wall caused by the line load. Use the modified equation given in Section 10.13.

Solution   We are given $H = 10$ ft, $q = 800$ lb/ft, and

$$m = \frac{5}{10} = 0.5 > 0.4$$

So Eq. (10.78) will apply:

$$\sigma_x = \frac{4q}{\pi H} \frac{m^2 n}{(m^2 + n^2)^2}$$

Now the following table can be prepared:

$n = \dfrac{z}{H}$	$\dfrac{4q}{\pi H}$	$\dfrac{m^2 n}{(m^2 + n^2)^2}$	σ_x (lb/ft^2)
0	101.86	0	0
0.2	101.86	0.595	60.61
0.4	101.86	0.595	60.61
0.6	101.86	0.403	41.05
0.8	101.86	0.252	25.67
1.0	101.86	0.16	16.3

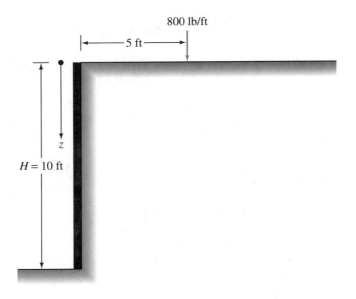

(a)

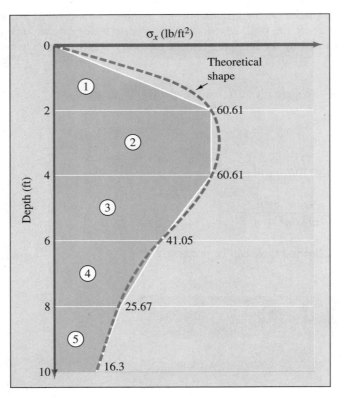

(b)

▼ FIGURE 10.36

Refer to the diagram in Figure 10.36b.

Area no.	Area	
1	$\left(\dfrac{1}{2}\right)(2)(60.61) =$	60.61 lb/ft
2	$\left(\dfrac{1}{2}\right)(2)(60.61 + 60.61) =$	121.22 lb/ft
3	$\left(\dfrac{1}{2}\right)(2)(60.61 + 41.05) =$	101.66 lb/ft
4	$\left(\dfrac{1}{2}\right)(2)(41.05 + 25.67) =$	66.72 lb/ft
5	$\left(\dfrac{1}{2}\right)(2)(25.67 + 16.3) =$	41.97 lb/ft

$$\text{Total} = \textbf{392.18 lb/ft}$$
$$\approx \textbf{390 lb/ft}$$

▼

▼ **EXAMPLE 10.10**

Refer to Figure 10.35. For a strip loading, we are given $q = 1000$ lb/ft², $m_1 = 10$ ft, $m_2 = 5$ ft, and $H = 10$ ft. Determine the force P per unit length of the wall caused by a surcharge load.

Solution   From Eqs. (10.82), (10.83), and (10.84),

$$P = \frac{q}{90}\,[H(\theta_2 - \theta_1)]$$

$$\theta_1 = \tan^{-1}\left(\frac{m_1}{H}\right) = \tan^{-1}\left(\frac{10}{10}\right) = 45°$$

$$\theta_2 = \tan^{-1}\left(\frac{m_1 + m_2}{H}\right) = \tan^{-1}\left(\frac{15}{10}\right) = 56.31°$$

Hence

$$P = \frac{1000}{90}\,[10(56.31 - 45)] = \textbf{1256 lb/ft}\quad ▼$$

10.14 BRACED CUTS

It frequently happens during the construction of foundations or utilities (such as sewers) that open trenches with vertical soil slopes are excavated. Although most of these are temporary, the sides of the cuts have to be supported by proper bracing systems.

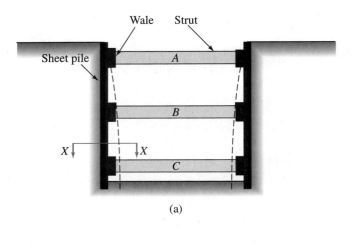

(a)

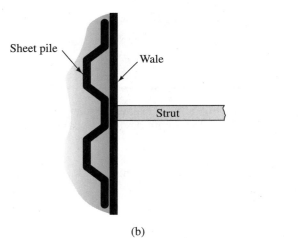

(b)

▼ **FIGURE 10.37** Braced cut: (a) cross-section; (b) plan (section at $X - X$)

Figure 10.37 shows one of several bracing systems commonly adopted in construction practice. The bracing consists of sheet piles, wales, and struts.

Proper design of these elements requires a knowledge of the lateral earth pressure exerted on the braced walls. The magnitude of the lateral earth pressure at various depths of the cut is very much influenced by the deformation condition of the sheeting. To understand the nature of the deformation of the braced walls, one needs to follow the sequence of construction. Construction of the unit begins with driving the sheetings. The top row of the wales and struts (marked A in Figure 10.37a) is emplaced immediately after a small cut is made. This must be done immediately so that there is no time for the soil mass outside the cut to deform and cause the sheetings to yield. As the sequence of driving the sheetings, excavating the soil, and placing rows of wales and struts (see B and C in Figure 10.37) continues, the sheetings move inward at greater depths. This action is caused by greater earth pressure exerted by the soil outside the cut. The deformation of the braced walls is shown by the broken lines in Figure 10.37a. Essentially, the problem models a condition where the walls are rotating about the level of the top row of struts.

The deformation of a braced wall differs from the deformation condition of a retaining wall in that, in a braced wall, the rotation is about the top. For this reason, neither Coulomb's nor Rankine's theory will give the actual earth pressure distribution. This fact is illustrated in Figure 10.38, in which AB is a frictionless wall with a granular soil backfill. When the wall deforms to position AB', failure surface BC develops. Since the upper portion of the soil mass in the zone ABC does not undergo sufficient deformation, it does not pass into Rankine's active state. The sliding surface BC intersects the ground surface almost at 90°. The corresponding earth pressure will be somewhat parabolic like acb shown in Figure 10.38b. With this type of pressure distribution, the point of application of the resultant active thrust P_a will be at a height of $n_a H$ measured from the bottom of the wall with $n_a > 1/3$ (for triangular pressure distribution $n_a = 1/3$). Theoretical evaluation and field measurements have shown that n_a could be as high as 0.55.

Figure 10.39 shows the laboratory observations of Sherif and Fang (1984) related to the distribution of the horizontal component of the lateral earth pressure on a model retaining wall with a dry granular backfill rotating about the top. This clearly demonstrates the nonhydrostatic distribution of the lateral earth pressure for this type of wall movement.

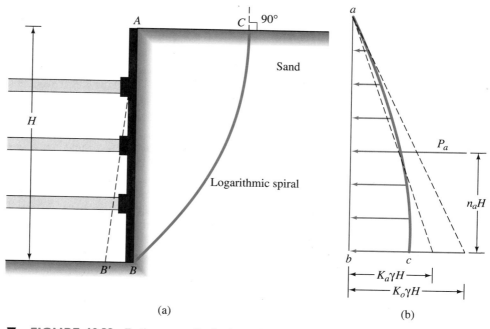

(a)　　　　　　　　　　　(b)

▼ **FIGURE 10.38** Earth pressure distribution against a wall with rotation about the top

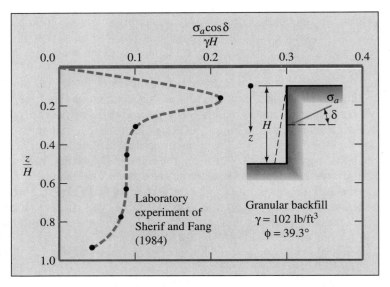

▼ **FIGURE 10.39**   Laboratory observation of the distribution of horizontal component of lateral earth pressure on retaining wall rotating about the top (*Note:* The value of δ was constant.)

Determination of Active Thrust on Bracing Systems of Open Cuts in Granular Soil

The theoretical estimation of the active thrust on the bracing system of open cuts can be done by means of trial wedges using Terzaghi's general wedge theory (1941). The general procedure for determination of the active thrust follows.

Figure 10.40a shows a braced wall AB of height H that deforms by rotating about its top. The wall is assumed to be rough, with the angle of wall friction equal to δ. The point of application of the active thrust (that is, $n_a H$) is assumed to be known. The curve of sliding is assumed to be an arc of a logarithmic spiral. As we discussed in the preceding section, the curve of sliding intersects the horizontal ground surface at 90°. To proceed with the trial wedge solution, a point b_1 is selected. From b_1, a line $b_1 b_1'$ that makes an angle ϕ with the ground surface is drawn. (Note that ϕ = angle of friction of the soil.) The arc of the logarithmic spiral, $b_1 B$, which defines the curve of sliding for this trial, can now be drawn with the center of the spiral (point O_1) located on the line $b_1 b_1'$. This can be traced out using a trial-and-error procedure by superimposing the worksheet on another sheet of paper on which a logarithmic spiral is drawn. Note that the equation for the logarithmic spiral is given by $r_1 = r_o e^{\theta_1 \tan \phi}$ and, in this case, $\overline{O_1 b_1} = r_o$ and $\overline{O_1 B} = r_1$. Also, it is interesting to see that the horizontal line that represents the ground surface is the normal to the curve of sliding at the point b_1, and $O_1 b_1$ is a radial line. The angle between them is equal to ϕ, which agrees with the property of the spiral.

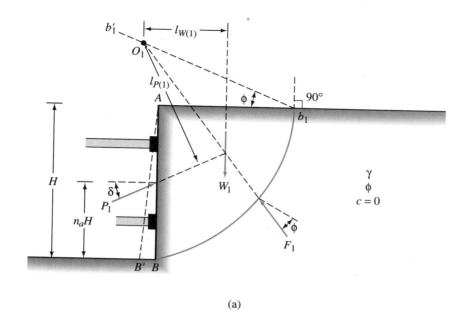

(a)

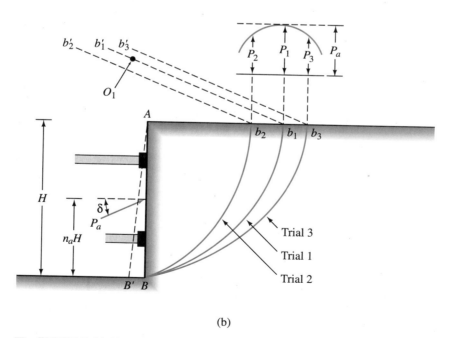

(b)

▼ **FIGURE 10.40** Determination of active force on bracing system of open cut in cohesionless soil

To look at the equilibrium of the failure wedge, the following forces per unit length of the braced wall are considered:

1. W_1 = the weight of the wedge ABb_1 = (area of ABb_1) $\times$ (γ) $\times$ (1).
2. P_1 = the active thrust acting at a point $n_a H$ measured vertically upward from the bottom of the cut and inclined at an angle δ with the horizontal.
3. F_1 = the resultant of the shear and normal forces that act along the trial failure surface. The line of action of the force F_1 will pass through the point O_1.

Now, taking the moments of the above forces about O_1, we have

$$W_1[l_{w(1)}] + F_1(0) - P_1[l_{p(1)}] = 0$$

or

$$P_1 = \frac{W_1 l_{w(1)}}{l_{p(1)}} \tag{10.85}$$

where $l_{w(1)}$ and $l_{p(1)}$ are the moment arms for the forces W_1 and P_1, respectively.

The value of $l_{p(1)}$ can be found graphically. To determine $l_{w(1)}$, the center of gravity of the section ABb_1 has to be determined. This can be done by trimming a cardboard the size of ABb_1 and hanging it by a thread at its corners. Once the center of gravity is known, $l_{w(1)}$ can be scaled out.

This procedure of finding the active thrust can now be repeated for several wedges such as ABb_2, ABb_3, ..., ABb_n (Figure 10.40b). Note that the centers of the logarithmic spiral arcs will lie on lines $b_2 b_2'$, $b_3 b_3'$, ..., $b_n b_n'$, respectively. The active thrusts P_1, P_2, P_3, ..., P_n derived from the trial wedges are plotted to some scale in the upper portion of Figure 10.40b. The maximum point of the smooth curve drawn through these points will yield the desired maximum active thrust, P_a, on the braced wall.

Kim and Preber (1969) determined the values of $P_a/0.5\gamma H^2$ for braced excavations for various values of ϕ, δ, and n_a. These are given in Appendix F (Table F.1). In general, the average magnitude of P_a is about 10% greater when the wall rotation is about the top as compared with the value obtained by Coulomb's active earth pressure theory.

Active Thrust on Bracing Systems for Cuts in Cohesive Soil ($\phi = 0$)

To determine the active thrust against the braced wall of an open cut in saturated cohesive soil, a trial wedge procedure similar to that described earlier can be adopted. However, for undrained conditions—that is, $\phi = 0$—the equation of the logarithmic spiral, $r_1 = r_o^{\theta_1 \tan \phi}$, yields $r_1 = r_o$, which is the equation of a circle.

Figure 10.41a shows a trial wedge for a braced wall of height H. The curve of sliding Bb_1 is an arc of a circle, the center of which is located at the level of the ground surface. Considering the unit length of the wall, we find that the forces for equilibrium of the wedge ABb_1 are:

1. W_1 = weight of the soil wedge
2. P_1 = active thrust that acts at a height of $n_a H$ measured from the bottom of the wall
3. F_1 = resultant of the normal forces that act along the surface of sliding

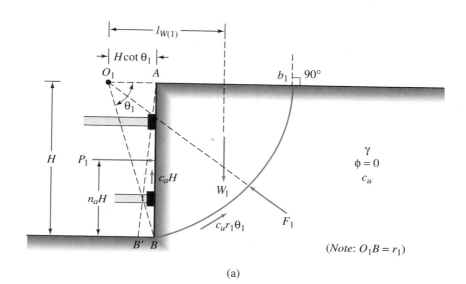

(a)

(b)

FIGURE 10.41 Determination of active force on bracing system of open cut in cohesive soil ($\phi = 0$ concept)

4. $c_u r_1 \theta_1$ = force from cohesion acting along the surface of sliding
5. $c_a H$ = force from adhesion between the soil and the sheeting (c_a = adhesion between the soil and the material used for sheeting; $c_a \leq c_u$)

Taking the moments of the preceding forces about O_1, we have

$$W_1[l_{w(1)}] + F_1[0] - c_u r_1 \theta_1[r_1] - c_a H[H \cot \theta_1] - P_1[(1 - n_a)H] = 0$$

or

$$P_1 = \frac{1}{H(1 - n_a)} [W_1 l_{w(1)} - c_u r_1^2 \theta_1 - c_a H^2 \cot \theta_1] \tag{10.86}$$

The trial active thrusts, such as $P_1, P_2, P_3, \ldots, P_n$, can be obtained from several trial wedges. If these are plotted to some scale as shown in Figure 10.41b, the highest point of the smooth curve drawn through these points will give the desired value of active thrust, P_a.

Using the preceding procedure, Das and Seeley (1975) showed that

$$P_a = \frac{1}{2(1 - n_a)} (0.677 - K N_c) \gamma H^2 \tag{10.87}$$

where $N_c = \left(\frac{c_u}{\gamma H}\right)$ (10.88)

$$K = f\left(\frac{c_a}{c_u}\right) \tag{10.89}$$

The values of K are as follows:

$\left(\dfrac{c_a}{c_u}\right)$	K
0	2.762
0.5	3.056
1.0	3.143

Active Thrust on Bracing Systems for Cuts in c-φ Soil

Using the principles of general wedge theory, we can also determine the active thrust on bracing systems for cuts made in $c\text{-}\phi$ soil. Table F-2 in Appendix F gives the variation of P_a in a nondimensional form for various values of ϕ, δ, n_a, and $c/\gamma H$.

10.15 DYNAMIC EARTH PRESSURE DISTRIBUTION BEHIND A WALL ROTATING ABOUT THE TOP

Based on laboratory model test results, Sherif and Fang (1984) reported the dynamic earth pressure distribution behind a rigid retaining wall ($H = 1$ m) with dense sand as backfill material and rotation about its top. Figure 10.42 shows a plot of $\sigma_a \cos \delta$ versus depth for various values of k_h (for $k_v = 0$). The magnitude of active thrust, P_{ae}, can be obtained as

$$P_{ae} \cos \delta = \int_0^H (\sigma_a \cos \delta) \, dz$$

or

$$P_{ae} = \frac{1}{\cos \delta} \int_0^H (\sigma_a \cos \delta) \, dz \qquad (10.90)$$

For a given value of k_h, the magnitude of P_{ae} is 15% to 20% higher compared with that obtained by using Eq. (10.56) (that is, the case of wall rotation about the bottom).

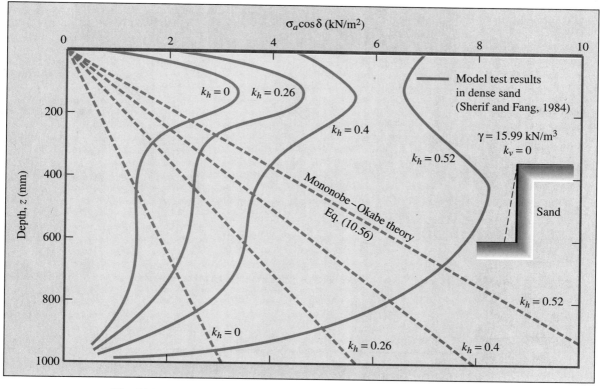

▼ **FIGURE 10.42** Dynamic lateral earth pressure distribution behind a rigid model retaining wall rotating about the top

10.16 PRESSURE VARIATION FOR DESIGN OF SHEETINGS, STRUTS, AND WALES

The active thrust against sheeting in a braced cut, calculated using the general wedge theory, does not explain the variation of the earth pressure with depth that is necessary for design work. An important difference between bracings in open cuts and retaining walls is that retaining walls fail as single units, whereas bracings in an open cut undergo progressive failure where one or more struts fail at one time.

Empirical lateral pressure diagrams against sheetings for the design of bracing systems have been given by Peck (1969). These pressure diagrams for cuts in loose sand, soft to medium clay, and stiff clay are given in Figure 10.43. Strut loads may be determined by assuming that the vertical members are hinged at each strut level except the topmost and bottommost ones (Figure 10.44). Example 10.11 illustrates the procedure for the calculation of strut loads.

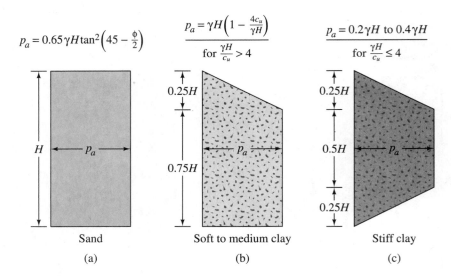

▼ **FIGURE 10.43** Peck's pressure diagrams for design of bracing systems

▼ **EXAMPLE 10.11**

A 7-m deep braced cut in sand is shown in Figure 10.45. In plan, the struts are placed at $s = 2$ m center-to-center. Using Peck's empirical pressure diagram, calculate the design strut loads.

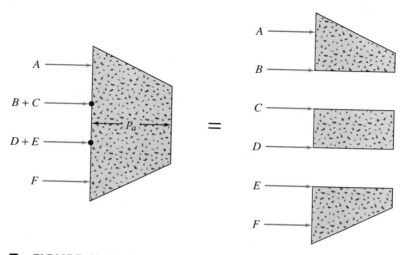

▼ **FIGURE 10.44** Determination of strut loads from empirical lateral pressure diagrams

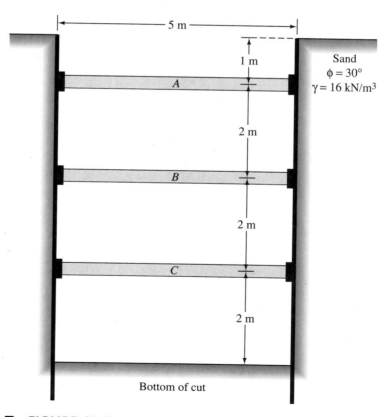

▼ **FIGURE 10.45**

Solution   Refer to Figure 10.43a. For the lateral earth pressure diagram,

$$p_a = 0.65\gamma H \tan^2\left(45 - \frac{\phi}{2}\right) = (0.65)(16)(7)\tan^2\left(45 - \frac{30}{2}\right) = 24.27 \text{ kN/m}^2$$

Assume the sheeting is hinged at strut level B. Now refer to the diagram in Figure 10.46. We need to find reactions at A, B_1, B_2, and C. Taking the moment about B_1, we have

$$2A = (24.27)(3)\left(\frac{3}{2}\right); \qquad A = 54.61 \text{ kN/m}$$

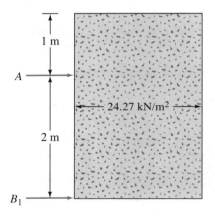

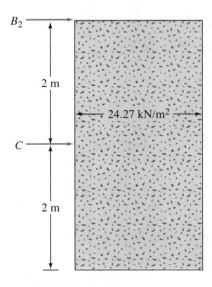

▼ **FIGURE 10.46**

Hence,

$$B_1 = (24.27)(3) - 54.61 = 18.2 \text{ kN/m}$$

Again, taking the moment about B_2, we have

$$2C = (24.27)(4)\left(\frac{4}{2}\right)$$

$$C = 97.08 \text{ kN/m}$$

So

$$B_2 = (24.27)(4) - 97.08 = 0$$

The strut loads are as follows:

at level A: $(A)(s) = (54.61)(2) = \mathbf{109.22 \ kN}$

at level B: $(B_1 + B_2)(s) = (18.2 + 0)(2) = \mathbf{36.4 \ kN}$

at level C: $(C)(s) = (97.08)(2) = \mathbf{194.16 \ kN}$ ▼

10.17 GENERAL COMMENTS

In this chapter, the concepts and relationships for at-rest, active, and passive pressure on retaining walls were developed. Also, the nature and distribution of lateral earth pressure for braced cuts in granular and cohesive soils were discussed.

For design, it is important to realize that the lateral active pressure on a retaining wall can be calculated using Rankine's theory only when the wall moves *sufficiently* outward either by rotation about the toe of the footing or by deflection of the wall. If sufficient wall movement cannot, or is not allowed to, occur, then the lateral earth pressure will be greater than the Rankine active pressure and sometimes may be closer to the at-rest earth pressure. Hence, proper selection of the lateral earth pressure coefficient is important for safe and proper design. It is a general practice to assume a value for the soil friction angle (ϕ) of the backfill in order to calculate the Rankine active pressure distribution, ignoring the contribution of the cohesion (c). The general range of ϕ used for retaining wall design is given in the table.

Soil type	Soil friction angle, ϕ (deg)
Soft clay	0–15
Compacted clay	20–30
Dry sand and gravel	30–40
Silty sand	20–30

In Section 10.4, we saw that the lateral earth pressure on a retaining wall is greatly increased in the presence of a water table above the base of the wall. Most retaining walls are not designed to withstand full hydrostatic pressure; hence, it is important that adequate drainage facilities are provided to ensure that the backfill soil does not become fully saturated. This can be achieved by providing weepholes at regular intervals along the length of the wall.

During the last 20 years or so, many retaining walls have been constructed with *reinforced earth* backfill. Reinforced earth walls are flexible walls. Their main components are:

1. *Backfill*, which is a free-draining granular soil
2. *Reinforcing strips*, which are thin, wide galvanized steel strips placed at regular intervals
3. A *cover* on the front face, which is referred to as the *skin*

Figure 10.47 is a schematic diagram of a reinforced earth wall. Note that, at any given depth, the reinforcing strips are placed with a horizontal spacing of S_H center-to-

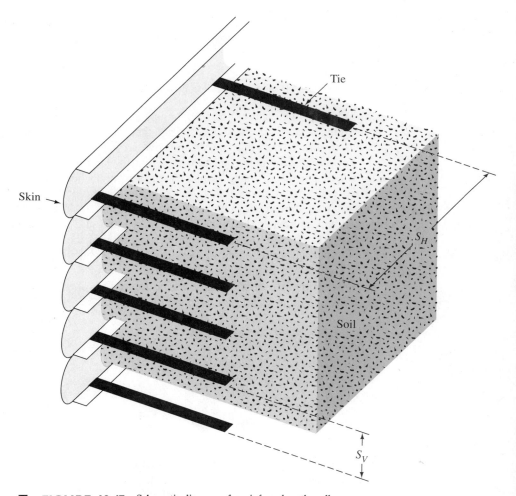

▼ **FIGURE 10.47** Schematic diagram of a reinforced earth wall

center; the vertical spacing of the strips (center-to-center) is S_V. The skin can be constructed with sections of relatively flexible thin material. In some cases, precast concrete slabs can also be used as skin. The slabs are grooved to fit each other so that soil does not flow out between the joints. When metal skins are used, they are bolted together, and reinforcing strips are placed between the skins. Reinforced earth walls are now used in the construction of retaining walls, bridge abutments, waterfront walls, and so forth. There are three basic ways to design ties that will resist the lateral earth pressure:

1. The Rankine method
2. The Coulomb force method
3. The Coulomb moment method

The Rankine method is the simplest and most widely used of the three design methods. For more details, refer to Das (1990).

    The design and construction of braced cuts (Sections 10.14 and 10.16) are highly related to safety issues. Hence, all health and safety recommendations provided by state and federal agencies should be followed.

PROBLEMS

10.1 Assuming that the wall shown in Figure 10.48 is restrained from yielding, find the magnitude and location of the resultant lateral force per unit length of the wall for the following cases:

    a. $H = 10$ ft, $\gamma = 110$ lb/ft³, $\phi = 34°$
    b. $H = 12$ ft, $\gamma = 105$ lb/ft³, $\phi = 36°$
    c. $H = 5$ m, $\gamma = 14.4$ kN/m³, $\phi = 31°$
    d. $H = 4$ m, $\gamma = 13.4$ kN/m³, $\phi = 28°$

10.2 Figure 10.48 shows a retaining wall with cohesionless soil backfill. For the following cases, determine the total active force per unit length of the wall for Rankine's state, the

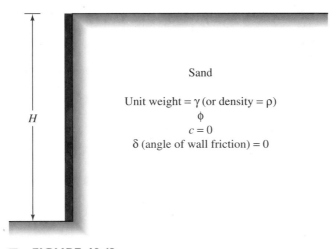

Sand

Unit weight = γ (or density = ρ)
ϕ
$c = 0$
δ (angle of wall friction) = 0

H

▼ **FIGURE 10.48**

location of the resultant, and the variation of the active pressure with depth.
a. $H = 15$ ft, $\gamma = 110$ lb/ft³, $\phi = 30°$
b. $H = 20$ ft, $\gamma = 98$ lb/ft³, $\phi = 28°$
c. $H = 18$ ft, $\gamma = 108$ lb/ft³, $\phi = 36°$
d. $H = 16.5$ ft, $\gamma = 90$ lb/ft³, $\phi = 30°$
e. $H = 4.5$ m, $\gamma = 17.6$ kN/m³, $\phi = 36°$
f. $H = 5$ m, $\gamma = 17.0$ kN/m³, $\phi = 38°$
g. $H = 4$ m, $\gamma = 19.95$ kN/m³, $\phi = 42°$

10.3 From Figure 10.48, determine the passive force, P_p, per unit length of the wall for Rankine's case. Also state Rankine's passive pressure at the bottom of the wall. Consider the following cases:
a. $H = 10$ ft, $\gamma = 110$ lb/ft³, $\phi = 30°$
b. $H = 14$ ft, $\gamma = 120$ lb/ft³, $\phi = 36°$
c. $H = 2.45$ m, $\gamma = 16.67$ kN/m³, $\phi = 33°$
d. $H = 4$ m, $\rho = 1800$ kg/m³, $\phi = 38°$

10.4 A retaining wall is shown in Figure 10.49. Determine Rankine's active force, P_a, per unit length of the wall and the location of the resultant for each of the following cases:
a. $H = 10$ ft, $H_1 = 4$ ft, $\gamma_1 = 110$ lb/ft³, $\gamma_2 = 122$ lb/ft³, $\phi_1 = 32°$, $\phi_2 = 32°$, $q = 0$
b. $H = 15$ ft, $H_1 = 5$ ft, $\gamma_1 = 110$ lb/ft³, $\gamma_2 = 120$ lb/ft³, $\phi_1 = 36°$, $\phi_2 = 36°$, $q = 500$ lb/ft²
c. $H = 6$ m, $H_1 = 2$ m, $\gamma_1 = 16$ kN/m³, $\gamma_2 = 19$ kN/m³, $\phi_1 = 32°$, $\phi_2 = 36°$, $q = 15$ kN/m²
d. $H = 5$ m, $H_1 = 1.5$ m, $\gamma_1 = 17.2$ kN/m³, $\gamma_2 = 20.4$ kN/m³, $\phi_1 = 30°$, $\phi_2 = 34°$, $q = 19.15$ kN/m²

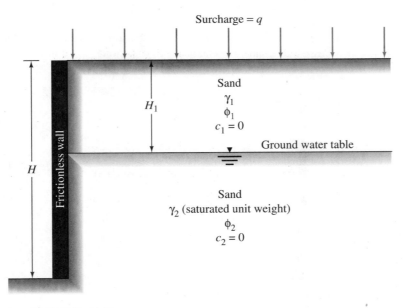

▼ **FIGURE 10.49**

10.5 Refer to Figure 10.49. Determine Rankine's passive force, P_p, per unit length of the wall for the following cases. Also find the location of the resultant for each case.
  a. $H = 15$ ft, $H_1 = 5$ ft, $\gamma_1 = 105$ lb/ft³, $\gamma_2 = 120$ lb/ft³, $\phi_1 = 30°$, $\phi_2 = 36°$, $q = 0$
  b. $H = 20$ ft, $H_1 = 6$ ft, $\gamma_1 = 110$ lb/ft³, $\gamma_2 = 126$ lb/ft³, $\phi_1 = 34°$, $\phi_2 = 34°$, $q = 300$ lb/ft²

10.6 A retaining wall 20 ft high with a vertical back face retains a homogeneous saturated soft clay. The saturated unit weight of the clay is 120 lb/ft³. Laboratory tests showed that the undrained shear strength, c_u, of the clay is 350 lb/ft².
  a. Do the necessary calculations and draw the variation of Rankine's active pressure on the wall with depth.
  b. Find the depth up to which a tensile crack can occur.
  c. Determine the total active force per unit length of the wall before the tensile crack occurs.
  d. Determine the total active force per unit length of the wall after the tensile crack. Also find the location of the resultant.

10.7 Redo Problem 10.6, assuming that the backfill is supporting a surcharge of 200 lb/ft².

10.8 Repeat Problem 10.6 with the following values:
  Height of wall = 6 m
  $\gamma_{sat} = 19.8$ kN/m³
  $c_u = 14.7$ kN/m²

10.9 A retaining wall 20 ft high with a vertical back face has a c-ϕ soil for backfill. For the backfill, $\gamma = 115$ lb/ft³, $c = 600$ lb/ft², and $\phi = 18°$. Taking the existence of the tensile crack into consideration, determine the active force, P_a, per unit length of the wall for Rankine's active state.

10.10 For the wall described in Problem 10.9, determine the passive force, P_p, per unit length for Rankine's passive state.

10.11 For the retaining wall shown in Figure 10.50, determine the active force, P_a, for Rankine's

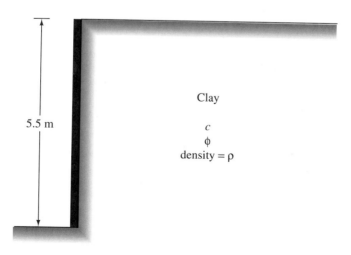

5.5 m

Clay

c

ϕ

density = ρ

▼ **FIGURE 10.50**

state. Also find the location of the resultant. Assume that the tensile crack exists.
   a. $\rho = 2300 \text{ kg/m}^3$, $\phi = 0°$, $c = c_u = 32 \text{ kN/m}^2$
   b. $\rho = 1850 \text{ kg/m}^3$, $\phi = 16°$, $c = 15 \text{ kN/m}^2$

10.12   A retaining wall is shown in Figure 10.51. The height of wall is 20 ft and the unit weight of the sand backfill is 120 lb/ft^3. Calculate the active force, P_a, on the wall using Coulomb's equation for the following values of the angle of wall friction:
   a. $\delta = 0°$
   b. $\delta = 10°$
   c. $\delta = 20°$
Comment on the direction and location of the resultant.

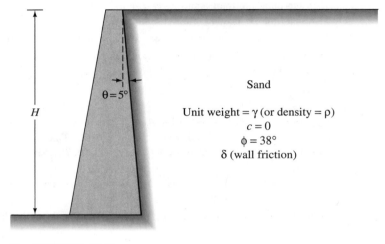

The figure shows a retaining wall of height H with a wall face inclined at $\theta = 5°$, retaining sand with:
Sand
Unit weight = γ (or density = ρ)
$c = 0$
$\phi = 38°$
δ (wall friction)

▼   **FIGURE 10.51**

10.13   For the retaining wall described in Problem 10.12, determine the passive force, P_p, per unit length of the wall using Coulomb's equation for the following values of the angle of wall friction:
   a. $\delta = 0°$
   b. $\delta = 10°$
   c. $\delta = 20°$

10.14   Draw a logarithmic spiral according to the equation $r_1 = r_o e^{\theta_1 \tan \phi}$ with θ_1 varying from 0° to 180°. Use $\phi = 35°$ and $r_o = 1$ in.

10.15   Refer to Figure 10.51. If $H = 5$ m, the density of soil $\rho = 1900 \text{ kg/m}^3$, and the angle of wall friction $\delta = 19°$, determine the passive force, P_p, per unit length of the wall. Use the graphic trial wedge technique.

10.16   Refer to Problem 10.15. If other quantities remain the same and if $\theta = 0$, what will be the passive force, P_p, per unit length of the wall? Use Table 10.6.

10.17 Referring to Figure 10.52, determine Coulomb's active force, P_a, per unit length of the wall for the following cases. Use Culmann's graphic construction procedure.

  a. $H = 15$ ft, $\beta = 85°$, $n = 1$, $H_1 = 20$ ft, $\gamma = 128$ lb/ft^3, $\phi = 38°$, $\delta = 10°$
  b. $H = 18$ ft, $\beta = 90°$, $n = 2$, $H_1 = 22$ ft, $\gamma = 116$ lb/ft^3, $\phi = 34°$, $\delta = 17°$
  c. $H = 5.5$ m, $\beta = 80°$, $n = 1$, $H_1 = 6.5$ m, $\rho = 1680$ kg/m^3, $\phi = 30°$, $\delta = 30°$

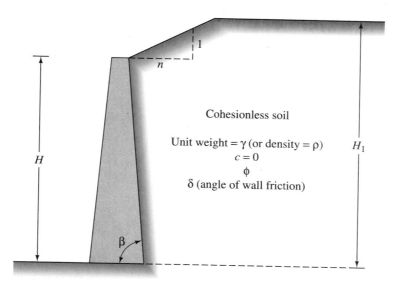

Cohesionless soil

Unit weight $= \gamma$ (or density $= \rho$)
$c = 0$
ϕ
δ (angle of wall friction)

H_1

H

β

n

1

▼ **FIGURE 10.52**

10.18 Refer to Figure 10.26. A vertical retaining wall ($\theta = 0$) is 15 ft high with a horizontal backfill (that is, $\alpha = 0$). It has a cohesionless soil as backfill. Given:

Unit weight of soil $= 95$ lb/ft^3
Angle of friction, $\phi = 30°$
$\delta = 15°$
$k_h = 0.35$
$k_v = 0$

determine the active force, P_{ae}, per unit length of the retaining wall.

10.19 Refer to Problem 10.18. Determine the location of the point of intersection of the resultant force, P_{ae}, with the back face of the retaining wall.

10.20 Consider a retaining wall 15 ft high that has a vertical back face with a horizontal backfill. A vertical point load of 4000 lb is placed on the ground surface at a distance of 6 ft from the wall. Calculate the increase in the lateral pressure on the wall for the section that contains the point load. Plot the variation of the pressure increase with depth. Use the modified equation given in Section 10.13.

10.21 A line load of 50 kN/m is located at a distance of 3 m from the back face of a retaining wall 5 m high. The retaining wall has a vertical back face and a horizontal backfill. Determine the lateral force per unit length of the wall caused by the line load.

10.22   Using the graphic procedure described in the section on general wedge theory, determine the active thrust, P_a, for the braced wall shown in Figure 10.53.

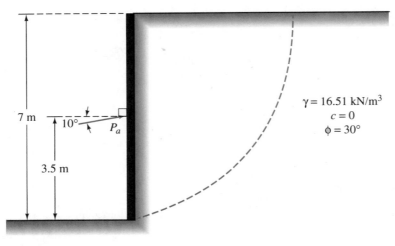

$\gamma = 16.51 \text{ kN/m}^3$
$c = 0$
$\phi = 30°$

▼ **FIGURE 10.53**

10.23   The elevation and plan of a bracing system for an open cut in sand are shown in Figure 10.54. Assuming $\gamma_{sand} = 105 \text{ lb/ft}^3$ and $\phi = 35°$, determine the strut loads.

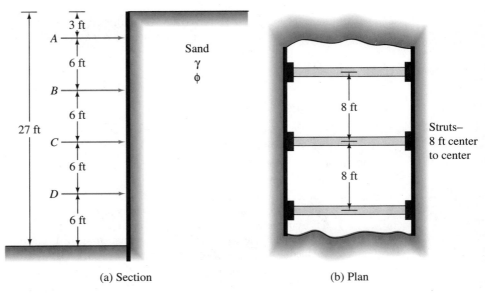

(a) Section   (b) Plan

▼ **FIGURE 10.54**

REFERENCES

Applied Technology Council (1978). "Tentative Provisions for the Development of Seismic Regulations for Buildings," *Publication ATC-3-06*.

Caquot, A., and Kerisel, J. (1948). *Tables for the Calculation of Passive Pressure, Active Pressure, and Bearing Capacity of Foundations*, Gauthier-Villars, Paris.

Coulomb, C. A. (1776). "Essai sur une Application des Règles de Maximis et Minimis à quelques Problèmes de Statique, relatifs a l'Architecture," *Mem. Roy. des Sciences*, Paris, Vol. 3, 38.

Culmann, C. (1875). *Die graphische Statik*, Meyer and Zeller, Zurich.

Das, B. M. (1990). *Principles of Foundation Engineering*, 2nd ed., PWS–Kent, Boston.

Das, B. M., and Seeley, G. R. (1975). "Active Thrust on Braced Cut in Clay," *Journal of the Construction Division*, ASCE, Vol. 101, No. CO4, 945–949.

Davies, T. G., Richards, R., and Chen, K. H. (1986). "Passive Pressure During Seismic Loading," *Journal of Geotechnical Engineering*, American Society of Civil Engineers, Vol. 112, No. GT4, 479–484.

Gerber, E. (1929). *Untersuchungen über die Druckverteilung im Örlich belasteten Sand*, Technische Hochschule, Zurich.

Hijab, W. (1956). "A Note on the Centroid of a Logarithmic Spiral Sector," *Geotechnique*, Vol. 4, No. 2, 96–99.

Jaky, J. (1944). "The Coefficient of Earth Pressure at Rest," *Journal of the Society of Hungarian Architects and Engineers*, Vol. 7, 355–358.

Janbu, N. (1957). "Earth Pressure and Bearing Capacity Calculations by Generalized Procedure of Slices," *Proceedings*, 4th International Conference on Soil Mechanics and Foundation Engineering, Vol. 2, 207–213.

Jarquio, R. (1981). "Total Lateral Surcharge Pressure Due to a Strip Load," *Journal of the Geotechnical Engineering Division*, ASCE, Vol. 107, No. GT10, 1424–1428.

Kapila, J. P. (1962). "Earthquake Resistant Design of Retaining Walls," *Proceedings*, Second Earthquake Symposium, Roorkee, India.

Kim, J. S., and Preber, T. (1969). "Earth Pressure Against Braced Excavations," *Journal of the Soil Mechanics and Foundations Division*, ASCE, Vol. 95, No. SM6, 1581–1584.

Massarsch, K. R. (1979). "Lateral Earth Pressure in Normally Consolidated Clay," *Proceedings of the Seventh European Conference on Soil Mechanics and Foundation Engineering*, Brighton, England, Vol. 2, 245–250.

Mononobe, N. (1929). "On the Determination of Earth Pressures During Earthquakes," *Proceedings*, World Engineering Conference, Vol. 9, 274–280.

Okabe, S. (1926). "General Theory of Earth Pressure," *Journal of the Japanese Society of Civil Engineers*, Tokyo, Vol. 12, No. 1.

Packshaw, S. (1969). "Earth Pressure and Earth Resistance," *A Century of Soil Mechanics*, The Institution of Civil Engineers, London, England, 409–435.

Peck, R. B. (1969). "Deep Excavation and Tunneling in Soft Ground," *Proceedings*, 7th International Conference on Soil Mechanics and Foundation Engineering, Mexico City, State-of-the-Art Vol., 225–290.

Rankine, W. M. J. (1857). "On Stability on Loose Earth," *Philosophic Transactions of Royal Society*, London, Part I, 9–27.

Richards, R., and Elms, D. G. (1979). "Seismic Behavior of Gravity Retaining Walls," *Journal of the Geotechnical Engineering Division*, ASCE, Vol. 105, No. GT4, 449–464.

Seed, H. B., and Whitman, R. V. (1970). "Design of Earth Retaining Structures for Dynamic

Loads," *Proceedings*, Specialty Conference on Lateral Stresses in the Ground and Design of Earth Retaining Structures, ASCE, 103–147.

Sherif, M. A., and Fang, Y. S. (1984). "Dynamic Earth Pressure on Walls Rotating About the Top," *Soils and Foundations*, Vol. 24, No. 4, 109–117.

Sherif, M. A., Fang, Y. S., and Sherif, R. I. (1984). "K_A and K_O Behind Rotating and Non-Yielding Walls," *Journal of Geotechnical Engineering*, American Society of Civil Engineers, Vol. 110, No. GT1, 41–56.

Shields, D. H., and Tolunay, A. Z. (1973). "Passive Pressure Coefficients by Method of Slices," *Journal of the Soil Mechanics and Foundations Division*, ASCE, Vol. 99, No. SM12, 1043–1053.

Spangler, M. G. (1938). "Horizontal Pressures on Retaining Walls Due to Concentrated Surface Loads," Iowa State University Engineering Experiment Station, *Bulletin*, No. 140.

Terzaghi, K. (1941). "General Wedge Theory of Earth Pressure," *Transactions*, ASCE, Vol. 106, 68–97.

Terzaghi, K., and Peck, R. B. (1967). *Soil Mechanics in Engineering Practice*, Wiley, New York.

CHAPTER ELEVEN

SOIL-BEARING CAPACITY FOR SHALLOW FOUNDATIONS

The lowest part of a structure is generally referred to as the *foundation*. Its function is to transfer the load of the structure to the soil on which it is resting. A properly designed foundation is one that transfers the load throughout the soil without overstressing the soil. Overstressing the soil can result in either excessive settlement or shear failure of the soil, both of which cause damage to the structure. Thus, geotechnical and structural engineers who design foundations must evaluate the bearing capacity of soils.

Depending on the structure and soil encountered, various types of foundations are used. Figure 11.1 shows the most common types of foundations. A *spread footing* is simply an enlargement of a load-bearing wall or column that makes it possible to spread the load of the structure over a larger area of the soil. In soil with low load-bearing capacity, the size of the spread footings required is impractically large. In that case, it is more economical to construct the entire structure over a concrete pad. This is called a *mat foundation*.

Pile and *drilled shaft foundations* are used for heavier structures when great depth is required for supporting the load. Piles are structural members made of timber, concrete, or steel that transmit the load of the superstructure to the lower layers of the soil. According to how they transmit their load into the subsoil, piles can be divided into two categories: friction piles and end-bearing piles. In the case of friction piles, the superstructure load is resisted by the shear stresses generated along the surface of the pile. In the end-bearing pile, the load carried by the pile is transmitted at its tip to a firm stratum.

In the case of drilled shafts, a shaft is drilled into the subsoil and is then filled with concrete. A metal casing may be used while the shaft is being drilled. It may be left in place or may be withdrawn during the placing of concrete. Generally, the diameter of a drilled shaft is much larger than that of a pile. The distinction between piles and drilled shafts becomes hazy at an approximate diameter of 3 ft (0.9 m), and the definitions and nomenclature are inaccurate.

465

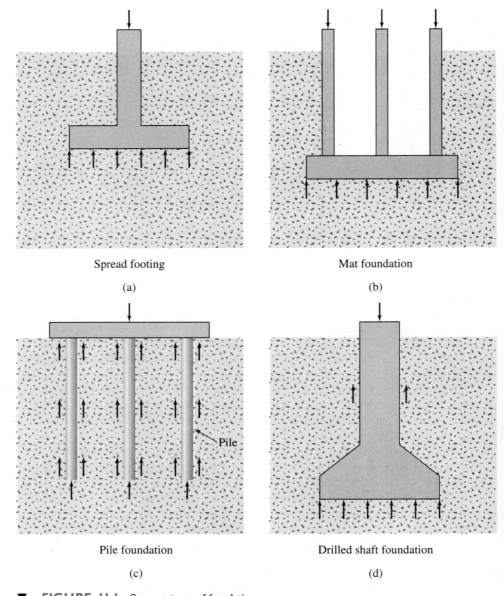

Spread footing

(a)

Mat foundation

(b)

Pile foundation

(c)

Drilled shaft foundation

(d)

▼   **FIGURE 11.1**   Common types of foundations

Spread footings and mat foundations are generally referred to as shallow foundations, and pile and drilled shaft foundations are classified as deep foundations.

In a more general sense, shallow foundations are those foundations that have a depth-of-embedment-to-width ratio of approximately less than four. When the depth-of-embedment-to-width ratio of a foundation is greater than four, it may be classified as a deep foundation.

In this chapter, we discuss the soil-bearing capacity for shallow foundations. As mentioned before, for a foundation to function properly, (1) the settlement of soil caused by the load must be within the tolerable limit, and (2) shear failure of the soil supporting the foundation must not occur. Compressibility of soil—consolidation and elasticity theory—was introduced in Chapter 8. This chapter introduces the load-carrying capacity of shallow foundations based on the criteria of shear failure in soil.

11.1 ULTIMATE SOIL-BEARING CAPACITY FOR SHALLOW FOUNDATIONS

To understand the concept of the ultimate soil-bearing capacity and the mode of shear failure in soil, let us consider the case of a long rectangular footing of width B located at the surface of a dense sand layer (or stiff soil) shown in Figure 11.2a. When a uniformly distributed load of q per unit area is applied to the footing, it will settle. If the uniformly distributed load (q) is increased, the settlement of the footing will gradually increase. When the value of $q = q_u$ is reached (Figure 11.2b), bearing capacity failure will occur;

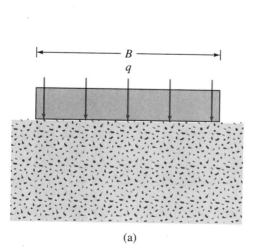

(a)

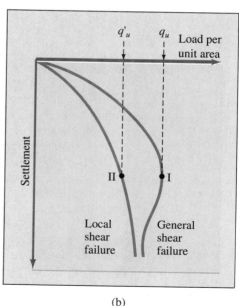

(b)

FIGURE 11.2   Ultimate soil-bearing capacity for shallow foundation: (a) model footing; (b) load-settlement relationship

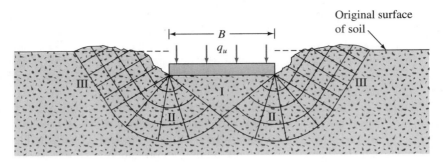

(a) General shear failure of soil

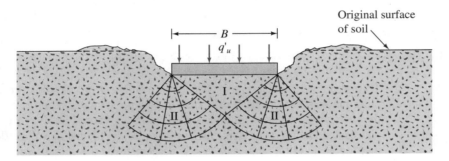

(b) Local shear failure of soil

▼   **FIGURE 11.3**   Modes of bearing capacity failure in soil

the footing will undergo a very large settlement without any further increase of q. The soil on one or both sides of the foundation will bulge and the slip surface will extend to the ground surface. The load-settlement relationship will be like curve I shown in Figure 11.2b. In this case, q_u is defined as the ultimate bearing capacity of soil.

The bearing-capacity failure described above is called a *general shear failure* and can be explained with reference to Figure 11.3a. When the foundation settles under the application of a load, a triangular wedge-shaped zone of soil (marked I) is pushed down and, in turn, it presses the zones marked II and III sideways and then upward. At the ultimate pressure, q_u, the soil passes into a state of plastic equilibrium and failure takes place by sliding.

If the footing test described above had been conducted in a loose to medium dense sand, the load-settlement relationship would have been like curve II. Beyond a certain value of $q = q'_u$, the load-settlement relationship becomes steep and straight. In this case, q'_u is defined as the ultimate bearing capacity of soil. This type of soil failure is referred to as *local shear failure* and is shown in Figure 11.3b. The triangular wedge-shaped zone (marked I) below the footing moves downward, but, unlike general shear failure, the slip surfaces end somewhere inside the soil. Some signs of soil bulging are seen, however.

11.2 TERZAGHI'S ULTIMATE BEARING CAPACITY EQUATION

In 1921, Prandtl published results of his study on the penetration of hard bodies, such as metal punches, into a softer material. Terzaghi (1943) extended the plastic failure theory of Prandtl to evaluate the bearing capacity of soils for shallow strip footings. For practical considerations, a long wall footing (length-to-width ratio more than about 5) may be called a *strip footing*. According to Terzaghi, a foundation may be defined as a shallow foundation if the depth, D_f, is less than or equal to its width, B (Figure 11.4). He also assumed that for ultimate soil-bearing-capacity calculations, the weight of soil above the base of the footing may be replaced by a uniform surcharge, $q = \gamma D_f$.

The failure mechanism assumed by Terzaghi for determining the ultimate soil-bearing-capacity (general shear failure) for a rough strip footing located at a depth D_f measured from the ground surface is shown in Figure 11.5a. The soil wedge ABJ (zone I) is an elastic zone. Both AJ and BJ make an angle ϕ with the horizontal. Zones marked II (AJE and BJD) are the radial shear zones, and zones marked III are the Rankine passive zones. The rupture lines JD and JE are arcs of a logarithmic spiral, and DF and EG are straight lines. AE, BD, EG, and DF make angles of $45 - \phi/2$ degrees with the horizontal. The equation of the arcs of the logarithmic spirals JD and JE may be given as

$$r = r_o e^{\theta \tan \phi}$$

If the load per unit area, q_u, is applied to the footing and general shear failure occurs, the passive force P_p is acting on each of the faces of the soil wedge ABJ. This is easy to conceive of if we imagine that AJ and BJ are two walls that are pushing the soil wedges $AJEG$ and $BJDF$, respectively, to cause passive failure. P_p should be inclined at an angle δ (which is the angle of wall friction) to the perpendicular drawn to the wedge faces (that is, AJ and BJ). In this case, δ should be equal to the angle of friction of soil, ϕ. Since AJ and BJ are inclined at an angle ϕ to the horizontal, the direction of P_p should be vertical.

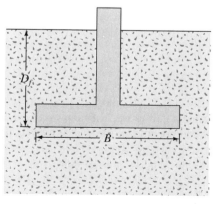

Unit weight of soil $= \gamma$
$q = \gamma D_f$
$D_f \leq B$

▼ **FIGURE 11.4**   Shallow strip footing

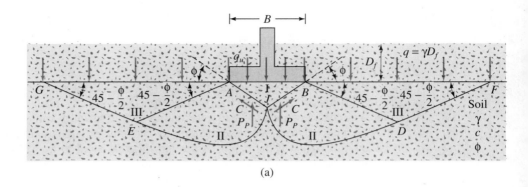

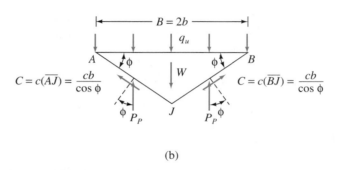

(b)

▼ **FIGURE 11.5**   Terzaghi's bearing capacity analysis

Now let us consider the free body diagram of the wedge ABJ as shown in Figure 11.5b. Considering the unit length of the footing, we have for equilibrium

$$(q_u)(2b)(1) = -W + 2C \sin \phi + 2P_p \tag{11.1}$$

where $b = B/2$

W = weight of soil wedge $ABJ = \gamma b^2 \tan \phi$
C = cohesive force acting along each face, AJ and BJ, that is equal to the unit cohesion times the length of each face = $cb/(\cos \phi)$

Thus,

$$2bq_u = 2P_p + 2bc \tan \phi - \gamma b^2 \tan \phi \tag{11.2}$$

The passive pressure in Eq. (11.2) is the sum of the contribution of the weight of soil, γ; cohesion, c; and surcharge, q. Figure 11.6 shows the distribution of passive pressure from each one of these components on the wedge face BJ. Thus, we can write

$$P_p = \frac{1}{2} \gamma(b \tan \phi)^2 K_\gamma + c(b \tan \phi)K_c + q(b \tan \phi)K_q \tag{11.3}$$

where K_γ, K_c, and K_q are earth pressure coefficients that are functions of the soil friction angle, ϕ.

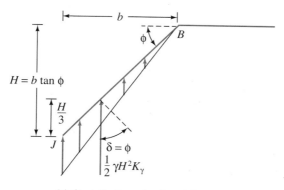

(a) Contribution of soil weight, γ

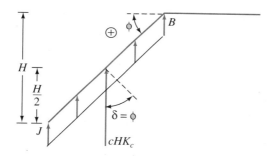

(b) Contribution of cohesion, c

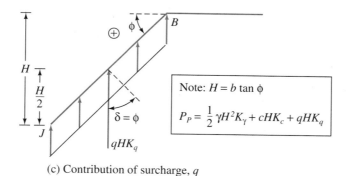

(c) Contribution of surcharge, q

Note: $H = b \tan \phi$

$$P_P = \frac{1}{2}\gamma H^2 K_\gamma + cHK_c + qHK_q$$

▼ **FIGURE 11.6** Passive force distribution on the wedge face BJ shown in Figure 11.5

Combining Eqs. (11.2) and (11.3), we obtain

$$q_u = cN_c + qN_q + \frac{1}{2}\gamma BN_\gamma \tag{11.4}$$

where

$$N_c = \tan\phi\,(K_c + 1) \tag{11.5}$$

$$N_q = K_q \tan\phi \tag{11.6}$$

$$N_\gamma = \frac{1}{2}\tan\phi(K_\gamma \tan\phi - 1) \tag{11.7}$$

The terms N_c, N_q, and N_γ in Eq. (11.4) are, respectively, the contributions of cohesion, surcharge, and the unit weight of soil to the ultimate load-bearing capacity. It is extremely tedious to evaluate K_c, K_q, and K_γ. For that reason, Terzaghi used an approximate method to determine the ultimate bearing capacity, q_u. The principle of this approximation follows:

1. If $c = 0$ and surcharge $q = 0$ (that is, $D_f = 0$), then from Eq. (11.4),

$$q_u = q_\gamma = \frac{1}{2}\gamma BN_\gamma \tag{11.8}$$

2. If $\gamma = 0$ (that is, weightless soil) and $q = 0$, then from Eq. (11.4),

$$q_u = q_c = cN_c \tag{11.9}$$

3. If $\gamma = 0$ (weightless soil) and $c = 0$, then

$$q_u = q_q = qN_q \tag{11.10}$$

By the method of superimposition, when the effects of the unit weight of soil, cohesion, and surcharge are taken into consideration, we have

$$q_u = q_c + q_q + q_\gamma = cN_c + qN_q + \frac{1}{2}\gamma BN_\gamma \tag{11.11}$$

Equation (11.11) is referred to as *Terzaghi's bearing capacity equation*. The terms N_c, N_q, and N_γ are called the *bearing capacity factors*. The values of these bearing capacity factors are given in Figure 11.7.

For square and circular footings, Terzaghi suggested the following equations for ultimate soil-bearing capacity:

Square footing:

$$q_u = 1.3cN_c + qN_q + 0.4\gamma BN_\gamma \tag{11.12}$$

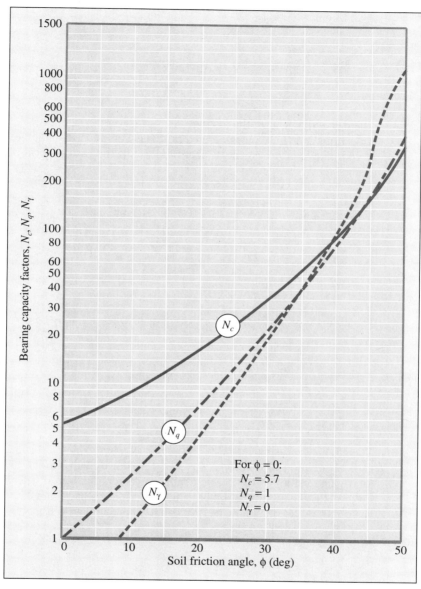

For φ = 0:
$N_c = 5.7$
$N_q = 1$
$N_\gamma = 0$

▼ **FIGURE 11.7** Terzaghi's bearing capacity factors for general shear failure

Circular footing:

$$q_u = 1.3cN_c + qN_q + 0.3\gamma BN_\gamma$$

(11.13)

where B = diameter of the footing.

Equation (11.11) was derived on the assumption that the bearing capacity failure of soil takes place by general shear failure. In the case of local shear failure, we may assume that

$$c' = \frac{2}{3}c$$

(11.14)

and

$$\tan \phi' = \frac{2}{3}\tan \phi$$

(11.15)

The ultimate bearing capacity of soil for a strip footing may be given by

$$q_u' = c'N_c' + qN_q' + \frac{1}{2}\gamma BN_\gamma'$$

(11.16)

The modified bearing capacity factors N_c', N_q', and N_γ' are calculated by using the same general equation as that for N_c, N_q, and N_γ but by substituting $\phi' = \tan^{-1}(2/3\tan\phi)$ for ϕ. The values of the bearing capacity factors for a local shear failure are given in Figure 11.8. The ultimate soil-bearing capacity for square and circular footings for local shear failure case may now be given as follows [similar to Eqs. (11.12) and (11.13)]:

Square footing:

$$q_u' = 1.3c'N_c' + qN_q' + 0.4\gamma BN_\gamma'$$

(11.17)

Circular footing:

$$q_u' = 1.3c'N_c' + qN_q' + 0.3\gamma BN_\gamma'$$

(11.18)

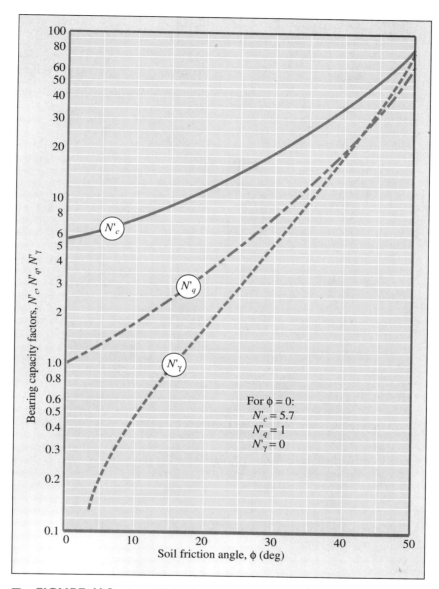

For $\phi = 0$:
$N'_c = 5.7$
$N'_q = 1$
$N'_\gamma = 0$

▼ **FIGURE 11.8** Terzaghi's bearing capacity factors for local shear failure

11.3 EFFECT OF GROUND WATER TABLE

In developing the bearing capacity equations given in the preceding section, it was assumed that the ground water table is located at a depth much greater than the width, B, of the footing. However, if the ground water table is close to the footing, some changes are required in the second and third terms of Eqs. (11.11) to (11.13) and Eqs.

(11.16) to (11.18). Three different conditions can arise regarding the location of the ground water table with respect to the bottom of the foundation. They are shown in Figure 11.9. Each of these conditions is briefly described below.

▶   *Case I (Figure 11.9a):* If the ground water table is located at a distance D above the bottom of the foundation, then the magnitude of q in the second

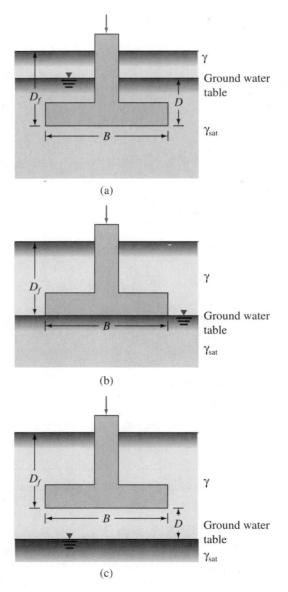

(a)

(b)

(c)

▼   **FIGURE 11.9**   Effect of the location of ground water table on the bearing capacity of shallow foundations: (a) case I; (b) case II; (c) case III

term of the bearing capacity equation should be calculated as

$$q = \gamma(D_f - D) + \gamma'D \tag{11.19}$$

where $\gamma' = \gamma_{sat} - \gamma_w$ = effective unit weight of soil. Also, the unit weight of soil γ that appears in the third term of the bearing capacity equations should be replaced by γ'.

▶ *Case II (Figure 11.9b)*: If the ground water table coincides with the bottom of the foundation, then the magnitude of q is equal to γD_f. However, the unit weight γ in the third term of the bearing capacity equations should be replaced by γ'.

▶ *Case III (Figure 11.9c)*: When the ground water table is at a depth D below the bottom of the foundation, $q = \gamma D_f$. The magnitude of γ in the third term of the bearing capacity equations should be replaced by γ_{av}.

$$\gamma_{av} = \frac{1}{B}\left[\gamma D + \gamma'(B - D)\right] \qquad \text{(for } D \leq B) \tag{11.20a}$$

$$\gamma_{av} = \gamma \qquad \text{(for } D > B) \tag{11.20b}$$

11.4 FACTOR OF SAFETY

Generally, a factor of safety, F_s, of about 3 or more is applied to the ultimate soil-bearing capacity to arrive at the value of the allowable bearing capacity. This is not considered to be too conservative. In nature, soils are neither homogeneous nor isotropic. A great deal of uncertainty is involved in evaluating the basic shear-strength parameters of soil.

There are three different definitions of the allowable bearing capacity of shallow foundations. They are gross allowable bearing capacity, net allowable bearing capacity, and gross allowable bearing capacity with a factor of safety with respect to shear failure.

The *gross allowable bearing capacity, q_{all}*, can be calculated as

$$q_{all} = \frac{q_u}{F_s} \tag{11.21}$$

q_{all} as defined by Eq. (11.21) is the allowable load per unit area to which the soil under the foundation should be subjected to avoid any chance of bearing capacity failure. It includes the contribution (Figure 11.10) of (a) the dead and live loads above the ground surface, $W_{(D+L)}$; (b) the self-weight of the foundation, W_F; and (c) the weight of the soil located immediately above the foundation, W_S. Thus,

$$q_{all} = \frac{q_u}{F_s} = \left[\frac{W_{(D+L)} + W_F + W_S}{A}\right]\frac{1}{F_s} \tag{11.22}$$

where A = area of the foundation.

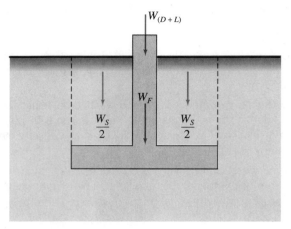

▼   **FIGURE 11.10**

The *net allowable bearing capacity* is the allowable load per unit area of the foundation in excess of the existing vertical effective stress at the level of the foundation. The vertical effective stress at the foundation level is equal to $q = \gamma D_f$. So, the net ultimate load is

$$q_{u(net)} = q_u - q \qquad (11.23)$$

Hence,

$$q_{all(net)} = \frac{q_{u(net)}}{F_s} = \frac{q_u - q}{F_s} \qquad (11.24)$$

If we assume that the weights of the soil and of the concrete from which the foundation is made are approximately the same, then

$$q = \gamma D_f \simeq \frac{W_S + W_F}{A}$$

Hence,

$$q_{all(net)} = \frac{W_{(D+L)}}{A} = \frac{q_u - q}{F_s} \qquad (11.25)$$

The *gross allowable bearing capacity* can be calculated by using a factor of safety, F_s, on the shear strength of soil. The procedure is as follows:

1.   Calculate

$$c_d = \frac{c}{F_s} \qquad (11.26)$$

where c_d is the developed cohesion.

2. Calculate

$$\tan \phi_d = \frac{\tan \phi}{F_s} \tag{11.27}$$

where ϕ_d is the developed angle of friction of soil.

3. Calculate the gross allowable bearing capacity of the foundation by using equations such as (11.11), (11.12), and (11.13) with c_d and ϕ_d as the shear-strength parameters of the soil. For example, the gross allowable bearing capacity of a continuous foundation according to Terzaghi's equation may be written as

$$q_{all} = c_d N_c + q N_q + \frac{1}{2} \gamma B N_\gamma \tag{11.28}$$

where N_c, N_q, and N_γ are the bearing capacity factors for friction angle ϕ_d.

In many situations, a factor of safety of 3 to 4 against the gross or net ultimate bearing capacity, along with a factor of safety of 2 to 3 with respect to shear failure, is considered satisfactory.

Another factor that must also be kept in mind is the tolerable settlement of a footing. The settlement of footings obtained at ultimate load, q_u (or q'_u), may range from about 5% to 25% of B (footing width) in sandy soils and from about 3% to 15% of B in clay soils. Hence, for large footings, the settlement may be too large for the safety of the structure at the ultimate load.

▼ **EXAMPLE II.I**

A continuous footing is shown in Figure 11.11. Using Terzaghi's bearing capacity factors, determine the gross allowable load per unit area (q_{all}) that the footing can carry.

$$\gamma = 115 \text{ lb/ft}^3$$
$$c = 300 \text{ lb/ft}^3$$
$$\phi = 28°$$
$$D_f = 2 \text{ ft}$$
$$B = 2.5 \text{ ft}$$

Factor of safety $= 4$

  a. Assume general shear failure.
  b. Assume local shear failure.

Solution

  a. From Eq. (11.11),

$$q_u = c N_c + q N_q + \frac{1}{2} \gamma B N_\gamma$$

From Figure 11.7, for $\phi = 28°$,

$$N_c = 32, \ N_q = 18, \text{ and } N_\gamma = 16$$

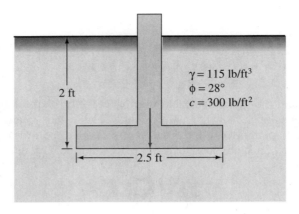

$\gamma = 115$ lb/ft^3
$\phi = 28°$
$c = 300$ lb/ft^2

2 ft

2.5 ft

Also

$$q = \gamma D_f = (115)(2) = 230 \text{ lb/ft}^2$$

So

$$q_u = (300)(32) + (230)(18) + \left(\frac{1}{2}\right)(115)(2.5)(16)$$

$$= 16{,}040 \text{ lb/ft}^2$$

$$q_{all} = \frac{q_u}{F_s} = \frac{16{,}040}{4} = \textbf{4010 lb/ft}^2$$

b.   From Eq. (11.16),

$$q'_u = c'N'_c + qN'_q + \frac{1}{2}\gamma BN'_\gamma$$

From Figure 11.8, for $\phi = 28°$,

$N'_c = 18$, $N'_q = 7$, and $N'_\gamma = 4.5$

So

$$q'_u = \left(\frac{2}{3}\right)(300)(18) + (230)(7) + \left(\frac{1}{2}\right)(115)(2.5)(4.5)$$

$$= 5857 \text{ lb/ft}^2$$

$$q_{all} = \frac{q'_u}{F_s} = \frac{5857}{4} \approx \textbf{1464 lb/ft}^2 \qquad ▼$$

▼   **EXAMPLE** 11.2

A square footing is shown in Figure 11.12. The footing will carry a gross mass of 30,000 kg. Using a factor of safety of 3, determine the size of the footing—that is, the size of B.

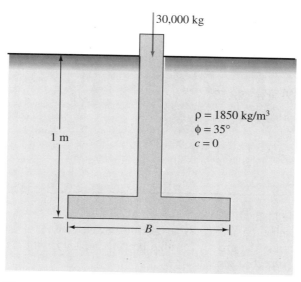

30,000 kg

$\rho = 1850$ kg/m³
$\phi = 35°$
$c = 0$

1 m

B

▼   **FIGURE** 11.12

Solution   We have soil density = 1850 kg/m³, so

$$\gamma = \frac{1850 \times 9.81}{1000} = 18.148 \text{ kN/m}^3$$

The total gross load to be supported by the footing is

$$\frac{(30,000)9.81}{1000} = 294.3 \text{ kN} = Q_{all}$$

From Eq. (11.12),

$$q_u = 1.3cN_c + qN_q + 0.4\gamma BN_\gamma$$

With a factor of safety of 3, we have

$$q_{all} = \frac{q_u}{3} = \frac{1}{3}(1.3cN_c + qN_q + 0.4\gamma BN_\gamma)$$

Also

$$q_{all} = \frac{Q_{all}}{B^2} = \frac{294.3}{B^2}$$

Thus,

$$\frac{294.3}{B^2} = \frac{1}{3}(1.3cN_c + qN_q + 0.4\gamma BN_\gamma)$$

From Figure 11.7, for $\phi = 35°$, $N_c = 57.8$, $N_q = 41.4$, and $N_\gamma = 42.4$. Substituting these values into the above equation, we get

$$\frac{294.3}{B^2} = \frac{1}{3}[(1.3)(0)(57.8) + (18.148 \times 1)(41.4) + 0.4(18.148)(B)(42.4)]$$

or

$$\frac{294.3}{B^2} = 250.44 + 102.6B$$

By trial and error,

$$B \simeq \mathbf{0.95\ m} \qquad \text{(say, 1 m)} \qquad \blacktriangledown$$

▼ **EXAMPLE 11.3**

Refer to Example 11.1.
  a.   Determine the net allowable bearing capacity with a factor of safety of 4.
  b.   Determine the gross allowable bearing capacity with a factor of safety of 4 with respect to shear failure.

Solution
  a.   From Example 11.1, $q_u = 16,040$ lb/ft^2, so

$$q_{u(\text{net})} = q_u - q = 16,040 - 230 = 15,810 \text{ lb/ft}^2$$

$$q_{\text{all(net)}} = \frac{q_{u(\text{net})}}{F_s} = \frac{15,810}{4} = \mathbf{3953\ lb/ft^2}$$

  b.   Using $c = 300$ lb/ft^2 and $\phi = 28°$ with Eqs. (11.26) and (11.27), we have

$$c_d = \frac{c}{F_s} = \frac{300}{4} = 75 \text{ lb/ft}^2$$

$$\phi_d = \tan^{-1}\frac{\tan \phi}{F_s} = \tan^{-1}\left[\frac{\tan 28°}{4}\right] = 7.57°$$

For $\phi = 7.57°$, from Figure 11.7, $N_c \approx 8.5$, $N_q \approx 2.2$, and $N_\gamma \approx 0.9$ (by extrapolation). From Eq. (11.28),

$$q_{\text{all}} = c_d N_c + qN_q + \frac{1}{2}\gamma BN_\gamma$$

$$= (75)(8.5) + (230)(2.2) + \left(\frac{1}{2}\right)(115)(2.5)(0.9)$$

$$= \mathbf{1273\ lb/ft^2} \qquad \blacktriangledown$$

11.5 GENERAL BEARING CAPACITY EQUATION

After the development of Terzaghi's bearing capacity equation, several investigators worked in this area and refined the solution (that is, Meyerhof, 1951, 1963; Lundgren and Mortensen, 1953; Balla, 1962). Different solutions show that the bearing capacity factors N_c and N_q do not change very much. However, for a given value of ϕ, the values of N_γ obtained by different investigators vary over a wide range. This is because of the variation of the assumption of the wedge shape of soil located directly below the footing, as explained in the following paragraph.

While deriving the bearing capacity equation for a strip footing, Terzaghi used the case of a rough footing and assumed that the sides AJ and BJ of the soil wedge ABJ (Figure 11.5a) make an angle ϕ with the horizontal. Later model tests (for example, DeBeer and Vesic, 1958) showed that Terzaghi's assumption of the general nature of the rupture surface in soil for bearing capacity failure is correct. However, tests have shown that the sides AJ and BJ of the soil wedge ABJ make angles of about $45 + \phi/2$ degrees instead of ϕ with the horizontal. This type of failure mechanism is shown in Figure 11.13. It consists of a Rankine active zone ABJ (zone I), two radial shear zones (zones II), and two Rankine passive zones (zones III). The curves JD and JE are arcs of a logarithmic spiral.

Based on this type of failure mechanism, the ultimate bearing capacity of a strip footing may be evaluated by the approximate method of superimposition described in Section 11.2 as

$$q_u = q_c + q_q + q_\gamma \tag{11.29}$$

where q_c, q_q, and q_γ are the contributions of cohesion, surcharge, and unit weight of soil, respectively.

Reissner (1924) expressed q_q as

$$q_q = qN_q \tag{11.30}$$

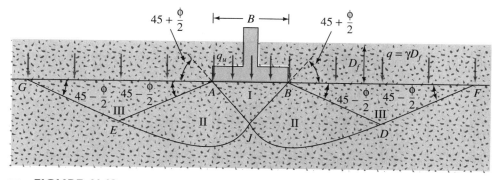

▼ FIGURE 11.13 Soil-bearing capacity calculation—general shear failure

where

$$N_q = e^{\pi \tan \phi} \tan^2 \left(45 + \frac{\phi}{2} \right)$$

(11.31)

Prandtl (1921) showed that

$$q_c = cN_c$$

(11.32)

where

$$N_c = (N_q - 1) \cot \phi$$
$$\uparrow$$
Eq. (11.31)

(11.33)

Caquot and Kerisel (1953) expressed q_γ as

$$q_\gamma = \frac{1}{2} B\gamma N_\gamma$$

(11.34)

The numerical values by Caquot and Kerisel can be approximated (Vesic, 1973) as

$$N_\gamma = 2(N_q + 1) \tan \phi$$
$$\uparrow$$
Eq. (11.31)

(11.35)

Combining Eqs. (11.29), (11.30), (11.32), and (11.34), we obtain

$$q_u = cN_c + qN_q + \frac{1}{2} \gamma BN_\gamma$$

(11.36)

This is in the same general form as that given by Terzaghi [Eq. (11.11)]; however, the values of the bearing capacity factors are not the same. The values of N_q, N_c, and N_γ, defined by Eqs. (11.31), (11.33), and (11.35), are given in Table 11.1, but, for all practical purposes, Terzaghi's bearing capacity factors will yield good results. Differences in bearing capacity factors are usually minor compared with the unknown soil parameters.

The soil-bearing capacity equation for a strip footing given by Eq. (11.36) can be modified for general use by incorporating the following factors:

1. Depth factor: to account for the shearing resistance developed along the failure surface in soil above the base of the footing
2. Shape factor: to determine the bearing capacity of rectangular and circular footings

3. Inclination factor: to determine the bearing capacity of a footing on which the direction of load application is inclined at a certain angle to the vertical

Thus, the modified general ultimate bearing capacity equation can be written as

$$q_u = c\lambda_{cs}\,\lambda_{cd}\,\lambda_{ci}\,N_c + q\lambda_{qs}\,\lambda_{qd}\,\lambda_{qi}\,N_q + \frac{1}{2}\,\lambda_{\gamma s}\,\lambda_{\gamma d}\,\lambda_{\gamma i}\,\gamma B N_\gamma \tag{11.37}$$

where λ_{cs}, λ_{qs}, and $\lambda_{\gamma s}$ = shape factors
λ_{cd}, λ_{qd}, and $\lambda_{\gamma d}$ = depth factors
λ_{ci}, λ_{qi}, and $\lambda_{\gamma i}$ = inclination factors

▼ **TABLE 11.1** **Bearing Capacity Factors* [Eqs. (11.31), (11.33), and (11.35)]**

ϕ (1)	N_c (2)	N_q (3)	N_γ (4)	N_q/N_c (5)	$\tan\phi$ (6)	ϕ (1)	N_c (2)	N_q (3)	N_γ (4)	N_q/N_c (5)	$\tan\phi$ (6)
0	5.14	1.00	0.00	0.20	0.00	26	22.25	11.85	12.54	0.53	0.49
1	5.38	1.09	0.07	0.20	0.02	27	23.94	13.20	14.47	0.55	0.51
2	5.63	1.20	0.15	0.21	0.03	28	25.80	14.72	16.72	0.57	0.53
3	5.90	1.31	0.24	0.22	0.05	29	27.86	16.44	19.34	0.59	0.55
4	6.19	1.43	0.34	0.23	0.07	30	30.14	18.40	22.40	0.61	0.58
5	6.49	1.57	0.45	0.24	0.09	31	32.67	20.63	25.99	0.63	0.60
6	6.81	1.72	0.57	0.25	0.11	32	35.49	23.18	30.22	0.65	0.62
7	7.16	1.88	0.71	0.26	0.12	33	38.64	26.09	35.19	0.68	0.65
8	7.53	2.06	0.86	0.27	0.14	34	42.16	29.44	41.06	0.70	0.67
9	7.92	2.25	1.03	0.28	0.16	35	46.12	33.30	48.03	0.72	0.70
10	8.35	2.47	1.22	0.30	0.18	36	50.59	37.75	56.31	0.75	0.73
11	8.80	2.71	1.44	0.31	0.19	37	55.63	42.92	66.19	0.77	0.75
12	9.28	2.97	1.69	0.32	0.21	38	61.35	48.93	78.03	0.80	0.78
13	9.81	3.26	1.97	0.33	0.23	39	67.87	55.96	92.25	0.82	0.81
14	10.37	3.59	2.29	0.35	0.25	40	75.31	64.20	109.41	0.85	0.84
15	10.98	3.94	2.65	0.36	0.27	41	83.86	73.90	130.22	0.88	0.87
16	11.63	4.34	3.06	0.37	0.29	42	93.71	85.38	155.55	0.91	0.90
17	12.34	4.77	3.53	0.39	0.31	43	105.11	99.02	186.54	0.94	0.93
18	13.10	5.26	4.07	0.40	0.32	44	118.37	115.31	224.64	0.97	0.97
19	13.93	5.80	4.68	0.42	0.34	45	133.88	134.88	271.76	1.01	1.00
20	14.83	6.40	5.39	0.43	0.36	46	152.10	158.51	330.35	1.04	1.04
21	15.82	7.07	6.20	0.45	0.38	47	173.64	187.21	403.67	1.08	1.07
22	16.88	7.82	7.13	0.46	0.40	48	199.26	222.31	496.01	1.12	1.11
23	18.05	8.66	8.20	0.48	0.42	49	229.93	265.51	613.16	1.15	1.15
24	19.32	9.60	9.44	0.50	0.45	50	266.89	319.07	762.89	1.20	1.19
25	20.72	10.66	10.88	0.51	0.47						

* After Vesic, 1973

The approximate values of the shape factors for rectangular, square, and circular footings were given by DeBeer (1970). Similarly, the approximate values for the depth factors and inclination factors were given by Hansen (1970) and Meyerhof (1963), respectively. These are empirical factors based on experimental observations. These shape, depth, and inclination factors are given in Table 11.2.

The effect of the ground water table has to be taken into consideration in determining the values of q and γ in Eq. (11.37). This can be done in the same manner as was explained in Section 11.3.

▼   **EXAMPLE 11.4**

A circular footing is shown in Figure 11.14. Determine the safe gross load (factor of safety of 3) that the footing can carry. Use Eq. (11.37).

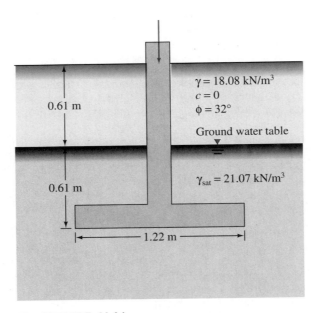

0.61 m

$\gamma = 18.08 \text{ kN/m}^3$
$c = 0$
$\phi = 32°$

Ground water table

0.61 m

$\gamma_{sat} = 21.07 \text{ kN/m}^3$

1.22 m

▼   **FIGURE 11.14**

Solution   From Eq. (11.37),

$$q_u = c\lambda_{cs}\,\lambda_{cd}\,N_c + q\lambda_{qs}\,\lambda_{qd}\,N_q + \frac{1}{2}\,\lambda_{\gamma s}\,\lambda_{\gamma d}\,BN_\gamma$$

(*Note:* λ_{ci}, λ_{qi}, and $\lambda_{\gamma i}$ are all equal to 1 because the load is vertical.) For $\phi = 32°$ (from Table 11.1), $N_c = 35.49$, $N_q = 23.18$, and $N_\gamma = 30.22$. From Table 11.2,

Shape factors for rectangular footing
(B = width of footing; L = length of footing)

$$\lambda_{cs} = 1 + \left(\frac{B}{L}\right)\left(\frac{N_q}{N_c}\right)$$

$$\lambda_{qs} = 1 + \left(\frac{B}{L}\right)(\tan \phi)$$

$$\lambda_{\gamma s} = 1 - 0.4\left(\frac{B}{L}\right)$$

Shape factors for square and circular footing

$$\lambda_{cs} = 1 + \frac{N_q}{N_c}$$

$$\lambda_{qs} = 1 + \tan \phi$$
$$\lambda_{\gamma s} = 0.6$$

Depth factors for $\dfrac{D_f}{B} \leq 1$

$$\lambda_{qd} = 1 + 2 \tan \phi (1 - \sin \phi)^2 \left(\frac{D_f}{B}\right)$$

$$\lambda_{cd} = \lambda_{qd} - \frac{1 - \lambda_{qd}}{N_q \tan \phi}$$

$$\lambda_{\gamma d} = 1$$

Depth factor for $\phi = 0$

$$\lambda_{\gamma d} = 1 + 0.4\left(\frac{D_f}{B}\right)$$

Depth factors for $\dfrac{D_f}{B} > 1$

$$\lambda_{qd} = 1 + 2 \tan \phi (1 - \sin \phi)^2 \tan^{-1}\left(\frac{D_f}{B}\right)$$

$$\lambda_{cd} = \lambda_{qd} - \frac{1 - \lambda_{qd}}{N_q \tan \phi}$$

$$\lambda_{\gamma d} = 1$$

Depth factor for $\phi = 0$

$$\lambda_{cd} = 1 + 0.4 \tan^{-1}\left(\frac{D_f}{B}\right)$$

Inclination factors

$$\lambda_{ci} = \left(1 - \frac{\alpha^\circ}{90^\circ}\right)^2$$

$$\lambda_{qi} = \left(1 - \frac{\alpha^\circ}{90^\circ}\right)^2$$

$$\lambda_{\gamma i} = \left(1 - \frac{\alpha^\circ}{\phi^\circ}\right)^2$$

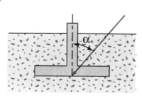

$$\lambda_{cs} = 1 + \frac{N_q}{N_c} = 1 + \frac{23.18}{35.49} = 1.65$$

$$\lambda_{qs} = 1 + \tan \phi = 1 + 0.62 = 1.62$$

$$\lambda_{\gamma s} = 0.6$$

$$\lambda_{qd} = 1 + 2 \tan \phi (1 - \sin \phi)^2 \left(\frac{D_f}{B}\right)$$

$$= 1 + (2)(0.62)(0.22)(1) = 1.273$$

$$\lambda_{\gamma d} = 1$$

$$\lambda_{cd} = \lambda_{qd} - \frac{1 - \lambda_{qd}}{N_q \tan \phi} = 1.273 - \frac{1 - 1.273}{(23.18)(0.62)} = 1.292$$

The ground water table is located above the bottom of the footing, so

$$q = 0.61(18.08) - 0.61(21.07 - 9.81) = 11.029 + 6.869$$

$$= 17.898 \ kN/m^2$$

Thus,

$$q_u = (17.898)(1.62)(1.273)(23.18) + (0.05)(0.6)(21.07 - 9.81)(1.22)(30.22)$$

$$= 855.58 + 124.54 \approx 980.12 \ kN/m^2$$

Thus,

$$q_{all} = \frac{q_u}{3} = \frac{980.12}{3} = 326.71 \ kN/m^2$$

Hence, the total gross load is

$$\left(\frac{\pi}{4} B^2\right) q_{all} = \frac{\pi}{4} (1.22)^2 (326.71) = \textbf{381.9 kN} \qquad \blacktriangledown$$

11.6 ULTIMATE LOAD FOR SHALLOW FOUNDATIONS UNDER ECCENTRIC LOAD

To calculate the bearing capacity of shallow foundations with eccentric loading, Meyerhof (1953) introduced the concept of effective width. This can be explained with reference to Figure 11.15, in which a footing of length L and width B is subjected to an eccentric load, Q_u. If Q_u is the ultimate load on the footing, it may be approximately calculated as follows. The load eccentricities are e_l and e_b with respect to the length and width as shown in Figure 11.15b.

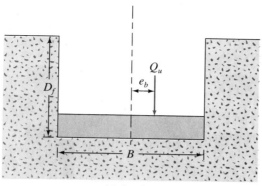

(a) Section

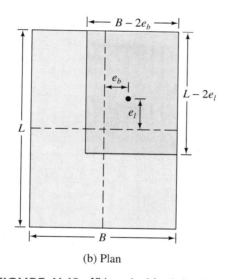

(b) Plan

▼ **FIGURE 11.15** Ultimate load for shallow foundations under eccentric load

The effective width of the footing may be given by

$$B' = B - 2e_b$$ (11.38)

and the effective length of the footing may be given by

$$L' = L - 2e_l$$ (11.39)

So the effective area is equal to B' times L'. Now, using the effective width, we can rewrite Eq. (11.37) as

$$q_u = c\lambda_{cs} \, \lambda_{cd} \, N_c + q\lambda_{qs} \, \lambda_{qd} \, N_q + \frac{1}{2} \, \lambda_{\gamma s} \, \lambda_{\gamma d} \, \gamma B' N_\gamma \tag{11.40}$$

Note that the preceding equation is obtained by substituting B' for B in Eq. (11.37). While computing the shape and depth factors, one should use B' for B, and L' for L. Note that, if $L - 2e_l$ is less than $B - 2e_b$, then $L - 2e_l = B'$ and $B - 2e_b = L'$.

Once the value of q_u is calculated from Eq. (11.40), we can obtain the total gross ultimate load as

$$Q_u = q_u(B'L')$$

Example 11.5 demonstrates the use of this procedure.

▼ EXAMPLE 11.5

A rectangular footing 5 ft × 2.5 ft is shown in Figure 11.16. Determine the magnitude of the gross ultimate load applied eccentrically for bearing capacity failure in soil.

Solution   The effective width $B' = 2.5 - 2(0.2) = 2.1$ ft. The effective length $L' = 5 - 2(0.4) = 4.2$ ft. Substituting $c = 0$ in Eq. (11.40), we have

$$q_u = q\lambda_{qs} \, \lambda_{qd} \, N_q + \frac{1}{2} \, \lambda_{\gamma s} \, \gamma_{\gamma d} \, \gamma B' N_\gamma$$

For $\phi = 30°$, $N_q = 18.4$, and $N_\gamma = 22.4$,

$$\lambda_{qs} = 1 + \left(\frac{B'}{L'}\right) \tan \phi = 1 + \left(\frac{2.1}{4.2}\right)(0.58) = 1.29$$

$$\lambda_{\gamma s} = 1 - 0.4\left(\frac{B'}{L'}\right) = 1 - 0.4\left(\frac{2.1}{4.2}\right) = 0.8$$

$$\lambda_{dq} = 1 + 2 \tan \phi(1 - \sin \phi^2)\left(\frac{D_f}{B'}\right)$$

$$= 1 + 2(\tan 30)(1 - \sin 30)^2\left(\frac{2}{2.1}\right) = 1.275$$

$$\lambda_{\gamma d} = 1$$

So

$$q_u = (2 \times 115)(1.29)(1.275)(18.4) + \frac{1}{2}(0.8)(1)(115)(2.1)(22.4)$$

$$= 9986.9 + 2163.8 = 12{,}150.7 \text{ lb/ft}^2$$

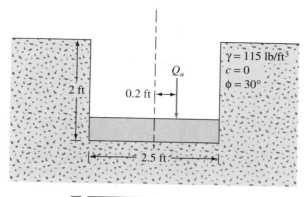

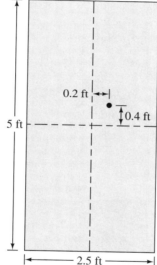

▼   **FIGURE 11.16**

Hence,

$$Q_u = q_u(B'L') = (12,150.7)(2.1 \times 4.2) = \textbf{197,169 lb} \qquad \blacktriangledown$$

11.7 SHALLOW FOUNDATIONS ON LAYERED SOIL

So far in this chapter we have considered only the load-bearing capacity of homogeneous soils that support shallow foundations. However, if a footing is placed on a stratified soil deposit, and the top layer on which the base of the foundation is located is relatively thin, then the rupture line in soil at ultimate load may pass through the lower layer of soil. In that situation, the soil properties of both upper and lower layers have to be taken into consideration. There have been few studies regarding the ultimate bearing capacity of foundations on layered soils. Some of these cases are discussed below.

Foundations on Layered Sand—Dense over Loose

A simple theory for determining the ultimate bearing capacity of a foundation that rests on a layer of dense sand underlain by loose sand has been proposed by Meyerhof and Hanna (1978). The basic principle of this theory can be explained with the aid of Figure 11.17, which is for a strip foundation. When the top dense sand layer is relatively thick, as shown by the right-hand side of Figure 11.17, the failure surface in soil under the foundation will be fully located inside the dense sand. For this case,

$$q_u = q_{u(t)} = \gamma_1 D_f N_{q(1)} + \frac{1}{2}\gamma_1 B N_{\gamma(1)}$$

(for strip foundations)   (11.41)

$$q_u = q_{u(t)} = \gamma_1 D_f N_{q(1)} + 0.3\gamma_1 B N_{\gamma(1)}$$

(for circular or square foundations)   (11.42)

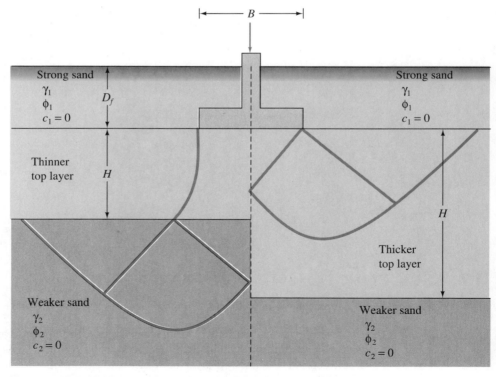

▼ **FIGURE 11.17** Bearing capacity in layered sand—strong sand underlain by weak sand

and

$$q_u = q_{u(t)} = \gamma_1 D_f N_{q(1)} + \frac{1}{2}\left[1 - 0.4\left(\frac{B}{L}\right)\right]\gamma_1 B N_{\gamma(1)}$$

(for rectangular foundations)

(11.43)

where
γ_1 = unit weight of top layer (dense sand in this case)
$N_{q(1)}$ and $N_{\gamma(1)}$ = bearing capacity factors with reference to the soil friction angle, ϕ_1 (Table 11.1)

Note that Eqs. (11.41), (11.42), and (11.43) are similar to Eq. (11.37). However, the depth factors have not been incorporated; they can be assumed to be somewhat conservative.

If the dense sand layer under the foundation H is relatively thin, the failure in soil would take place by *punching* in the dense sand layer followed by a general shear failure in the bottom (or weaker) sand layer, as shown in the left-hand side of Figure 11.17. For such a case, the ultimate bearing capacity for the foundation can be given as

$$q_u = q_{u(b)} + \gamma_1 H^2\left(1 + \frac{2D_f}{H}\right)K_s\frac{\tan\phi_1}{B} - \gamma_1 H \le q_{u(t)}$$

$\uparrow$

[Eq. (11.41)]

(for strip foundations)

(11.44)

$$q_u = q_{u(b)} + 2\gamma_1 H^2\left(1 + \frac{2D_f}{H}\right)\left(\frac{K_s\tan\phi_1}{B}\right)\lambda_s' - \gamma_1 H \le q_{u(t)}$$

$\uparrow$

[Eq. (11.42)]

(for square or circular foundations)

(11.45)

and

$$q_u = q_{u(b)} + \left(1 + \frac{B}{L}\right)\gamma_1 H^2\left(1 + \frac{2D_f}{H}\right)\left(\frac{K_s\tan\phi_1}{B}\right)\lambda_s' - \gamma_1 H \le q_{u(t)}$$

$\uparrow$

[Eq. (11.43)]

(11.46)

(for rectangular foundations)

where K_s = punching shear coefficient
λ_s' = shape factor
$q_{u(b)}$ = ultimate bearing capacity of the bottom soil layer

The value of the shape factor λ'_s can be taken to be approximately 1. The punching shear coefficient is

$$K_s = f(\gamma_1, \gamma_2, N_{\gamma(1)}, N_{\gamma(2)}) \tag{11.47}$$

where γ_2 = unit weight of the lower layer of sand
$N_{\gamma(2)}$ = bearing capacity factor for the soil friction angle, ϕ_2

The variation of K_s is shown in Figure 11.18. The term $q_{u(b)}$ in Eqs. (11.44), (11.45), and (11.46) is given by the following relationships:

$$q_{u(b)} = \gamma_1(D_f + H)N_{q(2)} + \frac{1}{2}\gamma_2 BN_{\gamma(2)}$$

$$\text{(for strip foundations)} \tag{11.48}$$

$$q_{u(b)} = \gamma_1(D_f + H)N_{q(2)} + 0.3\gamma_2 BN_{\gamma(2)}$$

$$\text{(for circular or square foundations)} \tag{11.49}$$

$$q_{u(b)} = \gamma_1(D_f + H)N_{q(2)} + \frac{1}{2}\left[1 - 0.4\left(\frac{B}{L}\right)\right]\gamma_2 BN_{\gamma(2)}$$

$$\text{(for rectangular foundations)} \tag{11.50}$$

For a given layered soil, the variation of q_u with H/B will be as shown in Figure 11.19.

Foundations on Layered Sand—Loose over Dense

Figure 11.20 shows a strip foundation supported by a loose sand layer underlain by a dense sand layer. Depending on the magnitude of H/B, the following two types of failure surface can be observed in the soil supporting the foundation:

1.   If the loose sand layer under the foundation (H) is thicker than the width of the foundation (B), then the failure surface in the soil would be completely in the weaker soil layer (right half of Figure 11.20). For that case,

$$q_u = q_{u(t')} = \gamma_1 D_f N_{q(1)} + \frac{1}{2}\gamma_1 BN_{\gamma(1)}$$

$$\text{(for strip foundations)} \tag{11.51}$$

$$q_y = q_{u(t')} = \gamma_1 D_f N_{q(1)} + 0.3\gamma_1 BN_{\gamma(1)}$$

$$\text{(for circular and square foundations)} \tag{11.52}$$

$$q_u = q_{u(t')} = \gamma_1 D_f N_{q(1)} + \frac{1}{2}\left[1 - 0.4\left(\frac{B}{L}\right)\right]\gamma_1 BN_{\gamma(1)}$$

$$\text{(for rectangular foundations)} \tag{11.53}$$

where γ_1 = unit weight of upper soil layer
$N_{q(1)}$ and $N_{\gamma(1)}$ = bearing capacity factors with respect to soil friction angle, ϕ_1

2. When H is thin compared with the foundation width B, the failure surface in soil will pass through the top and the bottom soil layer. This is shown in the left-hand side of Figure 11.20. With this condition (Meyerhof and Hanna, 1978),

$$q_u = q_{u(t')} + [q_{u(b')} - q_{u(t')}]\left(1 - \frac{H}{H_f}\right)^2 \qquad (11.54)$$

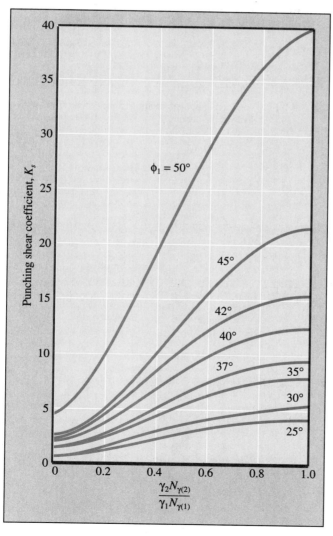

▼ **FIGURE 11.18**   Variation of K_s with $(\gamma_2 N_{\gamma(2)})/(\gamma_1 N_{\gamma(1)})$

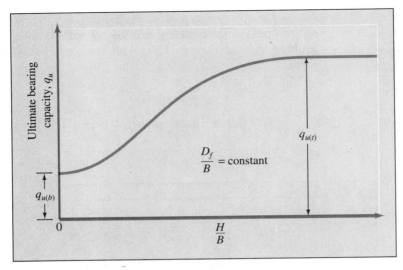

FIGURE 11.19   Variation of ultimate bearing capacity with H/B—layered sand (dense on top and loose at the bottom)

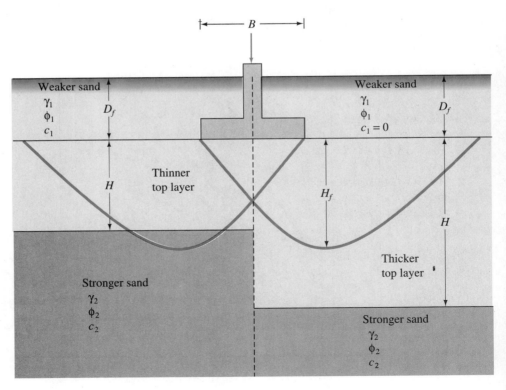

▼ **FIGURE 11.20**   Bearing capacity in layered sand—weak sand underlain by strong sand

where $q_{u(b')} = \gamma_2 D_f N_{q(2)} + \dfrac{1}{2} \gamma_2 B N_{\gamma(2)}$

(for strip foundations) (11.55)

$$q_{u(b')} = \gamma_2 D_f N_{q(2)} + 0.3\gamma_2 B N_{\gamma(2)}$$

(for square and circular foundations) (11.56)

$$q_{u(b')} = \gamma_2 D_f N_{q(2)} + \frac{1}{2}\left[1 - 0.4\left(\frac{B}{L}\right)\right]\gamma_2 B N_{\gamma(2)}$$

(for rectangular foundations) (11.57)

and

γ_2 = unit weight of the bottom layer of soil

$N_{q(2)}, N_{\gamma(2)}$ = bearing capacity factors with respect to the soil friction angle, ϕ_2

H_f = depth of the extent of failure surface under the foundation when it is located on a thick layer of loose sand

For all practical purposes, H_f can be taken as equal to $2B$. Note that Eq. (11.54) has the limits

$$q_{u(t')} \le q_u \le q_{u(b')}$$ (11.58)

The variation of q_u with H/B is shown in Figure 11.21.

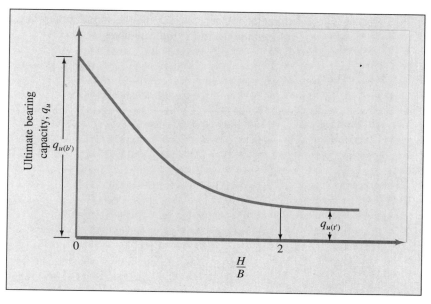

▼ **FIGURE 11.21** Variation of ultimate bearing capacity with H/B—layered sand (loose on top and dense at the bottom)

Foundation on Layered Clay ($\phi = 0$ condition)— Stronger over Weaker

The theory of Meyerhof and Hanna (1978) previously explained for the estimation of the ultimate bearing capacity of a foundation supported by dense sand underlain by loose sand can be extended to the bearing capacity calculation of a foundation on a stronger clay layer underlain by a weaker clay layer ($\phi = 0$ condition). This case is shown in Figure 11.22.

If the magnitude of H/B is relatively small, then failure in the soil at ultimate load will occur by punching in the stronger top soil layer, followed by a general shear failure in the weaker bottom soil layer. This is shown in the left-hand side of Figure 11.22. However, if H/B is relatively large, then, at ultimate load, the failure surface will be fully contained in the top soil layer. So the ultimate bearing capacity of a rectangular foundation can be given as (Meyerhof and Hanna, 1978):

$$q_u = \left[1 + 0.2\left(\frac{B}{L}\right)\right]c_{u(2)}N_c + \left(1 + \frac{B}{L}\right)\left(\frac{2c_a H}{B}\right) + \gamma_1 D_f \leq q_{u(t'')} \tag{11.59}$$

where c_a = adhesion along plane a–b as shown in the left-hand side of Figure 11.22
$c_{u(2)}$ = undrained cohesion of the bottom weaker clay layer
γ_1 = unit weight of top clay layer
N_c = bearing capacity factor (with $\phi = 0$) = 5.14
$q_{u(t'')}$ = ultimate bearing capacity of the foundation when the failure surface is fully contained in the top clay layer

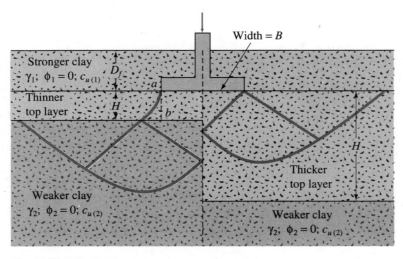

▼ **FIGURE 11.22**   Bearing capacity of layered clay—stronger over weaker

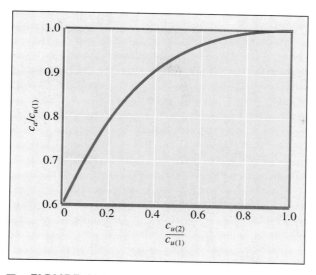

▼ FIGURE 11.23   Variation of $c_a/c_{u(1)}$ with $c_{u(2)}/c_{u(1)}$ according to Meyerhof and Hanna's theory

The term c_a is a function of $c_{u(2)}/c_{u(1)}$ [where $c_{u(1)}$ = undrained cohesion of the top clay layer], and

$$q_{u(t'')} = \left[1 + 0.2\left(\frac{B}{L}\right)\right]c_{u(1)}N_c + \gamma_1 D_f \tag{11.60}$$

Figure 11.23 shows the variation of $c_a/c_{u(1)}$ as a function of $c_{u(2)}/c_{u(1)}$. The nature of the variation of q_u with H/B is shown in Figure 11.24.

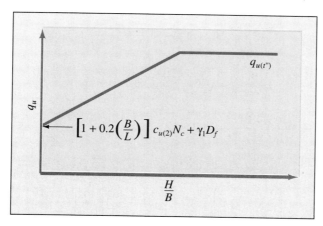

▼ FIGURE 11.24   Nature of variation of ultimate bearing capacity with H/B

▼   **EXAMPLE 11.6**

Figure 11.25 shows a rectangular foundation with $B = 4$ ft and $L = 6$ ft. Using a factor of safety of 3, determine the net allowable load the foundation can carry.

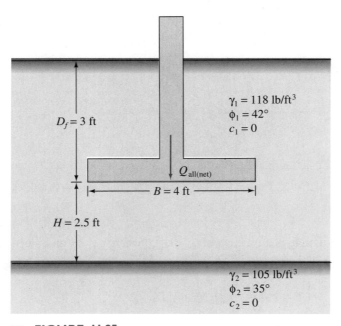

$\gamma_1 = 118$ lb/ft^3
$\phi_1 = 42°$
$c_1 = 0$

$D_f = 3$ ft

$Q_{all(net)}$

$B = 4$ ft

$H = 2.5$ ft

$\gamma_2 = 105$ lb/ft^3
$\phi_2 = 35°$
$c_2 = 0$

▼   **FIGURE 11.25**

Solution   The top layer of sand is dense since it has $\phi_1 = 42°$, which is greater than $\phi_2 = 35°$. Also, $\gamma_1 > \gamma_2$. So, Eq. (11.46) should be used to calculate q_u:

$$q_u = q_{u(b)} + \left(1 + \frac{B}{L}\right)\gamma_1 H^2\left(1 + \frac{2D_f}{H}\right)\left(\frac{K_s \tan \phi_1}{B}\right)\lambda_s' - \gamma_1 H$$

We are given $\gamma_1 = 118$ lb/ft^3, $\gamma_2 = 105$ lb/ft^3, $\phi_1 = 42°$, and $\phi_2 = 35°$. Also, from Table 11.1, $N_{\gamma(1)} = 155.55$ and $N_{\gamma(2)} = 48.03$. So

$$\frac{\gamma_2 N_{\gamma(2)}}{\gamma_1 N_{\gamma(1)}} = \frac{(105)(48.03)}{(118)(155.55)} = 0.275$$

From Figure 11.18, for $\phi_1 = 42°$, the value of K_s is 6.5. Thus, from Eq. (11.46),

$$q_u = q_{u(b)} + \left(1 + \frac{4}{6}\right)(118)(2.5)^2 + \left[1 + \frac{(2)(3)}{2.5}\right]\left[\frac{(6.5)(\tan 42°)}{4}\right](1) - (118)(2.5)$$

$$= q_{u(b)} + 5819$$

(a)

From Eq. (11.50),

$$q_{u(b)} = \gamma_1(D_f + H)N_{q(2)} + \frac{1}{2}\left[1 - 0.4\left(\frac{B}{L}\right)\right]\gamma_2 BN_{\gamma(2)}$$

From Table 11.1, for $\phi = 35°$, the values of $N_{q(2)} = 33.3$ and $N_{\gamma(2)} = 48.03$. Hence,

$$q_{u(b)} = (118)(3 + 2.5)(33.3) + \frac{1}{2}\left[1 - 0.4\left(\frac{4}{6}\right)\right](105)(4)(48.03) = 29,008 \qquad \text{(b)}$$

From Eqs. (a) and (b),

$$q_u = 29,008 + 5819 = 34,827 \text{ lb/ft}^2 \qquad \text{(c)}$$

We also need to check Eq. (11.43):

$$q_u = \gamma_1 D_f N_{q(1)} + \frac{1}{2}\left[1 - 0.4\left(\frac{B}{L}\right)\right]\gamma_1 BN_{\gamma(1)}$$

From Table 11.1, for $\phi_1 = 42°$, $N_{\gamma(1)} = 155.55$ and $N_{q(1)} = 85.38$, so

$$q_u = (118)(3)(85.38) + \left(\frac{1}{2}\right)\left[1 - 0.4\left(\frac{4}{6}\right)\right](118)(4)(155.55)$$

$$= 57,145 \text{ lb/ft}^2 \qquad \text{(d)}$$

Comparing Eqs. (c) and (d), $q_u = 34,827 \text{ lb/ft}^2$, we have

$$q_{u(net)} = q_u - \gamma_1 D_f = 34,827 - (3)(118) = 34,473 \text{ lb/ft}^2$$

$$Q_{all} = \frac{q_{u(net)}BL}{F_s} = \frac{(34,473)(4)(6)}{(1000)(3)} \approx \textbf{276 kips} \qquad \blacktriangledown$$

▼ **EXAMPLE 11.7**

Refer to Figure 11.26. Determine the gross allowable bearing capacity of the foundation. Use $F_s = 5$.

Solution　We have $\phi_1 < \phi_2$, and also $\gamma_1 < \gamma_2$. So, the top layer of sand is weaker. Hence, Eq. (11.54) will be used. From Eq. (11.57),

$$q_{u(b')} = \gamma_2 D_f N_{q(2)} + \frac{1}{2}\left[1 - 0.4\left(\frac{B}{L}\right)\right]\gamma_2 BN_{\gamma(2)}$$

For $\phi_2 = 38°$, Table 11.1 gives $N_{q(2)} = 48.93$ and $N_{\gamma(2)} = 78.03$. Hence,

$$q_{u(b')} = (108)(3)(48.93) + \frac{1}{2}\left[1 - (0.4)\left(\frac{4}{6}\right)\right](108)(4)(78.03)$$

$$= 15,853.3 + 12,360 = 28,213.3 \text{ lb/ft}^2 = 14.11 \text{ ton/ft}^2$$

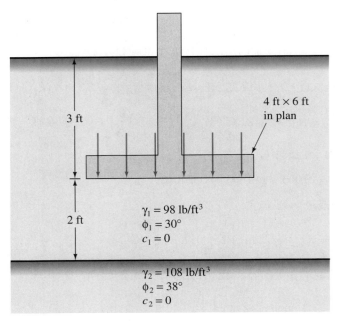

$\gamma_1 = 98 \text{ lb/ft}^3$
$\phi_1 = 30°$
$c_1 = 0$

$\gamma_2 = 108 \text{ lb/ft}^3$
$\phi_2 = 38°$
$c_2 = 0$

▼   **FIGURE 11.26**

Again, from Eq. (11.53),

$$q_{u(t')} = \gamma_1 D_f N_{q(1)} + \frac{1}{2}\left[1 - 0.4\left(\frac{B}{L}\right)\right]\gamma_1 B N_{\gamma(1)}$$

For $\phi_1 = 30°$, $N_{q(1)} = 18.4$ and $N_{\gamma(1)} = 22.4$, so

$$q_{u(t')} = (98)(3)(18.4) + \frac{1}{2}\left[1 - 0.4\left(\frac{4}{6}\right)\right](98)(4)(22.4)$$

$$= 5409.6 + 3219.6 = 8629.2 \text{ lb/ft}^2 = 4.31 \text{ ton/ft}^2$$

Now, using Eq. (11.54) (*note:* $H_f \simeq 2B$), we have

$$q_u = q_{u(t')} + [q_{u(b')} - q_{u(t')}]\left[1 - \frac{H}{H_f}\right]^2$$

$$= 4.31 + (14.11 - 4.31)\left[1 - \frac{2}{(2)(4)}\right]^2 = 9.82 \text{ ton/ft}^2$$

So

$$q_{\text{all}} = \frac{q_u}{F_s} = \frac{9.82}{5} = \textbf{1.96 ton/ft}^2 \qquad \blacktriangledown$$

11.8 FIELD LOAD TEST

In some cases, it is desirable to conduct field load tests to determine the soil-bearing capacity of foundations. The standard method for a field load test is given by the American Society for Testing and Materials under Designation D-1194. Circular steel bearing plates 6 in. to 30 in. (152.4 mm to 762 mm) in diameter and 1 ft × 1 ft (304.8 mm × 304.8 mm) square plates are used for this type of test.

A diagram of the load test is shown in Figure 11.27. To conduct the test, a pit of depth D_f is excavated. The width of the test pit should be at least four times the width of the bearing plate to be used for the test. The bearing plate is placed on the soil at the bottom of the pit, and an incremental load on the bearing plate is applied. After the application of an incremental load, enough time is allowed for settlement to take place. When the settlement of the bearing plate becomes negligible, another incremental load is applied. In this manner, a load-settlement plot can be obtained, as shown in Figure 11.28.

From the results of field load tests, the ultimate soil-bearing capacity of actual footings can be approximately calculated as follows:

For clays:

$$q_{u(\text{footing})} = q_{u(\text{plate})} \tag{11.61}$$

For sandy soils:

$$q_{u(\text{footing})} = q_{u(\text{plate})} \frac{B_{(\text{footing})}}{B_{(\text{plate})}} \tag{11.62}$$

For a given intensity of load, q, the settlement of the actual footing can also be approximately calculated from the following equations:

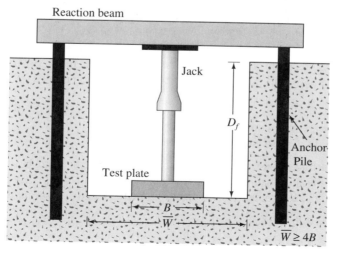

▼ **FIGURE 11.27**   Diagram of field plate load test

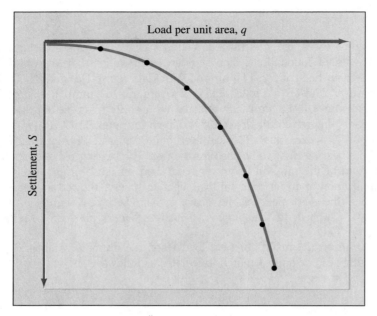

Settlement, S

▼   **FIGURE 11.28**   Typical load-settlement curve obtained from plate load test

In clay:

$$S_{(footing)} = S_{(plate)} \frac{B_{(footing)}}{B_{(plate)}}$$ (11.63)

In sandy soil:

$$S_{(footing)} = S_{(plate)} \left[\frac{2B_{(footing)}}{B_{(footing)} + B_{(plate)}} \right]^2$$ (11.64)

Housel (1929) also proposed a method for obtaining the soil-bearing capacity of a footing that rests on a cohesive soil for a given settlement, S. According to this procedure, the total load carried by a footing of area A and perimeter P can be given by

$$Q = Aq + Ps$$ (11.65)

where q = compression stress below the footing
     s = unit shear stress at the perimeter

Note that q and s are the two unknowns that have to be determined from the results of the field load tests conducted on two different-size plates. If Q_1 and Q_2 are the loads required to produce a settlement S in plates 1 and 2, respectively, then

$$Q_1 = A_1 q + P_1 s$$ (11.66)

and

$$Q_2 = A_2 q + P_2 s$$ (11.67)

Solution of Eqs. (11.66) and (11.67) will give the values of q and s. Housel's method is not used very much in practice now.

▼ EXAMPLE 11.8

The ultimate bearing capacity of a 30-in. diameter plate as determined from field load tests is 6000 lb/ft². Estimate the ultimate bearing capacity of a circular footing with a diameter of 4.5 ft. The soil is sandy.

Solution From Eq. (11.62),

$$q_{u(\text{footing})} = q_{u(\text{plate})} \frac{B_{(\text{footing})}}{B_{(\text{plate})}} = 6000\left(\frac{4.5}{2.5}\right)$$

$$= \mathbf{10{,}800 \ lb/ft^2} \quad ▼$$

▼ EXAMPLE 11.9

Following are the results of two plate load tests in a cohesive soil:

Plate size (ft)	Settlement (in.)	Total load, Q (lb)
1.5 × 1.5	0.5	15,750
2.5 × 2.5	0.5	33,750

If a square footing 5.75 ft × 5.75 ft is to be constructed and the allowable settlement is 0.5 in., what should be the magnitude of the total load it can carry?

Solution From Eq. (11.65),

$$Q = Aq + Ps$$

So

$$15{,}750 = (1.5)^2 q + (4 \times 1.5)s \qquad \text{(a)}$$

$$33{,}750 = (2.5)^2 q + (4 \times 2.5)s \qquad \text{(b)}$$

From Eqs. (a) and (b),

$$q = 3000 \ \text{lb/ft}^2 \quad \text{and} \quad s = 1500 \ \text{lb/ft}$$

$$Q = Aq + Ps = (5.75)^2(3000) + (4 \times 5.75)(1500)$$

$$= 133{,}687.5 \ \text{lb} = \mathbf{133.68 \ kip} \quad ▼$$

11.9 BEARING CAPACITY OF SAND BASED ON SETTLEMENT

It is usually difficult to obtain undisturbed specimens of cohesionless sand during a soil exploration program. For that reason, the results of standard penetration tests (SPT) performed during subsurface exploration are commonly used to predict the allowable soil-bearing capacity of foundations on sand. (The procedure for conducting standard penetration tests is discussed in detail in Chapter 14.)

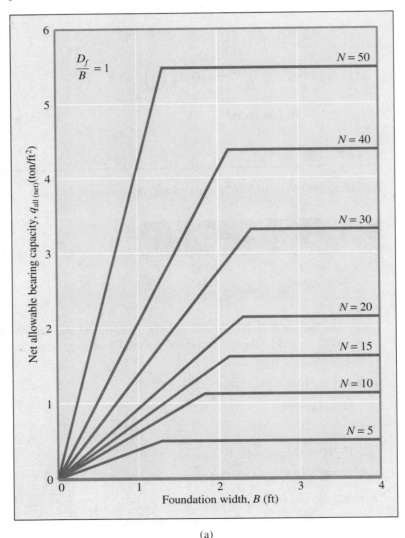

(a)

▼ **FIGURE 11.29**   Correlation of net allowable bearing capacity with standard penetration number for foundation settlements not exceeding 1 in. (25.4 mm) (after Peck, Hanson, and Thornburn, 1974)

Based on the corrected standard penetration numbers, Peck, Hanson, and Thornburn (1974) gave a design chart for the net allowable load per unit area of a foundation. This is shown in Figure 11.29. The principles behind the development of Figure 11.29 are as follows.

From Eqs. (11.61)–(11.64), it can be seen that the settlement at ultimate load increases with the increase of the foundation width B. However, most building codes require that the foundation should not undergo settlements of more than 1 in. (25.4 mm). For foundations with a smaller width B, the ultimate load may be reached before a settlement of 1 in. (24.5 mm) occurs. For those cases,

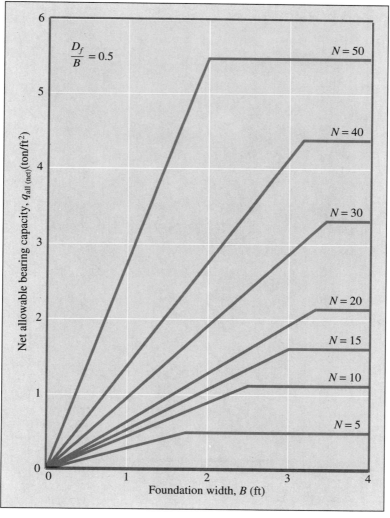

(b)

▼ **FIGURE 11.29** *(Continued)*

$$q_{all(net)} = \frac{q_{u(net)}}{F_s}$$

The sloping lines in the left-hand sides of Figures 11.29a, b, and c correspond to the preceding equation with $F_s = 2$. For larger values of B, the ultimate load will occur at a settlement greater than 1 in. (25.4 mm). For those cases,

$$q_{all(net)} = q_{net(at\ settlement\ of\ 1\ in.)}$$

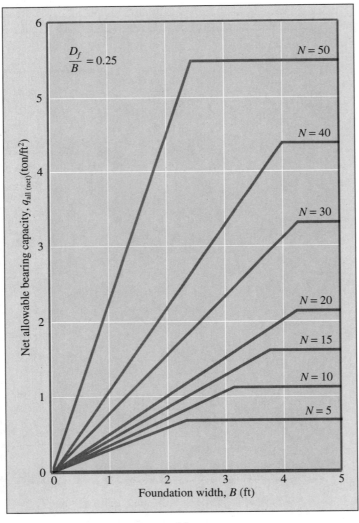

(c)

▼   **FIGURE 11.29**   *(Continued)*

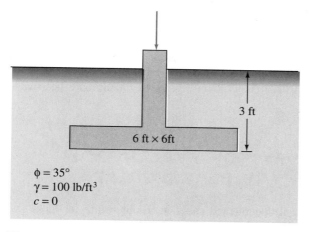

The horizontal lines in the right-hand sides of Figures 11.29a, b, and c correspond to the preceding condition.

Section 11.4 pointed out that the settlement of footings at ultimate load may be as high as 20% of the footing width B. Hence, for large footings, the magnitude of $q_{all} = (q_u/F_s)$ may be too large to satisfy the criteria for tolerable settlement. For this type of condition, the allowable bearing capacity is based on tolerable settlement. This fact can be demonstrated by the following example.

Figure 11.30 shows a 6-ft square footing. The allowable bearing capacity of this footing can be calculated conservatively from Eq. (11.12) as

$$q_u = qN_q + 0.4\gamma BN_\gamma$$

(*Note:* $c = 0$.) For $\phi = 35°$, $N_q = 41.4$ and $N_\gamma = 42.4$ (Figure 11.7), so

$$q_u = (100 \times 3)(41.4) + (0.4)(100)(6)(42.4)$$

$$= 12,420 + 10,176 = 22,596 \text{ lb/ft}^2$$

If $F_s = 4$ is used, we have

$$q_{all} = \frac{q_u}{F_s} = \frac{22,596}{4} = 5649 \text{ lb/ft}^2$$

Table 14.3 shows that for $\phi = 35°$, the standard penetration number (N) will be about 10. For this problem,

$$\frac{D_f}{B} = \frac{3 \text{ ft}}{6 \text{ ft}} = 0.5$$

Now, referring to Figure 11.29b, for $N = 10$ and $B = 6$ ft,

$$q_{all} \simeq 1.1 \text{ ton/ft}^2 = 2200 \text{ lb/ft}^2$$

The preceding value of $q_{all} = 2200$ lb/ft^2 is for a settlement of 1 in. (25.4 mm). So, if a tolerable settlement of 1 in. is a design criterion for the foundation, q_{all} based on settlement ($= 2200$ lb/ft^2) rather than q_{all} based on ultimate bearing capacity (5649 lb/ft^2) will control.

11.10   A CASE HISTORY FOR EVALUATION OF THE ULTIMATE BEARING CAPACITY

There are several documented cases in the literature for large-scale field load tests to determine the ultimate bearing capacity of shallow foundations. One of those field load tests is discussed in this section. The results of the field load tests will be compared with the theory presented in this chapter.

Skempton (1942) reported a field load test in clay for a large foundation with $B = 8$ ft (2.44 m) and $L = 9$ ft (2.74 m). This test was also reported by Bishop and Bjerrum (1960). Figure 11.31 shows the foundation and the soil profile. The variation of

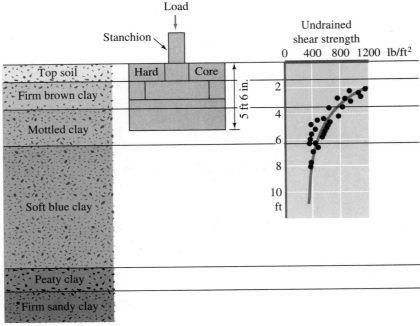

Net foundation pressure at failure $= 2500$ lb/ft^2

▼ **FIGURE 11.31**   Skempton's field load test on a foundation supported by a saturated clay (after Bishop and Bjerrum, 1960)

the undrained cohesion (c_u) of the soil profile is also shown in Figure 11.31. The average moisture content, liquid limit, and plastic limit of the clay underlying the foundation were 50%, 70%, and 28%, respectively. The foundation was loaded to failure immediately after construction. The net ultimate bearing capacity was determined to be 2500 lb/ft² ($\simeq 119.79$ kN/m²).

The net ultimate bearing capacity was defined in Eq. (11.23) as

$$q_{u(net)} = q_u - q$$

From Eq. (11.37), for the vertical loading condition and $\phi = 0$ (*Note:* $N_q = 1$, $N_\gamma = 0$, $\lambda_{qs} = 1$, and $\lambda_{qd} = 1$),

$$q_u = c_u \lambda_{cs} \lambda_{cd} N_c + q$$

So

$$q_{u(net)} = (c_u \lambda_{cs} \lambda_{cd} N_c + q) - q = c_u \lambda_{cs} \lambda_{cd} N_c$$

For the case under consideration, $c_u \simeq 350$ lb/ft², $N_c = 5.14$ (Table 11.1),

$$\lambda_{cs} = 1 + \frac{B}{L} \frac{N_q}{N_c} = 1 + \left(\frac{8}{9}\right)\left(\frac{1}{5.14}\right) = 1.173 \quad \text{(Table 11.2)}$$

$$\lambda_{cd} = \lambda_{qd} - \frac{1 - \lambda_{qd}}{N_q \tan \phi} \quad \text{(Table 11.2)}$$

With $\phi = 0$, $\lambda_{qd} = 1$, so $\lambda_{cd} = 1$. Hence,

$$q_{u(net)} = (350)(1.173)(1)(5.14) = 2110.2 \text{ lb/ft}^2$$

So, for this field load test,

$$\frac{q_{u(net\text{-}theory)}}{q_{u(net\text{-}actual)}} = \frac{2110.2}{2500} \simeq 0.844$$

This is fairly good agreement between the theoretical estimate and the field load test result. The slight variation between the theory and the field test result may be because of the estimation of the average value of c_u.

Bishop and Bjerrum (1960) cited several end-of-construction failures of footings on saturated clay. These are given in Table 11.3. It can be seen from this table that, in all cases, $q_{u(net\text{-}theory)}/q_{u(net\text{-}actual)}$ is approximately 1. This confirms that the design of shallow foundations based on the net ultimate bearing capacity is a reliable technique.

▼ **TABLE 11.3   End-of-Construction Failures of Footings—Saturated Clay Foundation: $\phi = 0$ Condition***

Locality	Data of clay					$\dfrac{q_{u(net\text{-}theory)}}{q_{u(net\text{-}actual)}}$
	w (%)	LL	PL	PI	$\dfrac{w - PL}{PI}$	
Loading test, Marmorerá	10	35	25	20	−0.25	0.92
Kensal Green	—	—	—	—	—	1.02
Silo, Transcona	50	110	30	80	0.25	1.09
Kippen	50	70	28	42	0.52	0.95
Screw pile, Lock Ryan	—	—	—	—	—	1.05
Screw pile, Newport	—	—	—	—	—	1.07
Oil tank, Fredrikstad	45	55	25	30	0.67	1.08
Oil tank A, Shellhaven	70	87	25	62	0.73	1.03
Oil tank B, Shellhaven	—	—	—	—	—	1.05
Silo, U.S.A.	40	—	—	—	—	0.98
Loading test, Moss	9	—	—	—	—	1.10
Loading test, Hagalund	68	55	20	35	1.37	0.93
Loading test, Torp	27	24	16	8	1.39	0.96
Loading test, Rygge	45	37	19	18	1.44	0.95

Note: w = moisture content; LL = liquid limit; PL = plastic limit; PI = plasticity index
* After Bishop and Bjerrum (1960)

11.11   GENERAL COMMENTS

In this chapter, theories for estimating the ultimate and allowable bearing capacities of shallow foundations were presented. Procedures for field load tests and estimation of the allowable bearing capacity of granular soil based on limited settlement criteria were briefly discussed.

Several building codes now used in the United States and elsewhere provide presumptive bearing capacities for various types of soil. It is extremely important to realize that they are *approximate values only*. The bearing capacity of foundations depends on several factors:

1. Subsoil stratification
2. Shear strength parameters of the subsoil
3. Location of the ground water table
4. Environmental factors
5. Building size and weight
6. Depth of excavation
7. Type of structure.

Hence, it is important that the allowable bearing capacity at a given site be determined based on the findings of soil exploration at that site, past experience of foundation construction, and fundamentals of the geotechnical engineering theories for bearing capacity.

The allowable bearing capacity relationships based on settlement considerations such as those given in Section 11.9 do not take into account the settlement caused by consolidation of the clay layers. Excessive settlement usually causes the building to crack, which may ultimately lead to structural failure. Uniform settlement of a structure does not produce cracking; on the other hand, differential settlement may produce cracks and damage to a building.

PROBLEMS   **11.1**   For the continuous footing shown in Figure 11.32, determine the gross allowable bearing capacity. Use Terzaghi's bearing capacity factors and a factor of safety of 4. Assume general bearing capacity failure.

   a. $\gamma = 120 \text{ lb/ft}^3$, $c = 0$, $\phi = 40°$, $D_f = 3$ ft, $B = 3.5$ ft
   b. $\gamma = 115 \text{ lb/ft}^3$, $c = 600 \text{ lb/ft}^2$, $\phi = 25°$, $D_f = 3.5$ ft, $B = 4$ ft
   c. $\gamma = 17.5 \text{ kN/m}^3$, $c = 14 \text{ kN/m}^2$, $\phi = 20°$, $D_f = 1.0$ m, $B = 1.2$ m
   d. $\gamma = 118 \text{ lb/ft}^3$, $c = 450 \text{ lb/ft}^2$, $\phi = 28°$, $D_f = 4$ ft, $B = 4$ ft
   e. $\gamma = 17.7 \text{ kN/m}^3$, $c = 48 \text{ kN/m}^2$, $\phi = 0°$, $D_f = 0.6$ m, $B = 0.8$ m

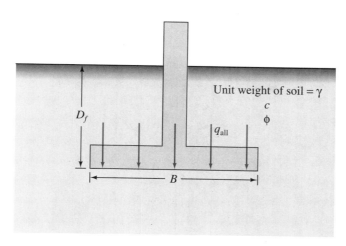

Unit weight of soil $= \gamma$
c
ϕ

D_f

q_{all}

B

▼   **FIGURE 11.32**

   11.2   Repeat Problem 11.1 assuming local shear failure.

   11.3   Redo Problem 11.1 using the Prandtl–Reissner and Caquot–Kerisel bearing capacity factors given in Table 11.1 and Eq. (11.37).

   11.4   A square footing has the following values:
   Gross allowable load $= 42,260$ lb
   Factor of safety $= 3$
   $D_f = 3$ ft
   Soil properties: $\gamma = 110 \text{ lb/ft}^3$
         $\phi = 20°$
         $c = 200 \text{ lb/ft}^2$
   Use Eq. (11.12) to determine the size of the footing.

11.5 Repeat Problem 11.4 for the following data:
Gross allowable load = 92.5 kip
Factor of safety = 3
$D_f = 2$ ft
$\gamma = 115$ lb/ft^3
$c = 0$
$\phi = 35°$

11.6 A square footing is shown in Figure 11.33. Use Terzaghi's equation for general shear failure and a factor of safety of 3. Determine the safe gross allowable load.
$\gamma = 100$ lb/ft^3
$\gamma_{sat} = 115$ lb/ft^3
$c = 0$
$\phi = 30°$
$B = 4$ ft
$D_f = 3.5$ ft
$h = 2$ ft

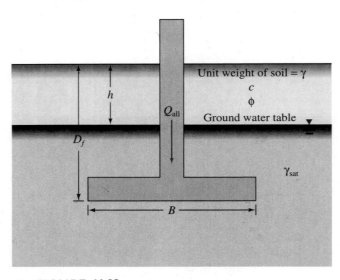

▼ **FIGURE 11.33**

11.7 Redo Problem 11.6 using Eq. (11.37).

11.8 Solve Problem 11.6 with the following data:
$\gamma = 115$ lb/ft^3
$\gamma_{sat} = 122.4$ lb/ft^3
$c = 100$ lb/ft^2
$\phi = 30°$
$B = 4$ ft
$D_f = 3$ ft
$h = 4$ ft

11.9 Redo Problem 11.8 using Eq. (11.37).

11.10   Solve Problem 11.6 with the following values:

$$\gamma = 15.72 \text{ kN/m}^3$$
$$\gamma_{sat} = 18.55 \text{ kN/m}^3$$
$$c = 0$$
$$\phi = 35°$$
$$B = 1.53 \text{ m}$$
$$D_f = 1.22 \text{ m}$$
$$h = 0.61 \text{ m}$$

11.11   Redo Problem 11.10 using Eq. (11.37).

11.12   The square footing shown in Figure 11.34 is subjected to an eccentric load. For the following cases, determine the gross allowable load that the footing could carry. Use $F_s = 3$.

    a. $\gamma = 105 \text{ lb/ft}^3$, $c = 0$, $\phi = 30°$, $B = 4.5$ ft, $D_f = 3.5$ ft, $x = 0.5$ ft, $y = 0$
    b. $\gamma = 120 \text{ lb/ft}^3$, $c = 400 \text{ lb/ft}^2$, $\phi = 25°$, $B = 6$ ft, $D_f = 4.5$ ft, $x = 0.6$ ft, $y = 0.5$ ft
    c. $\rho = 2000 \text{ kg/m}^3$, $c = 0$, $\phi = 42°$, $B = 2.5$ m, $D_f = 1.5$ m, $x = 0.2$ m, $y = 0.2$ m
    d. $\rho = 1950 \text{ kg/m}^3$, $c = 0$, $\phi = 36°$, $B = 3$ m, $D_f = 1.4$ m, $x = 0.3$ m, $y = 0$

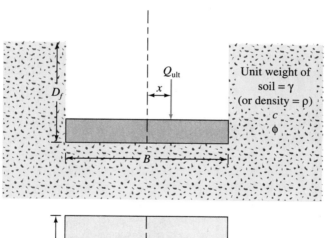

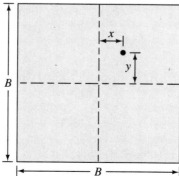

▼   **FIGURE 11.34**

11.13 Figure 11.35 shows a footing on layered sand. For the following conditions, determine the net allowable load it can carry:

Square footing; $B = 5$ ft
Factor of safety required $= 4$
$D_f = 3.5$ ft
$H = 2$ ft
$\gamma_1 = 118$ lb/ft^3
$\gamma_2 = 105$ lb/ft^3
$\phi_1 = 40°$
$\phi_2 = 30°$

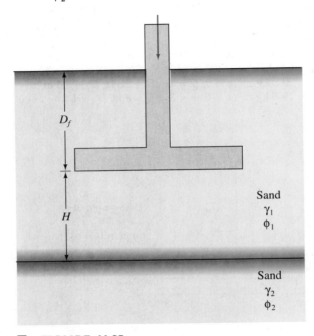

Sand
γ_1
ϕ_1

Sand
γ_2
ϕ_2

▼   **FIGURE 11.35**

11.14 Redo Problem 11.13 with the following values:
Rectangular footing; $B = 1$ m, $L = 1.5$ m
Factor of safety required $= 3$
$D_f = 1$ m
$H = 0.6$ m
$\gamma_1 = 17.5$ kN/m^3
$\gamma_2 = 15$ kN/m^3
$\phi_1 = 42°$
$\phi_2 = 34°$

11.15 Redo Problem 11.13 with the following data:
Circular footing; diameter $= 2$ m
$D_f = 1.5$ m
$H = 1.25$ m

$$\gamma_1 = 16.3 \text{ kN/m}^3$$
$$\gamma_2 = 18.2 \text{ kN/m}^3$$
$$\phi_1 = 32°$$
$$\phi_2 = 42°$$

11.16 Redo Problem 11.13 with the following data:
Square footing; $B = 3$ ft
$D_f = 3$ ft
$H = 1.5$ ft
$\gamma_1 = 100 \text{ lb/ft}^3$
$\gamma_2 = 115 \text{ lb/ft}^3$
$\phi_1 = 30°$
$\phi_2 = 36°$

11.17 Figure 11.36 shows a footing on layered saturated clay. For the following conditions, estimate the gross allowable load it can carry:
Square footing; $B = 4.5$ ft
$D_f = 4$ ft
$H = 3$ ft
$\gamma_1 = 120 \text{ lb/ft}^3$
$\gamma_2 = 117 \text{ lb/ft}^3$
$c_{u(1)} = 1500 \text{ lb/ft}^3$
$c_{u(2)} = 700 \text{ lb/ft}^3$
Factor of safety $= 3$

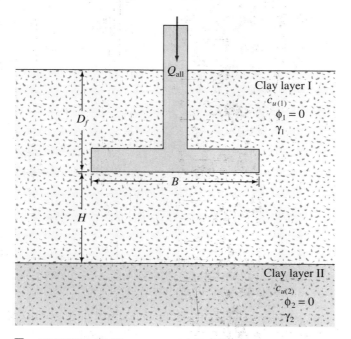

▼ **FIGURE 11.36**

11.18   Repeat Problem 11.17 for the following case:
Square footing; $B = 2$ m

$D_f = 1.5$ m
$H = 1.5$ m
$\gamma_1 = 20$ kN/m^3
$\gamma_2 = 16$ kN/m^3
$c_{u(1)} = 50$ kN/m^2
$c_{u(2)} = 28$ kN/m^2

11.19   Refer to Figure 11.36. Assume that the footing is rectangular with $B = 4$ ft and $L = 5.8$ ft. Given $D_f = 2$ ft, $c_{u(1)} = 1100$ lb/ft^2, $c_{u(2)} = 400$ lb/ft^2, $\gamma_1 = 120$ lb/ft^3, and $\gamma_2 = 112$ lb/ft^3, determine the minimum value of H at which the failure surface in soil at ultimate load will be fully contained in the top layer.

11.20   A plate load test was conducted in a sandy soil in which the size of the bearing plate was 1 ft × 1 ft. The ultimate load per unit area (q_u) for the test was found to be 4200 lb/ft^2. Estimate the total allowable load (Q_{all}) for a footing of size 5.5 ft × 5.5 ft. Use a factor of safety of 4.

11.21   A plate load test (bearing plate of 762 mm diameter) was conducted in clay. The ultimate load per unit area, q_u, for the test was found to be 248.9 kN/m^2. What should be the total allowable load, Q_{all}, for a column footing 2 m in diameter? Use a factor of safety of 3.

11.22   The results of two field load tests in a clay soil are given in the table. Based on these results, determine the size of a square footing that will carry a total load of 300 kN with a maximum settlement of 15 mm.

Plate diameter (mm)	Settlement (mm)	Total load (kN)
204.8	15	49.5
457.2	15	133.1

REFERENCES

American Society for Testing and Materials (1991). *Annual Book of Standards*, Vol. 04–08, Philadelphia, Pa.

Balla, A. (1962). "Bearing Capacity of Foundations," *Journal of the Soil Mechanics and Foundations Division*, ASCE, Vol. 89, No. SM5, 12–34.

Bishop, A. W., and Bjerrum, L. (1960). "The Relevance of the Triaxial Test to the Solution of Stability Problems," *Proceedings*, Research Conference on Shear Strength of Cohesive Soils, ASCE, 437–501.

Caquot, A., and Kerisel, J. (1953). "Sur le terme de surface dans le calcul des fondations en milieu pulverulent," *Proceedings*, 3rd International Conference on Soil Mechanics and Foundation Engineering, Vol. I, 336–337.

DeBeer, E. E. (1970). "Experimental Determination of Shape Factor and Bearing Capacity Factors of Sand," *Geotechnique*, Vol. 20, No. 4, 387–411.

DeBeer, E. E., and Vesic, A. S. (1958). "Etude Experimentale de la Capacité Portante du Sable Sous des Fondations Directes Etablies en Surface," *Ann. Trav. Publics Belg.*, Vol. 59, No. 3.

Hansen, J. B. (1970). "A Revised and Extended Formula for Bearing Capacity," Danish Geotechnical Institute, *Bulletin No. 28*, Copenhagen.

Housel, W. S. (1929). "A Practical Method for the Selection of Foundations Based on Fundamental Research in Soil Mechanics," University of Michigan Research Station, *Bulletin No. 13*, Ann Arbor.

Lundgren, H., and Mortensen, K. (1953). "Determination by the Theory of Elasticity of the Bearing Capacity of Continuous Footing on Sand," *Proceedings*, 3rd International Conference on Soil Mechanics and Foundation Engineering, Vol. I, 409–412.

Meyerhof, G. G. (1951). "The Ultimate Bearing Capacity of Foundations," *Geotechnique*, Vol. 2, No. 4, 301–331.

Meyerhof, G. G. (1953). "The Bearing Capacity of Foundations under Eccentric and Inclined Loads," *Proceedings*, 3rd International Conference on Soil Mechanics and Foundation Engineering, Vol. 1, 440–445.

Meyerhof, G. G. (1963). "Some Recent Research on the Bearing Capacity of Foundations," *Canadian Geotechnical Journal*, Vol. 1, 16–26.

Meyerhof, G. G., and Hanna, A. M. (1978). "Ultimate Bearing Capacity of Foundations on Layered Soil under Inclined Load," *Canadian Geotechnical Journal*, Vol. 15, No. 4, 565–572.

Peck, R. B., Hanson, W. E., and Thornburn, T. H. (1974). *Foundation Engineering*, 2nd ed., Wiley, New York.

Prandtl, L. (1921). "Über die Eindringungsfestigkeit (Harte) plasticher Baustoffe und die Festigkeit von Schneiden," *Zeitschrift für Angewandte Mathematik und Mechanik*, Vol. 1, No. 1, 15–20, Basel, Switzerland.

Reissner, H. (1924). "Zum Erddruckproblem," *Proceedings*, 1st International Congress of Applied Mechanics, 295–311.

Skempton, A. W. (1942). "An Investigation of the Bearing Capacity of a Soft Clay Soil." *Journal of the Institute of Civil Engineers*, London, Vol. 18, 307–321.

Terzaghi, K. (1943). *Theoretical Soil Mechanics*, Wiley, New York.

Vesic, A. S. (1973). "Analysis of Ultimate Loads on Shallow Foundations," *Journal of the Soil Mechanics and Foundations Division*, ASCE, Vol. 99, No. SM1, 45–73.

CHAPTER

TWELVE

SLOPE STABILITY

An exposed ground surface that stands at an angle with the horizontal is called an *unrestrained slope*. The slope can be natural or man-made. If the ground surface is not horizontal, a component of gravity will tend to move the soil downward as shown in Figure 12.1. If the component of gravity is large enough, slope failure can occur, that is, the soil mass in zone *abcdea* can slide downward. The driving force overcomes the resistance from the shear strength of the soil along the rupture surface.

In many cases, civil engineers are expected to make calculations to check the safety of natural slopes, slopes of excavations, and compacted embankments. This check involves determining and comparing the shear stress developed along the most likely rupture surface with the shear strength of the soil. This process is called *slope stability analysis*.

The stability analysis of a slope is not an easy task. Evaluation of variables such as the soil stratification and its in-place shear strength parameters may prove to be a formidable task. Seepage through the slope and the choice of a potential slip surface add to the complexity of the problem.

This chapter explains the basic principles involved in slope stability analysis.

12.1 FACTOR OF SAFETY

The task of the engineer charged with analyzing slope stability is to determine the factor of safety. Generally, the factor of safety is defined as

$$F_s = \frac{\tau_f}{\tau_d} \tag{12.1}$$

where F_s = factor of safety with respect to strength
  τ_f = average shear strength of the soil
  τ_d = average shear stress developed along the potential failure surface

The shear strength of a soil consists of two components, cohesion and friction, and may be written as

$$\tau_f = c + \sigma \tan \phi \tag{12.2}$$

where c = cohesion
  ϕ = angle of friction
  σ = normal stress on the potential failure surface

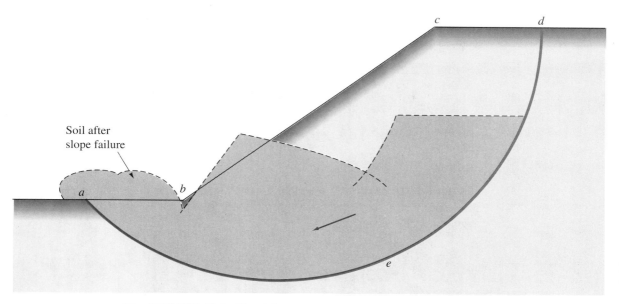

▼   **FIGURE 12.1**   Slope failure

In a similar manner, we can also write

$$\tau_d = c_d + \sigma \tan \phi_d \tag{12.3}$$

where c_d and ϕ_d are, respectively, the cohesion and the angle of friction that develop along the potential failure surface. Substituting Eqs. (12.2) and (12.3) into Eq. (12.1), we get

$$F_s = \frac{c + \sigma \tan \phi}{c_d + \sigma \tan \phi_d} \tag{12.4}$$

Now we can introduce some other aspects of the factor of safety—that is, the factor of safety with respect to cohesion, F_c, and the factor of safety with respect to friction, F_ϕ. They are defined as follows:

$$F_c = \frac{c}{c_d} \tag{12.5}$$

and

$$F_\phi = \frac{\tan \phi}{\tan \phi_d} \tag{12.6}$$

When Eqs. (12.4), (12.5), and (12.6) are compared, it becomes obvious that when F_c becomes equal to F_ϕ, it gives the factor of safety with respect to strength. Or, if

$$\frac{c}{c_d} = \frac{\tan \phi}{\tan \phi_d}$$

we can write

$$F_s = F_c = F_\phi \tag{12.7}$$

When F_s is equal to 1, the slope is in a state of impending failure. Generally, a value of 1.5 for the factor of safety with respect to strength is quite acceptable for the design of a stable slope.

12.2 STABILITY OF INFINITE SLOPES WITHOUT SEEPAGE

In considering the problem of slope stability, we may start with the case of an infinite slope as shown in Figure 12.2. The shear strength of the soil may be given by [Eq. (12.2)]

$$\tau_f = c + \sigma \tan \phi$$

Assuming that the pore water pressure is 0, we will evaluate the factor of safety against a possible slope failure along a plane AB located at a depth H below the ground surface. The slope failure can occur by the movement of soil above the plane AB from right to left.

Let us consider a slope element, $abcd$, that has a unit length perpendicular to the plane of the section shown. The forces, F, that act on the faces ab and cd are equal and opposite and may be ignored. The weight of the soil element is

$$W = \text{(volume of the soil element)} \times \text{(unit weight of soil)} = \gamma LH \tag{12.8}$$

The weight, W, can be resolved into two components:

1. Force perpendicular to the plane $AB = N_a = W \cos \beta = \gamma LH \cos \beta$.
2. Force parallel to the plane $AB = T_a = W \sin \beta = \gamma LH \sin \beta$. Note that this is the force that tends to cause the slip along the plane.

Thus, the normal stress σ and the shear stress τ at the base of the slope element can be given as

$$\sigma = \frac{N_a}{\text{area of the base}} = \frac{\gamma LH \cos \beta}{\left(\dfrac{L}{\cos \beta}\right)} = \gamma H \cos^2 \beta \tag{12.9}$$

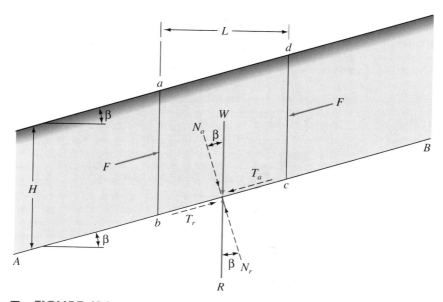

▼ FIGURE 12.2 Analysis of infinite slope (without seepage)

and

$$\tau = \frac{T_a}{\text{area of the base}} = \frac{\gamma LH \sin \beta}{\left(\dfrac{L}{\cos \beta}\right)} = \gamma H \cos \beta \sin \beta \qquad (12.10)$$

The reaction to the weight W is an equal and opposite force R. The normal and tangential components of R with respect to the plane AB are N_r and T_r:

$$N_r = R \cos \beta = W \cos \beta \qquad (12.11)$$

$$T_r = R \sin \beta = W \sin \beta \qquad (12.12)$$

For equilibrium, the resistive shear stress that develops at the base of the element is equal to $(T_r)/(\text{area of the base}) = \gamma H \sin \beta \cos \beta$. This may also be written in the form [Eq. (12.3)]

$$\tau_d = c_d + \sigma \tan \phi_d$$

The value of the normal stress is given by Eq. (12.9). Substitution of Eq. (12.9) into Eq. (12.3) yields

$$\tau_d = c_d + \gamma H \cos^2 \beta \tan \phi_d \qquad (12.13)$$

Thus,

$$\gamma H \sin \beta \cos \beta = c_d + \gamma H \cos^2 \beta \tan \phi_d$$

or

$$\frac{c_d}{\gamma H} = \sin \beta \cos \beta - \cos^2 \beta \tan \phi_d$$

$$= \cos^2 \beta (\tan \beta - \tan \phi_d) \qquad (12.14)$$

The factor of safety with respect to strength has been defined in Eq. (12.7), from which

$$\tan \phi_d = \frac{\tan \phi}{F_s} \quad \text{and} \quad c_d = \frac{c}{F_s}$$

Substituting the preceding relationships into Eq. (12.14), we obtain

$$F_s = \frac{c}{\gamma H \cos^2 \beta \tan \beta} + \frac{\tan \phi}{\tan \beta} \qquad (12.15)$$

For granular soils, $c = 0$, and the factor of safety, F_s, becomes equal to $(\tan \phi)/(\tan \beta)$. This indicates that in an infinite slope in sand, the value of F_s is independent of the height H, and the slope is stable as long as $\beta < \phi$.

If a soil possesses cohesion and friction, the depth of the plane along which critical equilibrium occurs may be determined by substituting $F_s = 1$ and $H = H_{cr}$ into Eq. (12.15). Thus,

$$H_{cr} = \frac{c}{\gamma} \frac{1}{\cos^2 \beta (\tan \beta - \tan \phi)} \qquad (12.16)$$

▼ **EXAMPLE 12.1**

For the infinite slope shown in Figure 12.3, determine:

    a.   The factor of safety against sliding along the soil-rock interface given $H = 8$ ft.

    b.   The height, H, that will give a factor of safety (F_s) of 2 against sliding along the soil-rock interface.

Solution

    a.   From Eq. (12.15),

$$F_s = \frac{c}{\gamma H \cos^2 \beta \tan \beta} + \frac{\tan \phi}{\tan \beta}$$

Given $c = 200 \text{ lb/ft}^2$, $\gamma = 100 \text{ lb/ft}^3$, $\phi = 15°$, $\beta = 25°$, and $H = 8$ ft, we have

$$F_s = \frac{200}{(100)(8)(\cos^2 25)(\tan 25)} + \frac{\tan 15}{\tan 25} = \mathbf{1.23}$$

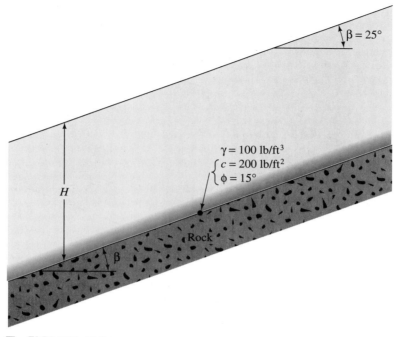

$\gamma = 100 \ \text{lb/ft}^3$
$\begin{cases} c = 200 \ \text{lb/ft}^2 \\ \phi = 15° \end{cases}$

$\beta = 25°$

Rock

b. $F_s = \dfrac{c}{\gamma H \cos^2 \beta \ \tan \beta} + \dfrac{\tan \phi}{\tan \beta}$

$2 = \dfrac{200}{(100)(H)(\cos^2 25)(\tan 25)} + \dfrac{\tan 15}{\tan 25}$

$H = \textbf{3.66 ft}$ ▼

12.3   STABILITY OF INFINITE SLOPES WITH SEEPAGE

Figure 12.4a shows an infinite slope. It is assumed that there is seepage through the soil and that the ground water level coincides with the ground surface. The shear strength of the soil is given by

$$\tau_f = c + \sigma' \tan \phi \tag{12.17}$$

Note that, unlike Eq. (12.2), σ' is used in the preceding equation to differentiate between total stress and effective stress.

To determine the factor of safety against failure along the plane AB, consider the slope element $abcd$. The forces that act on the vertical faces ab and cd are equal and

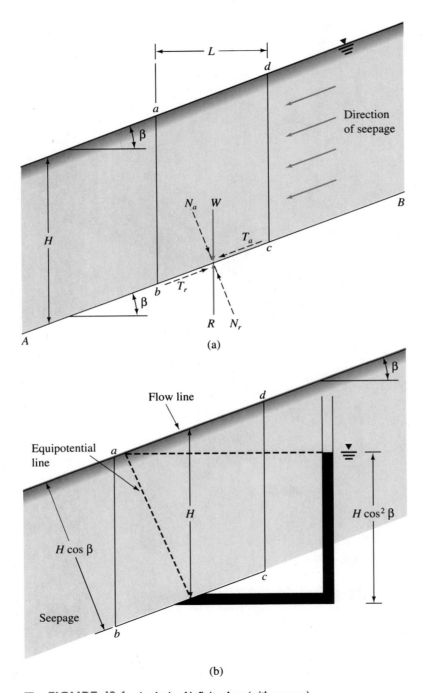

▼   **FIGURE 12.4**   Analysis of infinite slope (with seepage)

opposite. The total weight of the slope element of unit length is

$$W = \gamma_{\text{sat}} LH \tag{12.18}$$

The components of W in the directions normal and parallel to plane AB are

$$N_a = W \cos \beta = \gamma_{\text{sat}} LH \cos \beta \tag{12.19}$$

and

$$T_a = W \sin \beta = \gamma_{\text{sat}} LH \sin \beta \tag{12.20}$$

The reaction to weight W is equal to R. Thus,

$$N_r = R \cos \beta = W \cos \beta = \gamma_{\text{sat}} LH \cos \beta \tag{12.21}$$

and

$$T_r = R \sin \beta = W \sin \beta = \gamma_{\text{sat}} LH \sin \beta \tag{12.22}$$

The total normal stress and the shear stress at the base of the element are as follows:

Total normal stress:

$$\sigma = \frac{N_r}{\left(\dfrac{L}{\cos \beta} \right)} = \gamma_{\text{sat}} H \cos^2 \beta \tag{12.23}$$

Shear stress:

$$\tau = \frac{T_r}{\left(\dfrac{L}{\cos \beta} \right)} = \gamma_{\text{sat}} H \cos \beta \sin \beta \tag{12.24}$$

The resistive shear stress developed at the base of the element can also be given by

$$\tau_d = c_d + \sigma' \tan \phi_d = c_d + (\sigma - u) \tan \phi_d \tag{12.25}$$

where u = the pore water pressure = $\gamma_w H \cos^2 \beta$ (see Figure 12.4b).

Substituting the value of σ [Eq. (12.23)] and u into Eq. (12.25), we get

$$\tau_d = c_d + (\gamma_{\text{sat}} H \cos^2 \beta - \gamma_w H \cos^2 \beta) \tan \phi_d$$

$$= c_d + \gamma' H \cos^2 \beta \tan \phi_d \tag{12.26}$$

Now, setting the right-hand sides of Eqs. (12.24) and (12.26) equal to each other gives

$$\gamma_{\text{sat}} H \cos \beta \sin \beta = c_d + \gamma' H \cos^2 \beta \tan \phi_d$$

or

$$\frac{c_d}{\gamma_{\text{sat}} H} = \cos^2 \beta \left(\tan \beta - \frac{\gamma'}{\gamma_{\text{sat}}} \tan \phi_d \right) \tag{12.27}$$

The factor of safety with respect to strength can be found by substituting $\tan \phi_d = (\tan \phi)/F_s$ and $c_d = c/F_s$ into Eq. (12.27), or

$$F_s = \frac{c}{\gamma_{\text{sat}} H \cos^2 \beta \tan \beta} + \frac{\gamma'}{\gamma_{\text{sat}}} \frac{\tan \phi}{\tan \beta} \qquad (12.28)$$

▼   **EXAMPLE 12.2**

Refer to Figure 12.3. If there is seepage through the soil and the ground water table coincides with the ground surface, what is the factor of safety, F_s, given $H = 3.66$ ft and $\gamma_{\text{sat}} = 118 \text{ lb/ft}^3$?

Solution   From Eq. (12.28),

$$F_s = \frac{c}{\gamma_{\text{sat}} H \cos^2 \beta \tan \beta} + \frac{\gamma'}{\gamma_{\text{sat}}} \frac{\tan \phi}{\tan \beta}$$

or

$$F_s = \frac{200}{(118)(3.66)(\cos^2 25)(\tan 25)} + \frac{(118 - 62.4)}{118}\left(\frac{\tan 15}{\tan 25}\right) = \mathbf{1.48} \qquad ▼$$

12.4 FINITE SLOPES—GENERAL

When the value of H_{cr} approaches the height of the slope, it may generally be considered a finite slope. When analyzing the stability of a finite slope in a homogeneous soil, for simplicity, we need to make an assumption about the general shape of the surface of potential failure. Although there is considerable evidence that slope failures usually occur on curved failure surfaces, Culmann (1875) approximated the surface of potential failure as a plane. The factor of safety F_s calculated by using Culmann's approximation gives fairly good results for near-vertical slopes only. After extensive investigation of slope failures in the 1920s, a Swedish geotechnical commission recommended that the actual surface of sliding may be approximated to be circularly cylindrical.

Since that time, most conventional stability analyses of slopes have been made by assuming that the curve of potential sliding is an arc of a circle. However, there are many circumstances (for example, zoned dams and foundations on weak strata) where stability analysis using plane failure of sliding is more appropriate and yields excellent results.

Analysis of Finite Slope with Plane Failure Surface (Culmann's Method)

This analysis is based on the assumption that the failure of a slope occurs along a plane when the average shearing stress tending to cause the slip is more than the shear strength of the soil. Also, the most critical plane is the one that has a minimum ratio of the average shearing stress that tends to cause failure to the shear strength of soil.

Figure 12.5 shows a slope of height H. The slope rises at an angle β with the horizontal. AC is a trial failure plane. Considering a unit length perpendicular to the section of the slope, the weight of the wedge $ABC = W$:

$$W = \frac{1}{2} (H)(\overline{BC})(1)(\gamma)$$

$$= \frac{1}{2} H(H \cot \theta - H \cot \beta)\gamma$$

$$= \frac{1}{2} \gamma H^2 \left[\frac{\sin (\beta - \theta)}{\sin \beta \sin \theta} \right] \tag{12.29}$$

The normal and tangential components of W with respect to the plane AC are as follows:

$$N_a = \text{normal component} = W \cos \theta$$

$$= \frac{1}{2} \gamma H^2 \left[\frac{\sin (\beta - \theta)}{\sin \beta \sin \theta} \right] \cos \theta \tag{12.30}$$

$$T_a = \text{tangential component} = W \sin \theta$$

$$= \frac{1}{2} \gamma H^2 \left[\frac{\sin (\beta - \theta)}{\sin \beta \sin \theta} \right] \sin \theta \tag{12.31}$$

The average normal stress and shear stress on the plane AC may be given by

$$\sigma = \text{average normal stress}$$

$$= \frac{N_a}{(\overline{AC})(1)} = \frac{N_a}{\left(\dfrac{H}{\sin \theta} \right)}$$

$$= \frac{1}{2} \gamma H \left[\frac{\sin (\beta - \theta)}{\sin \beta \sin \theta} \right] \cos \theta \sin \theta \tag{12.32}$$

and

$$\tau = \text{average shear stress}$$

$$= \frac{T_a}{(\overline{AC})(1)} = \frac{T_a}{\left(\dfrac{H}{\sin \theta} \right)}$$

$$= \frac{1}{2} \gamma H \left[\frac{\sin (\beta - \theta)}{\sin \beta \sin \theta} \right] \sin^2 \theta \tag{12.33}$$

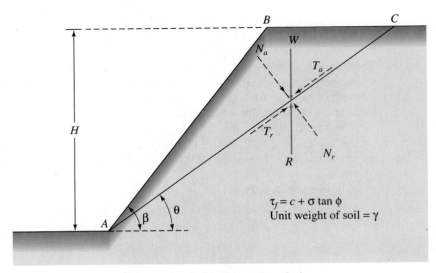

▼ **FIGURE 12.5**   Finite slope analysis—Culmann's method

The average resistive shearing stress developed along the plane AC may also be expressed as

$$\tau_d = c_d + \sigma \tan \phi_d$$

$$= c_d + \frac{1}{2} \gamma H \left[\frac{\sin (\beta - \theta)}{\sin \beta \sin \theta} \right] \cos \theta \sin \theta \tan \phi_d \qquad (12.34)$$

Now, from Eqs. (12.33) and (12.34),

$$\frac{1}{2} \gamma H \left[\frac{\sin (\beta - \theta)}{\sin \beta \sin \theta} \right] \sin^2 \theta = c_d + \frac{1}{2} \gamma H \left[\frac{\sin (\beta - \theta)}{\sin \beta \sin \theta} \right] \cos \theta \sin \theta \tan \phi_d \qquad (12.35)$$

or

$$c_d = \frac{1}{2} \gamma H \left[\frac{\sin (\beta - \theta)(\sin \theta - \cos \theta \tan \phi_d)}{\sin \beta} \right] \qquad (12.36)$$

The expression in Eq. (12.36) is derived for the trial failure plane AC. In an effort to determine the critical failure plane, the principle of maxima and minima is used (for a given value of ϕ_d) to find the angle θ where the developed cohesion would be maximum. Thus, the first derivative of c_d with respect to θ is set equal to 0, or

$$\frac{\partial c_d}{\partial \theta} = 0 \qquad (12.37)$$

Since γ, H, and β are constants in Eq. (12.36), we have

$$\frac{\partial}{\partial \theta} [\sin (\beta - \theta)(\sin \theta - \cos \theta \tan \phi_d)] = 0 \qquad (12.38)$$

Solution of Eq. (12.38) gives the critical value of θ, or

$$\theta_{cr} = \frac{\beta + \phi_d}{2} \tag{12.39}$$

Substitution of the value of $\theta = \theta_{cr}$ into Eq. (12.36) yields

$$c_d = \frac{\gamma H}{4} \left[\frac{1 - \cos(\beta - \phi_d)}{\sin \beta \cos \phi_d} \right] \tag{12.40}$$

The maximum height of the slope for which critical equilibrium occurs can be obtained by substituting $c_d = c$ and $\phi_d = \phi$ into Eq. (12.40). Thus,

$$H_{cr} = \frac{4c}{\gamma} \left[\frac{\sin \beta \cos \phi}{1 - \cos(\beta - \phi)} \right] \tag{12.41}$$

▼ **EXAMPLE 12.3**

A cut is to be made in a soil that has $\gamma = 105$ lb/ft^3, $c = 600$ lb/ft^2, and $\phi = 15°$. The side of the cut slope will make an angle of 45° with the horizontal. What should be the depth of the cut slope that will have a factor of safety, F_s, of 3?

Solution   We are given $\phi = 15°$ and $c = 600$ lb/ft^2. If $F_s = 3$, then F_c and F_ϕ should both be equal to 3:

$$F_c = \frac{c}{c_d}$$

or

$$c_d = \frac{c}{F_c} = \frac{c}{F_s} = \frac{600}{3} = 200 \text{ lb/ft}^2$$

Similarly,

$$F_\phi = \frac{\tan \phi}{\tan \phi_d}$$

$$\tan \phi_d = \frac{\tan \phi}{F_\phi} = \frac{\tan \phi}{F_s} = \frac{\tan 15}{3}$$

or

$$\phi_d = \tan^{-1} \left[\frac{\tan 15}{3} \right] = 5.1°$$

Substituting the preceding values of c_d and ϕ_d into Eq. (12.40) gives

$$H = \frac{4c_d}{\gamma}\left[\frac{\sin \beta \cos \phi_d}{1 - \cos (\beta - \phi_d)}\right] = \frac{4 \times 200}{105}\left[\frac{\sin 45 \cos 5.1}{1 - \cos (45 - 5.1)}\right]$$

$$= \textbf{23 ft} \qquad \blacktriangledown$$

12.5 ANALYSIS OF FINITE SLOPE WITH CIRCULARLY CYLINDRICAL FAILURE SURFACE—GENERAL

In general, slope failure occurs in one of the following modes (Figure 12.6):

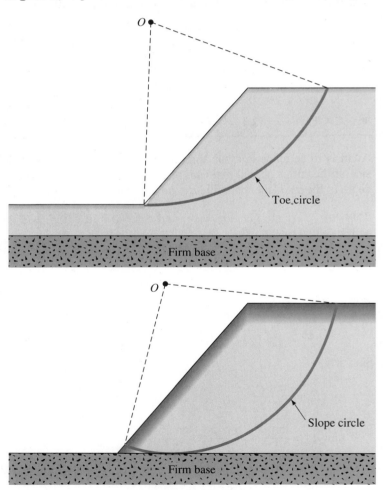

(a) Slope failure

▼ **FIGURE 12.6**   Modes of failure of finite slope

1. When the failure occurs in such a way that the surface of sliding intersects the slope at or above its toe, it is called a *slope failure* (Figure 12.6a). The failure circle is referred to as a *toe circle* if it passes through the toe of the slope and as a *slope circle* if it passes above the toe of the slope. Under certain circumstances, it is possible to have a shallow slope failure as shown in Figure 12.6b.

2. When the failure occurs in such a way that the surface of sliding passes at some distance below the toe of the slope, it is called a *base failure* (Figure 12.6c). The failure circle in the case of base failure is called a *midpoint circle*.

Various procedures of stability analysis may, in general, be divided into two major classes:

1. Mass procedure. In this case, the mass of the soil above the surface of sliding

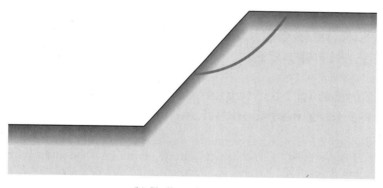

(b) Shallow slope failure

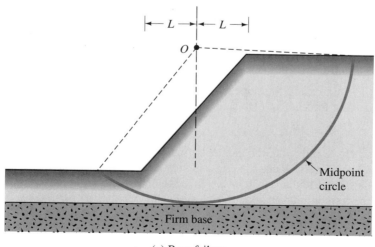

(c) Base failure

▼ **FIGURE 12.6** (*Continued*)

is taken as a unit. This procedure is useful when the soil that forms the slope is assumed to be homogeneous, although this is hardly the case in most natural slopes.

2. Method of slices. In this procedure, the soil above the surface of sliding is divided into a number of vertical parallel slices. The stability of each of the slices is calculated separately. This is a versatile technique in which the non-homogeneity of the soils and pore water pressure can be taken into consideration. It also accounts for the variation of the normal stress along the potential failure surface.

The fundamentals of the analysis of slope stability by mass procedure and method of slices are given in the following sections.

12.6 MASS PROCEDURE OF STABILITY ANALYSIS (CIRCULARLY CYLINDRICAL FAILURE SURFACE)

Slopes in Homogeneous Clay Soil with $\phi = 0$ (Undrained Condition)

Figure 12.7 shows a slope in a homogeneous soil. The undrained shear strength of the soil is assumed to be constant with depth and may be given by $\tau_f = c_u$. To make the stability analysis, we choose a trial potential curve of sliding AED, which is an arc of a circle that has a radius r. The center of the circle is located at O. Considering unit length perpendicular to the section of the slope, we can give the weight of the soil above the curve AED as $W = W_1 + W_2$, where

$$W_1 = (\text{area of } FCDEF)(\gamma)$$

and

$$W_2 = (\text{area of } ABFEA)(\gamma)$$

Failure of the slope may occur by the sliding of the soil mass. The moment of the driving force about O to cause slope instability is

$$M_d = W_1 l_1 - W_2 l_2 \tag{12.42}$$

where l_1 and l_2 are the moment arms.

The resistance to sliding is derived from the cohesion that acts along the potential surface of sliding. If c_d is the cohesion that needs to be developed, then the moment of the resisting forces about O is

$$M_R = c_d(\widehat{AED})(1)(r) = c_d r^2 \theta \tag{12.43}$$

For equilibrium, $M_R = M_d$; thus,

$$c_d r^2 \theta = W_1 l_1 - W_2 l_2$$

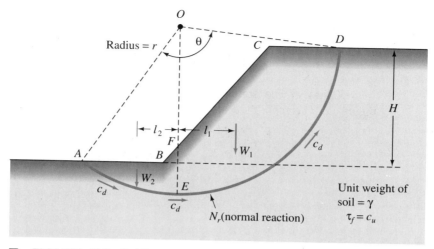

▼ **FIGURE 12.7** Stability analysis of slope in homogeneous clay soil ($\phi = 0$)

or

$$c_d = \frac{W_1 l_1 - W_2 l_2}{r^2 \theta} \qquad (12.44)$$

The factor of safety against sliding may now be found:

$$F_s = \frac{\tau_f}{c_d} = \frac{c_u}{c_d} \qquad (12.45)$$

Note that the potential curve of sliding, *AED*, was chosen arbitrarily. The critical surface is the one for which the ratio of c_u to c_d is a minimum. In other words, c_d is maximum. To find the critical surface for sliding, a number of trials are to be made for different trial circles. The minimum value of the factor of safety thus obtained is the factor of safety against sliding for the slope, and the corresponding circle is the critical circle.

Stability problems of this type have been solved analytically by Fellenius (1927) and Taylor (1937). For the case of *critical circles*, the developed cohesion can be expressed by the relationship

$$c_d = \gamma H m$$

or

$$\boxed{\frac{c_d}{\gamma H} = m} \qquad (12.46)$$

Note that the term *m* in the right-hand side of the preceding equation is non-dimensional and is referred to as the *stability number*. The critical height (that is,

$F_s = 1$) of the slope can be evaluated by substituting $H = H_{cr}$ and $c_d = c_u$ (full mobilization of the undrained shear strength) into the above equation. Thus,

$$H_{cr} = \frac{c_u}{\gamma m} \tag{12.47}$$

Values of the stability number m for various slope angles β are given in Figure 12.8. Terzaghi used the term $\gamma H/c_d$, the reciprocal of m, and called it the *stability*

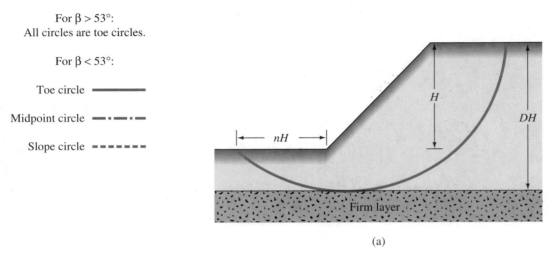

For $\beta > 53°$:
All circles are toe circles.

For $\beta < 53°$:

Toe circle ————

Midpoint circle —·—·—

Slope circle — — — —

(a)

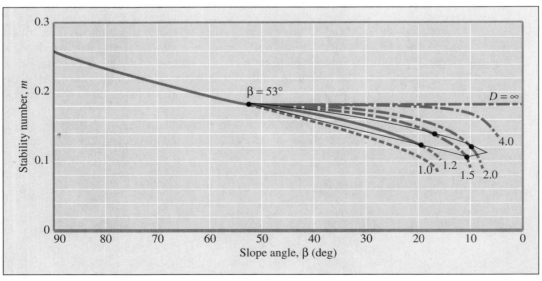

(b)

▼ **FIGURE 12.8** (a) Definition of parameters for midpoint circle-type failure; (b) plot of stability number against slope angle (redrawn from Terzaghi and Peck, 1967)

factor. Readers should be careful in using Figure 12.8 and note that it is valid for slopes of saturated clay and is applicable to only undrained conditions ($\phi = 0$).

In reference to Figure 12.8, the following need to be pointed out:

1. For slope angle β greater than 53°, the critical circle is always a toe circle. The location of the center of the critical toe circle may be found with the aid of Figure 12.9.

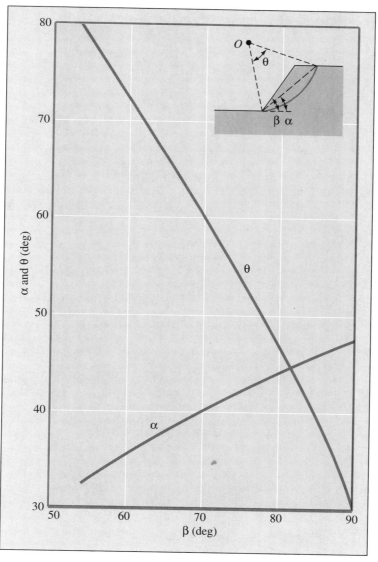

▼ **FIGURE 12.9**   Location of the center of critical circles for $\beta > 53°$

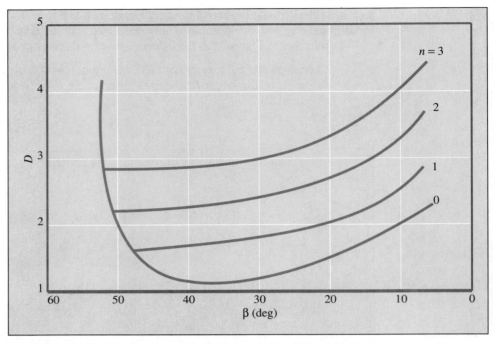

▼ **FIGURE 12.10**   Location of midpoint circles (after Terzaghi and Peck, 1967)

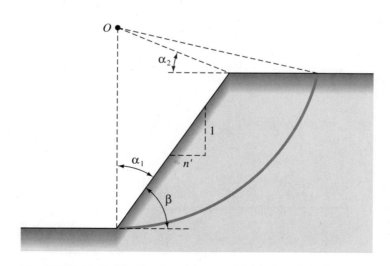

▼ **FIGURE 12.11**   Location of the center of critical toe circles for $\beta < 53°$

▼ **TABLE 12.1**   **Location of the Center of Critical Toe Circles ($\beta < 53°$)**

n	β (deg)	α_1 (deg)	α_2 (deg)
1.0	45	28	37
1.5	33.68	26	35
2.0	26.57	25	35
3.0	18.43	25	35
5.0	11.32	25	37

Note: For notations of n, β, α_1, and α_2, see Figure 12.11.

2. For $\beta < 53°$, the critical circle may be a toe, slope, or midpoint circle, depending on the location of the firm base under the slope. This is called the *depth function*, which is defined as

$$D = \frac{\text{vertical distance from the top of the slope to the firm base}}{\text{height of the slope}} \qquad (12.48)$$

3. When the critical circle is a midpoint circle (that is, the failure surface is tangent to the firm base), its position can be determined with the aid of Figure 12.10.

4. The maximum possible value of the stability number for failure at the midpoint circle is 0.181.

Fellenius (1927) also investigated the case of critical toe circles for slopes with $\beta < 53°$. The location of these can be determined with the use of Figure 12.11 and Table 12.1. Note that these critical toe circles are not necessarily the most critical circles that exist.

▼ **EXAMPLE 12.4**

A cut slope in saturated clay (Figure 12.12) makes an angle of 56° with the horizontal.

a. Determine the maximum depth up to which the cut could be made. Assume that the critical surface for sliding is circularly cylindrical. What will be the nature of the critical circle (that is, toe, slope, or midpoint)?
b. Referring to part (a), determine the distance of the point of intersection of the critical failure circle from the top edge of the slope.
c. How deep should the cut be made if a factor of safety of 2 against sliding is required?

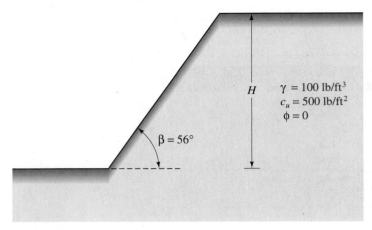

$\gamma = 100\ \text{lb/ft}^3$
$c_u = 500\ \text{lb/ft}^2$
$\phi = 0$

$\beta = 56°$

▼   **FIGURE 12.12**

Solution

a.   Since the slope angle $\beta = 56° > 53°$, the critical circle is a **toe circle**.
From Figure 12.8, for $\beta = 56°$, $m = 0.185$. Using Eq. (12.47), we have

$$H_{cr} = \frac{c_u}{\gamma m} = \frac{500}{(110)(0.185)} = \textbf{24.57 ft}$$

b.   Refer to Figure 12.13. For the critical circle,

$$\overline{BC} = \overline{EF} = \overline{AF} - \overline{AE} = H_{cr}(\cot \alpha - \cot 56°)$$

From Figure 12.9, for $\beta = 56°$, the magnitude of α is 33°, so

$$\overline{BC} = 24.57(\cot 33 - \cot 56) = \textbf{21.25 ft}$$

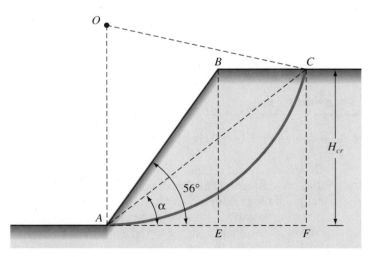

O

B C

H_{cr}

56°

α

A

E F

▼   **FIGURE 12.13**

c.   Developed cohesion is

$$c_d = \frac{c_u}{F_s} = \frac{500}{2} = 250 \text{ lb/ft}^2$$

From Figure 12.8, for $\beta = 56°$, $m = 0.185$. Thus, we have

$$H = \frac{c_d}{\gamma m} = \frac{250}{(110)(0.185)} = \textbf{12.29 ft} \quad \blacktriangledown$$

▼   **EXAMPLE 12.5**

A cut slope was excavated in a saturated clay. The slope made an angle of 40° with the horizontal. Slope failure occurred when the cut reached a depth of 6.1 m. Previous soil explorations showed that a rock layer was located at a depth of 9.15 m below the ground surface. Assume an undrained condition and $\gamma_{sat} = 17.29 \text{ kN/m}^3$.

    a.   Determine the undrained cohesion of the clay (use Taylor's chart, Figure 12.8).

    b.   What was the nature of the critical circle?

    c.   With reference to the toe of the slope, at what distance did the surface of sliding intersect the bottom of the excavation?

Solution

    a.   Referring to Figure 12.8, we find

$$D = \frac{9.15}{6.1} = 1.5$$

$$\gamma_{sat} = 17.29 \text{ kN/m}^3$$

$$H_{cr} = \frac{c_u}{\gamma m}$$

From Figure 12.8, for $\beta = 40°$ and $D = 1.5$, $m = 0.175$, so

$$c_u = (H_{cr})(\gamma)(m) = (6.15)(17.29)(0.175) = \textbf{18.6 kN/m}^2$$

    b.   **Midpoint circle**

    c.   Again, from Figure 12.10, for $D = 1.5$, $\beta = 40°$; $n = 0.9$, so

$$\text{distance} = (n)(H_{cr}) = (0.9)(6.1) = \textbf{5.49 m} \quad \blacktriangledown$$

Slopes in Homogeneous Soil with $\phi > 0$

A slope in a homogeneous soil is shown in Figure 12.14a. The shear strength of the soil is given by

$$\tau_f = c + \sigma \tan \phi$$

The pore water pressure is assumed to be 0. $\overset{\frown}{AC}$ is a trial circular arc that passes through the toe of the slope, and O is the center of the circle. Considering unit length

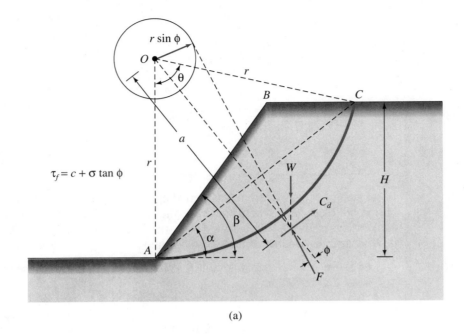

(a)

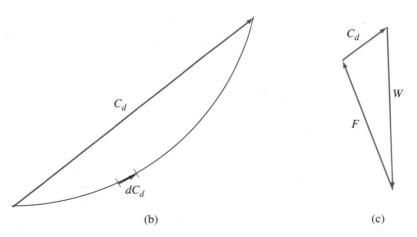

(b)   (c)

▼   **FIGURE 12.14**   Analysis of slopes in homogeneous soils with $\phi > 0$

perpendicular to the section of the slope, we find

weight of the soil wedge $ABC = W =$ (area of ABC)(γ)

For equilibrium, the following other forces are acting on the wedge:

1. C_d—resultant of the cohesive force that is equal to the unit cohesion developed times the length of the cord $\overline{AC}$. The magnitude of C_d is given by (Figure 12.14b).

$$C_d = c_d(\overline{AC}) \tag{12.49}$$

C_d acts in a direction parallel to the cord AC (Figure 12.14b) and at a distance a from the center of the circle O such that

$$C_d(a) = c_d(\widehat{AC})r$$

or

$$a = \frac{c_d(\widehat{AC})r}{C_d} = \frac{\widehat{AC}}{\overline{AC}}r \tag{12.50}$$

2. F—the resultant of the normal and frictional forces along the surface of sliding. For equilibrium, the line of action of F will pass through the point of intersection of the line of action of W and C_d.

Now, if we assume that full friction is mobilized ($\phi_d = \phi$ or $F_\phi = 1$), then the line of action of F will make an angle of ϕ with a normal to the arc, and thus it will be a tangent to a circle with its center at O and having a radius of $r \sin \phi$. This is called the *friction circle*. Actually, the radius of the friction circle is a little larger than $r \sin \phi$.

Since the directions of W, C_d, and F are known and the magnitude of W is known, a force polygon, as shown in Figure 12.14c, can be plotted. The magnitude of C_d can be determined from the force polygon. So the unit cohesion developed can be found:

$$c_d = \frac{C_d}{\overline{AC}}$$

Determination of the magnitude of c_d described previously is based on a trial surface of sliding. Several trials must be made to obtain the most critical sliding surface along which the developed cohesion is a maximum. So it is possible to express the maximum cohesion developed along the critical surface as

$$c_d = \gamma H[\,f(\alpha, \beta, \theta, \phi)\,] \tag{12.51}$$

For critical equilibrium—that is, $F_c = F_\phi = F_s = 1$—we can substitute $H = H_{cr}$ and $c_d = c$ into Eq. (12.51):

$$c = \gamma H_{cr}[\,f(\alpha, \beta, \theta, \phi)\,]$$

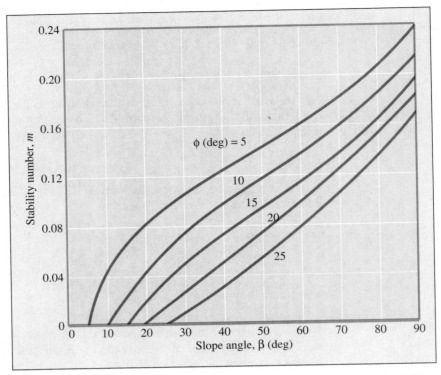

Stability number, m

Slope angle, β (deg)

ϕ (deg) = 5

10

15

20

25

▼   **FIGURE 12.15**   Plot of stability number with slope angle; $\phi > 0$ (after Taylor, 1937)

or

$$\frac{c}{\gamma H_{cr}} = f(\alpha, \beta, \theta, \phi) = m \qquad (12.52)$$

where m = stability number. The values of m for various values of ϕ and β are given in Figure 12.15. Example 12.6 illustrates the use of this chart.

Calculations have shown that, for ϕ greater than about 3°, the critical circles are all *toe circles*. Using Taylor's method of slope stability (as shown in Example 12.6), Singh (1970) provided graphs of equal factors of safety, F_s, for various slopes. These are given in Appendix G. In these charts, the pore water pressure was assumed to be 0.

▼   **EXAMPLE 12.6**

A slope with $\beta = 45°$ is to be constructed with a soil that has $\phi = 20°$ and $c = 23.95$ kN/m². The unit weight of the compacted soil will be 18.87 kN/m³.

a.   Find the critical height of the slope.

b. If the height of the slope is 10 m, determine the factor of safety with respect to strength.

Solution

a. $m = \dfrac{c}{\gamma H_{cr}}$

From Figure 12.15, for $\beta = 45°$ and $\phi = 20°$, $m = 0.06$. So

$$H_{cr} = \frac{c}{\gamma m} = \frac{23.95}{(18.87)(0.06)} = \textbf{21.15 m}$$

b. If we assume that full friction is mobilized, then, referring to Figure 12.15 (for $\beta = 45°$ and $\phi_d = \phi = 20°$), we have

$$m = 0.06 = \frac{c_d}{\gamma H}$$

or

$$c_d = (0.06)(18.87)(10) = 11.32 \text{ kN/m}^2$$

Thus,

$$F_\phi = \frac{\tan \phi}{\tan \phi_d} = \frac{\tan 20}{\tan 20} = 1$$

and

$$F_c = \frac{c}{c_d} = \frac{23.95}{11.32} = 2.12$$

Since $F_c \neq F_\phi$, this is not the factor of safety with respect to strength.

Now we can make another trial. Let the developed angle of friction, ϕ_d, be equal to 15°. For $\beta = 45°$ and the friction angle equal to 15°, we find

$$m = 0.085 = \frac{c_d}{\gamma H} \qquad \text{(Figure 12.15)}$$

or

$$c_d = (0.085)(18.87)(10) = 16.04 \text{ kN/m}^2$$

For this trial,

$$F_\phi = \frac{\tan \phi}{\tan \phi_d} = \frac{\tan 20}{\tan 15} = 1.36$$

and

$$F_c = \frac{c}{c_d} = \frac{23.95}{16.04} = 1.49$$

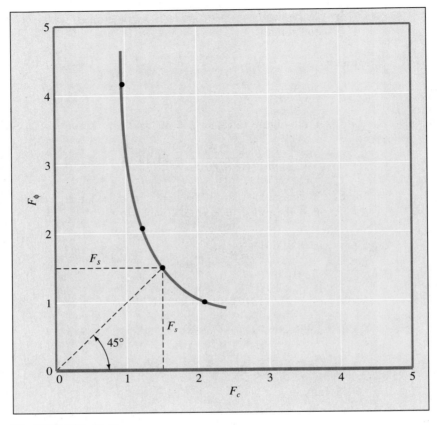

▼   **FIGURE 12.16**

Similar calculations of F_ϕ and F_c for various assumed values of ϕ_d can be made. They are given in the table.

ϕ_d	$\tan \phi_d$	F_ϕ	m	c_d (kN/m²)	F_c
20	0.364	1.0	0.06	11.32	2.12
15	0.268	1.36	0.085	16.04	1.49
10	0.176	2.07	0.11	20.75	1.15
5	0.0875	4.16	0.136	25.66	0.93

The values of F_ϕ are plotted against their corresponding values of F_c in Figure 12.16, from which we find

$$F_c = F_\phi = F_s = \mathbf{1.45} \qquad \blacktriangledown$$

12.7 MASS PROCEDURE FOR STABILITY OF CLAY SLOPE ($\phi = 0$ CONDITION) WITH EARTHQUAKE FORCES

The stability of saturated clay slopes ($\phi = 0$ condition) with earthquake forces has been analyzed by Koppula (1984). Figure 12.17 shows a clay slope with a potential curve of sliding *AED*, which is an arc of a circle that has radius *r*. The center of the circle is located at *O*. Considering unit length perpendicular to the slope, we consider these forces for stability analysis:

1. Weight of the soil wedge, W:

   $$W = (\text{area of } ABCDEA)(\gamma)$$

2. Horizontal inertia force, $k_h W$:

   $$k_h = \frac{\text{horizontal component of earthquake acceleration}}{g}$$

   where g = acceleration from gravity

3. Cohesive force along the surface of sliding, which will have a magnitude of $(\widehat{AED})c_u$

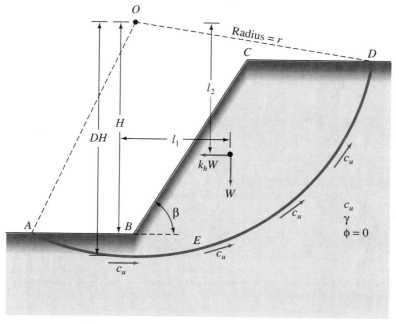

▼ **FIGURE 12.17** Stability analysis of slope in homogeneous clay with earthquake forces ($\phi = 0$ condition)

The moment of the driving forces about O can now be given as

$$M_d = Wl_1 + k_h Wl_2 \tag{12.53}$$

Similarly, the moment of the resisting forces about O is

$$M_r = (\widehat{AED})(c_u)r \tag{12.54}$$

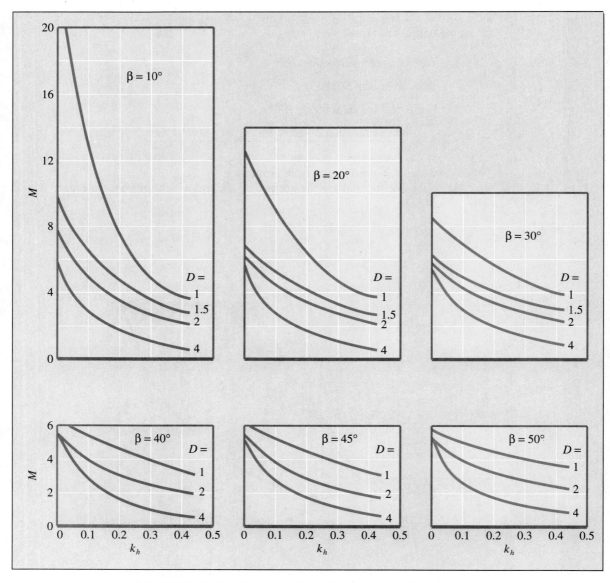

▼ **FIGURE 12.18**   Variation of M with k_h and β based on Koppula's analysis

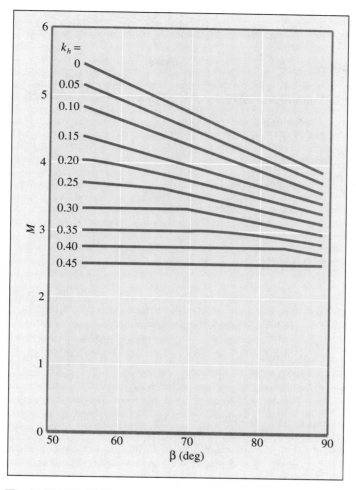

▼ **FIGURE 12.19** Variation of M with k_h based on Koppula's analysis (for $\beta \geq 55°$)

Thus, the factor of safety against sliding is

$$F_s = \frac{M_r}{M_d} = \frac{(\widehat{AED})(c_u)(r)}{Wl_1 + k_h Wl_2} = \frac{c_u}{\gamma H} M \qquad (12.55)$$

where M = stability factor.

The variations of the stability factor M with slope angle β and k_h based on Koppula's (1984) analysis are given in Figures 12.18 and 12.19.

▼ **EXAMPLE 12.7**

Solve parts (a) and (c) of Example 12.4 assuming $k_h = 0.25$.

Solution

a. For the critical height of the slopes, $F_s = 1$. So, from Eq. (12.55),

$$H_{cr} = \frac{c_u M}{\gamma}$$

From Figure 12.19, for $\beta = 56°$ and $k_h = 0.25$, $M = 3.66$. Thus,

$$H_{cr} = \frac{(500)(3.66)}{110} = \textbf{16.64 ft}$$

c. From Eq. (12.55),

$$H = \frac{c_u M}{\gamma F_s} = \frac{(500)(3.66)}{(110)(2)} = \textbf{8.32 ft} \qquad \blacktriangledown$$

12.8 METHOD OF SLICES

Stability analysis by using the method of slices can be explained with the use of Figure 12.20a, in which AC is an arc of a circle representing the trial failure surface. The soil above the trial failure surface is divided into several vertical slices. The width of each slice need not be the same. Considering unit length perpendicular to the cross-section shown, the forces that act on a typical slice (nth slice) are shown in Figure 12.20b. W_n is the weight of the slice. The forces N_r and T_r are the normal and tangential components of the reaction R, respectively. P_n and P_{n+1} are the normal forces that act on the sides of the slice. Similarly, the shearing forces that act on the sides of the slice are T_n and T_{n+1}. For simplicity, the pore water pressure is assumed to be 0. The forces P_n, P_{n+1}, T_n, and T_{n+1} are difficult to determine. However, we can make an approximate assumption that the resultants of P_n and T_n are equal in magnitude to the resultants of P_{n+1} and T_{n+1}, and also that their lines of action coincide.

For equilibrium consideration,

$$N_r = W_n \cos \alpha_n$$

The resisting shear force can be expressed as

$$T_r = \tau_d(\Delta L_n) = \frac{\tau_f(\Delta L_n)}{F_s} = \frac{1}{F_s} [c + \sigma \tan \phi] \, \Delta L_n \tag{12.56}$$

The normal stress σ in Eq. (12.56) is equal to

$$\frac{N_r}{\Delta L_n} = \frac{W_n \cos \alpha_n}{\Delta L_n}$$

For equilibrium of the trial wedge ABC, the moment of the driving force about O equals the moment of the resisting force about O, or

$$\sum_{n=1}^{n=p} W_n r \sin \alpha_n = \sum_{n=1}^{n=p} \frac{1}{F_s} \left(c + \frac{W_n \cos \alpha_n}{\Delta L_n} \tan \phi \right)(\Delta L_n)(r)$$

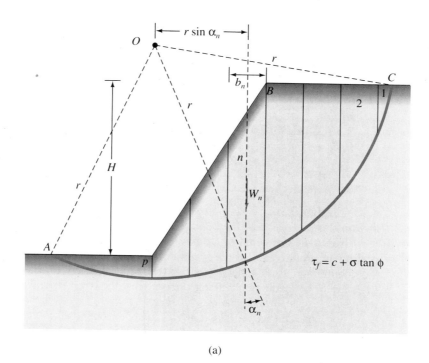

(a)

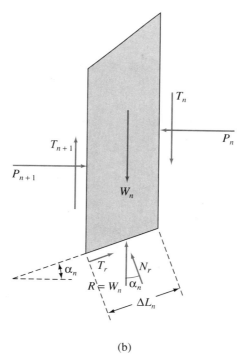

(b)

▼ **FIGURE 12.20** Stability analysis by ordinary method of slices: (a) trial failure surface; (b) forces acting on nth slice

or

$$F_s = \frac{\sum\limits_{n=1}^{n=p} (c \, \Delta L_n + W_n \cos \alpha_n \tan \phi)}{\sum\limits_{n=1}^{n=p} W_n \sin \alpha_n} \tag{12.57}$$

Note: ΔL_n in Eq. (12.57) is approximately equal to $(b_n)/(\cos \alpha_n)$, where $b_n =$ the width of the nth slice.

Note that the value of α_n may be either positive or negative. The value of α_n is positive when the slope of the arc is in the same quadrant as the ground slope. To find the minimum factor of safety—that is, the factor of safety for the critical circle—several trials are to be made by changing the center of the trial circle. This method is generally referred to as the *ordinary method of slices.*

For convenience, a slope in a homogeneous soil is shown in Figure 12.20. However, the method of slices can be extended to slopes with layered soil as shown in Figure 12.21. The general procedure of stability analysis is the same. However, some minor points should be kept in mind. When Eq. (12.57) is used for factor of safety calculation, the values of ϕ and c will not be the same for all slices. For example, for slice no. 3 (Figure 12.21), we have to use a friction angle of $\phi = \phi_3$ and cohesion $c = c_3$; similarly, for slice no. 2, $\phi = \phi_2$ and $c = c_2$.

Bishop's Simplified Method of Slices

In 1955, Bishop proposed a more refined solution to the ordinary method of slices. In this method, the effect of forces on the sides of each slice are accounted for to some

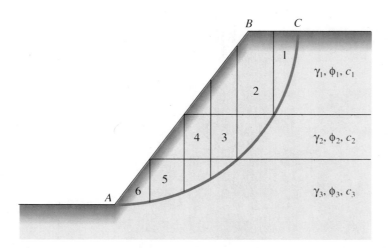

▼ **FIGURE 12.21**   Stability analysis, by ordinary method of slices, for slopes in layered soils

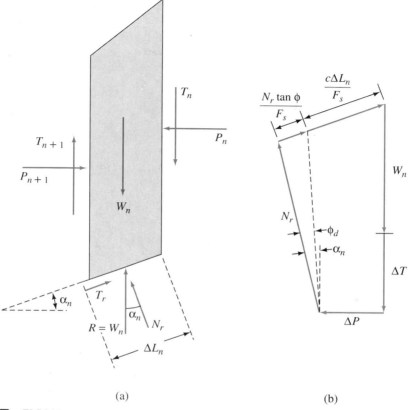

(a)   (b)

▼ **FIGURE 12.22**   Bishop's simplified method of slices: (a) forces acting on the nth slice; (b) force
polygon for equilibrium

degree. We can study this method by referring to the slope analysis presented in
Figure 12.20. The forces that act on the nth slice shown in Figure 12.20b have been
redrawn in Figure 12.22a. Now, let $P_n - P_{n+1} = \Delta P$ and $T_n - T_{n+1} = \Delta T$. Also, we
can write

$$T_r = N_r(\tan \phi_d) + c_d \, \Delta L_n = N_r\left(\frac{\tan \phi}{F_s}\right) + \frac{c \, \Delta L_n}{F_s} \tag{12.58}$$

Figure 12.22b shows the force polygon for equilibrium of the nth slice. Summing
the forces in the vertical direction gives

$$W_n + \Delta T = N_r \cos \alpha_n + \left[\frac{N_r \tan \phi}{F_s} + \frac{c \, \Delta L_n}{F_s}\right] \sin \alpha_n$$

or

$$N_r = \frac{W_n + \Delta T - \dfrac{c \, \Delta L_n}{F_s} \sin \alpha_n}{\cos \alpha_n + \dfrac{\tan \phi \sin \alpha_n}{F_s}} \tag{12.59}$$

For equilibrium of the wedge ABC (Figure 12.20a), taking the moment about O gives

$$\sum_{n=1}^{n=p} W_n r \sin \alpha_n = \sum_{n=1}^{n=p} T_r r \tag{12.60}$$

where $T_r = \dfrac{1}{F_s} (c + \sigma \tan \phi) \Delta L_n$

$$= \frac{1}{F_s} (c \Delta L_n + N_r \tan \phi) \tag{12.61}$$

Substitution of Eqs. (12.59) and (12.61) into Eq. (12.60) gives

$$F_s = \frac{\displaystyle\sum_{n=1}^{n=p} (cb_n + W_n \tan \phi + \Delta T \tan \phi) \dfrac{1}{m_{\alpha(n)}}}{\displaystyle\sum_{n=1}^{n=p} W_n \sin \alpha_n} \tag{12.62}$$

where

$$m_{\alpha(n)} = \cos \alpha_n + \frac{\tan \phi \sin \alpha_n}{F_s} \tag{12.63}$$

For simplicity, if we let $\Delta T = 0$, Eq. (12.62) becomes

$$F_s = \frac{\displaystyle\sum_{n=1}^{n=p} (cb_n' + W_n \tan \phi) \dfrac{1}{m_{\alpha(n)}}}{\displaystyle\sum_{n=1}^{n=p} W_n \sin \alpha_n} \tag{12.64}$$

Note that the term F_s is present on both sides of Eq. (12.64). Hence, a trial-and-error procedure needs to be adopted to find the value of F_s. As in the method of ordinary slices, a number of failure surfaces have to be investigated to find the critical surface that provides the minimum factor of safety.

Bishop's simplified method is probably the most widely used method. When incorporated into computer programs, it yields satisfactory results in most cases. The ordinary method of slices is presented in this chapter as a learning tool. It is rarely used now because it is too conservative.

▼ **EXAMPLE 12.8**

For the slope shown in Figure 12.23, find the factor of safety against sliding for the trial slip surface AC. Use the ordinary method of slices.

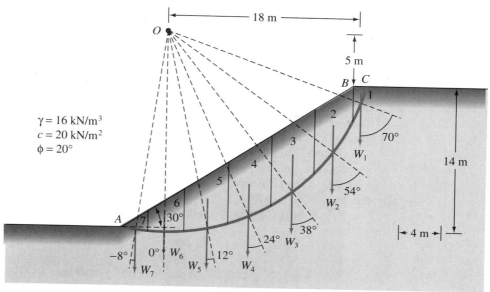

$\gamma = 16 \text{ kN/m}^3$
$c = 20 \text{ kN/m}^2$
$\phi = 20°$

▼ **FIGURE 12.23**

Solution   The sliding wedge is divided into seven slices. Other calculations are shown in the following table:

Slice no. (1)	W (kN/m) (2)	α_n (deg) (3)	$\sin \alpha_n$ (4)	$\cos \alpha_n$ (5)	ΔL_n (m) (6)	$W_n \sin \alpha_n$ (kN/m) (7)	$W_n \cos \alpha_n$ (kN/m) (8)
1	22.4	70	0.94	0.342	2.924	21.1	6.7
2	294.4	54	0.81	0.588	6.803	238.5	173.1
3	435.2	38	0.616	0.788	5.076	268.1	342.94
4	435.2	24	0.407	0.914	4.376	177.1	397.8
5	390.4	12	0.208	0.978	4.09	81.2	381.8
6	268.8	0	0	1	4	0	268.8
7	66.58	−8	−0.139	0.990	3.232	−9.25	65.9
					Σ Col. 6 = 30.501 m	Σ Col. 7 = 776.75 kN/m	Σ Col. 8 = 1638.04 kN/m

$$F_s = \frac{(\Sigma \text{ Col. 6})(c) + (\Sigma \text{ Col. 8}) \tan \phi}{\Sigma \text{ Col 7}}$$

$$= \frac{(30.501)(20) + (1638.04)(\tan 20)}{776.75} = \mathbf{1.55} \qquad \blacktriangledown$$

12.9   STABILITY ANALYSIS BY METHOD OF SLICES FOR STEADY STATE SEEPAGE

The fundamentals of the ordinary method of slices and Bishop's simplified method of slices were presented in Section 12.8, and we assumed the pore water pressure to be 0. However, for steady state seepage through slopes, as is the situation in many practical cases, the pore water pressure has to be taken into consideration when effective shear strength parameters are used. So we need to modify Eqs. (12.57) and (12.64) slightly.

Figure 12.24 shows a slope through which there is steady state seepage. For the nth slice, the average pore water pressure at the bottom of the slice is equal to $u_n = h_n \gamma_w$. The total force caused by the pore water pressure at the bottom of the nth slice is equal to $u_n \Delta L_n$.

Thus, Eq. (12.57) for the ordinary method of slices will be modified to read

$$F_s = \frac{\sum_{n=1}^{n=p} [c \, \Delta L_n + (W_n \cos \alpha_n - u_n \, \Delta L_n)] \tan \phi}{\sum_{n=1}^{n=p} W_n \sin \alpha_n} \qquad (12.65)$$

Similarly, Eq. (12.64) for Bishop's simplified method of slices will be modified to the form

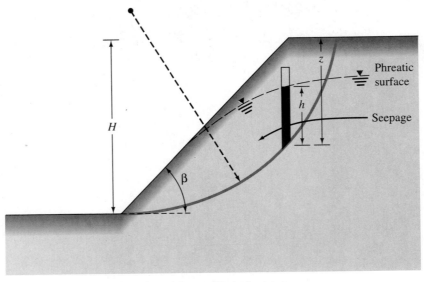

▼ **FIGURE 12.24**   Stability of slopes with steady state seepage

$$F_s = \frac{\sum\limits_{n=1}^{n=p} [cb_n + (W_n - u_n b_n) \tan \phi] \dfrac{1}{m_{(\alpha)n}}}{\sum\limits_{n=1}^{n=p} W_n \sin \alpha_n} \tag{12.66}$$

Note that W_n in Eqs. (12.65) and (12.66) is the total weight of the slice.

Using the method of slices, Bishop and Morgenstern (1960) and Spencer (1967) provided charts to determine the factor of safety of simple slopes that takes into account the effects of pore water pressure. These solutions are given in Sections 12.10 and 12.11.

12.10   BISHOP AND MORGENSTERN'S SOLUTION FOR STABILITY OF SIMPLE SLOPES WITH SEEPAGE

Using Eq. (12.66), Bishop and Morgenstern developed tables for the calculation of F_s for simple slopes. The principles of these developments can be explained as follows. In Eq. (12.66),

$$W_n = \gamma b_n z_n \tag{12.67}$$

where z_n = average height of the nth slice
$u_n = h_n \gamma_w$

So we can let

$$r_{u(n)} = \frac{u_n}{\gamma z_n} = \frac{h_n \gamma_w}{\gamma z_n} \tag{12.68}$$

Note that $r_{u(n)}$ is a nondimensional quantity. Substituting Eqs. (12.67) and (12.68) into Eq. (12.66) and simplifying, we obtain

$$F_s = \left[\frac{1}{\sum\limits_{n=1}^{n=p} \dfrac{b_n}{H} \dfrac{z_n}{H} \sin \alpha_n} \right] \times \sum\limits_{n=1}^{n=p} \left\{ \frac{\dfrac{c}{\gamma H} \dfrac{b_n}{H} + \dfrac{b_n}{H} \dfrac{z_n}{H} [1 - r_{u(n)}] \tan \phi}{m_{\alpha(n)}} \right\} \tag{12.69}$$

For a steady state seepage condition, a weighted average value of $r_{u(n)}$ can be taken, which is a constant. Let the weighted average value of $r_{u(n)}$ be r_u. For most practical cases, the value of r_u may range up to 0.5. So

$$F_s = \left[\frac{1}{\sum\limits_{n=1}^{n=p} \dfrac{b_n}{H} \dfrac{z_n}{H} \sin \alpha_n} \right] \times \sum_{n=1}^{n=p} \left\{ \frac{\left[\dfrac{c}{\gamma H} \dfrac{b_n}{H} + \dfrac{b_n}{H} \dfrac{z_n}{H} \overset{\text{Constant}}{\overset{\downarrow}{(1 - r_u)}} \tan \phi \right]}{m_{\alpha(n)}} \right\} \qquad (12.70)$$

The factor of safety based on the preceding equation can be solved and expressed in the form

$$\boxed{F_s = m' - n' r_u} \qquad (12.71)$$

where m' and $n' =$ stability coefficients. Table G.1 (Appendix G) gives the values of m' and n' for various combinations of $c/\gamma H$, D, ϕ, and β.

In order to determine F_s from Table G.1, the following step-by-step procedure should be used:

1. Obtain ϕ, β, and $c/\gamma H$.
2. Obtain r_u (weighted average value).
3. From Table G.1, obtain the values of m' and n' for $D = 1$, 1.25, and 1.5 (for the required parameters ϕ, β, r_u, and $c/\gamma H$).
4. Determine F_s using the values of m' and n' for each value of D.
5. The required value of F_s is the smallest one obtained in Step 4.

▼ EXAMPLE 12.9

Given:

Slope: 3 horizontal : 1 vertical
$H = 12.63$ m
$\phi = 25°$
$c = 12$ kN/m^2
$\gamma = 19$ kN/m^3
$r_u = 0.25$

determine the minimum factor of safety. Use Bishop and Morgenstern's method.

Solution   Given slope $= 3H : 1V$, $\phi = 25°$, and $r_u = 0.25$, we find

$$\frac{c}{\gamma H} = \frac{12}{(19)(12.63)} = 0.05$$

Using Tables G.1a, b, and c from Appendix G, we can prepare the following table:

D	m'	n'	$F_s = m' - n'r_u$
1	2.193	1.757	1.754
1.25	2.222	1.897	1.748
1.5	2.467	2.179	1.922

So the minimum factor of safety is **1.748**.   ▼

12.11   SPENCER'S SOLUTION FOR STABILITY OF SIMPLE SLOPES WITH SEEPAGE

Bishop's simplified method of slices described in Sections 12.8, 12.9, and 12.10 satisfies the equations of equilibrium with respect to the moment but not with respect to the forces. Spencer (1967) provided a method to determine the factor of safety (F_s) by taking into account the interslice forces (P_n, T_n, P_{n+1}, T_{n+1}, as shown in Figure 12.22), which does satisfy the equations of equilibrium with respect to moment and forces. The details of this method of analysis are beyond the scope of this text; however, the final results of Spencer's work is summarized in this section.

Consider a simple slope of height H as shown in Figure 12.20. Let the average values of unit weight, cohesion, and friction angle be γ, c, and ϕ, respectively, and let the slope be under steady state seepage. Figure 12.25 shows the variation of $c/F_s\gamma H$ for various values of the slope angle β, ϕ_d, and r_u from Spencer's analysis. Note that

$$\phi_d = \tan^{-1}\left(\frac{\tan\phi}{F_s}\right) \tag{12.72}$$

In order to use the charts given by Spencer (1967) and to determine the required value of F_s, the following step-by-step procedure needs to be used:

1. Determine c, γ, H, β, ϕ, and r_u for the given slope.
2. Assume a value of F_s.
3. Calculate $c/[F_{s\,(assumed)}\gamma H]$.
   ↑
   Step 2
4. With the value of $c/F_s\gamma H$ calculated in Step 3 and the slope angle β, enter the proper chart in Figure 12.25 to obtain ϕ_d. Note that Figures 12.25a, b, and c, are, respectively, for r_u values of 0, 0.25, and 0.5, respectively.
5. Calculate $F_s = \tan\phi/\tan\phi_d$.
   ↑
   Step 4

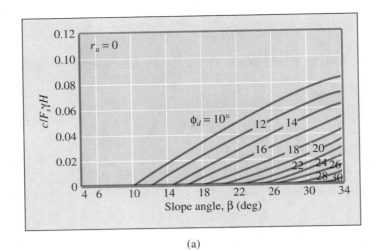

(a)

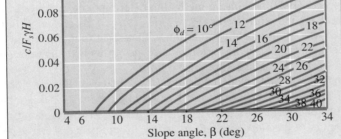

(b)

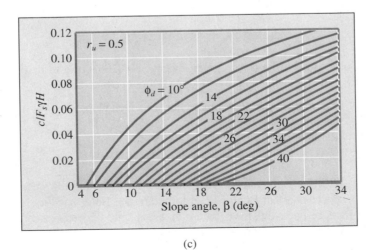

(c)

▼ **FIGURE 12.25** Plot of $c/F_s\gamma H$ against β for various values of ϕ_d (after Spencer, 1967)

6. If the values of F_s as assumed in Step 2 are not the same as those calculated in Step 5, repeat Steps 2, 3, 4, and 5 until they are the same.

▼ **EXAMPLE 12.10**

A given slope under steady state seepage has $H = 21.62$ m, $\phi = 25°$, slope = 2H : 1V, $c = 20$ kN/m², $\gamma = 18.5$ kN/m³, and $r_u = 0.25$. Determine the factor of safety, F_s. Use Spencer's method.

Solution β = slope angle = $\tan^{-1}\left(\dfrac{1}{2}\right) = 26.57°$

Now the following table can be prepared:

β (deg)	$F_{s\,(assumed)}$	$\dfrac{c}{F_{s\,(assumed)}\,\gamma H}$	ϕ_d (deg)[a]	$F_{s\,(calculated)} = \dfrac{\tan\phi}{\tan\phi_d}$
26.57	1.1	0.0455	18	1.435
26.57	1.2	0.0417	19	1.354
26.57	1.3	0.0385	20	1.281
26.57	1.4	0.0375	21	1.247

[a] From Figure 12.25b

Figure 12.26 shows a plot of $F_{s\,(assumed)}$ against $F_{s\,(calculated)}$, from which $F_s \simeq$ **1.3**.

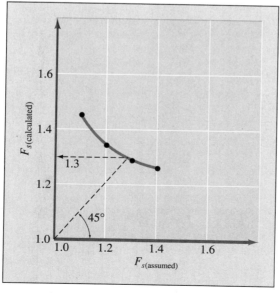

▼ **FIGURE 12.26**

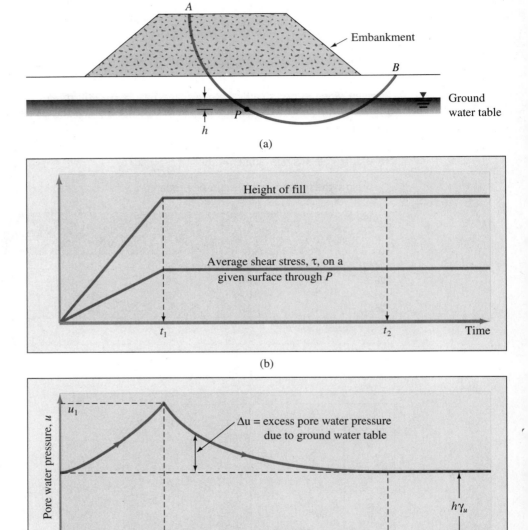

▼ **FIGURE 12.27**   Factor of safety variation with time for embankment on soft clay (redrawn after Bishop and Bjerrum, 1960)

12.12 FLUCTUATION OF FACTOR OF SAFETY OF SLOPES IN CLAY EMBANKMENT ON SATURATED CLAY

Figure 12.27 shows a low clay embankment constructed on a *saturated soft clay*. Let P be a point on a potential failure surface APB that is an arc of a circle. Before construction of the embankment, the pore water pressure at P can be expressed as

$$u = h\gamma_w \tag{12.73}$$

Under ideal conditions, let us assume that the height of the fill needed for the construction of the embankment is placed uniformly as shown in Figure 12.27b. At time $t = t_1$, the embankment height is equal to H, and it remains constant thereafter (that is, $t > t_1$). The average shear stress increase, τ, on the potential failure surface caused by the construction of the embankment is also shown in Figure 12.27b. The value of τ will increase linearly with time up to time $t = t_1$ and remain constant thereafter.

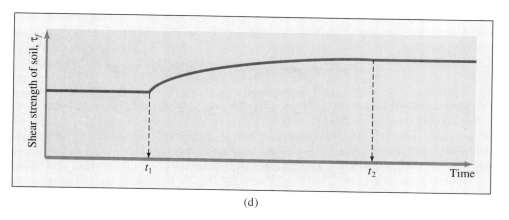

(d)

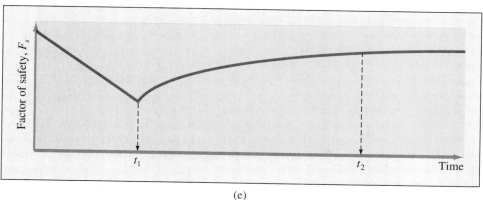

(e)

▼ **FIGURE 12.27** (*Continued*)

The pore water pressure at point P (Figure 12.27a) will continue to increase as construction of the embankment progresses, as shown in Figure 12.27c. At time $t = t_1$, $u = u_1 > h\gamma_w$. This is because of the slow rate of drainage from the clay layer. However, after construction of the embankment is completed (that is, $t > t_1$), the pore water pressure will gradually decrease with time as the drainage, and thus consolidation, progresses. At time $t \simeq t_2$,

$$u = h\gamma_w$$

For simplicity, if we assume that the construction of the embankment is rapid and practically no drainage takes place during the construction period, the average *shear strength* of the clay will remain constant from $t = 0$ to $t = t_1$, or $\tau_f = c_u$ (undrained shear strength). This is shown in Figure 12.27d. For time $t > t_1$, as consolidation progresses, the magnitude of the shear strength τ_f will gradually increase. At time $t \geq t_2$—that is, after consolidation is completed—the average shear strength of the clay will be equal to (Figure 12.27d) $\tau_f = c + \sigma' \tan \phi$ (drained shear strength). The factor of safety of the embankment along the potential surface of sliding can be given as

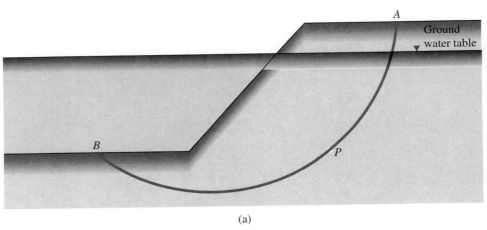

(a)

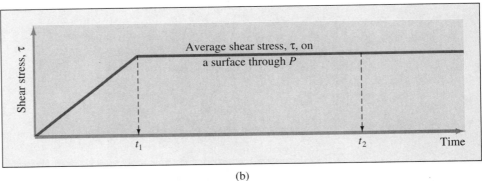

(b)

▼ **FIGURE 12.28**   Variation of factor of safety on cut slopes in soft clay (redrawn after Bishop and Bjerrum, 1960)

$$F_s = \frac{\text{average shear strength of clay along the sliding surface, } \tau_f \text{ (Figure 12.27d)}}{\text{average shear stress, } \tau, \text{ along the sliding surface (Figure 12.27b)}}$$

(12.74)

The general nature of the variation of the factor of safety F_s with time is shown in Figure 12.27e. As can be seen from this figure, the magnitude of F_s initially decreases

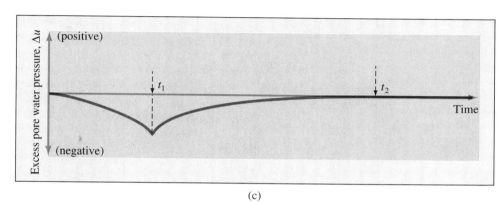

(c)

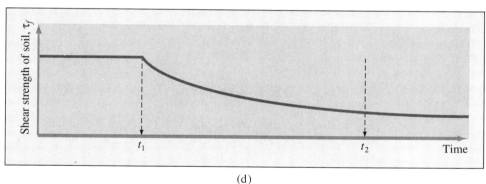

(d)

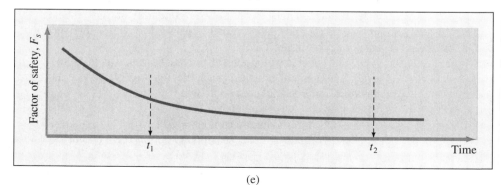

(e)

▼   **FIGURE 12.28**   *(Continued)*

with time. At the end of construction (time $t = t_1$), the value of the factor of safety is a minimum. Beyond this point, the value of F_s continues to increase with drainage up to time $t = t_2$.

Cuts in Saturated Clay

Figure 12.28a shows a cut slope in a saturated soft clay in which APB is a circular potential failure surface. During advancement of the cut, the average shear stress, τ, on the potential failure surface passing through P will increase. The maximum value of the average shear stress τ will be attained at the end of construction—that is, at time $t = t_1$. This is shown in Figure 12.28b.

Because of excavation of the soil, the effective overburden pressure at point P will decrease. This will induce a reduction in the pore water pressure. The variation of the net change of pore water pressure, Δu, is shown in Figure 12.28c. After excavation is complete (time $t > t_1$), the net negative excess pore water pressure will gradually dissipate. At time $t \geq t_2$, the magnitude of Δu will be equal to zero.

The variation of the average shear strength, τ_f, of the clay with time is shown in Figure 12.28d. Note that the shear strength of the soil after excavation gradually decreases. This is because of dissipation of the negative excess pore water pressure.

If the factor of safety of the cut slope F_s along the potential failure surface is defined by Eq. (12.74), its variation will be as shown in Figure 12.28e. Note that the magnitude of F_s decreases with time, and its minimum value is obtained at time $t \geq t_2$.

12.13 A CASE HISTORY OF SLOPE FAILURE

Ladd (1972) reported the results of a study of the failure of a slope that had been constructed over a sensitive clay. The study was conducted in relation to a major improvement program for Interstate Route 95 in Portsmouth, New Hampshire, which is 50 miles north of Boston on the coast. To study the stability of the slope, a test embankment was built to failure during the spring of 1968. The test embankment was heavily instrumented. The general subsoil condition at the test site, the section of the test embankment, and the instruments placed to monitor the performance of the test section are shown in Figure 12.29.

The ground water level at the test section was at an elevation of $+20$ ft (mean sea level). The general physical properties of the soft to very soft gray silty clay layer as shown in Figure 12.29 are as follows:

▶ Natural moisture content = $50 \pm 5\%$
▶ Undrained shear strength as obtained from field vane shear
   tests = 250 ± 50 lb/ft^2 ($\simeq 12 \pm 2.4$ kN/m^2)
▶ Remolded shear strength = 25 ± 5 lb/ft^2 ($\simeq 1.2 \pm 0.24$ kN/m^2)
▶ Liquid limit = 35 ± 5
▶ Plastic limit = 20 ± 2

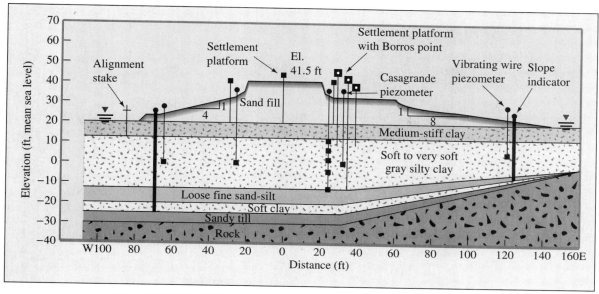

▼ **FIGURE 12.29**   Cross-section through the centerline of the experimental test section looking north (after Ladd, 1972)

During construction of the test embankment, fill was placed at a fairly uniform rate within a period of about one month. Failure of the slope (1 vertical : 4 horizontal) occurred on June 6, 1968, at night. The height of the embankment at failure was 21.5 ft (6.55 m). Figure 12.30 shows the actual failure surface of the slope. The rotated section shown in Figure 12.30 is the "before failure" section rotated through an angle of 13° about a point W 45 ft, El. 51 ft.

Ladd (1972) reported the total stress ($\phi = 0$ concept) stability analysis of the slope that failed by using Bishop's simplified method. The variation of the undrained shear strengths (c_u) used for the stability analysis is given in Table 12.2. Note that these values have not been corrected.

▼ **TABLE 12.2**   **Variation of Undrained Shear Strength with Depth**

Elevation (ft, mean sea level)	c_u as obtained from vane shear strength tests (lb/ft²)
20 to 15	1000
15 to 10	400
10 to 5	240
5 to 0	250
0 to −2.5	300
−2.5 to −5	235
−5 to −10	265
−10 to −13	300

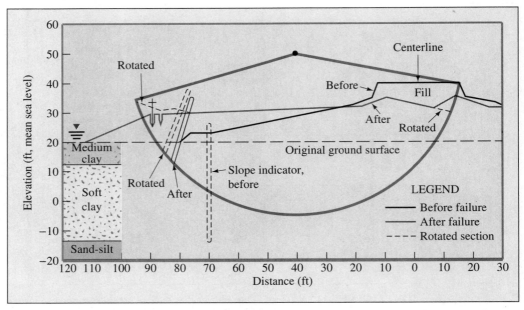

▼ **FIGURE 12.30**   Cross-section of the experimental test section before and after failure (after Ladd, 1972)

The factor of safety, F_s, as obtained from the stability analysis for the critical circle of sliding was 0.88. The critical circle of sliding is shown in Figure 12.31. The factor of safety for the *actual surface of sliding* as obtained by using Bishop's simplified method was 0.92. For comparison purposes, the actual surface of sliding is also shown in Figure 12.31. Note that the bottom of the actual failure surface is about 3 ft (0.91 m) above the theoretically determined critical failure surface.

Readers could check the factor of safety, F_s, for the actual failure surface as a homework problem. The general outline for doing so is as follows.

Figure 12.32 shows a section of the embankment and the actual failure surface. For simplicity, the soil layerings are not shown in this figure. In Figure 12.32, the bottom of slices no. $n = 1$ through m are in sand ($c = 0$). Similarly, the bottom of slices no. $n = m + 1$ to $n = p$ are in clay. Now, for total stress analysis by Bishop's simplified method, Eq. (12.64) may be used. Since, for slices no. 1 through m, $c = 0$, and, for slices no. $m + 1$ through p, $\phi = 0$ and $c = c_u$,

$$F_s = \frac{\sum\limits_{n=1}^{n=m} (W_n \tan \phi)\, \dfrac{1}{m_{(\alpha)n}} + \sum\limits_{n=m+1}^{n=p} (c_u b_n)\, \dfrac{1}{m_{(\alpha)n}}}{\sum\limits_{n=1}^{n=m} W_n \sin \alpha_n \qquad\quad \sum\limits_{n=m+1}^{n=p} W_n \sin \alpha_n} \tag{12.75}$$

Note that W_n in the preceding equation is the total weight of the slice under construction. For calculation of W_n, use the values of the saturated unit in Table 12.3. These values have been provided by Ladd (1972).

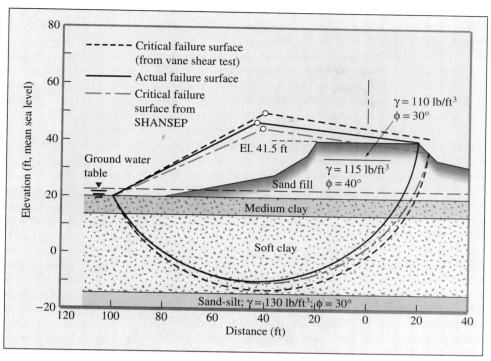

▼ **FIGURE 12.31** Results of total stress stability analysis (after Ladd, 1972) (*Note:* SHANSEP = Stress History and Normalized Soil Engineering Properties)

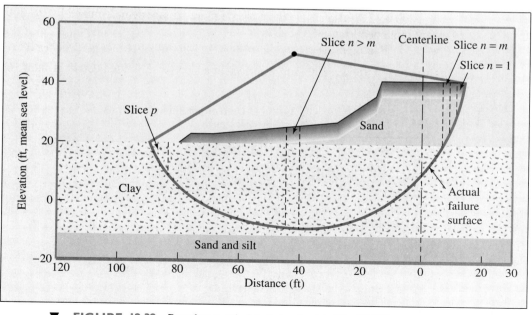

▼ **FIGURE 12.32** Procedure to calculate factor of safety [Eq. (12.75)]

▼ **TABLE 12.3**   **Variation of Unit Weight of Soil with Depth**

Elevation (ft, mean sea level)	Unit weight, γ (lb/ft^3)
20 to 15	118
12.5 to -5	109
-5 to $\simeq -14$	120

Ladd (1972) also reported the stability analysis of the slope based on the average undrained shear strength variation of the clay layer as determined by using the Stress History and Normalized Soil Engineering Properties (SHANSEP). The details of obtaining c_u by this procedure are beyond the scope of this text; however, the final results are given in Table 12.4.

Using the average values of c_u, Bishop's simplified method of stability analysis yields the following results:

Failure surface	Factor of safety, F_s
Actual failure surface	1.02
Critical failure surface	1.01

Figure 12.31 also shows the critical failure surface as determined by using the values of c_u obtained from SHANSEP.

Based on the preceding results, we can draw the following conclusions:

1. The actual failure surface of a slope with limited height is an arc of a circle.

▼ **TABLE 12.4**   **Variation of Undrained Shear Strength with Depth**

Elevation (ft, mean sea level)	Average c_u as obtained from SHANSEP (lb/ft^2)
20 to 15	1000
15 to 10	335
10 to 5	230
5 to 0	260
0 to -2.5	300
-2.5 to -5	320
-5 to -10	335
-10 to -13	400

2. The disagreement between the predicted critical failure surface and the actual failure surface is primarily caused by the shear strength assumptions. The c_u values obtained from SHANSEP give $F_s \simeq 1$, and the critical failure surface is practically the same as the actual failure surface.

This case study is another example that demonstrates the importance of the proper evaluation of soil parameters for predicting the stability of various structures.

12.14 GENERAL COMMENTS

In this chapter, the general principles of analyzing the stability of slopes were presented. Several general observations can be made regarding this subject.

1. Culmann's method of slope analysis using a planar failure surface passing through the toe of the slope was presented in Section 12.4. The accuracy of this method is good for steep slopes or vertical banks; however, it will underestimate the factor of safety for flatter slopes. A planar failure surface can be used to analyze slope stability when a sloping mass of soil rests on an inclined layer of stiff soil or rock (Figure 12.33). When water seeps from the top and then along the plane of discontinuity, it will weaken the soil along the plane of contact and thus may make the upper sloping soil mass unstable.

2. Several other mechanisms by which slopes may fail were not covered in this chapter. Figure 12.34a shows a *noncircular failure surface* in a slope. Another type of slope failure is shown in Figure 12.34b, which is called *sliding block failure*.

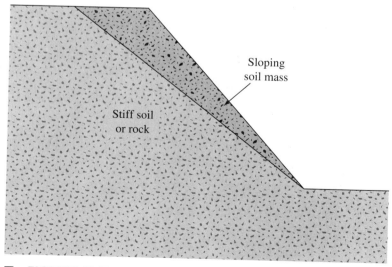

Sloping
soil mass

Stiff soil
or rock

▼ **FIGURE 12.33** Planar failure surface in slope stability analysis

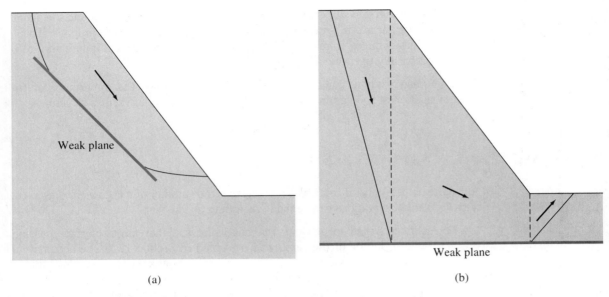

▼ **FIGURE 12.34**   (a) Noncircular failure surface in a slope; (b) sliding block failure

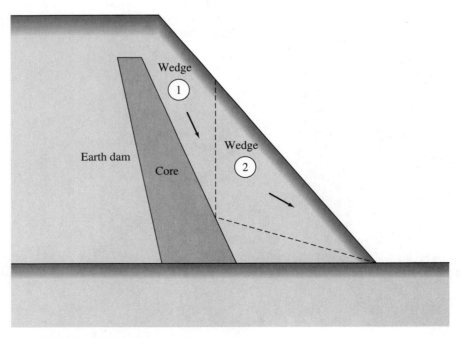

▼ **FIGURE 12.35**   Wedge analysis

In many situations, the potential failure surface can be approximated by several straight lines. An example of this configuration is shown in Figure 12.35. The soil mass in the potential failure zone is broken up into several wedges and the stability is then analyzed.

3. For a short-term stability analysis of nonfissured clay, the total stress method assuming $\phi = 0$ condition will be satisfactory. However, for long-term slope stability analysis, the effective stress method of analysis should be used. For overconsolidated clay slope analysis, the residual strength of the soil should be used. It was shown in Chapter 9 that, for overconsolidated clays, the stress-strain plots will be of the nature shown in Figure 12.36a. Figure 12.36b shows the peak and residual effective strength envelopes for an overconsolidated clay. Note that the residual strength can be approximately expressed as

$$\tau_r = \sigma' \tan \phi_r \tag{12.76}$$

For normally consolidated clays, the difference between peak and residual strengths is small.

4. Section 12.8 provided the procedure for stability analysis by the method of slices. Several computer programs are now available for quick routine analysis using Bishop's simplified method of slices.

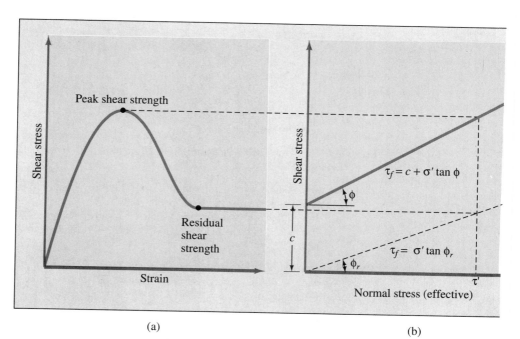

(a)   (b)

▼ **FIGURE 12.36**   Peak and residual strengths of overconsolidated clay

PROBLEMS **12.1** For the slope shown in Figure 12.37, find the height, H, for critical equilibrium given $\beta = 25°$.

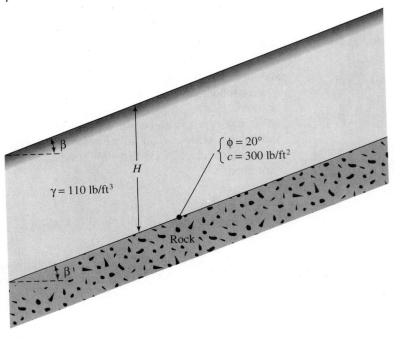

$\phi = 20°$
$c = 300$ lb/ft^2

H

$\gamma = 110$ lb/ft^3

Rock

▼ **FIGURE 12.37**

12.2 Refer to Figure 12.37.
  a. If $\beta = 25°$ and $H = 10$ ft, what is the factor of safety of the slope against sliding along the soil-rock interface?
  b. For $\beta = 30°$, find the height, H, that will have a factor of safety of 1.5 against sliding along the soil-rock interface.

12.3 Refer to Figure 12.37. Plot a graph of H_{cr} against the slope angle β (for β varying from 15° to 30°).

12.4 An infinite slope is shown in Figure 12.38. The shear strength parameters at the interface of soil and rock are $c = 18$ kN/m^2 and $\phi = 25°$.
  a. If $H = 8$ m and $\beta = 20°$, find the factor of safety against sliding on the rock surface.
  b. If $\beta = 30°$, find the height, H, for which $F_s = 1$. (Assume the pore water pressure to be 0.)

12.5 Refer to Figure 12.38. If there were seepage through the soil and the ground water table coincided with the ground surface, what would be the value of F_s? Use $H = 8$ m, $\rho_{sat} = 1900$ kg/m^3, and $\beta = 20°$.

12.6 For the infinite slope shown in Figure 12.39, find the factor of safety against sliding along the plane AB given $H = 10$ ft. Note that there is seepage through the soil, and the ground water table coincides with the ground surface.

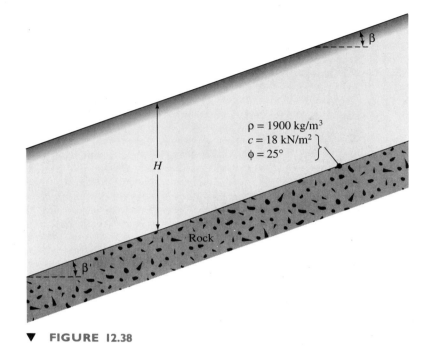

$\rho = 1900 \text{ kg/m}^3$
$c = 18 \text{ kN/m}^2$
$\phi = 25°$

Rock

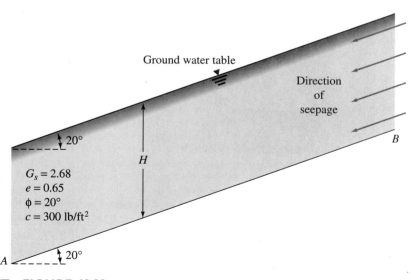

Ground water table

Direction
of
seepage

B

$20°$

H

$G_s = 2.68$
$e = 0.65$
$\phi = 20°$
$c = 300 \text{ lb/ft}^2$

$20°$

A

12.7   A slope is shown in Figure 12.40. AC represents a trial failure plane. For the wedge ABC, find the factor of safety against sliding.

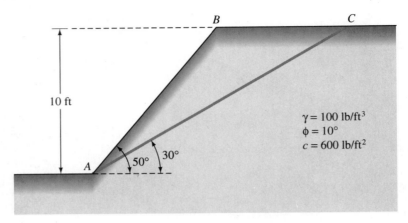

▼ **FIGURE 12.40**

12.8   A finite slope is shown in Figure 12.41. Assuming that the slope failure would occur along a plane (Culmann's assumption), find the height of the slope for critical equilibrium given $\phi = 10°$, $c = 250$ lb/ft^2, $\gamma = 110$ lb/ft^2, and $\beta = 50°$.

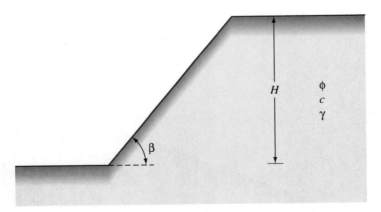

▼ **FIGURE 12.41**

12.9   Repeat Problem 12.8 with $\phi = 20°$, $c = 25$ kN/m^2, $\gamma = 18$ kN/m^3, and $\beta = 45°$.

12.10   Refer to Figure 12.41. Using the soil parameters given in Problem 12.8, find the height of the slope, H, that will have a factor of safety of 2.5 against sliding. Assume that the critical surface for sliding is a plane.

12.11   Refer to Figure 12.41. Given $\phi = 15°$, $c = 200$ lb/ft^2, $\gamma = 115$ lb/ft^3, $\beta = 60°$, and $H = 9$ ft, determine the factor of safety with respect to sliding. Assume that the critical surface for sliding is a plane.

12.12   Refer to Problem 12.11. Find the height of the slope, H, that will have $F_s = 1.5$. Assume that the critical surface for sliding is a plane.

12.13   A cut slope is to be made in a soft clay with its sides rising at an angle of 75° to the horizontal (Figure 12.42). Assume that $c_u = 650$ lb/ft^2 and $\gamma = 110$ lb/ft^3.

   a.   Determine the maximum depth up to which the excavation can be carried out.

   b.   Find the radius, r, of the critical circle when the factor of safety is equal to 1 [part (a)].

   c.   Find the distance $\overline{BC}$.

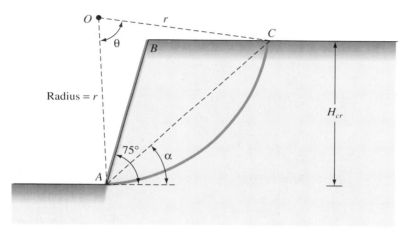

▼ **FIGURE 12.42**

12.14   If the cut described in Problem 12.13 is made to a depth of only 10 ft, what will be the factor of safety of the slope against sliding?

12.15   Using the graph given in Figure 12.8, determine the height of a slope, 1 vertical to $\frac{1}{2}$ horizontal, in saturated clay that has an undrained shear strength of 32.55 kN/m^2. The desired factor of safety against sliding is 2. Assume $\gamma = 18.9$ kN/m^3.

12.16   Refer to Problem 12.15. What should be the critical height of the slope? What will be the nature of the critical circle? Also find the radius of the critical circle.

12.17   For the slope shown in Figure 12.43, find the factor of safety against sliding for the trial surface $\widehat{AC}$.

12.18   A cut slope was excavated in a saturated clay. The slope angle β is equal to 40° with respect to the horizontal. Slope failure occurred when the cut reached a depth of 6.1 m. Previous soil explorations showed that a rock layer was located at a depth of 9.15 m below the ground surface. Assume an undrained condition and $\gamma_{sat} = 17.29$ kN/m^3.

   a.   Determine the undrained cohesion of the clay (use Taylor's chart for $\phi = 0$, Figure 12.8).

   b.   What was the nature of the critical circle?

   c.   With reference to the toe of the slope, at what distance did the surface of sliding intersect the bottom of the excavation?

12.19   If the cut slope described in Problem 12.18 is to be excavated in a manner such that $H_{cr} = 7.62$ m, what angle should the slope make with the horizontal? (Use Taylor's chart and the results of Problem 12.18a.)

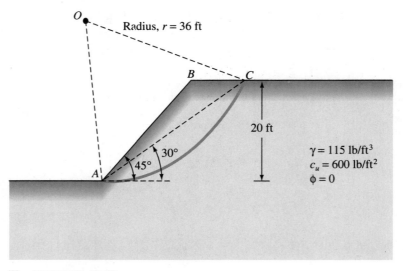

▼ **FIGURE 12.43**

12.20   Refer to Figure 12.44. Use Taylor's chart for $\phi > 0$ (Figure 12.15) to solve the following:
   a.   If $n' = 2$, $\phi = 15°$, $c = 650$ lb/ft^2, and $\gamma = 115$ lb/ft^3, find the critical height of the slope.
   b.   If $n' = 1$, $\phi = 25°$, $c = 500$ lb/ft^2, and $\gamma = 115$ lb/ft^3, find the critical height of the slope.
   c.   If $n' = 2.5$, $\phi = 12°$, $c = 25$ kN/m^2, and $\gamma = 17$ kN/m^3, find the critical height of the slope.
   d.   If $n' = 1.5$, $\phi = 18°$, $c = 18$ kN/m^2, and $\gamma = 16.5$ kN/m^3, find the critical height of the slope.

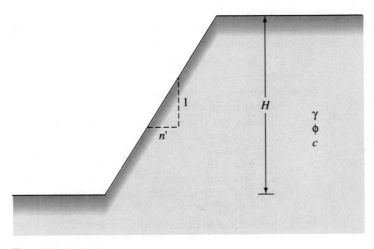

▼ **FIGURE 12.44**

12.21 Referring to Figure 12.44 and using Figure 12.15, find the factor of safety with respect to sliding for the following cases:

    a. $n' = 2$, $\phi = 15°$, $c = 700$ lb/ft², $\gamma = 115$ lb/ft³, and $H = 45$ ft

    b. $n' = 1$, $\phi = 20°$, $c = 400$ lb/ft², $\gamma = 115$ lb/ft³, and $H = 30$ ft

    c. $n' = 2.5$, $\phi = 12°$, $c = 23.94$ kN/m², $\gamma = 16.51$ kN/m³, and $H = 10$ m

    d. $n' = 1.5$, $\phi = 15°$, $c = 18$ kN/m², $\gamma = 16.5$ kN/m³, and $H = 6$ m

12.22 Refer to Figure 12.44 and Figure G.1 (Appendix G).

    a. If $n' = 2$, $\phi = 10°$, $c = 700$ lb/ft², and $\gamma = 110$ lb/ft³, draw a graph of the height of the slope, H, against F_s (varying from 1 to 3).

    b. If $n' = 1$, $\phi = 15°$, $c = 18$ kN/m², and $\gamma = 17.1$ kN/m³, draw a graph of the height of the slope, H, against F_s (varying from 1 to 3).

12.23 A clay slope is built over a layer of rock. For the slope:

Height = 16 m

Slope angle, $\beta = 30°$

Saturated unit weight of soil = 17 kN/m³

Undrained shear strength, $c_u = 50$ kN/m²

Determine the factor of safety if $k_h = 0.4$. (Use the procedure outlined in Section 12.7.)

12.24 For a slope in clay, $H = 50$ ft, $\gamma = 115$ lb/ft³, $\beta = 60°$, and $c_u = 1000$ lb/ft². Determine the factor of safety for $k_h = 0.3$. (Use the procedure outlined in Section 12.7.)

12.25 Referring to Figure 12.45 and using the ordinary method of slices, find the factor of safety against sliding for the following trial cases:

    a. $\beta = 45°$, $\phi = 20°$, $c = 400$ lb/ft², $\gamma = 115$ lb/ft³, $H = 40$ ft, $\alpha = 30°$, and $\theta = 70°$

    b. $\beta = 45°$, $\phi = 15°$, $c = 18$ kN/m², $\gamma = 17.1$ kN/m³, $H = 5$ m, $\alpha = 30°$, and $\theta = 80°$

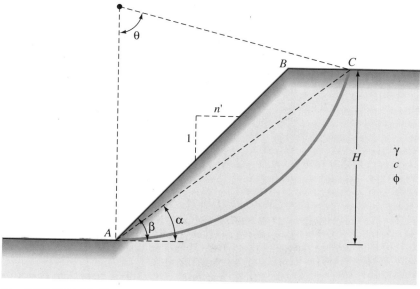

▼ **FIGURE 12.45**

12.26 Determine the minimum factor of safety of a slope with the following parameters: $H = 20$ ft, $\beta = 26.57°$, $\phi = 25°$, $c = 115$ lb/ft², $\gamma = 115$ lb/ft³, and $r_u = 0.5$. Use Bishop and Morgenstern's method.

12.27   Repeat Problem 12.26 with the following values:
Slope: 3 horizontal : 1 vertical
$H = 12.63$ m
$\phi = 25°$
$c = 12$ kN/m^2
$\gamma = 19$ kN/m^3
$r_u = 0.25$

12.28   Repeat Problem 12.26 using Spencer's chart.

12.29   Repeat Problem 12.27 using Spencer's chart.

12.30   Use Spencer's chart to determine the value of F_s for the given slope: $\beta = 20°$, $H = 15$ m, $\phi = 15°$, $c = 20$ kN/m^2, $\gamma = 17.5$ kN/m^3, $r_u = 0.5$.

REFERENCES

Bishop, A. W. (1955). "The Use of Slip Circle in the Stability Analysis of Earth Slopes," *Geotechnique*, Vol. 5, No. 1, 7–17.

Bishop, A. W., and Bjerrum, L. (1960). "The Relevance of the Triaxial Test to the Solution of Stability Problems," *Proceedings*, Research Conference on Shear Strength of Cohesive Soils, ASCE, 437–501.

Bishop, A. W., and Morgenstern, N. R. (1960). "Stability Coefficients for Earth Slopes," *Geotechnique*, Vol. 10, No. 4, 129–147.

Culmann, C. (1875). *Die Graphische Statik*, Meyer and Zeller, Zurich.

Fellenius, W. (1927). *Erdstatische Berechnungen*, revised edition, W. Ernst u. Sons, Berlin.

Koppula, S. D. (1984). "Pseudo-Static Analysis of Clay Slopes Subjected to Earthquakes," *Geotechnique*, Vol. 34, No. 1, 71–79.

Ladd, C. C. (1972). "Test Embankment on Sensitive Clay," *Proceedings*, Conference on Performance of Earth and Earth-Supported Structures, ASCE, Vol. 1, Part 1, 101–128.

Singh, A. (1970). "Shear Strength and Stability of Man-Made Slopes," *Journal of the Soil Mechanics and Foundations Division*, ASCE, Vol. 96, No. SM6, 1879–1892.

Spencer, E. (1967). "A Method of Analysis of the Stability of Embankments Assuming Parallel Inter-Slice Forces," *Geotechnique*, Vol. 17, No. 1, 11–26.

Taylor, D. W. (1937). "Stability of Earth Slopes," *Journal of the Boston Society of Civil Engineers*, Vol. 24, 197–246.

Terzaghi, K., and Peck, R. B. (1967). *Soil Mechanics in Engineering Practice*, 2nd ed., Wiley, New York.

Supplementary References for Further Study

Cousins, B. F. (1978). "Stability Charts for Simple Earth Slopes," *Journal of the Geotechnical Engineering Division*, ASCE, Vol. 104, No. GT2, 267–279.

Morgenstern, N. R. (1963). "Stability Charts for Earth Slopes During Rapid Drawdown," *Geotechnique*, Vol. 13, No. 2, 121–133.

Morgenstern, N. R., and Price, V. E. (1965). "The Analysis of the Stability of General Slip Surfaces," *Geotechnique*, Vol. 15, No. 1, 79–93.

O'Connor, M. J., and Mitchell, R. J. (1977). "An Extension of the Bishop and Morgenstern Slope Stability Charts," *Canadian Geotechnical Journal*, Vol. 14, No. 1, 144–151.

ENVIRONMENTAL GEOTECHNOLOGY

Enormous amounts of solid waste are generated every year in the United States and other industrialized countries. These waste materials can, in general, be classified into four major categories: (1) municipal waste, (2) industrial waste, (3) hazardous waste, and (4) low-level radioactive waste. Table 13.1 lists the waste material generated in 1984 in the United States in the above categories (Koerner, 1990).

The waste materials are generally placed in landfills. The landfill materials interact with moisture received from rainfall and snow to form a liquid called *leachate*. The chemical composition of leachates varies widely depending on the waste material involved. Leachates are a main source of ground water pollution; therefore, it is important that leachates be properly contained in all landfills, surface impoundments, and waste piles by using some type of liner system. In the following sections of this chapter, various types of liner systems and the materials used in them are discussed.

13.1 LANDFILL LINERS—OVERVIEW

Until about 1982, the predominant liner material used in landfills was clay. Proper clay liners will have a coefficient of permeability of about 10^{-7} cm/sec or less. In 1984, the U.S. Environmental Protection Agency's minimum technological requirements for hazardous waste landfill design and construction were introduced by the U.S. Congress in Hazardous and Solid Waste Amendments. In those amendments, Congress required that all new landfills should have double liners and systems for leachate collection and removal.

In order to understand the construction and functioning of the double liner system, it is necessary to review the general properties of the component materials involved in the double liner system—that is, clay soil and geosynthetics (such as geotextiles, geomembranes, and geonets). These materials are briefly described in the following sections.

▼ **TABLE 13.1   Waste Material Generation in the United States**

Waste type	Approximate quantity in 1984 (millions of metric tons)
Municipal	300
Industrial (building debris, degradable waste, nondegradable waste, and near hazardous)	600
Hazardous	150
Low-level radioactive	15

13.2   CLAY

The U.S. Environmental Protection Agency requires that clay liners have a coefficient of permeability of 10^{-7} cm/sec or less. In order to achieve this value, the soil should meet the following criteria (EPA, 1989):

1. The soil should have at least 20% fines (fine silt and clay-size particles).
2. The plasticity index (*PI*) should be greater than 10. Soils that have a *PI* greater than about 30 are difficult to work with in the field.
3. The soil should not have more than 10% gravel-size particles.
4. The soil should not contain any particles or chunks of rock that are larger than 1–2 in. ($\approx$ 25–50 mm).

In many instances, the soil found at the construction site may be somewhat non-plastic. Such soil may be blended with imported clay minerals like sodium bentonite to achieve the desired properties as a clay liner. Figure 13.1 shows the effect of sodium bentonite when used as an admixture in reducing the coefficient of permeability, k, of a silty sand.

The important factors that control the coefficient of permeability, k, of a compacted clay are as follows (Mitchell and Jaber, 1990):

I. Moisture Content and Dry Unit Weight of Compaction

Figure 13.2 shows the variation of the dry unit weight of compaction and the coefficient of permeability, k, with the molding water content for a clayey soil. This was also shown in Chapter 4. From this figure it can be seen that, when the *in situ* compaction is such that it results in a high dry unit weight with a moisture content on the wet side of optimum, the soil will have a low coefficient of permeability. It has also been observed that a heavy sheepsfoot roller that introduces large shear strain during compaction creates a more dispersed structure in the soil. This type of compacted soil will have a lower coefficient of permeability. It is advisable that small lifts be used during com-

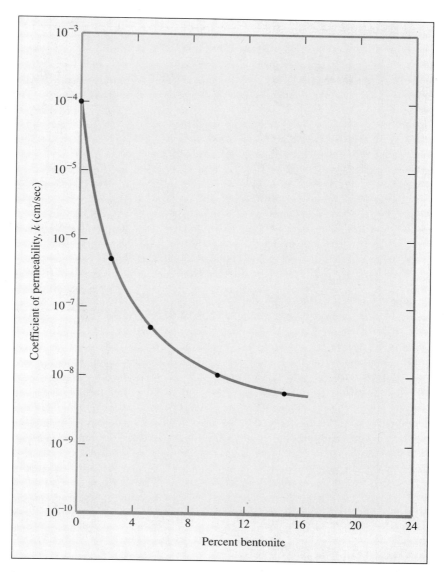

▼ **FIGURE 13.1** Effect of sodium bentonite on the coefficient of permeability of a silty sand (after Environmental Protection Agency, 1989)

paction, by which the feet of the compactor can penetrate the full depth of the lift (Figure 13.3).

2. Size of Clay Clods

The size of the clay clods has a great influence on the coefficient of permeability of a compacted clay liner. This fact is illustrated in Table 13.2 for a clay in Houston, Texas.

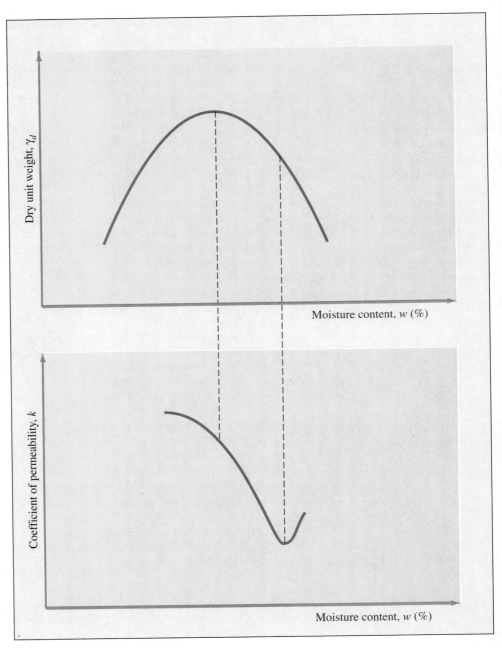

▼ **FIGURE 13.2**   Nature of variation of dry unit weight and coefficient of permeability with molding moisture content

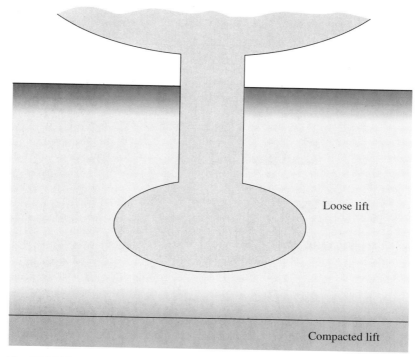

▼ FIGURE 13.3 Use of small lifts for compaction

From this table, it can be seen that larger clod sizes can increase the coefficient of permeability by several orders of magnitude. However, the degree of increase in k decreases with the increase in the molding moisture content. Hence, in the preparation of clay liners, it is important that clods are broken down mechanically to as small as possible. A very heavy roller used for compaction helps to break down the clods.

▼ TABLE 13.2 Influence of Clod Size on the Coefficient of Permeability of a Houston Clay*

Molding moisture content (%)	Coefficient of permeability, k (cm/sec)	
	0.2-in. ($\approx$5-mm) clods	0.75-in. ($\approx$19-mm) clods
12	2×10^{-8}	4×10^{-4}
16	2×10^{-9}	1×10^{-3}
18	1×10^{-9}	8×10^{-10}
20	2×10^{-9}	7×10^{-10}

* After Environmental Protection Agency (1989)

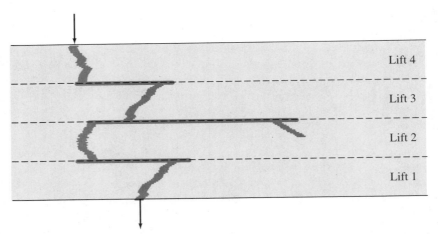

Lift 4

Lift 3

Lift 2

Lift 1

▼ **FIGURE 13.4** Pattern of flow through a clay liner with improper bonding between lifts (after Environmental Protection Agency, 1989)

3. Bonding Between Lifts

Bonding between successive lifts is also an important factor; otherwise, leachates can move through a vertical crack in the liner and then travel along the interface between two lifts until it finds another crack, as is schematically shown in Figure 13.4. This process can substantially reduce the overall coefficient of permeability of a clay liner. An example was seen in a trial pad construction in Houston in 1986. The trial pad was 3 ft (0.91 m) thick and built in six, 6-in. (152-mm) lifts. The results of the coefficient of permeability tests for the compact soil from the trial pad are given in Table 13.3. Note that, although the laboratory-determined values of k for various lifts are on the order of 10^{-7} to 10^{-9} cm/sec, the actual overall value of k increased to the order of 10^{-4}. For that reason, scarification and control of the moisture content after compaction of each lift are extremely important in constructing liners and achieving the desired coefficient of permeability.

▼ **TABLE 13.3 Hydraulic Conductivity from Houston Liner Tests***

Location	Sample	Laboratory k (cm/sec)
Lower lift	3-in. ($\approx$76-mm) tube	4×10^{-9}
Upper lift	3-in. ($\approx$76-mm) tube	1×10^{-9}
Lift interface	3-in. ($\approx$76-mm) tube	1×10^{-7}
Lower lift	Block	8×10^{-5}
Upper lift	Block	1×10^{-8}
Actual overall $k = 1 \times 10^{-4}$ cm/sec		

* After Environmental Protection Agency (1989)

13.3   GEOSYNTHETICS

In general, geosynthetics are fabriclike material made from polymers such as polyester, polyethylene, polypropylene, polyvinyl chloride, nylon, chlorinated polyethylene, and others. The term *geosynthetics* includes the following:

1. Geotextiles
2. Geomembranes
3. Geogrids
4. Geonets
5. Geocomposites

Each type of geosynthetic performs one or more of the following five major functions:

1. Separation
2. Reinforcement
3. Filtration
4. Drainage
5. Moisture barrier

The use of geosynthetics in civil engineering construction is about 15 years old and it is presently growing at a very rapid pace. It is not possible to provide detailed descriptions of manufacturing procedures, properties, and uses of all types of geosynthetics in this chapter. However, a broad overview of geotextiles, geomembranes, and geonets is given. For further information, readers are directed to a geosynthetics text (e.g. Koerner, 1990).

13.4   GEOTEXTILES

Geotextiles are textiles in the traditional sense; however, the fabrics are usually made from petroleum products such as polyester, polyethylene, and polypropylene. They may also be made from fiberglass. Geotextiles are not prepared from natural fabrics because these decay too quickly. They may be *woven, knitted,* or *nonwoven.*

Woven geotextiles are made of two sets of parallel filaments or strands of yarn systematically interlaced to form a planar structure. *Knitted geotextiles* are formed by interlocking a series of loops of one or more filaments or strands of yarn to form a planar structure. *Nonwoven geotextiles* are formed from filaments or short fibers arranged in an oriented or random pattern in a planar structure. These filaments, or short fibers, are first arranged into a loose web. They are then bonded using one or a combination of the following processes:

1. *Chemical bonding*—by glue, rubber, latex, cellulose derivative, and so forth
2. *Thermal bonding*—by heat for partial melting of filaments
3. *Mechanical bonding*—by needle punching

The *needle-punched nonwoven* geotextiles are thick and have high in-plane permeability.

There are four major uses of geotextiles:

1. *Drainage*: The fabrics can rapidly channel water from soil to various outlets.
2. *Filtration*: When placed between two soil layers, one coarse-grained and the other fine-grained, the fabric allows free seepage of water from one layer to the other. At the same time, it protects the fine-grained soil from being washed into the coarse-grained soil.
3. *Separation*: Geotextiles help keep various soil layers separate after construction. For example, in the construction of highways, a clayey subgrade can be kept separate from a granular base course.
4. *Reinforcement*: The tensile strength of geotextiles increases the load-bearing capacity of the soil.

Geotextiles presently available commercially have thicknesses that vary from about 0.01 to 0.3 in. (0.25 to 7.6 mm). The mass per unit area of these geotextiles ranges from about 150 to 700 g/cm^2.

One of the major functions of geotextiles is for filtration. For that, it is important that water be able to flow freely through the fabric of the geotextile (Figure 13.5). Hence, the *cross-plane permeability* is an important parameter for design purposes. It should be realized that geotextile fabrics are compressible, however, and their thickness may change depending on the effective normal stress to which they are being subjected. The change in thickness under normal stress will also change the cross-plane permeability of a geotextile. So, the cross-plane capability is generally expressed in terms of a quantity called *permittivity*, or

$$P = \frac{k_n}{t} \tag{13.1}$$

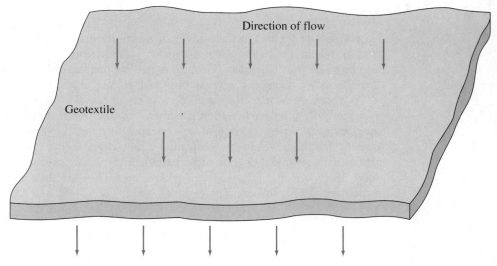

▼   **FIGURE 13.5**   Cross-plane flow through geotextile

where P = permittivity
$\quad k_p$ = coefficient of permeability for cross-plane flow
$\quad t$ = thickness of the geotextile

In a similar manner, in order to perform the function of drainage satisfactorily, geotextiles must possess excellent in-plane permeability. For reasons stated above, the in-plane coefficient of permeability is also dependent on the compressibility, and, hence, the thickness of the geotextile. The in-plane drainage capability can thus be expressed in terms of a quantity called *transmissivity*, which is expressed as

$$T = k_p t \tag{13.2}$$

where T = transmissivity
$\quad k_p$ = coefficient of permeability for in-plane flow (Figure 13.6)

The units of k_n and k_p are cm/sec or ft/min; however, the unit of permittivity, P, is sec^{-1} or min^{-1}. In a similar manner, the unit of transmissivity, T, is $\text{m}^3/\text{sec} \cdot \text{m}$ or $\text{ft}^3/\text{min} \cdot \text{ft}$. Depending on the type of geotextile, k_n and P, and k_p and T can vary over a wide range. Following are some typical values for k_n, P, k_p, and T:

► Coefficient of permeability, k_n: 1×10^{-3} to 2.5×10^{-1} cm/sec
► Permittivity, P: 2×10^{-2} to 2.0 sec^{-1}
► Coefficient of permeability, k_p:
  Nonwoven: 1×10^{-3} to 5×10^{-2} cm/sec
  Woven: 2×10^{-3} to 4×10^{-3} cm/sec
► Transmissivity, T:
  Nonwoven: 2×10^{-6} to $2 \times 10^{-9} \text{ m}^3/\text{sec} \cdot \text{m}$
  Woven: 1.5×10^{-8} to $2 \times 10^{-8} \text{ m}^3/\text{sec} \cdot \text{m}$

When a geotextile is being considered for use in the design and construction of landfill liners, certain properties need to be measured by tests on the geotextile to deter-

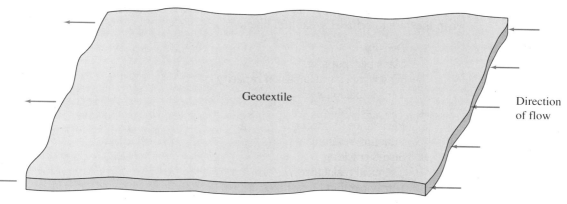

▼ **FIGURE 13.6** In-plane flow in geotextile

mine its applicability. A partial list of these tests follows:

1. Mass per unit area
2. Percent of open area
3. Equivalent opening size
4. Thickness
5. Ultraviolet resistivity
6. Permittivity
7. Transmissivity
8. Puncture resistance
9. Resistance to abrasion
10. Compressibility
11. Tensile strength and elongation properties
12. Chemical resistance

13.5 GEOMEMBRANES

Geomembranes are impermeable liquid or vapor barriers made primarily from continuous polymeric sheets that are flexible. The type of polymeric material used for geomembranes may be thermoplastic or thermoset. The thermoplastic polymers include polyvinyl chloride (PVC), polyethylene, chlorinated polyethylene, and polyamide. The thermoset polymers include ethylene vinyl acetate, polychloroprene, and isoprene-isobutylene. Although geomembranes are thought to be impermeable, they are not actually so. Water vapor transmission tests show that the coefficient of permeability of geomembranes is in the range of 10^{-10} to 10^{-13} cm/sec and, hence, they are "essentially impermeable."

Many scrim-reinforced geomembranes manufactured in single plies have thicknesses that range from 0.01 to about 0.016 in. (0.25 to 0.4 mm). These single plies of geomembranes can be laminated together to make thicker geomembranes. Some geomembranes made from PVC and polyethylene may be as thick as 0.18 to 0.2 in. (4.5 to 5 mm).

Following is a partial list of tests that should be conducted on geomembranes when they are to be used as landfill liners:

1. Density
2. Mass per unit area
3. Water vapor transmission capacity
4. Tensile behavior
5. Tear resistance
6. Resistance to impact
7. Puncture resistance
8. Stress cracking
9. Chemical resistance
10. Ultraviolet light resistance
11. Thermal properties
12. Behavior of seams

The most important aspect of construction with geomembranes is the preparation of seams. Otherwise, the basic reason for using geomembranes as a liquid or vapor barrier will be defeated. Geomembrane sheets are generally seamed together in the factory to prepare larger sheets. These larger sheets are field-seamed into their final position. There are several types of seams; some are briefly described here.

Lap Seam with Adhesive

▶ A solvent adhesive is used for this type of seam (Figure 13.7a). After application of the solvent, the two sheets of geomembrane are overlapped and then there is roller pressure.

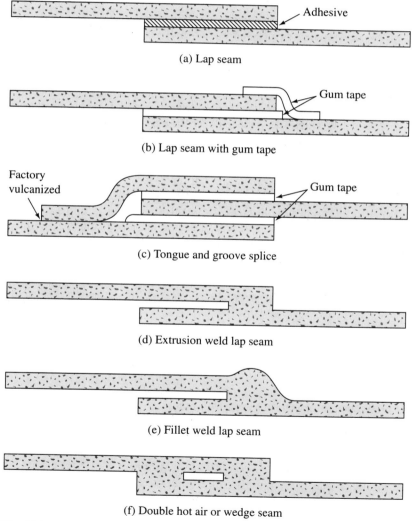

(a) Lap seam

(b) Lap seam with gum tape

(c) Tongue and groove splice

(d) Extrusion weld lap seam

(e) Fillet weld lap seam

(f) Double hot air or wedge seam

▼ **FIGURE 13.7** Configurations of filled geomembrane seams (after Environmental Protection Agency, 1989)

Lap Seam with Gum Tape

▶ This type of seam (Figure 13.7b) is used mostly in dense thermoset material such as isoprene-isobutylene.

Tongue and Groove Splice

▶ A schematic diagram of the tongue and groove splice is shown in Figure 13.7c. The tapes used for the splice are double-sided.

Extrusion Weld Lap Seam

▶ Extrusion or fusion welding is done on geomembranes made from polyethylene. A ribbon of molten polymer is extruded between the two surfaces to be joined (Figure 13.7d).

Fillet Weld Lap Seam

▶ This is similar to an extrusion weld lap seam; however, for fillet welding, the extrudate is placed over the edge of the seam (Figure 13.7e).

Double Hot Air or Wedge Seam

▶ In the hot air seam, hot air is blown to melt the two opposing surfaces. For melting, the temperatures should rise to about 500°F or more. After the opposite surfaces are melted, pressure is applied to form the seam (Figure 13.7f). For hot wedge seams, an electrically heated element like a blade is passed between the two opposing surfaces of the geomembrane. The heated element helps to melt the geomembrane, after which pressure is applied by a roller to form the seam.

13.6   GEONETS

Geonets are formed by the continuous extrusion of polymeric ribs at acute angles to each other. They have large openings in a netlike configuration. The primary function of geonets is drainage. Figure 13.8 is a photograph of a typical piece of geonet. Most of the geonets presently available are made of medium-density and high-density polyethylene. They are available in rolls with widths of 6 to 7 ft (≈ 1.8 to 2.1 m) and lengths of 100 to 300 ft (≈ 30 to 90 m). The approximate aperture sizes vary from 1.2 × 1.2 in. ($\approx 30 \times 30$ mm) to about 0.25 × 0.25 in. (≈ 6 mm × 6 mm). The thickness of geonets available commercially can vary from 0.15 to 0.3 in. (≈ 3.8 to 7.6 mm).

Seaming of geonets is somewhat more difficult. For that purpose, staples, threaded loops, and wire are sometimes used.

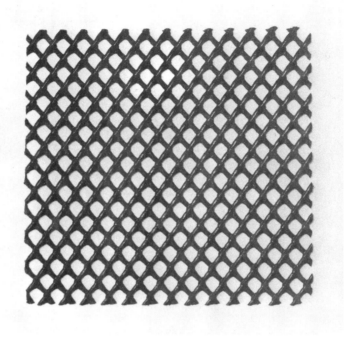

▼ **FIGURE 13.8** Geonet

13.7 SINGLE CLAY LINER AND SINGLE GEOMEMBRANE LINER SYSTEMS

Up until about 1982—that is, before the guidelines for the minimum technological requirements for hazardous waste landfill design and construction from the U.S. Environmental Protection Agency—most of the landfill liners were *single clay liners*. Figure 13.9 shows the cross-section of a single clay liner system for a landfill. It consists primarily of a compacted clay liner over the native foundation soil. The thickness of the compacted clay liner varies between 3 and 6 ft (0.9 and 1.8 m). The *maximum* required coefficient of permeability, k, is 10^{-7} cm/sec. Over the clay liner, there is a layer of gravel with perforated pipes for leachate collection and removal. Over the gravel layer, there is a layer of filter soil. The filter is used to protect the holes in the perforated pipes against the movement of fine soil particles. It is, in most cases, medium coarse to fine sandy soil. It is important to note that this system does not have any leak-detection system.

Around 1982, single layers of geomembranes were also used as a liner material for landfill sites. As shown in Figure 13.10, the geomembrane is laid over native foundation soil. Over the geomembrane, there is a layer of gravel with perforated pipes for leachate collection and removal. A layer of filter soil is placed between the solid waste material and the gravel. As in the single clay liner system, there is no provision for leak detection.

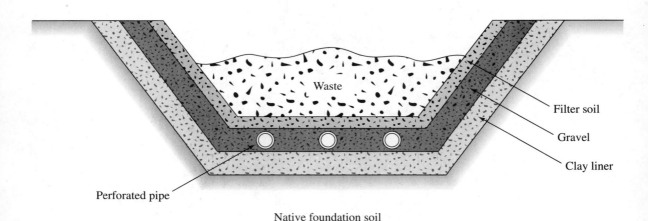

▼ FIGURE 13.9   Cross-section of single clay liner system for a landfill

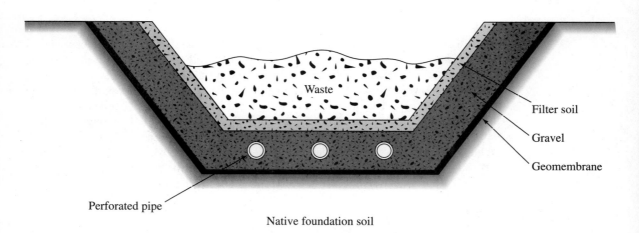

▼ FIGURE 13.10   Cross-section of single geomembrane liner system for a landfill

13.8 RECENT ADVANCES IN THE LINER SYSTEMS FOR LANDFILLS

Since 1984, most of the landfills developed for solid and hazardous wastes have double liners. The two liners are an upper primary liner and a lower secondary liner. Above the top liner there is a *primary* leachate collection and removal system. In general, the primary leachate collection system must be capable of maintaining a leachate head of 12 in. (≈ 0.3 m) or less. In between the primary and secondary liners, there is a system for leak detection, collection, and removal (LDCR) of leachates. The general guidelines for the primary leachate collection system, and the leak detection, collection, and removal systems are as follows:

1. It can be a granular drainage layer or a geosynthetic drainage material such as geonet.

2. If a granular drainage layer is used, it should have a minimum thickness of 12 in. (≈ 0.3 m).

3. The granular drainage layer (or the geosynthetic) should have a coefficient of permeability, k, greater than 10^{-2} cm/sec.

4. If a granular drainage layer is used, it should have a granular filter or a layer of geotextile over it to prevent clogging. A layer of geotextile is also required over the geonet when it is used as the drainage layer.

5. The granular drainage layer, when used, must be chemically resistant to the waste material and the leachate that are produced. It should also have a network of perforated pipes to collect the leachate effectively and efficiently.

In the design of the liner systems, the compacted clay layers should be at least 3 ft (≈ 0.9 m) thick with $k \le 10^{-7}$ cm/sec. Figures 13.11 and 13.12 show schematic dia-

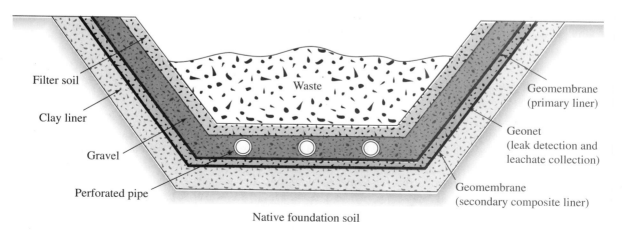

▼ **FIGURE 13.11** Cross-section of double-liner systems. (Note the secondary composite liner.)

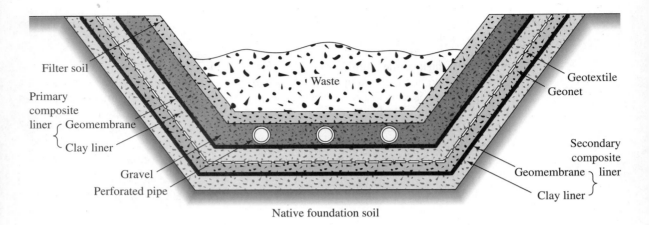

Filter soil

Primary
composite
liner ⎧ Geomembrane
⎩ Clay liner

Gravel
Perforated pipe

Waste

Geotextile
Geonet

Secondary
composite
Geomembrane ⎫ liner
Clay liner ⎭

Native foundation soil

▼   **FIGURE 13.12**   Cross-section of double-liner systems.
(Note the primary and secondary
composite liners.)

grams of two double-liner systems. In Figure 13.11, the primary leachate collection system is made of a granular material with perforated pipes and a filter system above it. The primary liner is a geomembrane. The LDCR system is made of geonet. The secondary liner is a *composite* liner made of a geomembrane with a compacted clay layer below it. In Figure 13.12, the primary leachate collection system is similar to that shown in Figure 13.11; however, the primary and secondary liners are both composite liners (geomembrane-clay). The LDCR system is a geonet with a layer of geotextile over it. The layer of geotextile acts as a filter and a separator.

The geomembranes used for landfill lining must have a minimum thickness of 0.03 in. (0.76 mm); however, all geomembranes that have a thickness of 0.03 in. (0.76 mm) may not be suitable in all situations. In practice, most geomembranes used as liners have thicknesses ranging from 0.07 to 0.1 in. (1.8 to 2.54 mm).

13.9   LEACHATE REMOVAL SYSTEMS

The bottom of a landfill needs to be properly graded so that the leachate collected from the primary collection system and the LDCR system will flow to a low point by gravity. Usually a grade of 2% or more is provided for large landfill sites. The low point of the leachate collection system ends at a sump. For primary leachate collection, a manhole is located at the sump, which rises through the waste material. Figure 13.13 shows a schematic diagram of the leachate removal system with a low-volume sump. A typical leachate removal system for high-volume sumps (for primary collection) is shown in Figure 13.14.

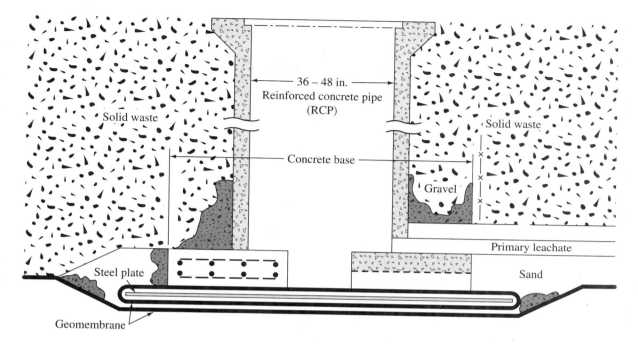

▼ **FIGURE 13.13**   Primary leachate removal system with a low-volume sump (after Environmental Protection Agency, 1989)

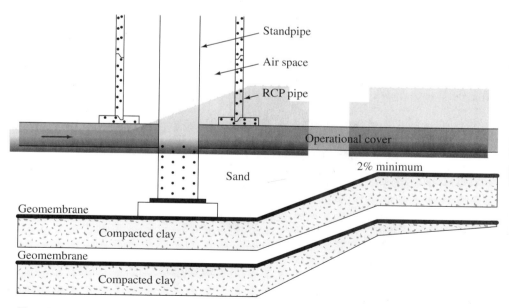

▼ **FIGURE 13.14**   Primary leachate removal system with a high-volume sump (after Environmental Protection Agency, 1989)

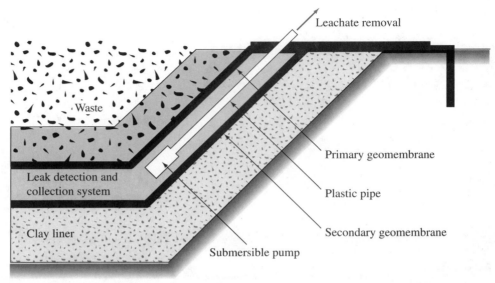

▼ **FIGURE 13.15** Secondary leak detection, collection, and removal system—via pumping. (*Note:* The plastic pipe penetrates the primary geomembrane.)

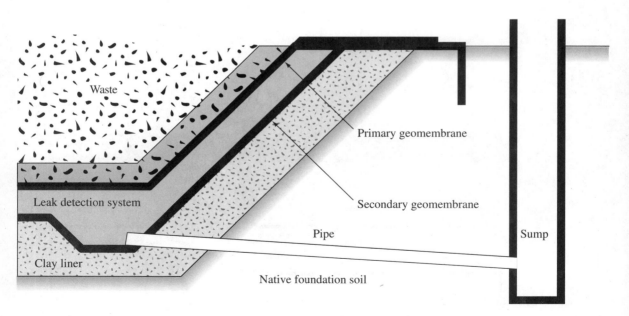

▼ **FIGURE 13.16** Secondary leak detection, collection, and removal system—via gravity monitoring. (*Note:* The plastic pipe penetrates the secondary geomembrane.)

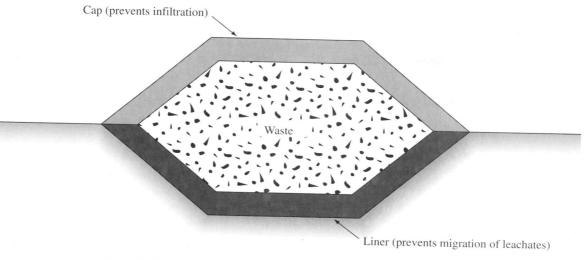

Cap (prevents infiltration)

Waste

Liner (prevents migration of leachates)

▼ **FIGURE 13.17**   Landfill with liner and cap

The removal of leachate from the leak detection and collection system can be done via pumping as shown in Figure 13.15 or via gravity monitoring as shown in Figure 13.16. When leachate is removed by pumping, the plastic pipe used for removal must penetrate the primary liner. On the other hand, if gravity monitoring is used, the pipe will penetrate the secondary liner.

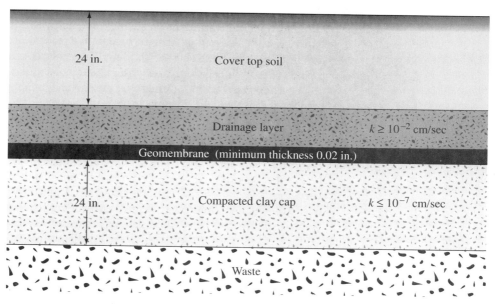

24 in.   Cover top soil

Drainage layer $k \geq 10^{-2}$ cm/sec

Geomembrane (minimum thickness 0.02 in.)

24 in.   Compacted clay cap $k \leq 10^{-7}$ cm/sec

Waste

▼ **FIGURE 13.18**   Schematic diagram of the layering system for landfill cap

13.10   CLOSURE OF LANDFILLS

When the landfill is complete and no more waste can be placed into it, a cap has to be put on the landfill (Figure 13.17). This cap will reduce and ultimately eliminate leachate generation. A schematic diagram of the layering system recommended by the U.S. Environmental Protection Agency (1979, 1986) and Koerner (1990) for hazardous waste landfills is shown in Figure 13.18. Essentially, it consists of a compacted clay cap over the solid waste, a geomembrane liner, a drainage layer, and a cover of top soil. The manhole used for leachate collection penetrates the landfill cover. Leachate removal continues until its generation is stopped. For hazardous waste landfill sites, the EPA (1989) recommends this period to be about 30 years.

13.11   GENERAL COMMENTS

This chapter provided a brief overview of the problems associated with solid and hazardous waste landfills. The general concepts for the construction of landfill liners using compacted clayey soil and geosynthetics (that is, geotextiles, geomembranes, and geonets) were discussed. Several areas were not addressed, however, since they are beyond the scope of the text. Some of these areas are:

1. *Selection of material.* The chemicals contained in leachates generated from hazardous and nonhazardous waste may interact with the liner materials. For that reason, it is essential that representative leachates are used to test the chemical compatibility so that the liner material remains intact during the periods of landfill operation and closure, and possibly longer. Selection of the proper leacheates becomes difficult because of the extreme variations that are encountered in the field. The mechanical properties of geomembranes are also very important. Properties such as workability, creep, stress cracking, and thermal coefficient of expansion should be investigated thoroughly.

2. *Stability of side slope liner.* The stability and slippage checks of the side slope liners of a landfill site are very important and complicated because of the variation of the frictional characters of the composite materials involved in liner construction. Interested readers are referred to any book on geosynthetics (e.g., Koerner, 1990) for a detailed treatment of this topic.

3. *Leak response action plan.* It is extremely important that any leaks or clogging of the drainage layer(s) in a given waste disposal site be detected as quickly as possible. Leaks or cloggings are likelihoods at a site even with good construction quality control. Each waste disposal facility should have a *leak response action plan.*

PROBLEMS **13.1** Refer to the cross-plane flow in a geotextile in Figure 13.19. If $\Delta h = 100$ mm, $L = 380$ mm, $w = 229$ mm, $t = 2.54$ mm, and flow rate $= 50$ cm³/min, calculate the transmissivity.

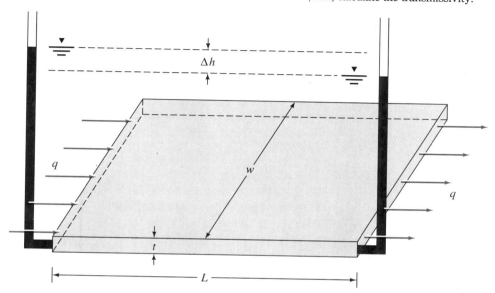

▼ **FIGURE 13.19**

13.2 Repeat Problem 13.1 with $\Delta h = 0.3$ in., $L = 3$ ft, $w = 12$ in., $t = 6.2$ mm, and $Q = 62$ cm³/min.

13.3 Similar to soils (Chapter 5), constant head permeability tests can be conducted on geotextile specimens to determine k_n (coefficient of permeability for cross-plane flow). Show that permittivity,

$$P = \frac{k_n}{t} = \frac{q}{(\Delta h)A}$$

where $q =$ flow rate
$\Delta h =$ loss of head
$A =$ area of the geotextile under test

13.4 Refer to Problem 13.3. Following are the results of a constant head permeability test on a geotextile:
Diameter of test specimen $= 70$ mm
Thickness of test specimen $= 5$ mm
Calculate the average permittivity (min⁻¹).

Δh (mm)	Flow rate, q (cm³/min)
80	350
100	700
120	822
140	997
160	1114

REFERENCES

Environmental Protection Agency (1989). *Requirements for Hazardous Waste Landfill Design, Construction, and Closure,* Publication No. EPA-625/4-89-022, Cincinnati, Ohio.

Environmental Protection Agency (1986). *Cover for Uncontrolled Hazardous Waste Sites,* Publication No. EPA-540/2-85-002, Cincinnati, Ohio.

Environmental Protection Agency (1979). *Design and Construction for Solid Waste Landfills,* Publication No. EPA-600/2-79-165, Cincinnati, Ohio.

Koerner, R. M. (1990). *Designing with Geosynthetics,* 2nd ed., Prentice-Hall, Englewood Cliffs, NJ.

Mitchell, J. K., and Jaber, M. (1990). "Factors Controlling the Long-Term Properties of Clay Liners," *Geotechnical Special Publication No. 25,* ASCE, 84–105.

FOURTEEN

SUBSOIL
EXPLORATION

The preceding chapters have reviewed the fundamental properties of soils and their behavior under stress and strain in idealized conditions. In practice, natural soil deposits are not homogeneous, elastic, or isotropic. In some places, the stratification of soil deposits may even change greatly within a horizontal distance of 50 to 100 ft ($\simeq$ 15 to 30 m). For foundation design and construction work, it is necessary to know the actual soil stratification at a given site, the laboratory test results of the soil samples obtained from various depths, and the observations made during the construction of other structures built under similar conditions. For most major structures, adequate subsoil exploration at the construction site must be conducted. The purposes of subsoil exploration include the following:

1. Determining the nature of soil at the site and its stratification
2. Obtaining disturbed and undisturbed soil samples for visual identification and appropriate laboratory tests
3. Determining the depth and nature of bedrock, if and when encountered
4. Performing some *in situ* field tests, such as permeability tests (Chapter 5), vane shear tests (Chapter 9), and standard penetration tests
5. Observing drainage conditions from and into the site
6. Assessing any special construction problems with respect to the existing structure(s) nearby

This chapter briefly summarizes subsoil exploration techniques. Readers are also encouraged to refer to the *Manual on Foundation Investigation* of the American Association of State Highway Transportation Officials (1967) for additional information.

14.1 PLANNING FOR SOIL EXPLORATION

A soil exploration program for a given structure can be broadly divided into four main categories:

1. *Compilation of the existing information regarding the structure:* This includes information such as the type of structure to be constructed and its

future use, the requirements of local building codes, and the column and load-bearing wall loads. If the exploration is for the construction of a bridge foundation, one must have an idea of the length of the span and the anticipated loads to which the piers and abutments will be subjected.

2. *Collection of existing information for the subsoil condition:* Considerable savings in the exploration program can sometimes be realized if the geotechnical engineer in charge of the project makes a thorough review of the existing information regarding the subsoil conditions at the site under consideration. Useful information can be obtained from the following sources:
   a. Geological survey maps
   b. County soil survey maps prepared by the U.S. Department of Agriculture and the Soil Conservation Service
   c. Soil manuals published by the state highway department
   d. Existing soil exploration reports prepared for the construction of nearby structures

   Information gathered from the preceding sources provides insight into the type of soil and problems that might be encountered during actual drilling operations.

3. *Reconnaissance of the proposed construction site:* The engineer should make a visual inspection of the site and the surrounding area. In many cases, the information gathered from such a trip is invaluable for future planning. The type of vegetation at a site may in some instances give an indication as to the type of subsoil that will be encountered. The accessibility of a site and the nature of drainage into and from it can also be determined. Open cuts near the site provide an indication about the subsoil stratification. Cracks in the walls of nearby structure(s) may indicate settlement from the possible existence of soft clay layers or the presence of expansive clay soils.

4. *Detailed site investigation:* This phase consists of making several test borings at the site and collecting disturbed and undisturbed soil samples from various depths for visual observation and for laboratory tests. There is no hard and fast rule for determining the number of borings or the depth to which the test borings are to be advanced. For most buildings, at least one boring at each corner and one at the center should provide a start. Depending on the uniformity of the subsoil, additional test borings may be made. Table 14.1 gives guidelines for initial planning of the spacing of boreholes.

The test borings should extend through unsuitable foundation materials to firm soil layers. Sowers and Sowers (1970) provided a rough estimate of the minimum depth of borings (unless bedrock is encountered) for multistory buildings. They can be given by the following equations:

Light steel or narrow concrete buildings:

$$z_b \text{ (ft)} = 10S^{0.7} \tag{14.1a}$$

$$z_b \text{ (m)} = 3S^{0.7} \tag{14.1b}$$

▼ **TABLE 14.1**   **Spacing of Borings**

Project	Boring spacings	
	ft	m
One-story buildings	75–100	23–30
Multistory buildings	50–75	15–23
Highways	750–1000	230–305
Earth dams	75–150	23–46
Residential subdivision planning	200–300	61–92

Heavy steel or wide concrete buildings:

$$z_b \text{ (ft)} = 20S^{0.7} \tag{14.2a}$$

$$z_b \text{ (m)} = 6S^{0.7} \tag{14.2b}$$

In Eqs. (14.1) and (14.2), z_b is the approximate depth of boring and S is the number of stories.

The American Society of Civil Engineers (1972) recommended the following rules of thumb for estimating the boring depths for buildings:

1. Estimate the variation of the net stress increase, Δp, that will result from the construction of the proposed structure with depth. This can be done by using the principles outlined in Chapter 7. Determine the depth, D_1, at which the value of Δp is equal to 10% of the average load per unit area of the structure.

2. Plot the variation of the effective vertical stress, σ', in the soil layer with depth. Compare this with the net stress increase variation, Δp, with depth as determined in Step 1. Determine the depth, D_2, at which $\Delta p = 0.05\sigma'$.

3. The smaller of the two depths, D_1 and D_2, is the approximate minimum depth of the boring.

When the soil exploration is for the construction of dams and embankments, the depth of boring may range from one-half to two times the embankment height.

The general techniques used for advancing test borings in the field and the procedure for the collection of soil samples are described in the following sections.

14.2 BORING METHODS

There are several methods by which the test boring can be advanced in the field. The simplest is the use of augers. Figure 14.1 shows two types of hand augers that can be used for making boreholes up to a depth of about 10 to 15 ft ($\simeq 3$ to 5 m). They can be used for soil exploration work for highways and small structures. Information regarding the types of soil present at various depths is obtained by noting the soil that holds to

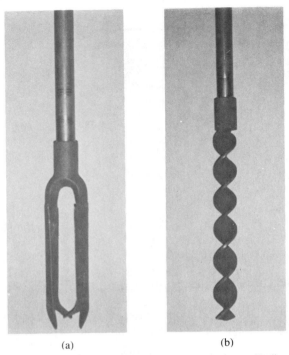

(a)   (b)

▼ **FIGURE 14.1**   Hand augers: (a) Iwan auger; (b) slip auger

the auger. The soil samples collected in this manner are disturbed, but they can be used to conduct laboratory tests such as grain-size determination and Atterberg's limits.

When the boreholes are to be advanced to greater depths, the most common method is to use continuous flight augers. These are power operated. The power for drilling is delivered by truck- or tractor-mounted drilling rigs. Continuous flight augers are commercially available in 3- to 5-ft sections. During the drilling operation, section after section of auger can be added and the hole extended downward. Continuous flight augers can be solid stem or hollow stem. Some of the commonly used solid-stem augers have outside diameters of $2\frac{5}{8}$ in. (66.68 mm), $3\frac{1}{4}$ in. (82.55 mm), 4 in. (101.6 mm), and $4\frac{1}{2}$ in. (114.3 mm). The inside and outside diameters of some hollow-stem augers are given in Table 14.2.

Flight augers bring the loose soil from the bottom of the hole to the surface. The driller can detect the change in soil type encountered by the change of speed and the sound of drilling. Figure 14.2 shows a drilling operation with flight augers. When solid-stem augers are used, it is necessary to withdraw the auger at regular intervals to obtain soil samples and also to conduct other operations such as standard penetration tests. Hollow-stem augers have a distinct advantage in this respect: they do not have to be removed at frequent intervals for sampling or other tests. As shown in Figure 14.3, the outside of the auger acts like a casing. A removable plug is attached to the bottom of the auger by means of a center rod. During the drilling, the plug can be pulled out with the auger in place, and soil sampling and standard penetration tests can be performed. When hollow-stem augers are used in sandy soils below the ground water table,

▼ **TABLE 14.2** **Dimensions of Commonly Used Hollow-Stem Augers**

| Inside diameter | | Outside diameter | |
in.	mm	in.	mm
2.5	63.5	6.25	158.75
2.75	69.85	7.0	177.8
3.0	76.2	8.0	203.2
3.5	88.9	9.0	228.6
4.0	101.6	10.0	254.0

it is possible that the sand might be pushed several feet into the stem of the auger by excess hydrostatic pressure immediately after the removal of the plug. In such conditions, the plug should not be used and, instead, water inside the hollow stem should be maintained at a higher level than the ground water table.

Rotary drilling is a procedure by which rapidly rotating drilling bits attached to the bottom of drilling rods cut and grind the soil and advance the borehole down. Several types of drilling bits are available for such work. Rotary drilling can be used in sand, clay, and rock (unless badly fissured). Water or drilling mud is forced down the

▼ **FIGURE 14.2** Drilling with flight augers (courtesy of Danny R. Anderson, El Paso, Texas)

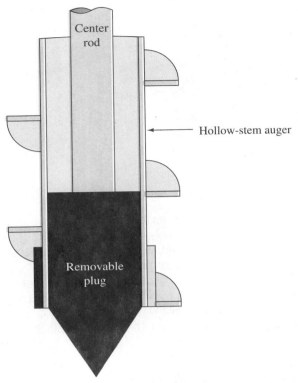

Center
rod

Hollow-stem auger

Removable
plug

▼ **FIGURE 14.3** Diagram of a hollow-stem auger with removable plug

drilling rods to the bits, and the return flow forces the cuttings to the surface. Drilling mud is a slurry prepared by mixing bentonite and water. Bentonite is a montmorillonite clay that is formed by the weathering of volcanic ash. Boreholes with diameters ranging from 2 to 8 in. (50.8 to 203.2 mm) can be made easily by this technique.

Wash boring is another method of advancing boreholes. In this method, a casing about 6 to 10 ft (2 to 3 m) long is driven into the ground. The soil inside the casing is then removed by means of a *chopping bit* that is attached to a drilling rod. Water is forced through the drilling rod, and it goes out at a very high velocity through the holes located at the bottom of the chopping bit (Figure 14.4). The water and the chopped soil particles rise upward in the drill hole and overflow at the top of the casing through a T-connection. The wash water is then collected in a container. The casing can be extended with additional pieces as the borehole progresses; however, it is not necessary if the borehole will stay open without caving in.

Percussion drilling is an alternate method of advancing a borehole, particularly through hard soil and rock. In this technique, a heavy drilling bit is raised and lowered to chop the hard soil. Casing for this type of drilling may be required. The chopped soil particles are brought up by the circulation of water.

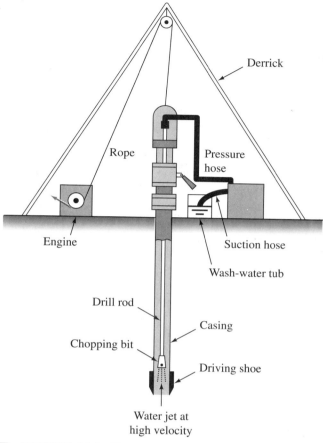

Derrick

Rope

Pressure hose

Engine

Suction hose

Wash-water tub

Drill rod

Casing

Chopping bit

Driving shoe

Water jet at
high velocity

▼ **FIGURE 14.4** Wash boring

14.3 COMMON METHODS OF SAMPLING

Various methods for advancing boreholes have been discussed in the preceding section. During the advancement of the boreholes, soil samples are collected at various depths for further analysis. This section briefly discusses some of the methods of sample collection.

Sampling by Standard Split Spoon

Figure 14.5 shows a diagram of a split-spoon sampler. It consists of a tool-steel driving shoe at the bottom, a steel tube (that is split longitudinally into halves) in the middle, and a coupling at the top. The steel tube in the middle has inside and outside diameters of $1\frac{3}{8}$ in. (34.93 mm) and 2 in. (50.8 mm), respectively. Figure 14.6 shows a photograph of a split-spoon sampler in an unassembled form.

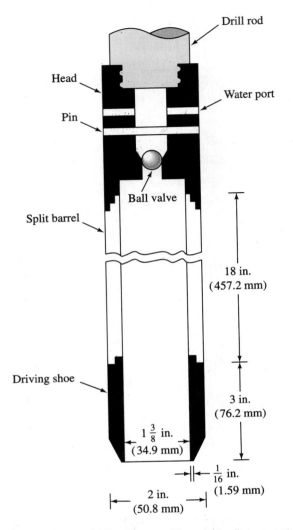

Drill rod

Head

Water port

Pin

Ball valve

Split barrel

18 in.
(457.2 mm)

Driving shoe

3 in.
(76.2 mm)

$1\frac{3}{8}$ in.
(34.9 mm)

$\frac{1}{16}$ in.
(1.59 mm)

2 in.
(50.8 mm)

▼ **FIGURE 14.5**   Diagram of a standard split-spoon sampler

When the borehole is advanced to a desired depth, the drilling tools are removed. The split-spoon sampler is attached to the drilling rod and then lowered to the bottom of the borehole. The sampler is driven into the soil at the bottom of the borehole by means of hammer blows. The hammer blows occur at the top of the drilling rod. The hammer weighs 140 lb (623 N). For each blow, the hammer drops a distance of 30 in. (0.762 m). The number of blows required for driving the sampler through three 6-in. (152.4-mm) intervals is recorded. The sum of the number of blows required for driving the last two 6-in. (152.4-mm) intervals is referred to as the *standard penetration number*, *N*. It is also commonly called the "blow count." The interpretation of the standard penetration number is given in Section 14.5. After driving is completed, the sampler is

▼ **FIGURE 14.6**   Split-spoon sampler, unassembled (courtesy of Soiltest, Inc., Lake Bluff, Illinois)

withdrawn and the shoe and coupling are removed. The soil sample collected inside the split tube is then removed and transported to the laboratory in small glass jars.

Determination of the standard penetration number and collection of split-spoon samples are usually done at 5-ft ($\simeq 1.5$-m) intervals.

Sampling by Thin Wall Tube

This is used for obtaining fairly undisturbed soil samples. The thin wall tubes are made of seamless thin tubes and are commonly referred to as *Shelby tubes* (Figure 14.7). To collect samples at a given depth in a borehole, the drilling tools are first removed. The sampler is attached to a drilling rod and lowered to the bottom of the borehole. After that, it is hydraulically pushed into the soil. It is then spun to shear off the base and pulled out. The sampler with the soil inside it is sealed and taken to the laboratory for testing. Most commonly used thin wall tube samplers have outside diameters of 2 in. (50.8 mm) and 3 in. (76.2 mm).

Sampling by Piston Sampler

Piston samplers are particularly useful when highly undisturbed samples are required. The cost of recovering such samples is, of course, higher. There are several types of

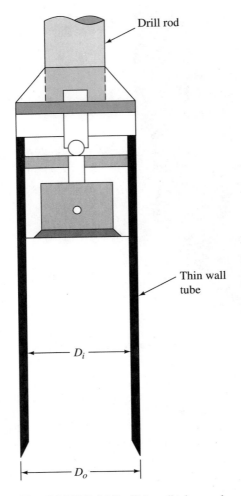

Drill rod

Thin wall
tube

D_i

D_o

▼ **FIGURE 14.7**   Thin wall tube sampler

piston samplers; however, the sampler proposed by Osterberg (1952) is the most advantageous (see Figure 14.8a and b). It consists of a thin wall tube with a piston. Initially, the piston closes the end of the thin wall tube. The sampler is first lowered to the bottom of the borehole (Figure 14.8a), and then the thin wall tube is pushed into the soil hydraulically past the piston. After that, the pressure is released through a hole in the piston rod (Figure 14.8b). The presence of the piston prevents distortion in the sample by neither letting the soil squeeze into the sampling tube very fast nor admitting excess soil. Samples obtained in this manner are consequently less disturbed than those obtained by Shelby tubes.

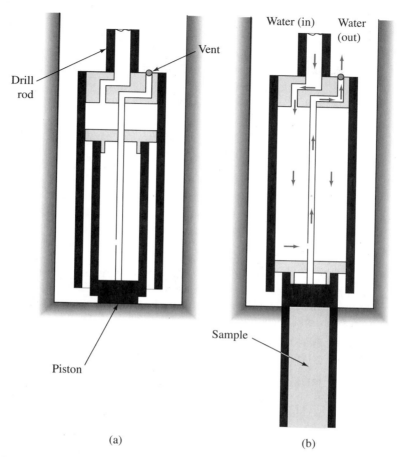

▼ **FIGURE 14.8** Piston sampler: (a) sampler lowered to bottom of borehole; (b) pressure released through hole in piston rod

14.4 SAMPLE DISTURBANCE

The degree of disturbance of the sample collected by various methods can be expressed by a term called *area ratio, A_r*, or

$$A_r(\%) = \frac{D_o^2 - D_i^2}{D_i^2} \times 100 \qquad (14.3)$$

where D_o = outside diameter of the sampler
D_i = inside diameter of the sampler

A soil sample can generally be considered undisturbed if the area ratio is less than or equal to 10%. Following is a calculation of A_r for a standard split-spoon sampler and a 2-in. (50.8-mm) Shelby tube.

Standard spit-spoon sampler: $D_i = 1.38$ in. and $D_o = 2$ in. Hence,

$$A_r(\%) = \frac{(2)^2 - (1.38)^2}{(1.38)^2} \times 100 = 110\%$$

Shelby tube sampler (2-in. diameter): $D_i = 1.875$ in. and $D_o = 2$ in. Hence,

$$A_r(\%) = \frac{(2)^2 - (1.875)^2}{(1.875)^2} \times 100 = 13.7\%$$

The preceding calculation indicates that the sample collected by split spoons is highly disturbed. The area ratio (A_r) of the 2-in. (50.8-mm) diameter Shelby tube samples is slightly higher than the 10% limit stated above. For practical purposes, it can be treated as an undisturbed sample. Note that the cost of recovery rapidly increases with samples of larger diameter.

The disturbed but representative soil samples recovered by split-spoon samplers can be used for laboratory tests such as grain-size distribution, liquid limit, plastic limit, and shrinkage limit. However, for performing tests such as consolidation, triaxial compression, and unconfined compression, undisturbed soil samples are necessary.

14.5 CORRELATIONS FOR STANDARD PENETRATION TEST

The procedure for conducting the standard penetration test was outlined in Section 14.3. The standard penetration number, N, is commonly used to correlate several physical parameters of soils. In Chapter 9, a qualitative description of the consistency for clay soils based on their unconfined compressive strengths (q_u) was presented (Table 9.3). The unconfined compression strength of clay soils can also be approximately correlated to the standard penetration number, N. Table 14.3 gives the approximate relationship among the standard penetration number at a given depth, the consistency, and the unconfined compression strength of clayey soils.

In granular soils, the standard penetration number is highly dependent on the effective overburden pressure, σ' (Chapter 6). This can be explained by means of Figure 14.9, in which an ideal condition of a homogeneous dry sand deposit is shown. At a depth of h_1, the vertical effective stress will be equal to

$$\sigma' = \sigma_1' = \gamma h_1 \tag{14.4}$$

where γ = unit weight of soil. In a similar manner, at a depth of h_2,

$$\sigma' = \sigma_2' = \gamma h_2 \tag{14.5}$$

Although the sand is homogeneous, having the same unit weight γ and hence the same relative density D_r and angle of friction ϕ, the higher effective overburden pres-

▼ **TABLE 14.3** **Approximate Correlation of Standard Penetration Number and Consistency of Clay**

Standard penetration number, N	Consistency	Unconfined compression strength, q_u (ton/ft²)
0		0
	Very soft	
2		0.25
	Soft	
4		0.5
	Medium stiff	
8		1
	Stiff	
16		2
	Very stiff	
32		4
>32	Hard	>4

Note: 1 ton/ft² = 95.76 kN/m²

sure (and hence higher lateral confining pressure) at depth h_2 will contribute to a higher standard penetration number. This fact was clearly demonstrated by Gibbs and Holtz (1957). The results of their findings are shown in Figure 14.10. As an example, one can see that at $D_r \simeq 80\%$, the standard penetration number is about 12 with $\sigma' = 0$ lb/in.². It increases to about 50 with $\sigma' = 40$ lb/in.² (276 kN/m²). For that reason, it is necessary to convert the standard penetration numbers obtained at various depths to reflect a constant effective overburden pressure.

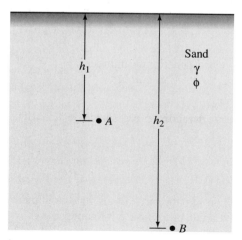

▼ **FIGURE 14.9**   Effect of effective overburden pressure on the standard penetration number

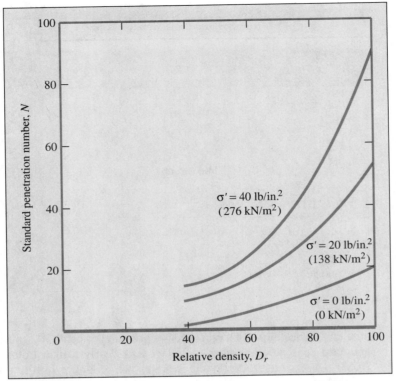

▼ FIGURE 14.10 Results of Gibbs and Holtz's study—variation of N with D_r and σ'

A number of empirical relationships have been proposed to convert the field standard penetration numbers to corrected penetration numbers that reflect a standard effective overburden pressure. The simplest of those relationships is that proposed by Liao and Whitman (1986), which can be expressed as

$$N' = C_N N_F \tag{14.6}$$

where N' = corrected standard penetration number
N_F = field standard penetration number
C_N = correction factor

For a standard effective overburden pressure $\sigma' = 1$ ton/ft^2,

$$C_N = \sqrt{\frac{1}{\sigma'}} \tag{14.7}$$

where σ' is in ton/ft^2.

Table 14.4 shows approximate correlations for the standard penetration number N', relative density D_r, and angle of friction ϕ for sands.

The standard penetration number is a very useful guideline in soil exploration and assessment of subsoil conditions, provided that the results are interpreted correctly.

▼ **TABLE 14.4** **Approximate Relation Among Corrected Standard Penetration Number, Angle of Friction, and Relative Density of Sand**

Corrected standard penetration number, N	Relative density, D_r (%)	Angle of friction, ϕ (degrees)
0–5	0–5	26–30
5–10	5–30	28–35
10–30	30–60	35–42
30–50	60–95	38–46

Note that all equations and correlations relating to the standard penetration numbers are approximate. Since soil is not homogeneous, a wide variation in the N-value may be obtained in the field. In soil deposits that contain large boulders and gravel, the standard penetration numbers may be erratic.

14.6 OTHER *IN SITU* TESTS

Depending on the type of project and the complexity of the subsoil, several types of *in situ* tests can be conducted during the exploration period. The soil properties evaluated from the *in situ* tests, in many cases, yield more representative values. This is primarily because the sample disturbance during soil exploration is eliminated. Following are some of the common tests that can be conducted in the field.

Vane Shear Test

The principles of the vane shear test have been discussed in Chapter 9. When soft clay is encountered during the advancement of a borehole, the undrained shear strength of clay, c_u, can be determined by conducting a vane shear test in the borehole. It provides valuable information about the strength in undisturbed clay.

Borehole Pressuremeter Test

The pressuremeter is a device that was originally developed by Menard in 1965 for *in situ* measurement of the stress-strain modulus. This device basically consists of a pressure cell and two guard cells (Figure 14.11). The test involves expanding the pressure cell inside a borehole and measuring the expansion of its volume. The Menard-type pressuremeter operates on the theory of expansion of an infinitely thick cylinder of soil. Figure 14.12 shows the variation of the pressure cell volume with changes in the cell pressure. In this figure, zone I represents the reloading portion, during which the soil around the borehole is pushed back to its initial state—that is, the state it was in before drilling. Zone II represents a pseudo-elastic zone in which the cell volume versus cell

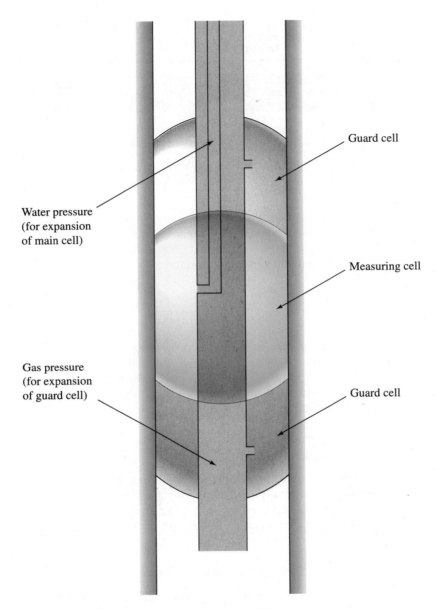

Water pressure
(for expansion
of main cell)

Gas pressure
(for expansion
of guard cell)

Guard cell

Measuring cell

Guard cell

▼   **FIGURE 14.11**   Diagram for pressuremeter test

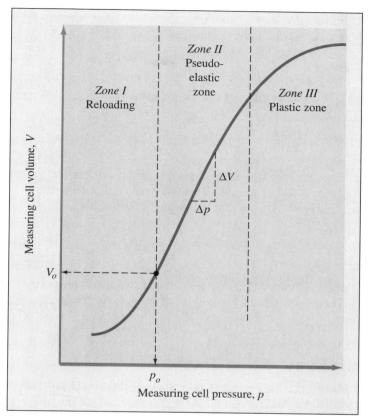

▼ FIGURE 14.12 Relationship between measuring pressure and measuring volume for Menard-type pressuremeter

pressure is practically linear. The zone marked III is the plastic zone. For the pseudo-elastic zone,

$$E = 2(1 + \mu)V_o \frac{\Delta p}{\Delta V}$$ (14.8)

where E = modulus of elasticity of soil
μ = Poisson's ratio
V_o = cell volume corresponding to pressure p_o (that is, the cell pressure corresponding to the beginning of zone II)

$$\frac{\Delta p}{\Delta V} = \frac{1}{\text{slope of straight line plot of zone II}}$$

Menard recommended a value of $\mu = 0.33$ for use in Eq. (14.8), or

$$E = 2.66V_o \frac{\Delta p}{\Delta V}$$ (14.9)

From the theory of elasticity, the relationship between the modulus of elasticity and the shear modulus can be given as

$$E = 2(1 + \mu)G \qquad (14.10)$$

where G = shear modulus of soil. Hence, combining Eqs. (14.8) and (14.10) gives

$$G = V_o \frac{\Delta p}{\Delta V} \qquad (14.11)$$

Pressuremeter test results can be used to determine the at-rest earth pressure coefficient, K_o (Chapter 10). This can be obtained from the ratio of p_o and σ' (σ' = effective vertical stress at the depth of the test), or

$$K_o = \frac{p_o}{\sigma'} \qquad (14.12)$$

Note that p_o (Figure 14.12) represents the *in situ* lateral pressure.

The pressuremeter tests are very sensitive to the conditions of a borehole before the test.

Cone Penetration Test

The Dutch cone penetrometer is a device by which a $60°$ cone with a base area of 1.54 in.^{2} (10 cm^2) (Figure 14.13) is continuously pushed into the soil at a rate of about 20 mm/sec, and the cone end resistance, q_c, to penetration is measured. Most cone penetrometers commonly used have friction sleeves that follow the point. This allows independent determination of the cone resistance and the frictional resistance of the soil above it. The friction sleeves have an exposed surface area of about 23.25 in.^{2} (150 cm^2).

In the past, the cone penetration test was used in Europe more commonly than in the United States. Recently, however, this test has attracted considerable interest in the United States. One of the major advantages of the cone penetration test is that boreholes are not necessary to conduct the test. Unlike the standard penetration test, however, soil samples cannot be recovered for visual observation and laboratory tests.

Figure 14.14 shows an approximate relationship between the vertical effective stress (σ'), q_c, and the peak soil friction angle (ϕ) for tests conducted in sand. Robertson, Campanella, and Wightman (1982) correlated the magnitude of q_c (kg/cm^2)/N (blows/ft) as a function of the mean grain size (D_{50}) of the soil, which covers a wide range of soil types. Kasim, Chu, and Jensen (1986) showed that the correlations of Robertson and colleagues are generally valid in spite of some scattering. This is shown in Figure 14.15.

The cone penetration resistance has also been correlated with the equivalent modulus of elasticity, E, of soils by various investigators. Schmertmann (1970) gave a simple correlation for sand as

$$E = 2q_c \qquad (14.13)$$

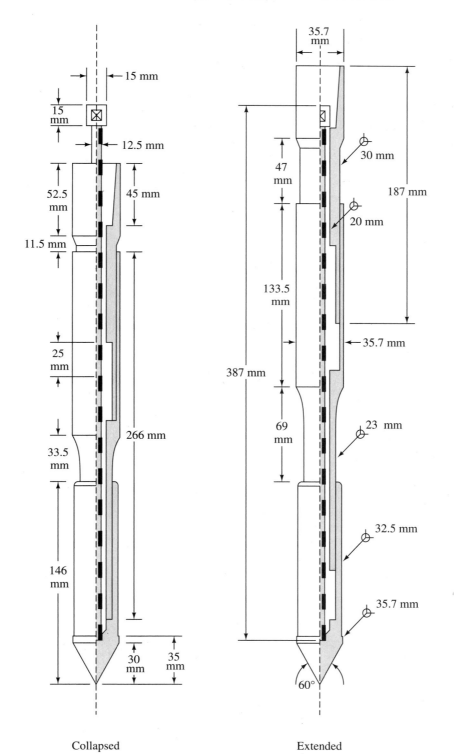

Collapsed   Extended

▼ **FIGURE 14.13**   Dutch cone penetrometer with friction sleeve (after American Society of Testing and Materials, 1991; copyright © ASTM; reprinted with permission).

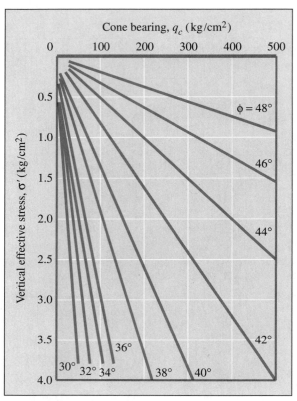

▼ **FIGURE 14.14** Variation of q_c with σ' and ϕ (after Robertson and Campanella, 1983)

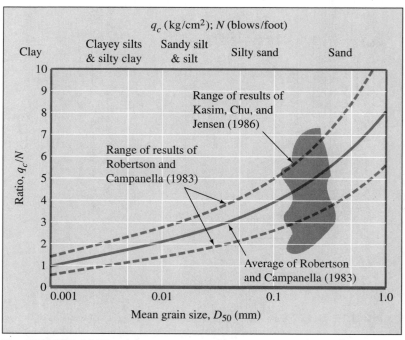

▼ **FIGURE 14.15** Variation of q_c (kg/cm²)/N with mean grain size

Trofimenkov (1974) also gave the following correlations for the stress-strain modulus in sand and clay:

$$E = 3q_c \quad \text{(for sands)} \tag{14.14}$$

$$E = 7q_c \quad \text{(for clays)} \tag{14.15}$$

Correlations such as Eqs. (14.13), (14.14), and (14.15) can be used in the calculation of the elastic settlement of foundations (Chapter 8).

Iowa Borehole Shear Test

The Iowa borehole shear test is a simple device used to determine the shear strength parameters of soil at a given depth during subsoil exploration. The shear device consists of two grooved plates that are pushed into the borehole (Figure 14.16). A controlled normal force, F, can be applied to each of the grooved plates. Shear failure in soil close

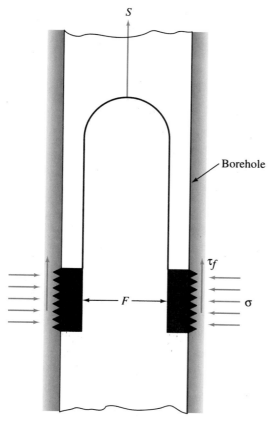

▼ **FIGURE 14.16**   Iowa borehole shear test

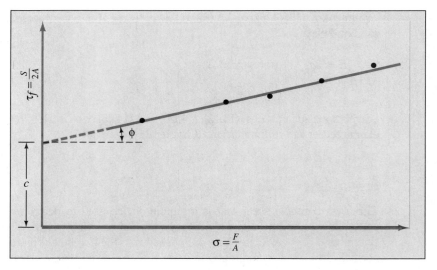

▼ **FIGURE 14.17**   Variation of τ_f with σ from Iowa borehole shear test

to the plates is induced by applying a vertical force S. So, the normal stress, σ, on the wall of the borehole can be given as

$$\sigma = \frac{F}{A} \tag{14.16}$$

where A = area of each plate in contact with the soil.

Similarly, the shear stress at failure, τ_f, is

$$\tau_f = \frac{S}{2A} \tag{14.17}$$

The test could be repeated with a number of increasing normal forces (F) without removing the shearing device. The results can be plotted in graphic form (Figure 14.17) to obtain the shear strength parameters (that is, cohesion c and angle of friction ϕ) of the soil. The shear strength parameters obtained in this manner are likely to represent those of a consolidated drained test (Chapter 9).

14.7 ROCK CORING

It may be necessary to core rock if bedrock is encountered at a certain depth during drilling. It is always desirable that coring be done for at least 10 ft ($\simeq 3$ m). If the bedrock is weathered or irregular, the coring may need to be extended to a greater depth. For coring, a core barrel is attached to the drilling rod. A coring bit is attached to the bottom of the core barrel. The cutting element in the bit may be diamond, tungsten, or carbide. The coring is advanced by rotary drilling. Water is circulated through the

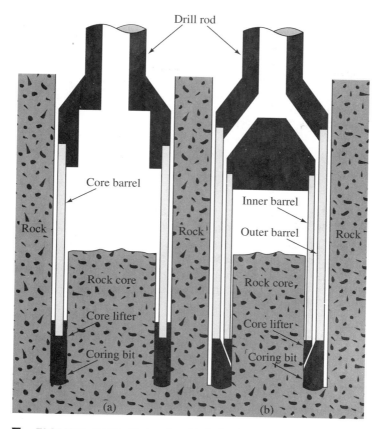

Rock coring: (a) single-tube core barrel; (b) double-tube core barrel

drilling rod during coring, and the cuttings are washed out. Figure 14.18 shows a diagram of rock coring by the use of a single-tube core barrel. Rock cores obtained by single-tube core barrels can be fractured because of torsion. To avoid this problem, double-tube core barrels can be used. Table 14.5 gives the details of various types of

▼ **TABLE 14.5   Details of Core Barrel Designations, Bits, and Core Samples**

Casing and core barrel designation	Outside diameter of core barrel bit (in.)	Diameter of core sample (in.)
EX	$1\frac{7}{16}$	$\frac{7}{8}$
AX	$1\frac{7}{8}$	$1\frac{1}{8}$
BX	$2\frac{5}{16}$	$1\frac{5}{8}$
NX	$2\frac{15}{16}$	$2\frac{1}{8}$

Note: 1 in. = 25.4 mm

▼ **TABLE 14.6**   **Qualitative Description Based on *RQD***

RQD	Rock quality
1–0.9	Excellent
0.9–0.75	Good
0.75–0.5	Fair
0.5–0.25	Poor
0.25–0	Very poor

casings and core barrels, diameters of core barrel bits, and diameters of core samples obtained. The core samples smaller than the BX size tend to break away during coring.

Based on the length of the rock core obtained from each run, the following quantities can be obtained for evaluation of the quality of rock:

1. $\text{Recovery ratio} = \dfrac{\text{length of rock core recovered}}{\text{length of coring}}$ (14.18)

2. $\text{Rock quality designation } (RQD) = \dfrac{\Sigma \text{ length of rock pieces recovered having lengths of 4 in. (101.6 mm) or more}}{\text{length of coring}}$ (14.19)

When the recovery ratio is equal to 1, it indicates intact rock. However, highly fractured rocks give a recovery ratio of 0.5 or less. Deere (1963) proposed the classification system in Table 14.6 for *in situ* rocks based on their *RQD*.

14.8 SOIL EXPLORATION REPORT

At the end of the soil exploration program, the soil and rock samples collected from the field are subjected to visual observation and laboratory tests. Then a soil exploration report is prepared for use by the planning and design office. Any soil exploration report should contain the following information:

1. Scope of investigation
2. General description of the proposed structure for which the exploration has been conducted
3. Geological conditions of the site
4. Drainage facilities at the site

5. Details of boring
6. Description of subsoil conditions as determined from the soil and rock samples collected
7. Ground water table as observed from the boreholes
8. Details of foundation recommendations and alternatives
9. Any anticipated construction problems
10. Limitations of the investigation

The following graphic presentations also need to be attached to the soil exploration report:

1. Site location map
2. Location of borings with respect to the proposed structure
3. Boring logs
4. Laboratory test results
5. Other special presentations

The boring log is the graphic presentation of the details gathered from each borehole. Figure 14.19 shows a typical boring log.

14.9 GENERAL COMMENTS

This chapter provided a brief overview of soil exploration methods generally used in the construction of foundations of various structures. A more elaborate treatment of this subject may be found in any foundation engineering text.

In many instances, owners view the soil exploration program as a waste of money, since there is nothing to "show for it." However, in some cases, it may save costly post-construction repair and maintenance. In any soil exploration program, the extent of boring and sampling depends on these factors:

1. Type of structure
2. Overall project cost
3. Requirements of local and state building codes
4. Uniformity of the subsoil conditions

In most cases, the soil exploration cost is about 0.5 to 1% of the total cost of the project. City and county officials may require soil exploration reports and recommendations for foundation construction before a construction permit is issued. In some critical construction projects, post-construction monitoring of building performance may be necessary.

It is also important for the geotechnical engineer who prepares the soil exploration report to be aware of the legal liabilities.

PROJECT TITLE ___Shopping center___

LOCATION ___Intersection Hill Street and Miner Street___   DATE ___June 7, 1983___

BORING NUMBER __4__   TYPE OF BORING ___Hollow stem auger___   GROUND ELEVATION ___132.2 ft___

DESCRIPTION OF SOIL	DEPTH (ft) AND SAMPLE NUMBER		STANDARD PENETRATION NUMBER, N	MOISTURE CONTENT, w (%)	COMMENTS
Tan sandy silt		1			
– – – – – – – –		2			
		3			
		4			
Light brown silty clay (CL)	SS-1	5 6	13	11	Liquid limit = 32 $PI = 9$
		7			
		8			
		9			
Ground water table June 14, 1983 ▼ – – – – ▽	SS-2	10 11	5	24	
		12			
		13			
		14			Liquid limit = 44 $PI = 26$ q_u = unconfined compression strength = 850 lb/ft²
Soft clay (CL)	ST-1	15 16	6	28	
		17			
		18			
		19			
Compact sand and gravel End of boring @ 22 ft	SS-3	20 21 22	32		

▼ **FIGURE 14.19**   Typical boring log (*Note:* SS = split-spoon sample; ST = Shelby tube sample)

PROBLEMS   **14.1**   Determine the area ratio of a Shelby tube sampler that has outside and inside diameters of 3.375 in. and 3.5 in., respectively.

14.2   Repeat Problem 14.1 with outside diameter = 3 in. and inside diameter = 2.875 in.

14.3   Following are the results of a standard penetration test in sand. Determine the corrected standard penetration numbers, N', at various depths. Note that the water table was not observed within a depth of 35 ft below the ground surface. Assume that the average unit weight of sand is 110 lb/ft^3.

Depth (ft)	N_F
5	8
10	7
15	12
20	14
25	13

14.4   The standard penetration number, N_F, of a clay at a certain depth is 12. Estimate its undrained shear strength.

14.5   Repeat Problem 14.4 for $N_F = 8$.

14.6   **a.**   From the results of Problem 14.3, estimate a design value of N' (corrected standard penetration number) for the construction of a shallow foundation.

   b.   Refer to Figure 11.29. For a 5-ft square column foundation in plan, what allowable load could the column carry? The bottom of the foundation is to be located at a depth of 2.5 ft below the ground surface. The maximum tolerable settlement is 1 in.

14.7   The undrained shear strength of a clay as determined from the field by a vane shear test is 900 lb/ft^2. The plasticity index of the clay is 16. What should be the corrected value of c_u for use in design work? [*Note:* Use the corrections suggested by Bjerrum and given in Eqs. (9.46) and (9.47).]

14.8   The average cone penetration resistance at a certain depth in a sandy soil is 205 kN/m^2. Estimate the modulus of elasticity of the soil at that depth.

14.9   During a field exploration program, rock was cored for a length of 4.2 ft. The length of the rock core recovered was 3.1 ft. Determine the recovery ratio.

REFERENCES

American Association of State Highway and Transportation Officials (1967). *Manual of Foundation Investigations*, National Press Building, Washington, D.C.

American Society of Civil Engineers (1972). "Subsurface Investigation for Design and Construction of Foundations of Buildings, Part I," *Journal of the Soil Mechanics and Foundations Division*, ASCE, Vol. 98, No. SM5, 481–490.

American Society of Testing and Materials (1991). *Annual Book of ASTM Standards*, Vol. 04.08, ASTM, Philadelphia.

Deere, D. U. (1963). "Technical Description of Rock Cores for Engineering Purposes," *Felsmechanik und Ingenieurgeologie*, Vol. 1, No. 1, 16–22.

Gibbs, H. J., and Holtz, W. G. (1957). "Research on Determining the Density of Sand by Spoon Penetration Testing," *Proceedings*, 4th International Conference on Soil Mechanics and Foundation Engineering, Vol. 1, 35–39, London.

Kasim, A. G., Chu, M., and Jensen, C. N. (1986). "Field Correlation of Cone and Standard Penetration Test," *Journal of Geotechnical Engineering*, ASCE, Vol. 112, No. 3, 368–372.

Liao, S., and Whitman, R. V. (1986). "Overburden Correction Factor for SPT in Sand," *Journal of Geotechnical Engineering*, ASCE, Vol. 112, No. 3, 373–377.

Menard, L. (1965). "Rules for Calculation of Bearing Capacity and Foundation Settlement Based on Pressuremeter Tests," *Proceedings*, 6th International Conference on Soil Mechanics and Foundation Engineering, Montreal, Canada, Vol. 2, 295–299.

Osterberg, J. O. (1952). "New Piston-Type Sampler," *Engineering News Solutions*, April 24.

Robertson, P. K., and Campanella, R. G. (1983). "Interpretation of Cone Penetration Tests. Part I: Sand," *Canadian Geotechnical Journal*, Vol. 20, No. 4, 718–733.

Robertson, P. K., Campanella, R. G., and Wightman, A. (1982). "SPT-CPT Correlations," University of British Columbia, Soil Mechanics Series No. 62, Canada.

Schmertmann, J. H. (1970). "Static Cone to Compute Static Settlement Over Sand," *Journal of the Soil Mechanics and Foundations Division*, ASCE, Vol. 96, No. SM3, 1011–1043.

Sowers, G. B., and Sowers, G. F. (1970). *Introductory Soil Mechanics and Foundations*, Macmillan, New York.

Trofimenkov, J. G. (1974). "General Reports: Eastern Europe," *Proceedings*, European Symposium of Penetration Testing, Stockholm, Sweden, Vol. 2.1, 24–39.

Supplementary References for Further Study

American Association of State Highway and Transportation Officials (1978). *Standard Specifications for Transportation Materials and Methods of Sampling and Testing*, Part II, Washington, D.C.

Brown, R. E. (1977). "Drill Rod Influence on Standard Penetration Test," *Journal of the Geotechnical Engineering Division*, ASCE, Vol. 103, No. GT11, 1332–1336.

Hvorslev, M. J. (1949). *Subsoil Exploration and Sampling of Soils for Civil Engineering Purposes*, Waterways Experiment Station, Vicksburg, Mississippi.

Marcuson, W. F., III, and Bieganousky, W. A. (1977). "SPT and Relative Density of Coarse Sands," *Journal of the Geotechnical Engineering Division*, ASCE, Vol. 103, No. GT11, 1295–1310.

CONVERSION FACTORS

A.1 CONVERSION FACTORS FROM ENGLISH TO SI UNITS

Length:
$$1 \text{ ft} = 0.3048 \text{ m}$$
$$1 \text{ ft} = 30.48 \text{ cm}$$
$$1 \text{ ft} = 304.8 \text{ mm}$$
$$1 \text{ in.} = 0.0254 \text{ m}$$
$$1 \text{ in.} = 2.54 \text{ cm}$$
$$1 \text{ in.} = 25.4 \text{ mm}$$

Area:
$$1 \text{ ft}^2 = 929.03 \times 10^{-4} \text{ m}^2$$
$$1 \text{ ft}^2 = 929.03 \text{ cm}^2$$
$$1 \text{ ft}^2 = 929.03 \times 10^2 \text{ mm}^2$$
$$1 \text{ in.}^2 = 6.452 \times 10^{-4} \text{ m}^2$$
$$1 \text{ in.}^2 = 6.452 \text{ cm}^2$$
$$1 \text{ in.}^2 = 645.16 \text{ mm}^2$$

Volume:
$$1 \text{ ft}^3 = 28.317 \times 10^{-3} \text{ m}^3$$
$$1 \text{ ft}^3 = 28.317 \text{ cm}^3 \times 10^3$$
$$1 \text{ in.}^3 = 16.387 \times 10^{-6} \text{ m}^3$$
$$1 \text{ in.}^3 = 16.387 \text{ cm}^3$$

Section modulus:
$$1 \text{ in.}^3 = 0.16387 \times 10^5 \text{ mm}^3$$
$$1 \text{ in.}^3 = 0.16387 \times 10^{-4} \text{ m}^3$$

Coefficient of permeability:
$$1 \text{ ft/min} = 0.3048 \text{ m/min}$$
$$1 \text{ ft/min} = 30.48 \text{ cm/min}$$
$$1 \text{ ft/min} = 304.8 \text{ mm/min}$$
$$1 \text{ ft/sec} = 0.3048 \text{ m/sec}$$
$$1 \text{ ft/sec} = 304.8 \text{ mm/sec}$$
$$1 \text{ in./min} = 0.0254 \text{ m/min}$$
$$1 \text{ in./sec} = 2.54 \text{ cm/sec}$$
$$1 \text{ in./sec} = 25.4 \text{ mm/sec}$$

Coefficient of consolidation:	$1 \text{ in.}^2/\text{sec} = 6.452 \text{ cm}^2/\text{sec}$
	$1 \text{ in.}^2/\text{sec} = 20.346 \times 10^3 \text{ m}^2/\text{yr}$
	$1 \text{ ft}^2/\text{sec} = 929.03 \text{ cm}^2/\text{sec}$

Force:

1 lb	$= 4.448 \text{ N}$
1 lb	$= 4.448 \times 10^{-3} \text{ kN}$
1 lb	$= 0.4536 \text{ kgf}$
1 kip	$= 4.448 \text{ kN}$
1 U.S. ton	$= 8.896 \text{ kN}$
1 lb	$= 0.4536 \times 10^{-3} \text{ metric ton}$
1 lb/ft	$= 14.593 \text{ N/m}$

Stress:

1 lb/ft^2	$= 47.88 \text{ N/m}^2$
1 lb/ft^2	$= 0.04788 \text{ kN/m}^2$
1 U.S. ton/ft^2	$= 95.76 \text{ kN/m}^2$
1 kip/ft^2	$= 47.88 \text{ kN/m}^2$
1 lb/in.^2	$= 6.895 \text{ kN/m}^2$

Unit weight:

1 lb/ft^3	$= 0.1572 \text{ kN/m}^3$
1 lb/in.^3	$= 271.43 \text{ kN/m}^3$

Moment:

1 lb-ft	$= 1.3558 \text{ N} \cdot \text{m}$
1 lb-in.	$= 0.11298 \text{ N} \cdot \text{m}$

Energy

1 ft-lb	$= 1.3558 \text{ J}$

Moment of inertia:

1 in.^4	$= 0.4162 \times 10^6 \text{ mm}^4$
1 in.^4	$= 0.4162 \times 10^{-6} \text{ m}^4$

A.2 CONVERSION FACTORS FROM SI TO ENGLISH UNITS

Length:

1 m	$= 3.281 \text{ ft}$
1 cm	$= 3.281 \times 10^{-2} \text{ ft}$
1 mm	$= 3.281 \times 10^{-3} \text{ ft}$
1 m	$= 39.37 \text{ in.}$
1 cm	$= 0.3937 \text{ in.}$
1 mm	$= 0.03937 \text{ in.}$

Area:

1 m^2	$= 10.764 \text{ ft}^2$
1 cm^2	$= 10.764 \times 10^{-4} \text{ ft}^2$
1 mm^2	$= 10.764 \times 10^{-6} \text{ ft}^2$
1 m^2	$= 1550 \text{ in.}^2$
1 cm^2	$= 0.155 \text{ in.}^2$
1 mm^2	$= 0.155 \times 10^{-2} \text{ in.}^2$

Volume:
$$1 \text{ m}^3 = 35.32 \text{ ft}^3$$
$$1 \text{ cm}^3 = 35.32 \times 10^{-4} \text{ ft}^3$$
$$1 \text{ m}^3 = 61{,}023.4 \text{ in.}^3$$
$$1 \text{ cm}^3 = 0.061023 \text{ in.}^3$$

Force:
$$1 \text{ N} = 0.2248 \text{ lb}$$
$$1 \text{ kN} = 224.8 \text{ lb}$$
$$1 \text{ kgf} = 2.2046 \text{ lb}$$
$$1 \text{ kN} = 0.2248 \text{ kip}$$
$$1 \text{ kN} = 0.1124 \text{ U.S. ton}$$
$$1 \text{ metric ton} = 2204.6 \text{ lb}$$
$$1 \text{ N/m} = 0.0685 \text{ lb/ft}$$

Stress:
$$1 \text{ N/m}^2 = 20.885 \times 10^{-3} \text{ lb/ft}^2$$
$$1 \text{ kN/m}^2 = 20.885 \text{ lb/ft}^2$$
$$1 \text{ kN/m}^2 = 0.01044 \text{ U.S. ton/ft}^2$$
$$1 \text{ kN/m}^2 = 20.885 \times 10^{-3} \text{ kip/ft}^2$$
$$1 \text{ kN/m}^2 = 0.145 \text{ lb/in.}^2$$

Unit weight:
$$1 \text{ kN/m}^3 = 6.361 \text{ lb/ft}^3$$
$$1 \text{ kN/m}^3 = 0.003682 \text{ lb/in.}^3$$

Moment:
$$1 \text{ N} \cdot \text{m} = 0.7375 \text{ lb-ft}$$
$$1 \text{ N} \cdot \text{m} = 8.851 \text{ lb-in.}$$

Energy:
$$1 \text{ J} = 0.7375 \text{ ft-lb}$$

Moment of inertia:
$$1 \text{ mm}^4 = 2.402 \times 10^{-6} \text{ in.}^4$$
$$1 \text{ m}^4 = 2.402 \times 10^6 \text{ in.}^4$$

Section modulus:
$$1 \text{ mm}^3 = 6.102 \times 10^{-5} \text{ in.}^3$$
$$1 \text{ m}^3 = 6.102 \times 10^4 \text{ in.}^3$$

Coefficient of permeability:
$$1 \text{ m/min} = 3.281 \text{ ft/min}$$
$$1 \text{ cm/min} = 0.03281 \text{ ft/min}$$
$$1 \text{ mm/min} = 0.003281 \text{ ft/min}$$
$$1 \text{ m/sec} = 3.281 \text{ ft/sec}$$
$$1 \text{ mm/sec} = 0.03281 \text{ ft/sec}$$
$$1 \text{ m/min} = 39.37 \text{ in./min}$$
$$1 \text{ cm/sec} = 0.3937 \text{ in./sec}$$
$$1 \text{ mm/sec} = 0.03937 \text{ in./sec}$$

Coefficient of consolidation:
$$1 \text{ cm}^2/\text{sec} = 0.155 \text{ in.}^2/\text{sec}$$
$$1 \text{ m}^2/\text{yr} = 4.915 \times 10^{-5} \text{ in.}^2/\text{sec}$$
$$1 \text{ cm}^2/\text{sec} = 1.0764 \times 10^{-3} \text{ ft}^2/\text{sec}$$

VARIOUS FORMS OF RELATIONSHIPS FOR γ, γ_d, AND γ_{sat}

Unit-weight relationship

$$\gamma = \frac{(1 + w)G_s \gamma_w}{1 + e}$$

$$\gamma = \frac{(G_s + Se)\gamma_w}{1 + e}$$

$$\gamma = \frac{(1 + w)G_s \gamma_w}{1 + \dfrac{wG_s}{S}}$$

$$\gamma = G_s \gamma_w (1 - n)(1 + w)$$

Dry unit weight

$$\gamma_d = \frac{\gamma}{1 + w}$$

$$\gamma_d = \frac{G_s \gamma_w}{1 + e}$$

$$\gamma_d = G_s \gamma_w (1 - n)$$

$$\gamma_d = \frac{G_s}{1 + \dfrac{wG_s}{S}} \gamma_w$$

$$\gamma_d = \frac{eS\gamma_w}{(1 + e)w}$$

$$\gamma_d = \gamma_{\text{sat}} - n\gamma_w$$

$$\gamma_d = \gamma_{\text{sat}} - \left(\frac{e}{1 + e}\right)\gamma_w$$

Saturated unit weight

$$\gamma_{\text{sat}} = \frac{(G_s + e)\gamma_w}{1 + e}$$

$$\gamma_{\text{sat}} = [(1 - n)G_s + n]\gamma_w$$

$$\gamma_{\text{sat}} = \left(\frac{1 + w}{1 + wG_s}\right)G_s \gamma_w$$

$$\gamma_{\text{sat}} = \left(\frac{e}{w}\right)\left(\frac{1 + w}{1 + e}\right)\gamma_w$$

$$\gamma_{\text{sat}} = \gamma_d + n\gamma_w$$

$$\gamma_{\text{sat}} = \gamma_d + \left(\frac{e}{1 + e}\right)\gamma_w$$

SUMMARY OF COMPACTION TEST SPECIFICATIONS

Description		ASTM D-698; AASHTO T-99				ASTM D-1557; AASHTO T-180			
		Method A	Method B	Method C	Method D	Method A	Method B	Method C	Method D
Mold: Volume	ft³	1/30	1/13.33	1/30	1/13.33	1/30	1/13.33	1/30	1/13.33
	cm³	943.9	2124.3	943.9	2124.3	943.9	2124.3	943.9	2124.3
Height	in.	4.58	4.58	4.58	4.58	4.58	4.58	4.58	4.58
	mm	116.33	116.33	116.33	116.33	116.33	116.33	116.33	116.33
Diameter	in.	4	6	4	6	4	6	4	6
	mm	101.6	152.4	101.6	152.4	101.6	152.4	101.6	152.4
Weight (mass) of hammer	lb	5.5	5.5	5.5	5.5	10	10	10	10
	kg	2.5	2.5	2.5	2.5	4.54	4.54	4.54	4.54
Height of drop of hammer	in.	12	12	12	12	18	18	18	18
	mm	304.8	304.8	304.8	304.8	457.2	457.2	457.2	457.2
Number of layers of soil		3	3	3	3	5	5	5	5
Number of blows per layers		25	56	25	56	25	56	25	56
Tested on soil fraction passing sieve		No. 4	No. 4	3/4 in.	3/4 in.	No. 4	No. 4	3/4 in.	3/4 in.

FILTER CRITERIA DEVELOPED FROM LABORATORY TESTING

Investigator	Year	Criteria developed
Bertram	1940	$\dfrac{D_{15(F)}}{D_{85(B)}} < 6;\ \dfrac{D_{15(F)}}{D_{85(B)}} < 9$
U.S. Corps of Engineers	1948	$\dfrac{D_{15(F)}}{D_{85(B)}} < 5;\ \dfrac{D_{50(F)}}{D_{50(B)}} < 25;$ $\dfrac{D_{15(F)}}{D_{15(B)}} < 20$
Sherman	1953	For $C_{u(base)} < 1.5$: $\dfrac{D_{15(F)}}{D_{15(B)}} < 6;\ \dfrac{D_{15(F)}}{D_{15(B)}} < 20;$ $\dfrac{D_{50(F)}}{D_{50(B)}} < 25$ For $1.5 < C_{u(base)} < 4.0$: $\dfrac{D_{15(F)}}{D_{85(B)}} < 5;\ \dfrac{D_{15(F)}}{D_{15(B)}} < 20;$ $\dfrac{D_{50(F)}}{D_{50(B)}} < 20$ For $C_{u(base)} > 4.0$: $\dfrac{D_{15(F)}}{D_{85(B)}} < 5;\ \dfrac{D_{15(F)}}{D_{85(B)}} < 40;$ $\dfrac{D_{15(F)}}{D_{85(B)}} < 25$
Leatherwood and Peterson	1954	$\dfrac{D_{15(F)}}{D_{85(B)}} < 4.1;\ \dfrac{D_{50(F)}}{D_{50(B)}} < 5.3$
Karpoff	1955	Uniform filter: $5 < \dfrac{D_{50(F)}}{D_{50(B)}} < 10$ Well-graded filter: $12 < \dfrac{D_{50(F)}}{D_{50(B)}} < 58;$ $12 < \dfrac{D_{15(F)}}{D_{15(B)}} < 40;$ and parallel grain-size curves
Zweck and Davidenkoff	1957	Base of medium and coarse uniform sand: $5 < \dfrac{D_{50(F)}}{D_{50(B)}} < 10$ Base of fine uniform sand: $5 < \dfrac{D_{50(F)}}{D_{50(B)}} < 15$ Base of well-graded fine sand: $5 < \dfrac{D_{50(F)}}{D_{50(B)}} < 25$

Note: $D_{50(F)}$ = diameter through which 50% of the filter passes; $D_{50(B)}$ = diameter through which 50% of the base material passes; C_u = uniformity coefficient

REFERENCES

Bertram, G. E. (1940). "An Experimental Investigation of Protective Filters," Soil Mechanics Series No. 7, *Publication No. 267*, Harvard University, Cambridge, Mass., 1–21.

Karpoff, K. P. (1955). "The Use of Laboratory Tests to Develop Design Criteria for Protective Filters," *Proceedings*, ASTM, Vol. 55, 1183–1198.

Leatherwood, F. N., and Peterson, D. F., Jr. (1954). "Hydraulic Head Loss at the Interface between Uniform Sands of Different Sizes," *Transactions*, American Geophysical Union, Vol. 35, No. 4, 588–594.

Sherman, W. C. (1953). "Filter Experiments and Design Criteria," U.S. Army Waterways Experiment Station, Vicksburg, Mississippi, NTIS AD 771076.

U.S. Corps of Engineers (1948). "Laboratory Investigation of Filters for Enid and Grenada Dams," U.S. Army Waterways Experiment Station, Vicksburg, Mississippi, Technical Memorandum No. 3–245.

Zweck, H., and Davidenkoff, R. (1957). "Etude Expérimentale des Filtres de Granulométrie Uniforme," *Proceedings*, Fourth International Conference on Soil Mechanics and Foundation Engineering, London, Vol. 2, 410–413.

SOIL COMPRESSIBILITY RELATIONSHIPS

E.1 EMPIRICAL PROCEDURE TO ESTIMATE VOID RATIO-PRESSURE RELATIONSHIPS

As has been shown in Figure 8.13, the void ratio-pressure relationships for an over-consolidated clay will follow the path *cbd* during consolidation (Figure E.1), the *in situ* void ratio and effective overburden pressure being e_O and p_O, respectively. Nagaraj and Murty (1985) provided empirical relationships for predicting the *e*-log *p* plot—that is, *cb* and *bd*. The relationship for line *cb* (that is, for overconsolidated clay) is as follows:

$$\frac{e}{e_L} = 1.122 - 0.188 \log p_c - 0.0463 \log p \qquad (E.1)$$

where e = void ratio at effective pressure p (kN/m²)
   e_L = void ratio of the soil at liquid limit
   p_c = preconsolidation pressure (kN/m²)

In Chapter 2 it was shown that, for saturated soil, $e = wG_s$. So

$$e_L = \left[\frac{LL(\%)}{100}\right]G_s \qquad (E.2)$$

If the *in situ* values e_O and p_O are known, then the magnitude of the preconsolidation pressure can be calculated as [from Eq. (E.1)]

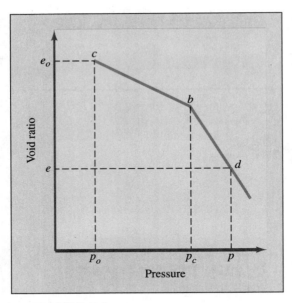

▼ **FIGURE E.I**

$$\log p_c = \frac{1.122 - \left(\dfrac{e_O}{e_L}\right) - 0.0463 \log p_O \;\overset{\text{kN/m}^2}{\downarrow}}{0.188}$$

$\underset{\uparrow}{}$
kN/m²

(E.3)

For line *bd* (that is, for normally consolidated clay), the relationship is

$$\frac{e}{e_L} = 1.122 - 0.2343 \log p$$
$$\underset{\text{kN/m}^2}{\uparrow}$$

(E.4)

or

$$e = \left[\frac{LL(\%)}{100}\right] G_s (1.122 - 0.2343 \log p)$$
$$\underset{\text{kN/m}^2}{\uparrow}$$

(E.5)

So, for a given overburden pressure, p, the void ratio in the field can be estimated if the liquid limit and the specific gravity of the soil solid are known.

E.2 CORRELATION FOR COMPRESSION INDEX

Several correlations for the compression index are available now. They have been developed by testing various clays. Some of these correlations are given in Table E.1. It is important to realize that they are for estimation purposes only.

▼ **TABLE E.I** Correlations for Compression Index, C_c*

Equation	Reference	Region of applicability
$C_c = 0.007(LL - 7)$	Skempton (1944)	Remolded clays
$C_c = 0.01w_N$		Chicago clays
$C_c = 1.15(e_o - 0.27)$	Nishida (1956)	All clays
$C_c = 0.30(e_o - 0.27)$	Hough (1957)	Inorganic cohesive soil: silt, silty clay, clay
$C_c = 0.0115w_N$		Organic soils, peats, organic silt, and clay
$C_c = 0.0046(LL - 9)$		Brazilian clays
$C_c = 0.75(e_o - 0.5)$		Soils with low plasticity
$C_c = 0.208e_o + 0.0083$		Chicago clays
$C_c = 0.156e_o + 0.0107$		All clays

* After Rendon-Herrero (1980)
Note: e_o = *in situ* void ratio; w_N = *in situ* water content

REFERENCES

Hough, B. K. (1957). *Basic Soils Engineering*. New York: Ronald Press.

Nagaraj, T., and Murty, B. R. S. (1985). "Prediction of the Preconsolidation Pressure and Recompression Index of Soils," *Geotechnical Testing Journal*, Vol. 8, No. 4, 199–202.

Nishida, Y. (1956). "A Brief Note on Compression Index of Soils," *Journal of the Soil Mechanics and Foundations Division*, ASCE, Vol. 82, No. SM3, 1027-1–1027-14.

Rendon-Herrero, O. (1980). "Universal Compression Index Equation," *Journal of the Geotechnical Engineering Division*, ASCE, Vol. 106, No. GT11, 1179–1200.

Skempton, A. W. (1944). "Notes on the Compressibility of Clays," *Quarterly Journal of the Geological Society of London*, Vol. 100, 119–135.

BRACED CUTS

▼ **TABLE F.I** $P_a/0.5\gamma H^2$ Against ϕ, δ, and $n_a(c = 0)$ for Braced Cuts (Section 10.14; See Figure 10.38 for notation)*

ϕ (deg)	δ (deg)	$P_a/0.5\gamma H^2$			
		$n_a = 0.3$	$n_a = 0.4$	$n_a = 0.5$	$n_a = 0.6$
10	0	0.653	0.734	0.840	0.983
	5	0.623	0.700	0.799	0.933
	10	0.610	0.685	0.783	0.916
15	0	0.542	0.602	0.679	0.778
	5	0.518	0.575	0.646	0.739
	10	0.505	0.559	0.629	0.719
	15	0.499	0.554	0.623	0.714
20	0	0.499	0.495	0.551	0.622
	5	0.430	0.473	0.526	0.593
	10	0.419	0.460	0.511	0.575
	15	0.413	0.454	0.504	0.568
	20	0.413	0.454	0.504	0.569
25	0	0.371	0.405	0.447	0.499
	5	0.356	0.389	0.428	0.477
	10	0.347	0.378	0.416	0.464
	15	0.342	0.373	0.410	0.457
	20	0.341	0.372	0.409	0.456
	25	0.344	0.375	0.413	0.461
30	0	0.304	0.330	0.361	0.400
	5	0.293	0.318	0.347	0.384
	10	0.286	0.310	0.339	0.374
	15	0.282	0.306	0.334	0.368
	20	0.281	0.305	0.332	0.367
	25	0.284	0.307	0.335	0.370
	30	0.289	0.313	0.341	0.377

ϕ (deg)	δ (deg)	$P_a/0.5\gamma H^2$			
		$n_a = 0.3$	$n_a = 0.4$	$n_a = 0.5$	$n_a = 0.6$
35	0	0.247	0.267	0.290	0.318
	5	0.239	0.258	0.280	0.318
	10	0.234	0.252	0.273	0.300
	15	0.231	0.249	0.270	0.296
	20	0.231	0.248	0.269	0.295
	25	0.232	0.250	0.271	0.297
	30	0.236	0.254	0.276	0.302
	35	0.243	0.262	0.284	0.312
40	0	0.198	0.213	0.230	0.252
	5	0.192	0.206	0.223	0.244
	10	0.189	0.202	0.219	0.238
	15	0.187	0.200	0.216	0.236
	20	0.187	0.200	0.216	0.235
	25	0.188	0.202	0.218	0.237
	30	0.192	0.205	0.222	0.241
	35	0.197	0.211	0.228	0.248
	40	0.205	0.220	0.237	0.259
45	0	0.156	0.167	0.180	0.196
	5	0.152	0.163	0.175	0.190
	10	0.150	0.160	0.172	0.187
	15	0.148	0.159	0.171	0.185
	20	0.149	0.159	0.171	0.185
	25	0.150	0.160	0.173	0.187
	30	0.153	0.164	0.176	0.190
	35	0.158	0.168	0.181	0.196
	40	0.164	0.175	0.188	0.204
	45	0.173	0.184	0.198	0.213

* After Kim and Preber (1969)

▼ **TABLE F.2** Values of $P_a/0.5\gamma H^2$ for Cuts in a c-ϕ Soil with the Assumption $c_a = c(\tan \delta/\tan \phi)$*

δ (deg)	$n_a = 0.3$ and $c/\gamma H = 0.1$	$n_a = 0.4$ and $c/\gamma H = 0.1$	$n_a = 0.5$ and $c/\gamma H = 0.1$
$\phi = 15°$			
0	0.254	0.285	0.322
5	0.214	0.240	0.270
10	0.187	0.210	0.238
15	0.169	0.191	0.218
$\phi = 20°$			
0	0.191	0.21	0.236
5	0.160	0.179	0.200
10	0.140	0.156	0.173
15	0.122	0.127	0.154
20	0.113	0.124	0.140
$\phi = 25°$			
0	0.138	0.150	0.167
5	0.116	0.128	0.141
10	0.099	0.110	0.122
15	0.085	0.095	0.106
20	0.074	0.083	0.093
25	0.065	0.074	0.083
$\phi = 30°$			
0	0.093	0.103	0.113
5	0.078	0.086	0.094
10	0.066	0.073	0.080
15	0.056	0.060	0.067
20	0.047	0.051	0.056
25	0.036	0.042	0.047
30	0.029	0.033	0.038

* After Kim and Preber (1969)

REFERENCE

Kim, J. S., and Preber, T. (1969). "Earth Pressure Against Braced Excavations," *Journal of the Soil Mechanics and Foundations Division*, ASCE, Vol. 95, No. SM6, 1581–1584.

FACTOR OF SAFETY FOR SLOPES

G.1 CONTOURS OF EQUAL FACTORS OF SAFETY USING TAYLOR'S METHOD

Using Taylor's method of slope stability (Figure 12.15), Singh (1970) provided contours of equal factors of safety, F_s, for various slopes. In preparing those, it was assumed that the pore water pressure in the soil is 0. Figure G.1 shows the plots of these contours of F_s for various slope angles, β. (*Note:* In Figure G.1, c = cohesion, γ = unit weight of soil, H = slope height, and ϕ = soil friction angle.)

G.2 BISHOP AND MORGENSTERN'S METHOD

According to Bishop and Morgenstern's method (Section 12.10), the factor of safety of a slope with seepage can be evaluated by using the equation

$$F_s = m' - n'r_u \tag{12.71}$$

where m' and n' are stability coefficients. Table G.1 provides the values of m' and n' for various combinations of $c/\gamma H$, D, ϕ, and β.

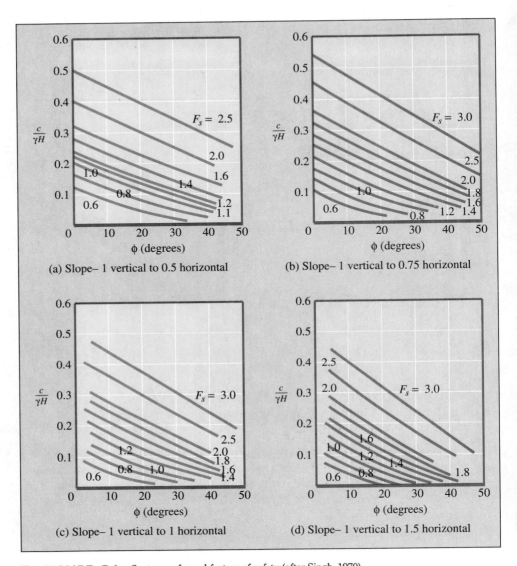

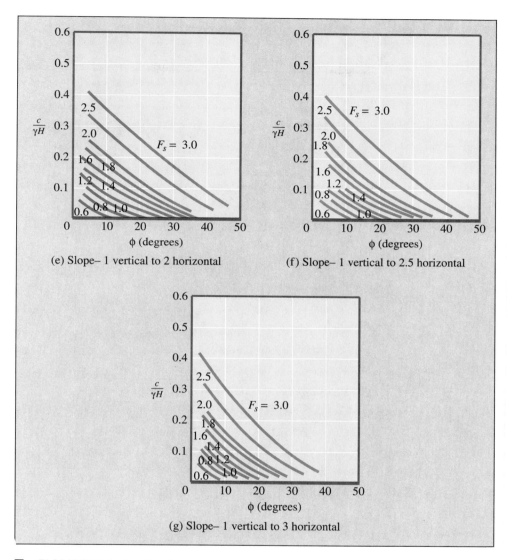

(e) Slope– 1 vertical to 2 horizontal

(f) Slope– 1 vertical to 2.5 horizontal

(g) Slope– 1 vertical to 3 horizontal

▼ **FIGURE G.1** (*continued*)

▼ **TABLE G.I** **Values of _m'_ and _n'_***

---

a. *Stability coefficients m' and n' for c/γH = 0*

	Stability coefficients for earth slopes						
	Slope 2 : 1		Slope 3 : 1		Slope 4 : 1		Slope 5 : 1	
ϕ	m'	n'	m'	n'	m'	n'	m'	n'
10.0	0.353	0.441	0.529	0.588	0.705	0.749	0.882	0.917
12.5	0.443	0.554	0.665	0.739	0.887	0.943	1.109	1.153
15.0	0.536	0.670	0.804	0.893	1.072	1.139	1.340	1.393
17.5	0.631	0.789	0.946	1.051	1.261	1.340	1.577	1.639
20.0	0.728	0.910	1.092	1.213	1.456	1.547	1.820	1.892
22.5	0.828	1.035	1.243	1.381	1.657	1.761	2.071	2.153
25.0	0.933	1.166	1.399	1.554	1.865	1.982	2.332	2.424
27.5	1.041	1.301	1.562	1.736	2.082	2.213	2.603	2.706
30.0	1.155	1.444	1.732	1.924	2.309	2.454	2.887	3.001
32.5	1.274	1.593	1.911	2.123	2.548	2.708	3.185	3.311
35.0	1.400	1.750	2.101	2.334	2.801	2.977	3.501	3.639
37.5	1.535	1.919	2.302	2.558	3.069	3.261	3.837	3.989
40.0	1.678	2.098	2.517	2.797	3.356	3.566	4.196	4.362

---

b. *Stability coefficients m' and n' for c/γH = 0.025 and D = 1.00*

	Stability coefficients for earth slopes						
	Slope 2 : 1		Slope 3 : 1		Slope 4 : 1		Slope 5 : 1	
ϕ	m'	n'	m'	n'	m'	n'	m'	n'
10.0	0.678	0.534	0.906	0.683	1.130	0.846	1.365	1.031
12.5	0.790	0.655	1.066	0.849	1.337	1.061	1.620	1.282
15.0	0.901	0.776	1.224	1.014	1.544	1.273	1.868	1.534
17.5	1.012	0.898	1.380	1.179	1.751	1.485	2.121	1.789
20.0	1.124	1.022	1.542	1.347	1.962	1.698	2.380	2.050
22.5	1.239	1.150	1.705	1.518	2.177	1.916	2.646	2.317
25.0	1.356	1.282	1.875	1.696	2.400	2.141	2.921	2.596
27.5	1.478	1.421	2.050	1.882	2.631	2.375	3.207	2.886
30.0	1.606	1.567	2.235	2.078	2.873	2.622	3.508	3.191
32.5	1.739	1.721	2.431	2.285	3.127	2.883	3.823	3.511
35.0	1.880	1.885	2.635	2.505	3.396	3.160	4.156	3.849
37.5	2.030	2.060	2.855	2.741	3.681	3.458	4.510	4.209
40.0	2.190	2.247	3.090	2.993	3.984	3.778	4.885	4.592

▼ **TABLE G.1** (*Continued*)

c. *Stability coefficients m' and n' for c/γH = 0.025 and D = 1.25*

	Stability coefficients for earth slopes							
	Slope 2 : 1		Slope 3 : 1		Slope 4 : 1		Slope 5 : 1	
ϕ	m'	n'	m'	n'	m'	n'	m'	n'
10.0	0.737	0.614	0.901	0.726	1.085	0.867	1.285	1.014
12.5	0.878	0.759	1.076	0.908	1.299	1.098	1.543	1.278
15.0	1.019	0.907	1.253	1.093	1.515	1.311	1.803	1.545
17.5	1.162	1.059	1.433	1.282	1.736	1.541	2.065	1.814
20.0	1.309	1.216	1.618	1.478	1.961	1.775	2.334	2.090
22.5	1.461	1.379	1.808	1.680	2.194	2.017	2.610	2.373
25.0	1.619	1.547	2.007	1.891	2.437	2.269	2.879	2.669
27.5	1.783	1.728	2.213	2.111	2.689	2.531	3.196	2.976
30.0	1.956	1.915	2.431	2.342	2.953	2.806	3.511	3.299
32.5	2.139	2.112	2.659	2.686	3.231	3.095	3.841	3.638
35.0	2.331	2.321	2.901	2.841	3.524	3.400	4.191	3.998
37.5	2.536	2.541	3.158	3.112	3.835	3.723	4.563	4.379
40.0	2.753	2.775	3.431	3.399	4.164	4.064	4.958	4.784

d. *Stability coefficients m' and n' for c/γH = 0.05 and D = 1.00*

	Stability coefficients for earth slopes							
	Slope 2 : 1		Slope 3 : 1		Slope 4 : 1		Slope 5 : 1	
ϕ	m'	n'	m'	n'	m'	n'	m'	n'
10.0	0.913	0.563	1.181	0.717	1.469	0.910	1.733	1.069
12.5	1.030	0.690	1.343	0.878	1.688	1.136	1.995	1.316
15.0	1.145	0.816	1.506	1.043	1.904	1.353	2.256	1.567
17.5	1.262	0.942	1.671	1.212	2.117	1.565	2.517	1.825
20.0	1.380	1.071	1.840	1.387	2.333	1.776	2.783	2.091
22.5	1.500	1.202	2.014	1.568	2.551	1.989	3.055	2.365
25.0	1.624	1.338	2.193	1.757	2.778	2.211	3.336	2.651
27.5	1.753	1.480	1.380	1.952	3.013	2.444	3.628	2.948
30.0	1.888	1.630	2.574	2.157	3.261	2.693	3.934	3.259
32.5	2.029	1.789	2.777	2.370	3.523	2.961	4.256	3.585
35.0	2.178	1.958	2.990	2.592	3.803	3.253	4.597	3.927
37.5	2.336	2.138	3.215	2.826	4.103	3.574	4.959	4.288
40.0	2.505	2.332	3.451	3.071	4.425	3.926	5.344	4.668

▼ **TABLE G.I**   *(Continued)*

e.   Stability coefficients m′ and n′ for c/γH = 0.05 and D = 1.25

	Stability coefficients for earth slopes							
	Slope 2 : 1		Slope 3 : 1		Slope 4 : 1		Slope 5 : 1	
ϕ	m'	n'	m'	n'	m'	n'	m'	n'
10.0	0.919	0.633	1.119	0.766	1.344	0.886	1.594	1.042
12.5	1.065	0.792	1.294	0.941	1.563	1.112	1.850	1.300
15.0	1.211	0.950	1.471	1.119	1.782	1.338	2.109	1.562
17.5	1.359	1.108	1.650	1.303	2.004	1.567	2.373	1.831
20.0	1.509	1.266	1.834	1.493	2.230	1.799	2.643	2.107
22.5	1.663	1.428	2.024	1.690	2.463	2.038	2.921	2.392
25.0	1.822	1.595	2.222	1.897	2.705	2.287	3.211	2.690
27.5	1.988	1.769	2.428	2.113	2.957	2.546	3.513	2.999
30.0	2.161	1.950	2.645	2.342	3.221	2.819	3.829	3.324
32.5	2.343	2.141	2.873	2.583	3.500	3.107	4.161	3.665
35.0	2.535	2.344	3.114	2.839	3.795	3.413	4.511	4.025
37.5	2.738	2.560	3.370	3.111	4.109	3.740	4.881	4.405
40.0	2.953	2.791	3.642	3.400	4.442	4.090	5.273	4.806

f.   Stability coefficients m′ and n′ for c/γH = 0.05 and D = 1.50

	Stability coefficients for earth slopes							
	Slope 2 : 1		Slope 3 : 1		Slope 4 : 1		Slope 5 : 1	
ϕ	m'	n'	m'	n'	m'	n'	m'	n'
10.0	1.022	0.751	1.170	0.828	1.343	0.974	1.547	1.108
12.5	1.202	0.936	1.376	1.043	1.589	1.227	1.829	1.399
15.0	1.383	1.122	1.583	1.260	1.835	1.480	2.112	1.690
17.5	1.565	1.309	1.795	1.480	2.084	1.734	2.398	1.983
20.0	1.752	1.501	2.011	1.705	2.337	1.993	2.690	2.280
22.5	1.943	1.698	2.234	1.937	2.597	2.258	2.990	2.585
25.0	2.143	1.903	2.467	2.179	2.867	2.534	3.302	2.902
27.5	2.350	2.117	2.709	2.431	3.148	2.820	3.626	3.231
30.0	2.568	2.342	2.964	2.696	3.443	3.120	3.967	3.577
32.5	2.798	2.580	3.232	2.975	3.753	3.436	4.326	3.940
35.0	3.041	2.832	3.515	3.269	4.082	3.771	4.707	4.325
37.5	3.299	3.102	3.817	3.583	4.431	4.128	5.112	4.735
40.0	3.574	3.389	4.136	3.915	4.803	4.507	5.543	5.171

* After Bishop and Morgenstern (1960)

REFERENCES

Bishop, A. W., and Morgenstern, N. R. (1960). "Stability Coefficients for Earth Slopes," *Geotechnique*, Vol. 20, No. 4, 129–147.

Singh, A. (1970). "Shear Strength and Stability of Man-Made Slopes," *Journal of the Soil Mechanics and Foundations Division*, ASCE, Vol. 96, No. SM6, 1879–1892.

ANSWERS TO PROBLEMS

Chapter 1

1.1 $C_u = 3.45; C_c = 0.86$

1.2 $C_u = 4.33; C_c = 0.73$

1.3   **b.** $D_{10} = 0.15$ mm; $D_{30} = 0.17$ mm; $D_{60} = 0.27$ mm
        c.   1.8   **d.**   0.714

1.4   **b.** $D_{10} = 0.12$ mm; $D_{30} = 0.22$ mm; $D_{60} = 0.4$ mm
        c.   3.33   **d.**   1.01

1.5   **b.** $D_{10} = 0.14$ mm; $D_{30} = 0.26$ mm; $D_{60} = 0.50$ mm
        c.   3.57   **d.**   0.97

1.6   **b.** $D_{10} = 0.17$ mm; $D_{30} = 0.39$ mm; $D_{60} = 0.46$ mm
        c.   2.71   **d.**   1.95

1.7   **a.**   Gravel: 0%; sand: 6%; silt: 52%; clay: 42%
        b.   Gravel: 0%; sand: 3%; silt: 55%; clay: 42%

1.8   **a.**   Gravel: 0%; sand: 17%; silt: 57%; clay: 26%
        b.   Gravel: 0%; sand: 11%; silt: 63%; clay: 26%

1.9   **a.**   Gravel: 0%; sand: 39%; silt: 27%; clay: 34%
        b.   Gravel: 0%; sand: 29%; silt: 37%; clay: 34%

1.10   **a.**   Gravel: 0%; sand: 17%; silt: 72%; clay: 11%
        b.   Gravel: 0%; sand: 8%; silt: 81%; clay: 11%

1.11   0.0052 mm

1.12   0.0042 mm

Chapter 2

2.1 $w = 15.6\%, \gamma = 121$ lb/ft^3; $\gamma_d = 104.7$ lb/ft^3; $e = 0.59$; $n = 0.37$; $S = 70.6\%$

2.2 $w = 14.94\%$; $\gamma = 120$ lb/ft^3; $\gamma_d = 104.4$ lb/ft^3; $e = 0.572$; $n = 0.364$; $S = 68.7\%$

2.3   **a.**   115 lb/ft^3 **b.**   103.6 lb/ft^3
        c.   0.626   **d.**   0.385
        e.   47.4%   **f.**   0.0365 ft^3

2.7   **a.**   106.6 lb/ft^3 **b.**   0.45   **c.**   2.48

2.8   **a.**   16.33 kN/m^3 **b.**   0.58   **c.**   2.05

2.9   **a.**   101.1 lb/ft^3 **b.**   0.648
        c.   0.39   **d.**   44.5%

2.10   **a.**   8.73 lb/ft^3 **b.**   13.6 lb/ft^3

2.11   **a.**   14.34 kN/m^3 **b.**   0.458
        c.   47.9%   **d.**   238.53 kg/m^3

2.12 19.55%

2.13 **a.** 20.59 kN/m^3 **b.** 1.99%

2.14 $\gamma_{sat} = 19.14$ kN/m^3; $\gamma_d = 14.95$ kN/m^3

2.15 **a.** 117.4 lb/ft^3 **b.** 96.6 lb/ft^3 **c.** 77.7%

2.16 **a.** 119.9 lb/ft^3 **b.** 103.3 lb/ft^3 **c.** 70.7%

2.17 $\gamma_d = 103.7$ lb/ft^3; $\gamma = 122.6$ lb/ft^3

2.18 **a.** 0.65   **b.** 126.7 lb/ft^3

2.19 **a.** 2.66   **b.** 0.81   **2.20** 69.75 lb/ft^3

2.21 16.87 kN/m^3 **2.22** 18.8 kN/m^3

2.23 109 lb/ft^3 **2.24** **a.** $LL = 39.7$ **b.** 21

2.25 53.8%   **2.26** **a.** $LL = 23.5$ **b.** 7

2.27 -24% **2.28** 19.4%

Chapter 3

3.1

Soil	Classification
A	Clay
B	Sandy clay
C	Loam
D	Sandy clay and sandy clay loam
E	Sandy loam
F	Silty loam
G	Clay loam
H	Clay
I	Silty clay
J	Loam

3.2

Soil	Classification
A	Gravelly sandy loam
B	Gravelly clay loam
C	Gravelly silty clay
D	Clay
E	Gravelly clay

3.3

Soil	Classification and group index
1	A-4(3)
2	A-7-6(28)
3	A-6(8)
4	A-4(1)
5	A-7-6(8)
6	A-2-4(0)
7	A-3(0)
8	A-6(10)
9	A-1-b(0)
10	A-7-5(33)

3.4

Soil	Classification and group index
A	A-1-a(0)
B	A-2-4(0)
C	A-2-6(0)
D	A-3(0)
E	A-2-5(0)

3.5

Soil	Classification and group index
A	A-1-b(0)
B	A-1-b(0)
C	A-7-5(23)
D	A-7-6(27)
E	A-6(5)

3.6

Soil	Group symbol	Group name
1	ML	Sandy silt
2	CH	Fat clay with sand
3	CL	Sandy lean clay
4	SC	Clayey sand with gravel
5	ML	Sandy silt
6	SC	Clayey sand

3.7

Soil	Group symbol	Group name
A	SP	Poorly graded sand
B	SW-SM	Well-graded sand with silt
C	MH	Elastic silt with sand
D	CH	Fat clay
E	SC	Clayey sand

3.8

Soil	Group symbol	Group name
A	SC	Clayey sand
B	GM-GC	Silty clayey gravel with sand
C	CH	Fat clay with sand
D	ML	Sandy silt
E	SM	Silty sand with gravel

Chapter 4

4.1

w (%)	γ_{zav} (lb/ft^3)
5	147.5
8	137.7
10	131.9
12	126.6
15	119.3

4.2

w (%)	γ_{zav} (lb/ft^3)
5	140.6
10	126.4
15	114.8
20	105.1

4.3 **a.** $\gamma_d = \dfrac{S\gamma_w}{w + \dfrac{S}{G_s}}$

  b.

w (%)	γ_d (lb/ft^3)
5	141.8
10	125.9
15	113.2
20	102.9

4.4 102.5 lb/ft^3

4.5 $\gamma_{d(max)} = 106$ lb/ft^3; optimum moisture $= 14.4\%$

4.6 $\gamma_{d(max)} = 125$ lb/ft^3; optimum moisture $= 12.2\%$

4.7 **a.** 86.95%   **b.** 50.4%

4.8 97.5%

4.9 $R = 96\%$; $\gamma_{d(field)} = 15.84$ kN/m^3

4.10 **a.** 15.23 kN/m^3 **b.** 27%   **c.** 16.45 kN/m^3

4.11 **a.** 18.56 kN/m^3 **b.** 97.7%

4.12 18; good

4.13 20.65; fair

Chapter 5

5.1 0.7276 ft^3/hr/ft

5.2 0.924 m^3/hr/m

5.3 **a.** 0.175×10^{-2} in./min
  b. 0.315×10^{-2} in./min
  c. 1×10^{-2} in./min

5.4 **a.** 4.178×10^{-3} cm/sec
  b. 5.57×10^{-3} cm/sec
  c. 15.7×10^{-3} cm/sec

5.5

h (mm)	q (cm^3/sec)
800	2.08
700	1.82
600	1.56
500	1.30
400	1.04

5.6 0.149 in./min

5.7 24.5 in.

5.8 3.61×10^{-3} cm/sec

5.9 0.15 in.^{2}

5.10 8.52×10^{-15} m^2

5.11 0.144 ft/min

5.12 0.167 ft/min

5.13 **a.** 0.0175 cm/sec   **b.** 0.0124 cm/sec

5.14 5.67×10^{-2} cm/sec

5.15 0.72×10^{-2} cm/sec

5.16 8.08×10^{-3} cm/sec

5.17 0.787×10^{-6} cm/sec

5.18 0.309×10^{-6} cm/sec

5.19 0.803×10^{-6} cm/sec

5.20 0.306×10^{-6} cm/sec

5.21 0.1×10^{-6} cm/sec

5.22 290.52 cm³/hr

5.24 22.45

5.26 1.6×10^{-4} ft³/ft/sec

5.28 14.286×10^{-3} ft³/ft/min

5.30 0.0057 cm/sec

5.23 $h_A = 287.88$ mm; $h_B = 247.55$ mm

5.25 0.074 ft/min

5.27 13.98×10^{-6} m³/m/sec

5.29 84.79 ton/ft

5.31 0.973 ft³/day/ft

Chapter 6

6.1

Point	σ (lb/ft²)	u (lb/ft²)	σ' (lb/ft²)
A	0	0	0
B	440	0	440
C	1040	312	728
D	1790	686.4	1103.6

6.2

Point	σ (lb/ft²)	u (lb/ft²)	σ' (lb/ft²)
A	0	0	0
B	427.5	0	427.5
C	1587.5	624	963.5
D	2624.5	1154.4	1470.1

6.3

Point	σ (kN/m²)	u (kN/m²)	σ' (kN/m²)
A	0	0	0
B	31.2	0	31.2
C	79.8	29.43	50.37
D	151	68.67	82.33

6.4

Point	σ (kN/m²)	u (kN/m²)	σ' (kN/m²)
A	0	0	0
B	55.08	0	55.08
C	135.52	39.24	96.28
D	173.9	58.86	115.04

6.5

Point	σ (lb/ft²)	u (lb/ft²)	σ' (lb/ft)
A	0	0	0
B	826.8	0	826.8
C	1614.96	374.7	1240.56
D	1958.16	561.6	1396.6

6.6

Depth (m)	σ (kN/m²)	u (kN/m²)	σ' (kN/m²)
0	0	0	0
4	67.72	0	67.72
7	122.89	29.43	93.46

6.7

Depth (ft)	σ (lb/ft²)	u (lb/ft²)	σ' (lb/ft)
0	0	0	0
8	861.36	0	861.36
30	3435.36	1372.8	2062.56

6.8   **a.**

Point	σ (kN/m²)	u (kN/m²)	σ' (kN/m²)
A	0	0	0
B	64.8	0	64.8
C	169.2	49.05	120.15

    b.   2.65 m

6.9

e	i_{cr}
0.35	1.23
0.45	1.14
0.55	1.07
0.7	0.976
0.8	0.92

6.10   19.68 ft

6.12 7.88×10^{-3} ft³/sec

6.14   113.6 to 568.2 mm

6.11   0.89 m

6.13   1.063 m

6.15

Depth (ft)	σ (lb/ft²)	u (lb/ft²)	σ' (lb/ft²)
0	0	0	0
6−	663.9	0	663.9
6+	663.9	−104.83	768.73
10	1095.34	0	1095.34
19	2152.3	561.6	1590.7

6.16

Depth (m)	σ (kN/m²)	u (kN/m²)	σ' (kN/m²)
0	0	0	0
1.5−	26.1	0	26.1
1.5+	26.1	−7.21	33.31
3.25	55.76	0	55.76
6.25	111.14	29.43	81.71

6.17   3.79

6.18   4.89

Chapter 7

	σ_1	σ_3	σ_n	τ_n (magnitude)
7.1	129.24 lb/ft²	30.76 lb/ft²	51.03 lb/ft²	39.82 lb/ft²
7.2	922.3 lb/ft²	227.7 lb/ft²	200 lb/ft² *wrong*	175 lb/ft²
7.3	161.1 kN/m²	68.9 kN/m²	138.5 kN/m²	39.7 kN/m²
7.4	315.85 lb/ft²	109.15 lb/ft²	180.8 lb/ft²	98.4 lb/ft²
wrong → **7.5**	36.54 lb/in.²	−1.54 lb/in.²	36.5 lb/in.²	1.13 lb/in.²
7.6	95 kN/m²	30 kN/m²	94.2 kN/m²	7.1 kN/m²
7.7	109.1 kN/m²	25.9 kN/m²	29.1 kN/m²	16.08 kN/m²
7.8	13.9 lb/ft²			
7.9	8.48 lb/ft²			

7.10

x (m)	Δp (kN/m^2)
0	10.61
± 2	5.08
± 4	1.38
± 6	0.41
± 8	0.16

7.11 1.33 ft

7.12 3.3 kN/m^2

7.13 22.73 lb/ft^2

7.14 17.3 kN/m^2

7.15 226 lb/ft^2

7.16 2457 kN/m^2

7.17

x (ft)	Δp (lb/ft^2)
7.5	367.5
5	1001
-5	56
-7.5	8.75

7.18 2582 lb/ft^2

7.19 184.4 kN/m^2

7.20
  a. 358.2 lb/ft^2
  b. 847.8 lb/ft^2
  c. 117 lb/ft^2

7.21
  a. 360 lb/ft^2
  b. 855 lb/ft^2
  c. 117 lb/ft^2

7.22 89.4 kN/m^2

7.23 1384.2 lb/ft^2

Chapter 8

8.1 5.49 in.

8.2 152 mm

8.3 196 mm

8.4 128 mm

8.5 1.91 in.

8.6 **a.**

p (lb/ft^2)	e
500	0.575
1,000	0.553
2,000	0.52
4,000	0.478
8,000	0.442
16,000	0.418

  b. 940 lb/ft^2 **c** 0.133

8.7 **b.** 3.1 ton/ft^2 **c.** 0.53

8.8 1.327

8.9 0.87

8.10 6.2 in.

8.11 **a.** 5.08×10^{-4} m^2/kN   **b.** 1.145×10^{-7} cm^2/sec

8.12 3.57 days

8.13 90.1 days

8.14 **a.** 386.8 days   **b.** 671.6 days

8.15 3.04×10^{-3} cm^2/sec

8.16 647.3 sec

8.17 **a.** 31.25%   **b.** 158.87 days   **c.** 39.72 days

8.18 5.733×10^{-7} ft/min

8.19 **a.** 20 min   **b.** 53.8 min
  c. 3.178×10^{-5} in.^{2}/sec   **d.** 5.086×10^{-5} in.^{2}/sec

8.20 0.00242 in.^{2}/min

8.21

Point	Δq (lb/ft²)
A	2200
B	220
C	890

8.22 5.36 in.

8.23 0.39 in.

8.24 11.56 mm

Chapter 9

9.1 $\phi = 34°$, shear force = 32.36 lb

9.2 0.246 lb

9.3 37.5 lb

9.4 0.292 kN

9.5 23.5°

9.6 37°

9.7 51.1 lb/in.²

9.8 **a.** 61.55° **b.** $\sigma' = 42.08$ lb/in.²; $\tau = 15.63$ lb/in.²

9.9 15 lb/in.²

9.10 **a.** 65.6° **b.** $\sigma' = 33.82$ lb/in.²; $\tau = 26.88$ lb/in.²

9.11 28.57°

9.12 15 lb/in.²

9.13 **a.** 19.45° **b.** 54.73° **c.** $\sigma' = 368.03$ kN/m²; $\tau = 130.12$ kN/m²

9.14 **a.** 414 kN/m²

9.15 $\phi = 18°$; $c = 55$ kN/m²

9.16 530.28 kN/m²

9.17 0.98 ton/ft²

9.18 $\phi_{cu} = 18°$; $\phi = 29.8°$

9.19 $\phi_{cu} = 15°$; $\phi = 23.3°$

9.20 **a.** 18.4° **b.** 50.2 kN/m²

9.21 185.8 kN/m²

9.22 $\Delta\sigma_{d(f)} = 13.4$ lb/in.²; $\Delta u_{d(f)} = 7.43$ lb/in.²

9.23 $\Delta\sigma_{d(f)} = 115.85$ kN/m²; $\Delta u_{d(f)} = 54.54$ kN/m²

9.24 1900 lb/ft²

9.25 -100.25 kN/m²

9.26 -0.854 ton/ft²

9.27 290 lb/ft²

9.28 1398 lb/ft²

9.29 **a.** $m = 10$ lb/in.²; $\alpha = 15°$
b. $c = 10.38$ lb/in.²; $\phi = 15.54°$

Chapter 10

10.1

Part	P_o	$\bar{z}$
a	2425.5 lb/ft	3.33 ft
b	3114.7 lb/ft	4 ft
c	87.3 kN/m	1.67 m
d	56.87 kN/m	1.33 m

10.2

Part	P_a	$\bar{z}$
a	4121 lb/ft	5 ft
b	7076 lb/ft	6.67 ft
c	4549 lb/ft	6 ft
d	4080 lb/ft	5.5 ft
e	46.33 kN/m	1.5 m
f	50.58 kN/m	1.67 m
g	31.6 kN/m	1.33 m

10.3

Part	P_p	σ_p
a	16,500 lb/ft	3300 lb/ft^2
b	45,276 lb/ft	6468 lb/ft^2
c	169.7 kN/m	138.5 kN/m^2
d	593.6 kN/m	296.8 kN/m^2

10.4

Part	P_a	$\bar{z}$
a	2533 lb/ft	2.89 ft
b	7606 lb/ft	5.11 ft
c	165.5 kN/m	1.93 m
d	139 kN/m	1.7 m

10.5

Part	P_p	$\bar{z}$
a	38,374 lb/ft	5.07 ft
b	89,067 lb/ft	9.35 ft

10.6 **a.** $z = 0$ ft, $\sigma_a = -700$ lb/ft^2; $z = 20$ ft, $\sigma_a = 1700$ lb/ft^2
 b. 5.83 ft   **c.** 10,000 lb/ft   **d.** $P_a = 12,045$ lb/ft, $\bar{z} = 4.2$ ft

10.7 **a.** $z = 0$ ft, $\sigma_a = -500$ lb/ft^2; $z = 20$ ft, $\sigma_a = 1900$ lb/ft^2
 b. 4.17 ft   **c.** 14,000 lb/ft   **d.** $P_a = 15,039$ lb/ft, $\bar{z} = 5.28$ ft

10.8 **a.** $z = 0$ m, $\sigma_a = -29.4$ kN/m^2; $z = 6$ m, $\sigma_a = 89.4$ kN/m^2
 b. 1.48 m   **c.** 180 kN/m   **d.** $P_a = 201.83$ kN/m, $\bar{z} = 1.5$ m

10.9 958 lb/ft

10.10 76,586 lb/ft

10.11 **a.** $P_a = 80$ kN/m; $\bar{z} = 0.888$ m   **b.** $P_a = 56.3$ kN/m; $\bar{z} = 1.1$ m

10.12

Part	P_a (lb/ft)
a	6523
b	6170
c	6055

10.13

Part	P_p (lb/ft)
a	88,990
b	126,050
c	192,960

10.14

θ (deg)	r (in.)
0	1
30	1.44
60	2.08
90	3.00
120	4.33
180	9.02

10.15 1748 kN/m

10.16 1631 kN/m

10.17

Part	P_a
a	5.2 kip/ft
b	6.3 kip/ft
c	125 kN/m

10.18 6840 lb/ft

10.20 209.3 kN/m

10.22 137 kN/m

10.23

Strut	Load (lb)
A	26,965
B	20,973
C	11,985
D	47,939

10.19 7.12 ft from the bottom of the wall

10.21 22.73 kN/m

Chapter 11

11.1 **a.** 12,588 lb/ft²   **b.** 5533 lb/ft²
       c. 104 kN/m²   **d.** 6668 lb/ft²
       e. 71 kN/m²

11.2 **a.** 2835 lb/ft²   **b.** 2240 lb/ft²
       c. 49 kN/m²   **d.** 2442 lb/ft²
       e. 48.3 kN/m²

11.3 **a.** 12,585 lb/ft² **b.** 6111 lb/ft²
       c. 120.9 kN/m²   **d.** 7134 lb/ft²
       e. 82.8 kN/m²

11.4 4 ft

11.6 42.3 kip

11.8 91.6 kip

11.10 659 kN

11.12 a. 81.2 kip   **b.** 235.3 kip
        c. 10,827 kN **d.** 6498 kN

11.13 108.2 kip

11.15 1853 kN

11.17 57,764 lb

11.19 5.04 ft

11.21 260.65 kN

11.5 4 ft

11.7 61.58 kip

11.9 124.3 kip

11.11 941 kN

11.14 624 kN

11.16 31.28 kip

11.18 451.2 kN

11.20 175 kip

11.22 0.7 m

Chapter 12

12.1 32.4 ft

12.3

β (deg)	H_{cr} (ft)
20	∞
25	32.4
30	17.0
35	12.0
40	9.78

12.2 a. 1.49   **b.** 7.24 ft

12.4 a. 1.66   **b.** 11.6 m   **12.5** 0.98

12.6 1.24   **12.7** 5.68

12.8 29.3 ft   **12.9** 39.4 m

12.10 9.11 ft   **12.11** 1.76

12.12 11.1 ft

                                            12.13 a. 26.98 ft   **b.** 46.34 ft
                                                   c. 22.95 ft

12.14 2.7   **12.15** 4.39 m

12.16 12.94 m; toe circle   **12.17** 1.7

12.18 a. 18.5 kN/m²   **b.** Midpoint circle   **c.** 5.49 m

12.19 36°

12.20 a. 141.3 ft   **b.** 96.6 ft   **12.21 a.** 1.9   **b.** 1.36
       c. 35.9 m   **d.** 20.6 m   **c.** 2   **d.** 1.9

12.22 a.

F_s	H (ft)
3.0	18.18
2.5	27.73
2.0	31.8
1.8	35.35
1.4	53
1.0	127.3

b.

F_s	H (m)
3.0	2.63
2.5	3.19
2.0	4.3
1.6	6.19
1.0	12.38

12.23 0.69   **12.24** 0.57

12.25 a. 1.3   **b.** 1.86   **12.26** 0.955

12.27 1.748   **12.28** 1.04

12.29 1.67   **12.30** 1.07

Chapter 13

13.1 8.3 cm²/m   **13.2** 244 cm²/m

13.4 1.68 min⁻¹

Chapter 14

14.1 7.54%   **14.2** 8.88%

14.3

Depth (ft)	N'
5	15
10	9
15	13
20	13
25	11

14.4 0.75 ton/ft²   **14.5** 0.5 ton/ft²

14.6 a. 12   **b.** 1.3 ton/ft²   **14.7** 945 lb/ft²

14.8 410 to 615 kN/m²   **14.9** 73.8%

INDEX